Practical Field Surveying
and Computations

Practical Field Surveying and Computations

A. L. ALLAN M.A., A.R.I.C.S.,
J. R. HOLLWEY BSc., F.R.I.C.S.,
J. H. B. MAYNES BSc., A.M.I.C.E.
A.E.I.Mun.E., C.Eng.

Waltham Forest Technical College

AMERICAN ELSEVIER PUBLISHING
COMPANY, INC.

52 Vanderbilt Ave., New York, N.Y. 10017

William Heinemann Ltd
LONDON   MELBOURNE   TORONTO
CAPE TOWN   AUCKLAND

First published 1968
© A. L. Allan, J. R. Hollwey, and J. H. B. Maynes 1968
434   90061   3

Printed in Northern Ireland at the Universities Press, Belfast

To our wives: Daphne, Mavis, and Winifred.

Preface

In this book we have drawn on our collective professional survey experiences in various parts of the world to provide a selection of the essential principles, observational and computational techniques, and mathematical proofs required by the ground surveyor today. The contents should be of considerable value to land, mining and geodetic surveyors; civil engineers; municipal and county engineers; town planners, and the many others, such as archaeologists, geographers, and geologists, who from time to time engage in surveying.

The detailed presentation of the subject matter is based on our more recent experiences of teaching surveying to students preparing for the following examinations: the first, intermediate, and final examinations of the Land Surveying Section of the Royal Institution of Chartered Surveyors; the final examinations of the Institution of Civil Engineers; BSc. external degree in civil engineering of the University of London; and Higher National Certificates in civil and structural engineering. It is hoped that the book will also prove to be suitable for other similar courses.

Even in a book of this size, considerable selection of subject matter has been necessary, and inevitably some traditional topics have been greatly curtailed or omitted completely. We have attempted to give mathematical proofs which are sufficient to explain the theory, and have rejected unnecessary terms at all stages of development. Since we firmly believe that the competent surveyor must have a thorough grasp both of theory and practice, and an understanding of their interdependence, no attempt has been made to subdivide the subject into these two categories.

The descriptions of particular instruments are intended to give typical examples of technique, and to illustrate the basic principles common to groups of similar instruments. We have placed considerable emphasis on computation by machine and tables of natural trigonometrical functions, and have given suggested layouts for computations. Where relevant, the theory is illustrated by fully-worked examples, most of which have been obtained from actual surveys, although in a few cases, simple figures have been selected to enable a student to follow a method unhampered by complicated arithmetic. Consideration has been given to those aspects of aerial mapping which affect the ground surveyor, though no attempt has been made to consider, even briefly, the other aspects of this vast specialist field. The chapter on field astronomy has been restricted to a short account of the determination of azimuth.

We wish to acknowledge the important contribution to our knowledge which has been generated by the many and varied questions posed by our students. Many of the ideas and methods expressed in this book have been developed from discussions with other members of the Land Surveying Section at

Waltham Forest Technical College. Although every care has been exercised to ensure that no errors appear in the text, we shall be grateful to learn of any that may have escaped our attention.

Waltham Forest Technical College A. L. Allan
J. R. Hollwey
J. H. B. Maynes

Acknowledgements

The authors wish to express their gratitude to the following organizations and persons who have assisted them by discussion and factual help in the preparation of this book, though the responsibility for the views expressed rests with the authors.

Waltham Forest Technical College: (Formerly South West Essex Technical College)

The Principal; Messrs F. J. Batson, G. E. Harrison, A. Haugh, J. J. Loader, Z. M. Michalski, R. J. Smith; B. R. Jones, R. H. Leppard, and P. M. Ball; and Miss J. V. L. Jenkins.

The Directorate of Overseas Surveys, England.

The Deputy Director Mr. J. W. Wright, Messrs A. A. Allum, H. H. Brazier, B. E. Furmston, H. F. Rainsford and W. Watson, and Dr. W. R. Logan.

The Directorate of Military Surveys, England.

Messrs T. Bassett, J. F. Bell, and J. A. Weightman.

The Ordnance Survey, England.

The former Director General, Major General A. H. Dowson C.B., C.B.E., Major R. E. J. Lower R.E.; Major T. A. Linley, R.E.; and Messrs D. K. Black and J. K. Holt.

University of Glasgow

Mr G. Petrie.

University College London.

Lt. Col. C. A. Biddle.

University College Swansea.

Mr. C. Tomlinson.

The Northern Nigeria Survey Department.

Messrs J. Ashton, and D. J. Moss.

Huntings Surveys Ltd., England.

Messrs C. C. Brown and I. Mathieson.

Tellurometer (U.K.) Ltd.
AGA (Signals) Ltd., England.
Messrs H. Wild and Co. Ltd., Switzerland.
Messrs Kern Ltd., Switzerland.
Messrs Carl Zeiss, Jena, E. Germany.
Messrs Hilger and Watts Ltd., England.
Vickers Instruments (Cooke, Troughton and Simms), England.
The Department of the Army, United States of America.
The Astronomer Royal, England.
The Editor, The Survey Review, England.

Contents

Useful Mathematical Formulae	xv

1. INTRODUCTION TO FIELD WORK 1

   Plane table surveying: exposition of basic survey methods–Reconnaissance

2. INTRODUCTION TO COMPUTATION 37

   The calculating machines–Basic processes in machine calculation–Types of calculating machine–Mathematical tables–Logarithmic methods of computation–The slide rule–Conversion of small angles from sexagesimal values to circular values–Spherical trigonometry–Interpolation–Coordinate transformation–Computation in the plane rectangular system.

3. THE THEODOLITE AND LEVEL 81

   The theodolite–Component parts of the theodolite–The telescope–Use of the telescope–The spirit level–The circles–The reading mechanism–Adjustments of the theodolite–Effects of malconstruction of the theodolite–Effect of maladjustment of the theodolite–Procedure to minimise the effects of malconstruction and maladjustment of the theodolite–Types of theodolite–Components of the level–The geodetic level–The automatic level–Adjustment of the level.

4. THE SPHEROID AND PROJECTIONS 149

   The spheroid–Survey projections.

5. THEORY OF ERRORS AND SURVEY ADJUSTMENT . 200

   Introduction–Systematic errors–Random errors–Principle of least squares–Adjustment of observations by the method of least squares–Observation equations–Solution of normal equations–The adjustment of observations of different weight–Condition equations–Observation and condition equations methods.

6. ANGULAR MEASUREMENT AND COMPUTATION . . 250

   Introduction–Preliminaries–Beacons and signals–Survey towers–Observation of horizontal angles–The adjustment of angles–Number of condition equations in a free network of triangulation–Formation of the condition equations in a network–Computation of intersections (analytical methods)–Analytical methods of resection computation–Variation of coordinates–The inaccessible base or two point problem.

7. TRAVERSES 323

Introduction–Sources of error–Equipment for traversing–Field work in theodolite traversing–The cadastral or town survey control traverse–Traverse computation–Permissible misclosures and expressions of accuracy–Miscellaneous problems in traversing–Adjustment of a traverse network–Subtense traversing–Precise traversing–Propagation of error in theodolite traverses.

8. VERTICAL CONTROL 384

Introduction–Methods of determining relative heights–Spirit levelling–Trigonometrical heighting–Miscellaneous aspects of heighting–Barometric heighting–Practical altimetry.

9. DETAIL SURVEY 437

Introduction–Definitions–Representation of detail–Chain surveying–Simple tacheometry–Self-reducing tacheometers with vertical staff–Large scale plane table equipment–Horizontal bar tacheometry–Completion and field examination.

10. CURVE DESIGN AND SETTING OUT 480

Setting out of works–General–Types of horizontal curve–Basic curve geometry–Computation of curve components–Field setting out–Design and fitting of horizontal curves–Other mathematical curves used as transition curves–Vertical curves.

11. VOLUMES OF EARTHWORKS 557

Preliminary note on prismoidal and end-area rules–Determination of volume from contours–Determination of volume from spot heights–Determination of volume from cross-sections–Mass-haul diagrams.

12. CADASTRAL SURVEYING 588

Introduction–The description–Urban layouts–Determination of area.

13. ELECTROMAGNETIC DISTANCE MEASUREMENT . . 608

Introduction–Geodimeter Model IV–Field operation of the Geodimeter Model IV–The Tellurometer–Practical hints on Tellurometer work–Other models of the Tellurometer–Other distance measuring apparatus–Application of electromagnetic distance measurement to surveying–Reduction of slant ranges to the spheroid–The least squares adjustment of measured lengths by condition equations–The computation of trilateration coordinates.

14. THE DETERMINATION OF ASTRONOMICAL AZIMUTH 666

   Introduction–Azimuth by simultaneous observation of horizontal and vertical angles–The determination of azimuth from observed horizontal angles and recorded time–Azimuth from circumpolar stars–Identification of stars.

INDEX 683

Useful Mathematical Formulae

The binomial expansion

$$(a + x)^n = x^n + nx^{n-1}a + \frac{n(n-1)}{2!}x^{n-2}a^2$$
$$+ \cdots + \frac{n(n-1)(n-2)\cdots(n-r+1)}{r!}x^{n-r}a^r$$

Taylor's theorem for a single variable

$$f(x + h) = f(x) + hf'(x) + \frac{h^2}{2!}f''(x) + \frac{h^3}{3!}f'''(x) + \cdots$$

Exponential series

$$e^x = 1 + x + x^2/2! + x^3/3! + \cdots$$
$$e = 2.718281828$$

Hyperbolic functions

$$\cosh x = \tfrac{1}{2}(e^x + e^{-x})$$
$$\sinh x = \tfrac{1}{2}(e^x - e^{-x})$$

Trigonometrical and other series

$$\sin \theta = \theta - \theta^3/3! + \theta^5/5! - \theta^7/7! \cdots$$
$$\cos \theta = 1 - \theta^2/2! + \theta^4/4! - \theta^6/6! \cdots$$
$$\sec \theta = 1 + \theta^2/2 + 5\theta^4/24 + 61\theta^6/720 \cdots$$
$$\tan \theta = \theta + \theta^3/3 + 2\theta^5/15 + 17\theta^7/315 + 62\theta^9/2835 + \cdots$$
$$\theta = \tan \theta - \tfrac{1}{3}\tan^3 \theta + \tfrac{1}{5}\tan^5 \theta - \cdots$$

$$1 + 2 + 3 + 4 + \cdots + n = \tfrac{1}{2}n(n+1)$$
$$1^2 + 2^2 + 3^2 + 4^2 + \cdots + n^2 = \tfrac{1}{6}n(n+1)(2n+1)$$

Radius of curvature of a plane curve
(a) Cartesian coordinates (x, y)

$$\rho = \frac{\{1 + (dy/dx)^2\}^{\frac{3}{2}}}{d^2y/dx^2}$$

(b) Polar coordinates (r, θ)

$$\rho = \frac{\{r^2 + (dr/d\theta)^2\}^{\frac{3}{2}}}{r^2 + 2(dr/d\theta)^2 - r(d^2r/d\theta^2)}$$

Useful Mathematical Formulae

$$1 \text{ radian} = 57°\cdot295\ 779\ 513$$
$$= 3437'\cdot746\ 770\ 785$$
$$= 206\ 264''\cdot806\ 247\ 096$$

1 metre = 3·2808 4558 British feet (Sears-Johnson-Jolly)
1928

1 *Introduction to Field Work*

1.1. PLANE TABLE SURVEYING: EXPOSITION OF BASIC SURVEY METHODS

In plane table surveying a map is drawn in the field on the flat surface of some drawing medium affixed to a small portable table resting on a tripod. Very little computation is required, fewer gross errors are likely since the plotting is carried out in the field, and in certain instances it is a very rapid method of producing a completed map. Bad weather makes progress difficult though not impossible if the drawing surface is durable, such as enamel plate, and if the surveyor is protected by an umbrella. Specially designed survey umbrellas are available not only to protect the surveyor and his work from rain, but to shield instruments from strong sunlight in tropical countries. The detail of topographical maps was formerly surveyed by plane table methods, and reconnaissance work was, and still is to some extent, based on rapid plane table diagrams. Today, map detail is almost invariably plotted from aerial photography and reconnaissance is assisted by *mosaics* constructed from such photography. At scales larger than 1:5000 however, the plane table method has much to recommend it on grounds of economy, speed, and accuracy. Perhaps the greatest use of plane tabling today is its application to training, since it is a cheap, individual process which not only initiates the beginner to survey techniques, but also enables him to gain self confidence in his own individual work. In later sections of this chapter, plane table surveying with simple equipment is explained, both for its own sake, and to illustrate basic survey principles and problems.

1.1.1. SIMPLE EQUIPMENT

In figure 1.1 the essential simple equipment is illustrated, and comprises:

1.1.1.1. *The board* usually 18 × 24 inches, attached to the tripod by a central screw (see figure 1.2) about which it rotates. The board can be clamped in position. Levelling of the table is usually by eye, or in large scale work with the aid of a small bubble placed on the table.

1.1.1.2. The *sight rule* or *alidade* enables one to sight on a distant object and draw a line on the board parallel to this line of sight. In fact, the drawing edge need not be parallel to the line of sight, but the two must bear a constant relationship to each other. A parallel rule facilitates drawing, though is by no means essential. For steep sights, a thread is fixed between the tops of the sighting vanes. In drawing, the pencil is held in a vertical position to minimize errors arising from a variation in the angle at which it is held. A sharp, hard point very lightly handled will give good results.

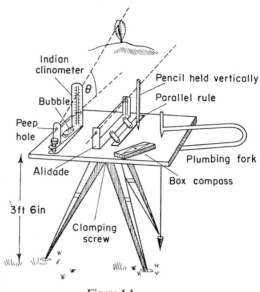

Figure 1.1

1.1.1.3. *The Indian clinometer* is used in heighting. Either vertical angles or their tangents may be read against a horizontal datum indicated by a small bubble. Before using the clinometer in the field it must be tested for vertical index error, that is one must know the position of the bubble when the line joining the peep hole to the zero mark of the graduations is horizontal. This can be found by observing an angle of slope in both directions. The mean angle or tangent will be correct. At one end of the slope this mean reading is set by the levelling screw and the bubble position noted and marked by a piece of paper stuck to the bubble tube. If desired, the bubble may be brought exactly to the centre of its run by its adjusting screws, though this adjustment is difficult in practice. Provided the bubble is not so far out of adjustment that it cannot run freely, the former course of marking the tube is preferable. Tangents can be read directly to 0·005 and estimated to 0·001, which limits the heighting precision to three significant figures.

1.1.1.4. The *box* or *trough compass* consists of a magnetic compass needle housed in a small box. The compass is used for approximate orientation of the board at new points. At some point where the board is in correct orientation, usually a control point, the compass needle is allowed to set itself in the magnetic meridian. When the needle is steady and central the outline of the box is drawn on the board, and thus the angle between true, or grid North and magnetic North is known. To orient the board at a new position the compass is placed in the outline of its box and the board twisted until the needle is again steady and central. This orientation is only approximate since local attraction of the compass needle is common, and the precision of the needle is usually inadequate. Many surveyors find the compass unnecessary.

Introduction to Field Work

1.1.1.5. *Ancillary equipment* may also be necessary and useful. The *plumbing fork* enables a point on the board to be plumbed exactly over a ground mark. This will only be necessary if the dimensions at the board are plottable at the scale of the survey. A *slide rule* is useful in height calculation, a *scale* of that of the survey or convenient multiple, *notebooks*, *pencils*, *rubber erasers*, *inks*, etc are all required, and if control coordinates are supplied, an *accurate scale*, *straight edge*, and *beam compass* are essential for plotting.

1.1.1.6. *The drawing medium:* A wide variety of drawing media is now available in addition to the traditional paper; plastics, paper–zinc sandwich, enamelled zinc are a few. A good 90 lb hot pressed paper provides a reasonable drawing

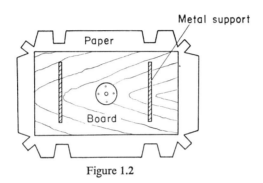

Figure 1.2

surface which will withstand the erasures of plane tabling. A linen-backed paper maintains its dimensional stability better than paper alone. Paper is wet mounted to the board as follows: The paper is cut to the shape shown in figure 1.2, or to some other convenient shape, moistened until limp, affixed to the underside of the board with a flour paste, held temporarily with drawing pins, and left to dry for 24 hours, after which time a very taut smooth drawing surface is obtained.

1.1.2. PLANE TABLE SURVEYING

1.1.2.1. *Preliminaries*

Before work may begin in the field the following preliminary work is carried out. The surveyor must be briefed as to the *scale* of the survey, its *scope*, *equipment* and *personnel* available, *funds* and *transport* permitted, and procedures concerning *access to property* and payment for any *damage* that may be done. He must also be briefed on the kind of *permanent ground marks* he should leave for possible future use, what type of *record* he should keep of these marks, how *duplicate copies* of essential survey data are to be made, and finally how this data is to be *conveyed* to the survey headquarters without fear of loss. In large surveys involving several surveyors a system of *unique numbering* of field sheets, etc. must be adopted to avoid confusion. The basic rule is that all work carried out by one surveyor should be documented in such a manner that a different surveyor should be able to follow exactly what was done, and be able to relocate ground marks after a considerable period of time. Many excellent surveys have been rendered virtually useless because of poor documentation.

1.1.2.2. *Estimated Accuracy of the Survey*

The technique adopted by the surveyor depends on the *precision* required for the immediate task, and if possible for any likely requirements in the future. In turn the methods chosen depend on the *equipment* available, and most important of all, on the *funds* available. A final decision is reached in the light of these factors and the extent to which the requirements of one are permitted to affect the other. In plane table surveying, the basic factor determining accuracy is the plottable error on the drawing surface and the dimensional stability of the drawing medium. The plottable accuracy is usually taken as 0·1 mm which is equivalent to 2·5 metres at a scale of 1:25 000, 5 cm at 1:500 scale, etc. Hence working at 1:25 000, a curve deviating from a straight line by less than 2·5 m will plot exactly straight, and significant features smaller than this are plotted by conventional signs. Adjustment for dimensional change of the drawing medium is made with reference to the base grid drawn on the map.

1.1.2.3. *Conventional Signs*

Many important features such as roads, buildings, etc. cannot be plotted at the scale of a survey, or at any rate, may plot too small for clarity, therefore conventional signs are adopted to depict these features. Wherever possible the centre of the sign is taken as the exact map position of the feature. For example, if a road is shown on a map at 1:25 000 scale by two lines 1·5 mm apart, the corresponding ground width is 37·5 metres. The centre line of the road is drawn in pencil on the field sheet and the final conventional sign placed on either side of and equidistant from this plotted centre line. A house 10 metres wide at the edge of the road, would have to be shown in the middle of the conventional sign for the road, if it is plotted in its correct position. To avoid possible confusion, the house is moved from its correct position to the edge of the road sign. In congested areas this procedure often means that individual buildings cannot be plotted and are therefore shown as a solid map sign.

1.1.2.4. *Providing Scale to the Map*

To give *scale* to a map at least one line is measured on the ground by chain, or tape (see §7.1.1), pacing, or electronic distance measurement. This line is called the *base line*, i.e. the line on which the scale is based. As a check, several base lines are normally measured in surveys covering a wide area. Suppose the line AB is measured. It is plotted of length ab = kAB on the plane table board where k is the map scale, and in such a position that as much of the area as possible may be drawn on the sheet. In large surveys, the area is divided into separate sections which are ultimately *brought* together, or a *compilation* is made of all the field sheets. (The same procedure is applied to photogrammetric plots.) Since ab is a short line, once the alidade has been removed from its original plotting position, it cannot be replaced with the same precision. For

Introduction to Field Work 5

this reason the direction of a short line on the plane table is marked by two short extensions called *repère marks*, thus permitting the alidade to be replaced more accurately.

1.1.2.5. *Method of Intersection*

Suppose a point C on the ground, c on the board, has to be fixed. The board is set up at A, the alidade drawing edge placed along ab, and the board turned carefully until the line of sight bisects an object, such as a ranging pole, placed at B, i.e. until abB is a straight line. The board is then in correct *orientation* or the board is *set*. The board is then clamped in position taking care that the line of sight still bisects the pole at B. The point C is then sighted and a line aC drawn on the board (See figure 1.3b). The board is removed to B, where the procedure is repeated giving the plotted position of c, by *intersection* from A and B.

1.1.2.6. *Triangulation*

In figure 1.3, if the board is taken to C and orientated on A, the plotted angle acb can be checked. If all three angles of the triangle are surveyed the process is called *triangulation*. In a large network, further points can be plotted from any two of points a, b, and c, and where possible more than two would be employed to check the work. Hence a *network* or *framework* of triangles can be plotted which will serve as the basis of the map. It is essential that all these *control points* are correctly positioned.

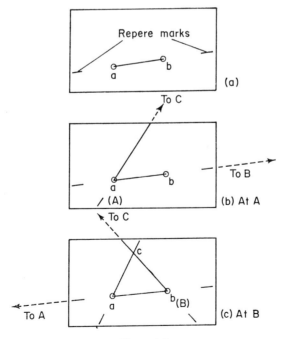

Figure 1.3

1.1.2.7. Radiation

The point c could have been fixed from A by the *direction* from a and the *distance* ac = kAC, where AC is measured on the ground. Fixation by direction and distance is called *radiation*. The method is used to fix points of detail from controls, the distances being measured by pacing, taping, or tacheometry as

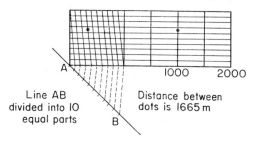

Figure 1.4

convenient. To assist plotting distances a *diagonal scale* in the unit of measurement is drawn on the board and distances transferred by bow dividers. Such a scale of metres is shown in figure 1.4.

1.1.2.8. Trilateration

The point c could also be fixed from a and b if the two sides ac = kAC and bc = kBC are known, by describing arcs with the beam compass centred on a and b equal to ac and bc respectively. Because c has been fixed from three sides the process is called *trilateration*.

1.1.2.9. Traverse

The traverse method of surveying is illustrated in figure 1.5.

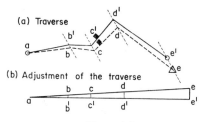

Figure 1.5

It consists of covering a considerable distance by a chain of points fixed from each other by a series of radiations. Since any error in one fixation, e.g. of b from a (figure 1.5), affects the position of all subsequent points, all small errors along the traverse will tend to accumulate and therefore a *traverse misclosure* such as ee' is common. In figure 1.5 the correct plotted position of E is e, and that brought through the traverse from a is e'. This illustrates the general principle that traverses should always start from and finish on points whose positions are known. If two fixed points are not available, the traverse may form a closed

Introduction to Field Work

loop, though this is not as foolproof as the former. To adjust the traverse, (see figure 1.5b) a line ab'c'd'e' is plotted equal to the total length of the traverse. At e', a line e'e is erected perpendicular to ae' and equal to the traverse misclosure. The line ae is joined and perpendiculars erected to meet it from the intermediate points. The distances bb', cc' etc are the adjustments to be made to the provisional positions on the board. The direction of these adjustments is parallel to that of the original misclosure at e. The final adjusted positions of the points are then given as in the figure 1.5a.

If a long traverse is contemplated, it pays to draw it on a separate sheet of paper so that only the adjusted traverse will be transferred to the plane table sheet. The traverse may also be plotted at twice the scale of the plane table map, and reduced down to fit it later. Any details plotted from the traverse will have to be moved slightly to accord with the new positions of the traverse points, e.g. the building fixed from C figure 1.5 is moved to accord with the new position of c' at c.

1.1.2.10. Resection

The *resection* method consists of fixing the position of a point P from three points A, B, and C visible from P, and whose positions a, b, and c are plotted on the board. In plane tabling this is a most useful method since it is only when one is at some point of detail that one realises which point should be fixed. The alternative to resection is to place some mark such as a ranging pole at P, and return to the control points to fix it by intersection, which may involve walking distances of several miles. The fixation of a point from three control points is possible in all circumstances except when all four points lie on the circumference of a circle called the *danger circle*. Two methods of resection are now described.

(a) The trial and error method (Lehmann's Method)

The essential problem of resection is one of orientation. This method is best explained by reversing the process of a practical solution. In figure 1.6a the plotted position of P resected from A, B, and C is given at p. The directions from A, B, and C to P are drawn on the board as full lines meeting at a point p. This is the correct position since the figures pab, etc. and PAB, etc. are similar in the mathematical sense. If the board is now turned clockwise through a small angle, a moves to a', b moves to b', and c moves to c'. If rays are pulled in from A, B, and C through their respective plotted positions a', b', and c' the dotted lines of the figure are obtained which do not now meet at a point, but form a triangle of error the size of which depends on the error of orientation. An enlarged view of this error triangle is shown in figure 1.6b. In practice the problem is to establish the position of p from the error figure. Let the perpendicular distances from p to Aa', Bb', and Cc' be d_a, d_b, and d_c respectively. Then

$$d_a:d_b:d_c = \text{pa}:\text{pb}:\text{pc} \tag{1}$$

Also, since the points on the board are rigidly fixed together a rotation of the board will move the rays such as Aa', Bb', and Cc' either *all clockwise* or *all anticlockwise*. Thus the position of p must lie either to the right-hand side of all

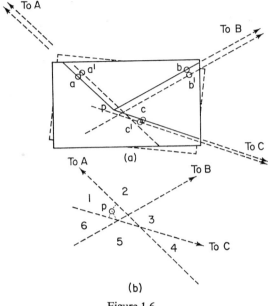

Figure 1.6

the rays looking towards the respective controls from which they were drawn, or to their left-hand sides. Thus the position of p must lie either in sector 1 of figure 1.6b or in sector 4. Equation (1) indicates which of these alternatives is correct since the ratios cannot be satisfied if p lies in sector 4. The position of p is plotted by estimation, bearing in mind equation (1). Then the board is orientated from this position of p and the most distant control point B. The lines of the solution are rubbed out and the process of pulling in rays is repeated. If they meet at a point p the solution is complete; if not, the error triangle is solved again and a second position obtained for p, and so on until the three rays meet at a point. With practice, resection by this method is very quick and not more than two approximations are normally required. As an absolute check, the ray to a fourth visible point should be checked, because a solution can always be obtained from three points even if the triangle on the ground is not similar to the plotted triangle. If P lies on one of the sides of the control triangle ABC faulty orientation will not give an error triangle, but two parallel lines from the collinear points cut by a line from the third control point. The correct position of p is found in the same way as above. If P lies inside the control triangle ABC, the position of p lies inside the error triangle.

(b) *The Collins' point method (Bessel's solution)*

The Collins' point or Bessel's method is attractive both in theory and practice, since it requires no successive approximation, and once mastered it is probably the quickest method. Consider the ground triangle ABC with respect to which the point P is required (see figure 1.7). AB is chosen as the base and the circle

Introduction to Field Work

ABP drawn. CP is produced to meet this circle at the Collin's point I. Thus if the position of I can be located the direction of IC and therefore of PC is known and the table can be set. In practice it is the positions of these points plotted on the board that are used, i.e. the position of i is found with respect to a, b, and c. The upper case letters refer to the ground and the lower case to the board. The circle is not drawn but is required to establish the theoretical basis of the method. In figure 1.7

$$A\hat{P}C = \alpha, \qquad B\hat{P}C = \beta$$
$$B\hat{A}I = B\hat{P}I = 180 - \beta$$
$$\therefore A'AB = \beta$$

Similarly
$$B'BA = \alpha$$

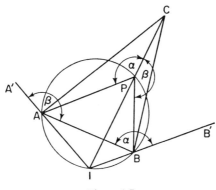

Figure 1.7

Therefore the position of i will be given by the intersection of the lines aa' and bb'. These lines are obtained by plotting β at a and α at b. This is simply achieved as follows: (Remember that the board is at P)

(i) With the alidade pointing along ab turn the board until B is sighted and clamp it, i.e. abB is a straight line.
(ii) Pull in the ray from that point not used as base, i.e. from C through the position a. The plotted point c is not used until the last stage. Thus the line a'a is drawn.
(iii) With the alidade along ba sight on A and clamp the board, i.e. baA is a straight line. Pull in the ray from C through the point b. This is the line b'b.
(iv) a'a and b'b are produced to meet at i.
(v) The alidade is layed along ic and the board rotated to sight C and clamped. The board is now set.
(vi) Rays are pulled back from A through a, and from B through b, and should intersect ic at the position p.

A check on a fourth point is desirable.

The only disadvantage of the method is that the board has to be rotated twice and unless this and the clamping is done with care, a poor fixation will result.

A little experience is required to choose the correct base, not only to give the best fix but also to ensure that the point i falls on the board. An interesting case arises when CP is a tangent to the circle ABP, in which case i and p coincide, though a solution is still possible.

1.1.2.11. *The Inaccessible Base or Two Point Resection* (see figure 1.8)

Occasionally, only two plotted control points A and B are visible from the point P whose position is required, thus simple resection is impossible. An auxiliary point Q is chosen as the other end of a base PQ whose length need not be known, and from which the two controls A and B are fixed. The plotted points on the board are ignored meantime. A line pq is plotted any convenient

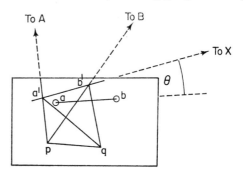

Figure 1.8

length and a' and b' intersected from P and Q which are both visited and marked by a ranging pole or bush. Thus the figure pqb'a', which is mathematically similar to the figure PQBA is drawn. To set the board, the alidade is placed along a'b' and some object X sighted in the background, or a pole is placed on line. The alidade is placed along the plotted positions ab and the board is rotated until X is again sighted. Thus the board has been rotated through the error of orientation θ and is now correctly set. The correct position of p is then obtained by pulling in rays from A and B through a and b respectively.

1.1.2.12. *Fixation of Details*

Any of the above survey methods may be used to fix points at or close to points of detail to be plotted on the map. Curved boundaries such as roads are plotted by interpolating the shapes between points accurately fixed. Because the plotting is completed on the ground some allowance can be made for other than linear interpolation between detail controls which means that fewer such control points are required than for office plotting from field books. Details may also be fixed from a measured distance by the chain survey method of rectangular offsets though this method will only be used occasionally in plane tabling.

1.1.2.13. *Contouring*

A *contour line* is a line drawn on a map through points at equal height above some *datum* or reference point. If one imagines the land surfaces to be flooded

Introduction to Field Work

up to some height, say 100 ft above sea level, the shore line of the flooded area will trace out the 100 ft contour line when plotted on a map. This flooding concept is a useful idea to have in mind when surveying contours in the field. The datum point is normally *mean sea level* taken over a long period, or in an isolated survey the datum will be chosen at some theoretical point below the lowest part of the area. Points of known height to which surveys are referred are called *bench marks* (B.M.'s). In the United Kingdom the Ordnance Survey has constructed *fundamental* or *primary bench* marks at intervals of about 30 miles covering the whole country; these are further broken down to a network of *secondary* bench marks at intervals of about 10 miles, and finally there exists a coverage of *tertiary* bench marks at intervals of about ¼ mile according to terrain and locality. Since information concerning these marks is readily available there should be no need to use an assumed datum in the United Kingdom. The vertical height between each contour line is called the *vertical interval* or V.I. and is normally some integral number of feet or metres. British practice is still to retain the foot as a unit for heighting, though the decimal system is almost universally adopted instead of inches. In choosing the vertical interval to be employed on a map consideration is given to the following factors:

(i) The purpose and therefore precision of the survey.
(ii) The scale of the map on which the contours are to be shown.
(iii) The character of the land surface itself.

No hard and fast rules can be laid down concerning the V.I. though the following have been widely adopted in the past:

Map scale	V.I. (ft)
1:100 000	100 or 250
1:50 000	50
1:25 000	25
1:10 000	10 or 20
1:5 000	10
1:2 500	10 or 5
1:1 250	From 1 to 10 according to circumstances.

Normally the same V.I. is used throughout the complete map though on occasion it has to be changed if the map covers both undulating and mountainous terrain. As a general rule, the accuracy of a contour is guaranteed to half the vertical interval.

1.1.2.14. *Surveying Contours*

There are four main methods of surveying contours:
(i) Levelling along the contour on the ground.
(ii) Interpolating the contour between height points on a grid.
(iii) Interpolating contours between height points carefully chosen at changes of slope and changes in the direction of a contour.
(iv) Plotting the contour in a photogrammetric plotter.

In the first method, a point at the exact height of the contour is established on the ground with reference to a bench mark. Thereafter the levelling staffman places the staff at points exactly on the contour. This requires a certain amount of trial and error. The position of the level and staff are fixed with respect to plan details, hence the contour may be plotted on the map. This process is both the most tedious and the most accurate and is used only when circumstances necessitate the very highest accuracy.

In the second method, a grid is set out on the ground at intervals of 100 ft or some other distance, and the intersections heighted by level or tacheometry.

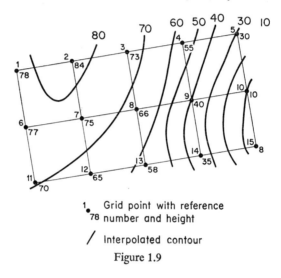

1̥ Grid point with reference
78 number and height

/ Interpolated contour

Figure 1.9

Since both the plan positions and heights of the grid are known the contours are interpolated between the grid points as shown in figure 1.9.

This method has the advantages of simplicity and gives three-dimensional coordinates which are simple to process mathematically, particularly in an electronic computer. It has the disadvantage of being very wasteful of surveying time.

In the third method, points such as A, B, C, D, E, F, and G of figure 1.10 are heighted, between which the contours are drawn. For greatest accuracy and speed this method should be carried out entirely in the field as in plane tabling. However it is much used in tacheometric surveys where, with the addition of a sketch, no ambiguity should arise when plotting in the office.

The photogrammetric method is beyond the scope of this book, except in so far as its requirement for ground control. The photogrammetrist requires a known height at each of four points situated well away from the base line of the photograph forming a stable platform on which he will construct a three-dimensional model of the ground. In detail, the area in which these *height control points* are supplied must be locally flat so that any uncertainty in the plan position of the point will not introduce a height error in the model. The plan position of the height control point is taken from the photograph on which

Introduction to Field Work

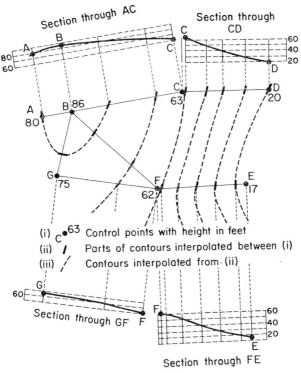

Figure 1.10

it is identified by the surveyor. Any recognized survey method will be used to fix the height of a point on the ground, provided it gives the required accuracy. For example a barometer will give an accuracy of about 5 feet which is sufficient to control 50 or 25 ft contours (see §8.6).

1.1.2.15. *Surveying Contours with the Indian Clinometer*

Contouring on the plane table is normally carried out by the method of interpolation between a few selected points whose heights have to be established by Indian clinometer, telescopic alidade, etc. We shall confine our consideration to the Indian clinometer since a more detailed account of large scale methods is given in §9.17.

In figure 1.11 a reading on the clinometer is taken from the plane table board

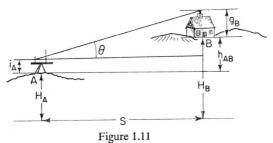

Figure 1.11

situated at point A to the chimney of a house situated at point B; H_A and H_B are the respective ground heights of A and B above the datum surface, i_A is the height of the instrument at A above the ground, g_B is the height of the observed point at B above ground, S is the horizontal distance between A and B, and θ the vertical angle at A to the observed point at B. The difference in height between A and B, h_{AB} is given by

$$h_{AB} = H_B - H_A = S \tan \theta + i_A - g_B \tag{2}$$

Tan θ is read directly on the clinometer, i_A and g_B are measured or estimated, and the distance S is scaled from the plotted positions of A and B on the board. Sometimes the position of a point such as A has to be resected for this purpose, though a forward ray to a stone in the middle of a field will often save much trouble. The value of g_B may have to be measured with the clinometer from a point close to B and whose distance from B is paced. The heights of prominent points such as B will be used to control the survey, hence these controls must first be established. Control heights should be fixed from at least two other points as a check before proceeding to use them.

Readings and computation by slide rule are noted in a small notebook as follows:

Station from and height		Station To	Tan T	S	ST	i g	Computed height	Notes
A	384	B	−0·100	1580	−158	4 −15	215	
C	279	B	−0·060	812	−49	4 −15	219	Accept mean 217

During the survey a trace is kept of all points and their heights from which the contours were drawn, so that any possible inconsistencies may be located, and the overall accuracy of the work may be assessed.

The method of tracing a contour may also be used effectively in plane tabling with the clinometer acting at a level. Points at which the contour cuts detail already plotted on the board are rayed in with the alidade.

1.1.2.16. Accuracy of Clinometer Heighting

Assuming that i_A and g_B can be measured to 0·1 ft with little trouble, the main sources of error are the distance S and the tangent T related by

$$h = ST \tag{3}$$

Differentiating

$$dh = S\,dT + T\,dS \tag{4}$$

By substituting values of dT and dS in (4) the corresponding error dh in h may be obtained. If the errors σ_T and σ_S are the probable errors in T and S respectively (see §5.3.7) the probable error σ_h in h becomes

$$\sigma_h^2 = S^2 \sigma_T^2 + T^2 \sigma_S^2 \tag{5}$$

A likely value of σ_T is ± 0.001 and for σ_S is ± 50 ft assuming a map at 1:25 000 scale. If θ is about 1°, $T = 1/57.3$, and S is 5 000 ft, σ_h is given by

$$\sigma_h^2 = (5 \times 10^3 \times 10^{-3})^2 + (50/57.3)^2 \doteq 25 + 1$$
$$\therefore \quad \sigma_h = \pm 5 \text{ feet}$$

It is apparent from (4) and (5) that the best results are obtained when S is small and $T = 0$; hence for the best results long lines and steep sights should be avoided. In work with the Indian clinometer, the correction for the Earth's curvature and the refraction of the light path are negligible, remembering that the precision of the clinometer is 0.001 in the tangent (see §8.4.3).

1.1.2.17. Surveying Details from Controls Already Supplied

In the above sections it has been assumed that the complete survey is carried out on the plane table. Very often this is not the case, since accurate coordinates of the control points fixed by theodolite, etc. are available. Before going

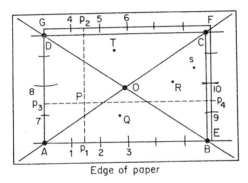

Figure 1.12

to the field, these coordinated points have to be plotted on the board. A grid is also drawn so that (a) any further controls that may come to hand may be plotted later, that (b) the edge of the area to be mapped may be established, and (c) any distortion in the drawing medium may be detected and its effects removed. The grid also serves as a convenient basis on which to reference points.

Figure 1.12 illustrates how the grid is plotted. The approximate centre of the field sheet is found by drawing the diagonals. From this centre O arcs are described by beam compass to cut the diagonals at points near the corners of the sheet, e.g. at A, B, C, and D. The quadrilateral ABCD should be a perfect rectangle. Before proceeding further a check is made to ensure that this is so, for AB should equal CD exactly, AD should equal BC, and AC should equal BD. If these respective dimensions are not exactly equal, the drawing should be repeated from the beginning. The needle points of the beam compass should be used for this purpose. A is chosen as origin of the grid. The dimensions of a grid square are calculated and with A as centre an arc A1 is described to cut AB at 1 such that A1 is the side of the grid square. A7 is also cut off on AD. The side length of two grid squares is calculated and arcs A2 and A8 cut off, and so on until AE and AG are drawn. It is essential that all grid lines are plotted from the origin at A. If point 2 is plotted from 1, etc, serious error will accumulate by the time E is reached. The point F is obtained by describing arcs from G and E equal to AE and AG respectively. Check that AF equals EG. Points 4, 5, 6, etc. are cut off from G, and 9, 10, etc. from E. As a further check the grid grid points 1, 2, . . . , 9, 10, etc. are also checked from the other end of the line from which they were originally plotted, e.g. point 1 is checked by describing arc E1 from E. No detectable error can be accepted in the positions of the grid points 1 $\cdots$ 10, etc. Relevant pairs of points are joined up to give the complete grid.

To plot a control point P, whose coordinates are E_P, N_P, the same procedure is adopted. The lengths of Ap_1 and Ap_3 are computed, and the position of P plotted at the intersection of p_1p_2 and p_3p_4. Coordinates plotted at the same time as the grid should always be plotted directly from A, and not from the grid line close to the point, because the former procedure should be more accurate. All control points are plotted on the field sheet in a similar manner. Suppose these points are Q, R, S, and T. As a check on the plotting, the lengths of certain distances between plotted points are compared with the respective distances computed from their coordinates corrected to the scale of the map. The distances chosen are those which give cuts at each point as close to right angles as possible, e.g. to check P we would choose the lines PT and PQ; to check T the lines PT and TR would be chosen, etc. It is absolutely essential that plotted coordinates are correct because the map is based on them. As a final check, the controls are tested in the field. The board is set up at P say, and rays are drawn to all other visible controls. These rays should pass through the plotted positions. This process is repeated from some other point giving good cuts at the observed points. Normally the plane tabler has to provide further controls by graphical methods to enable him to complete the map, but any error that arises will not accumulate since the base control is correct. This illustrates the basic survey principle that one should work 'from the whole to the part' and not *vice versa*, if errors are not to accumulate.

1.1.2.18. *Inking In*

The details on the field sheet should be inked up as soon as their positions are certain. This will normally mean that a little work will be inked on each evening of the survey. Inking is essential in plane tabling since much erasing of pencil work is necessary, both in the fixation of new detail and in surveying contours. A Pelican Graphos pen or similar type is recommended for the surveyor since the line thickness can be guaranteed and a steady flow of ink is ensured. Curved lines will be inked with the aid of curved templates or flexible moulded plastic rods specially designed for the purpose. In general it is easier to connect a straight line to a curved one than *vice versa*. If a mistake has to be erased, it should be left until all other inking is complete, since no ink will then require to be put on the damaged surface.

1.1.2.19. *Lettering*

Surveyors are advised to use Egyptian lettering on their field sheets since it is legible and simple to draw. It is easier to produce a neat effect if sloping letters are employed, since any small variation in slope will not be detected by the eye. Figure 1.13 gives an example of this lettering. Horizontal guide lines should be drawn and the words pencilled in to avoid missing out a letter. A few sloping guide lines also assist the eye when drawing the down strokes of the sloping letters, but no attempt should be made to draw 'boxes' into which a letter is placed. Each title should be thought of as a unit of shapes which is then split into smaller units—the words—and finally into the letters. The reason for this approach is that the size and spacing of each letter depends on those letters between which it lies. For example, when two t's come together as in the word S E T T I N G, the inside horizontal bars of the t's are shortened slightly to avoid a large gap between these letters. Round letters such as O, S, R, etc. are

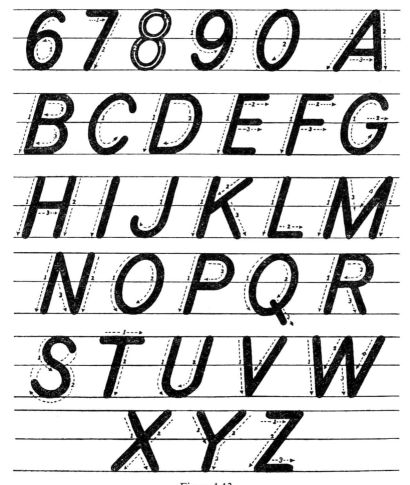

Figure 1.13

Figure 1.13 (continued)

drawn very slightly taller than the square letters such as E, F, etc to give the effect that all are of the same height. Natural draughtsmen are born, and most people have to practice diligently even to reach a fair standard. Practice is only useful if a serious attempt is made to produce the best results, and nothing is gained from a casual attitude to practice. The various stencil sets available do help the surveyor with his large titles, though they too require some practice if a pleasing effect is to be achieved. In this same connection, dry transfer lettering such as Lettraset is very effective. In large organizations, the field sheets will be fair drawn by draughtsmen using all forms of modern aids such as stick-on titles produced in Photonymograph machines. The surveyor is merely required to produce a neat legible document.

1.1.2.20. *Compilation and Map Production*

The individual field sheets are finally *compiled* into a complete map with the assistance of a master grid and the control points of the survey. The compilation is then photographed and *bromides* in blue are made. If a colour map is to be produced, all information to be in one colour is inked up on the bromide in a dense opaque ink, usually black. From this base the printing plate for that colour is made since the uninked blue portion will not photograph. A plate is made for each colour and the map printed by overprinting each colour on the next. Various combinations of colours are available by use of stipples, half-tone screens, and overprinting. For a more detailed account of these techniques the reader should refer to W. K. Kilford, *Elementary Air Survey* (London Pitman, 1963).

1.1.2.21. *Recent Developments and Trends*

In recent years, large organizations have supplanted ink work on paper with an etching process called *scribing*. In this process a hard coating on plastic or glass is etched out mechanically or photomechanically with a scribing tool consisting of a sapphire or hard steel point. Either a negative or a positive may be obtained directly. The line quality is of a very high standard and consistency, which will stand considerable enlargment. Work may be scribed directly from a photogrammetric plotter. To meet the tremendous increases in demand for maps of all kinds, and to keep pace with the greatly increased rates of production of both field survey techniques and photogrammetric plotting methods, there are indications that fully automated map production will become an economic reality, at least for the large mapping organizations. For further information on cartographic processes the reader should refer to the recently published *Cartographical Journal*†

1.2. RECONNAISSANCE

Once the purposes and precision of the survey have been decided upon, work commences on the survey itself. The first and perhaps the most important stage

† *The Cartographic Journal* (British Cartographic Society: London. 1963). Published twice a year.

is the *reconnaissance* or, more shortly, the *recce*. The most experienced surveyor is responsible for the recce, since only he understands the full technical and administrative problems involved, and often his wide experience saves the mistakes of the past being repeated. A few pounds spent on a good recce may save a great deal of money in the future.

The first stage is to analyse any information available concerning the area to be surveyed and any previous surveys that have been done. Very often, information about former surveys can only be obtained by the personal efforts of the surveyor responsible for the recce, and the very existence of old work may not be suspected until the area is visited on the ground. Documents to be studied include old maps, air photography, railway surveys, town plans, etc, and on the basis of this work a paper recce is produced in the form of a possible survey shown on a diagram. Such diagrams are usually drawn in pencil since many changes will be made when the ground is visited. At the preliminary stage, the datum of the survey is assessed. For example, the information about triangulation stations to which the survey will be tied is obtained and examined.

The second stage is to visit the area in question and decide where the points are to be located, the accessibility of stations, and the manner of fixation, etc. It may not be necessary to visit all proposed stations of a small survey involving short sides of a mile or so, but in large major control surveys with long sides, the stations are probably to be marked by pillars, and all lines should be checked for intervisibility.

1.2.1. RECCE DIAGRAMS

In the course of the recce, a diagram is drawn up showing possible stations and lines to be observed. Such a diagram should be made to scale, either based on an old map, or on a plane table survey made for the purpose, or on an air photo mosaic. When the final proposals are decided, the diagram should be drawn in ink so that copies may be made by Sun printing or by dyeline. A growth diagram capable of showing all stages of the field work through to completion should be initiated. An example of such a diagram is given in figure 1.14, which is the growth recce and observation diagram for a tellurometer control survey of the aerial photography of part of Somerset carried out on a field course by students. Since it was a training survey, a wide variety of methods was used. It will be seen that the diagram shows the state of the observations at the particular instant of time reached. Each stage is portrayed by a symbol created by overdrawing the symbol for the previous stage. As the work proceeds, copies may be taken off and sent with reports to headquarters. The student is advised to study figure 1.14 in some detail.

1.2.2. RECCE DESCRIPTIONS OF STATIONS

In a large survey to be carried out over a period of time by more than one surveyor, it is essential that a record of each proposed station is made. In a small survey a master diagram together with small sketches of the vicinity of each point will suffice, e.g. a diagram such as that of figure 1.14 and the 'vicinity sketch' of figure 1.16 will probably suffice, together with one general description

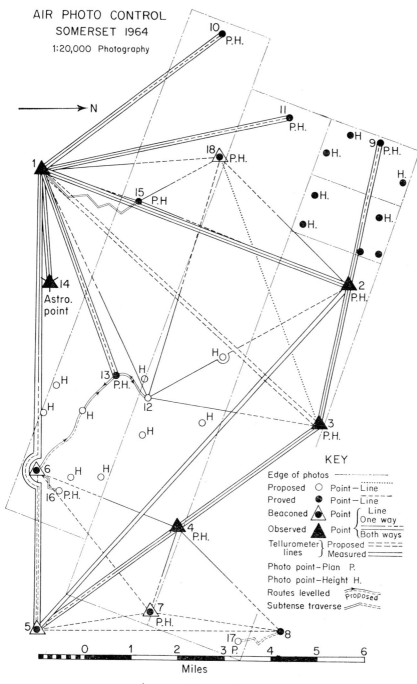

Figure 1.14

of the locality, services available, access, labour, etc. In major control schemes, a station description such as that of figures 1.15 and 1.16 will be required. By far the major part of the station information will be obtained on the recce: This will include the following:

(i) Access information and diagram, particularly if there are no roads, and pillars are to be constructed.
(ii) The approximate location of the point.
(iii) Any aerial photography available.
(iv) The approximate time taken to reach the station from the road, and the nature of the terrain crossed.
(v) The availability of materials to construct a permanent mark, i.e. sand, stones, and water.
(vi) A description of the nature of the ground on which the mark is to be constructed, e.g. sand or solid rock.
(vii) The likelihood of obtaining local labour and its source.
(viii) An accurate panoramic sketch of the view from the point such as that given in figure 1.15. Where the view is obscured by vegetation as in this example, an attempt should be made to complete the probable view from a point beyond the obstruction, and if necessary some clearing of timber may be required. An accurate panorama drawn with the aid of a theodolite is of inestimable value not only for the particular survey in hand, but also for any future work that may be contemplated. Such sketches avoid having to revisit the hill on a future recce, assist the observer to locate his points, and may even be used to control air photography if the hills to which observations were made can be identified on the photos with certainty. The sketch drawn by a surveyor is probably more useful than an all-round photograph, since it is more selective, though a photograph is a useful addition to the sketch.
(ix) Some object which may be identified from afar should be left at the station, e.g. a lone tree should be left if the hill is cleared.
(x) In thick bush country large daubs of paint on trees will last some years to indicate the route to the beacon, whilst a miniature beacon constructed from rough wood serves to mark the point where it leaves the road without being conspicuous.

1.2.3. RECCE FOR LEVELLING

The recce for levelling involves the following main points in addition to the general ones:

(i) The route should be perambulated so that steep slopes may be located and alternative routes adopted to avoid them.
(ii) Where possible a hard surface such as a made-up road should be chosen. If routes are not made-up, or open ground has to be crossed, pickets will have to be driven during the levelling. The recce should indicate this fact and suggest the type of picket required.

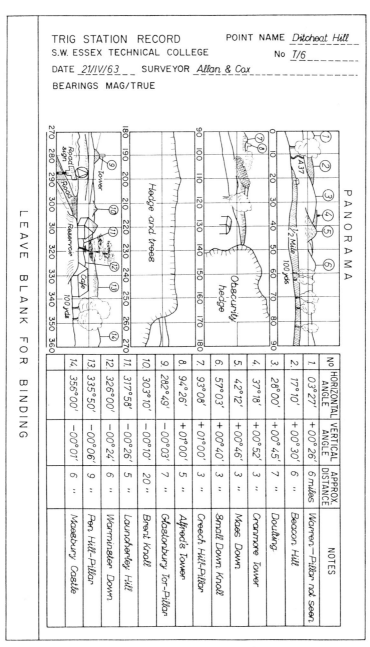

Figure 1.15

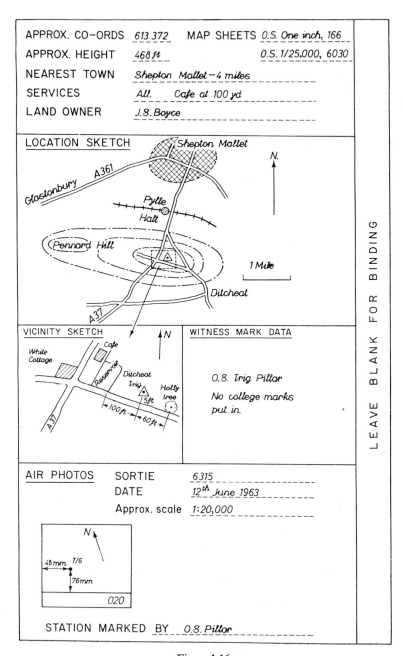

Figure 1.16

(iii) Particular care is required in the siting of permanent bench marks, and the levelling route may have to make a detour to include an area of solid rock.

1.2.4. CONSTRUCTION OF PERMANENT MARKS

In all classes of survey some mark giving the exact location and height of each main survey station should be left. These will vary from temporary ones, such as pegs used in setting out works, to permanent ones, such as triangulation pillars. The future usefulness of the survey depends on the precision with which these marks are located, their permanence, and the accuracy with which their position may be restored should they be removed or damaged. Generally, more

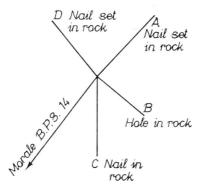

Figure 1.17

than one mark is placed to locate a survey station. Each main station is referenced to four *witness* or *reference marks* located close to it, but at slightly varying distances from it. Hence the chance of complete destruction of all five marks is slender, unless the destruction is deliberate. Figure 1.17 shows a typical layout of main and witness marks at a triangulation station, whilst table 1.1 gives the data observed from the main station. Since a bearing and distance to the main station may be computed for each witness mark only one of the latter is required to restablish the former. In practice all data would be used where possible, by

Table 1.1

Observations to witness marks

Point	Horizontal	Vertical	Slant distance from trunnion axis
morale	00° 00′	—	
D	113 05	−41° 06′	28·78 feet
A	181 59	−24 46	24·10 "
B	271 05	−22 28	13·08 "
C	343 42	21 48	19·23 "

describing arcs from the witness marks, and by semi-graphic resection at the main station, the drawing being on a horizontal board placed at the approximate position of the point to be relocated (see Chapter 6). In built-up areas, permanent features such as drain corners, etc. can be used as main stations as well as witness marks, though care must be exercised in describing the exact point used for the purpose. In engineering works, vital pegs are liable to be destroyed or moved by vehicles, diggers, etc, hence they must be referred to more permanent marks situated well back from the works. Re-establishment of pegs may occur several times per day in some types of work. Semi-permanent stations are easily marked on roads, pavements, etc. by road nails.

1.2.5. CONSTRUCTION OF A TRIANGULATION PILLAR

The construction of a triangulation pillar is now described as an example of the construction of permanent marks generally. Figure 1.18 shows details of a

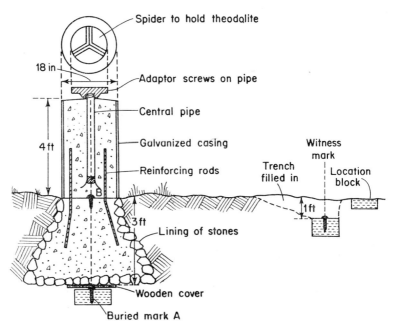

Figure 1.18

pillar set on a foundation in sand or soil. The hole for the foundation is dug roughly conical in shape so that the observer's weight will not rest directly on the foundation. A block of concrete made to hold the buried mark, which is either a brass bolt or an old cartridge case, is made in position of a two cement/four sand mix. A wooden cover is placed over the buried mark so that it is independent of the pillar foundation, and that this latter may be dug up in future without disturbing the mark. Four reference marks are also located at this time about one foot below ground and to which trenches are dug to enable the observer to view them from a theodolite set over the main mark. When these concrete blocks are dry, the theodolite is set up over the mark A and observations and measurements recorded as shown in table 1.1, Figure 1.17. A high trestle is then made and erected over the buried mark so that a nail may be fixed plumb above it. This trestle has to be about eight feet tall and must not be disturbed during the construction of the pillar. The concrete aggregate for the foundation is then mixed, i.e. two parts cement, four parts sand, and six parts small stones, which often have to be broken *in situ*. The inner surface of the foundation hole is lined with stones or paper and the aggregate poured into it. At ground level a mark B is plumbed down from the nail on the trestle. Three reinforcing rods are placed in the foundation at this stage. When the foundation is hard, the pillar mould consisting of a galvanized iron sheet moulded into a cylinder and bolted along the joining edge is placed in position, and a similar concrete aggregate is poured inside, taking care to locate the central pipe truly central and vertical with the aid of the trestle nail and a builder's level. Great care should be taken to tamp down the concrete to eliminate air bubbles. The pillar top is finally made of with a neat cement-sand mix and the name, number, and date of erection are printed deep into the top surface. Wooden dies cut specially are often used.

The protective pillar cap is well greased and screwed home on the central pipe. A pillar builder's report is made out giving the dimensions of the pillar, its foundation, and the observations to the witness marks. The outside is normally painted white with a foot-wide black band at the top. When the pillar is dry, a central vane may be left in position (see figure 6.2e). The triangulation observer will make observations to the witness marks from the centre of the pillar so that the relative positions of the buried permanent mark and the top of the pillar used in the survey can be computed if necessary. In some cases, the outside galvanized sheet is removed before the builder leaves the site, to prevent a thief doing likewise with less care. Pillars constructed on solid rock foundations have to be keyed directly on to the rock. Three holes are jumped at least one foot deep in the rock with a tungsten-tipped hand drill gently tapped with a hammer. The reinforcing rods are then concreted in position, and the pillar constructed thereafter in the normal way. In other types of foundation some ingenuity is required of the surveyor to try and locate a pillar as permanently as possible. One of the main advantages of the survey pillar in wild country, is not only its great observational stability but the ease with which it can be found. The Ordnance Survey of the United Kingdom, make a more elaborate pillar which has a permanently located spider to accept the theodolite, and has a flush bracket bench mark positioned on the side. The general principles of erecting ground blocks are identical to those described for the pillar.

1.2.6. PERMANENT BENCH MARKS

A *bench mark* is a point of known height to which lines of levels are tied. Figures 1.19, 1.20, and 1.21 show three types of bench marks used by the Ordnance Survey of Great Britain:

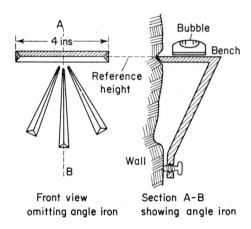

Figure 1.19 Cut bench mark

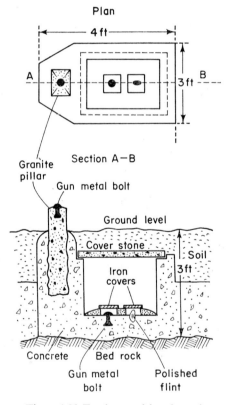

Figure 1.20 Fundamental bench mark

Introduction to Field Work

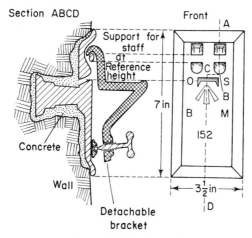

Figure 1.21 Flush bracket

1 2.6.1. *A Cut Bench Mark*

This is found on vertical faces of buildings, bridges, etc. The centre of the horizontal bar is the point to which heights are referred. An angle iron as indicated in the sketch serves as a 'bench' on which a levelling staff may be placed when taking readings.

1.2.6.2. *The Flush Bracket*

These are normally placed on a vertical wall of a building or a triangulation pillar. They are cast in brass and are concreted into a hole prepared in the wall. They are situated about every mile on a precise levelling line, and about 3 to 4 mile intervals on lines of secondary levelling.

1.2.6.3. *The Fundamental Bench Mark*

Such marks are situated at sites specially selected for their stability, i.e. on solid bed rock which is not liable to subsidence. In the U.K. they are located at about intervals of 30 miles. The open reference point is for the use of day to day levelling, whilst the buried marks are used only for special purposes.

Concreted bench marks should be emplaced at least six months in advance of the levelling operations from which their values will be derived so that their stability is ensured.

1.2.7. THE IDENTIFICATION OF CONTROL POINTS ON AERIAL PHOTOGRAPHY

In precise photogrammetric mapping, the positions of ground control points are required, and these points must be related to the photography with a very high degree of accuracy. Such points are termed *photo-points*. Points used to control the map in eastings and northings—planimetric photo-points—have to be selected in such a way that their positions may be surveyed accurately, that they may be identified on the photography, and that they appear as small distinct points when viewed in, the photogrammetric plotting machine. Height control points on the other hand are situated on areas of flat level terrain on which the dot of the plotting machine may be placed with certainty, and whose plan positions are invariably obtained from the photography itself. Accurate positioning of a height control is not required if the point is typical of the level area in which it is situated; any error in plan position will not affect its height. Where possible a control point will satisfy both requirements for plan and height, though many points satisfy only one of these functions. The amount of ground control required in photogrammetric mapping varies with the equipment available, the scale and type of photography, the accuracy required, and the time and money available for the job, a detailed consideration of which is beyond the scope of this book. However, it is essential that the field surveyor has a good working knowledge of basic photogrammetric methods, and that there is constant cooperation between the surveyor and the photogrammetrist, if the most efficient combination of field and office methods of map making is to be achieved.

1.2.7.1. PLANIMETRIC PHOTO-POINTS

Planimetric control points are of two kinds:
(i) The *post-identified* point; one which is identified on photograph after the photograph has been taken.
(ii) The *pre-identified* point; one which is marked off on the ground before the photography is done. This is generally known as pre-marking.

Generally speaking, for mapping at scales smaller than 1:10 000 either type of photo-point will suffice, and at scales larger than this, pre-marking is desirable and in some cases essential.

1.2.7.2. *Post-Identified Points*

The advantage of post-identified points is that the surveyor will carry a set of the photographs in the field, which he will use to select routes, and as the basis of sketches and diagrams; and on which he will name details to be shown on the map. Points selected as controls must be unambiguous, and must appear as definite points when viewed in the plotting machine, which may be at five times the scale of the field prints. Examples of photo-points are small bushes, rocks, one corner of large drain covers, etc. These points are tied into the survey control network by any suitable survey method. The method of radiation using electromagnetic distance measurement is much favoured today, and often permits the field surveyor to fix points where hitherto much time would have been spent in taped traverses. An accurate description of the control point is made, and its position pricked with a very fine needle on the print viewed stereoscopically. The position of the hole is circled both on the face and on the back of the print and suitably numbered. The description of the point is best carried out as in figure 1.22. The details contained within the circle drawn on the photograph are drawn to approximate scale within a large circle, and the photo-point is clearly marked. Only those details which appear on the photograph should be shown, since it may be from these alone that a future identification will have to be made if the original print is damaged or lost. If some detail of importance is not visible on the photograph it may be worth while drawing a second sketch on

Introduction to Field Work 31

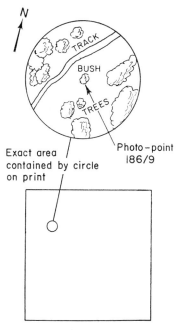

Figure 1.22

which this is shown. Identification of a point on a print may take several hours in bush country, and one may often require to pick one's way laboriously from tree to tree until one is satisfied that the correct identification has been made. If one is not sure of an identification, another area should be tried until one is sure.

1.2.7.3. *Pre-Marked Points*

The pre-marked point has the advantage over the post-identified point in that it is very much more exact and clear on the photography, if an efficient system of marking is used. Some map or photo-mosaic must exist to enable the positions of the controls to be carefully selected with accuracy, and the subsequent photography must also conform to the flight lines used as the basis for planning. The type of mark left on the ground depends on the particular ground in question and may vary in one place according to the seasons. In a recent photogrammetric experiment in Rhodesia† an elaborate mark was designed for very accurate photogrammetric work (see figure 1.23). However, less elaborate and less expensive marks, such as crosses of oil poured on the ground, and sheets of white calico three feet square, have been used effectively. The pre-marked points are also tied into the survey by any suitable method.

1.2.7.4. *Height Control Points*

A description of the procedure adopted for height controls is given in §8.7 dealing with altimetry, since this method of ground heighting is widely used.

† L. Eekhout, and F. E. Holmes, *An Investigation of Photogrammetric Methods for Cadastral Surveys*, Paper No. 29, Commonwealth Survey Officers' Conference 1963 (H.M.S.O: London).

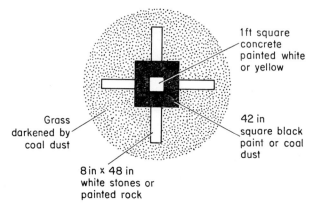

Figure 1.23

1.2.8. FIELD BOOKS

The actual manner in which field observations are recorded is much more important than many suppose. The basic rule is that all booking should be completely intelligible to some second person familiar with survey work. All figures must be clear and unambiguous, the unit of measurement clearly stated, and all documents should be signed so that their origin is known. Except in large government organizations in which the division of labour is large, printed field books are less convenient than a general page ruled in quarter or fifth of an inch squares. Possible exceptions to this are levelling, Geodimeter observations, and tacheometry in which further processing of the observations is readily carried out in the field book. Many surveyors prefer a small book, about 9 in. × 6 in., which may be placed in a pocket. Each book should carry a unique number, as should the pages of each book. If each page bears its full number, e.g. F.B64/20/P8, no difficulty is experienced in identifying pages when the book is split for photocopying. Pages are easily numbered with a repeating enumerator. All writing and figures should be clear; and any mistakes should be crossed out and the correction written above. On no account should figures be overwritten since the result is always ambiguous even to the person who wrote it. Opinions differ about the relative merits of ink, pencil, and ball-point pen, though the last seems to possess all the advantages, i.e. it is capable of taking a carbon copy, is difficult to erase without detection, and is comparatively permanent. Consideration should be given to the information placed in the page position most easily seen, e.g. the top right-hand corner. Normally the station number is placed here since this information is most wanted when working later with perhaps several hundred field books. The front cover of each book should list the contents of the book. If carbon copies are taken in the field, as may be the case if photocopying is impossible, a fresh carbon should be used for each book. If books are thereafter to be split, sufficient margins should be left. Actual booking in the field is best carried out with the book clipped to a piece of board and the pages prevented from turning in the wind by a large elastic band. Useful survey data can be written on the reverse side of the board, e.g. foot to metre conversions. If a large number of surveyors are working together on the same job, each surveyor should be assigned a series of numbers to be used solely by him if confusion is to be avoided. Should a field book become almost illegible due to immersion in water, etc., it will require to be copied by the surveyor. If this is so, the original must be retained for reference, and the copy must be clearly stated as such. Generally the copying of field books will be unnecessary, and is a thoroughly undesirable practice especially if only to cover up bad writing in the original.

1.2.9. ABSTRACTS

The abstract is a document giving the salient data of a field book which will be useful to a computor. For example, in the abstract of angles observed at a station, the means of faces for each pointing will be listed and the final mean for each pointing worked out. If several reference objects (R.O.) are used, all final means will be reduced to one R.O. Hence a computor does not require to work with the field books, and can obtain the information he requires from the abstract. In very important work such as primary triangulation, a separate abstract is made by each of two persons and the final results compared for identity.

1.2.10. CHECK COMPUTATION IN THE FIELD

In most types of survey, some degree of field check is advisable before leaving the field. Any inconsistencies and omissions may be detected and rectified whilst the surveyor is on the spot. The extent of field computation depends on circumstances. For example, if a complete survey of a distant country is carried out, computations in the field will reach a high stage of finality since the cost of returning to rectify omissions would be prohibitive. Check computations made by the surveyor in the field are very useful to the computor finally producing results, since they act as a check on the results, and set out the logical sequence of computation rather than the chronological order in which the work was observed.

1.2.11. RECORD KEEPING

All documents concerned with a survey should be retained for future reference and housed in a systematic manner that will make their future accessibility simple and speedy. The documents of an isolated survey are best stored together, whilst surveys of large continental areas will probably be easier to locate if stored according to map sheets or graticule lines. In a map series bounded by a graticule, a high degree of permanence of the reference system is obtained. The problem of storing important survey documents for a long period of time is one which is constantly occupying the minds of records officers. The mere problem of finding space to house the vast quantities of data is difficult to solve, though the copying of documents on microfilm may provide the solution.

1.2.12. STORES

A complete record of stores issued to each surveyor should be kept together with their signatures indicating responsibility. From time to time a complete check of equipment and stores should be made and losses made good. At periods of inactivity in the field, e.g. due to bad weather or whilst the surveyor is computing, opportunity should be taken to repair tents, oil tripods, and generally overhaul equipment. When transporting stores, all items should be housed in strong crates and should be clearly labelled with their destinations, preferably stencilled on with all-weather paint.

1.2.13. TRANSPORT

One of the most important aspects of a surveyor's activity is transportation, now-a-days by vehicle, though mules, camels, schooners, etc. are still in use. A surveyor must be able to drive a car and have a good working knowledge of its operation, maintenance, and repair. A wide range of spare parts should be carried together with ancilliary equipment such as special jacks, sheets for laying on sand, hawsers, etc. On arrival in a new territory, the surveyor should consult local people concerning the particular hazards of motoring in the country and any advice given should be conscienciously followed. For example in much of Africa, a wire mesh fitted to the front of the radiator prevents grass seed from entering and impairing its proper function. In some areas as many as three punctures a day are experienced, hence repair equipment must be carried. In many parts of the world, the roads become impassable in wet weather unless the vehicle is fitted with chains, used in the United Kingdom only in snow conditions. Regular maintenance of the vehicle is paramount and should have the personal attention of the surveyor.

The state of batteries should be watched constantly, petrol and oil consumption noted daily together with the mileage travelled. The actual driving of the vehicle in unfamiliar conditions can present problems which are very often solved by advice from locals with years of experience of these conditions. For example, the best gear to use in mud is third, and not second as the uninitiated suppose. The type of tyre best suited to various conditions is worth noting, whilst the importance of the track width of vehicles in sand country cannot be underrated, since a non-standard vehicle, which will not fit the ruts of the road, will easily turn over at quite slow speeds.

1.2.14. ACCESS TO LAND

The surveyor should familiarize himself with the statutory conditions pertaining to the access to property for the purpose of surveying, and act accordingly. At all times tact and good manners in dealing with owners is not only civilized but also essential if the work is to proceed unhindered. Should damage to crops or property be inevitable, a third party should be called in to adjudicate, preferably before the damage is done so that he will be able to give a proper assessment.

1.2.15. LABOUR

It is one of the normal duties of the field surveyor to recruit his labour force as required for the job, and to train them according to his requirements. Rates of pay should accord with local rates, advice concerning which may be obtained from various firms operating in the area. Additional payment will normally be required if the work is away from the labourer's home, and if conditions are arduous and at times lonely. The successful surveyor must maintain good labour relations, which can be obtained only as the result of fair, firm treatment tempered with understanding.

1.2.16. FINANCE

The surveyor will be responsible for the financial organization of his work. Prior to work in a new area, arrangements are made with a bank to receive and pay out the necessary funds. Normally an imprest account is opened in which there is sufficient money for the day-to-day running of the survey. Large amounts such as for the purchase of vehicles will usually be given special approval from headquarters and special financial arrangements will be made accordingly. The newcomer to a firm should make himself familiar with the system to be adopted and in particular he should make himself aware of his own particular responsibility, and the manner in which accounts are to be presented. Accounting in government departments is usually under the following heads:

(i) Ocean and air travel.
(ii) Local travel.
(iii) Labour.
(iv) Salaries to permanent staff.
(v) Transport.
(vi) Stores and equipment.
(vii) Miscellaneous items.

All payments must be accompanied by a receipt which is given a serial number and stored in a safe place. If accounts are regularly kept up to date much worry and wasting of time at the end of a month can be avoided.

1.2.17. REPORT WRITING

From time to time during the survey, and at its completion, the surveyor is normally required to submit reports of progress to his headquarters. As already stressed, the most effective and efficient way of indicating technical matters is by a growth diagram. The final survey report should contain all necessary and relevant information, but should be arranged in such a way that each person who will eventually be concerned is not required to read through the whole report to gain the information he wants. This means that the report is divided into sections, for example as follows: (*a*) Administration; (*b*) Finance, (*c*) Technical. A summary of the salient points of these three sections should be given so that a general picture of the survey can be obtained without excessive reading. The following are the main headings and information required in a report: Location of survey; time and dates of its execution; personnel involved; weather and other conditions; finance; technical (including datum used, unit of measurement, equipment, results of field computations, etc.); days lost due to weather; list of persons to whom the report is sent.

2 *Introduction to Computation*

2.1. THE CALCULATING MACHINES

2.1.1. INTRODUCTION

Essentially a calculating machine is designed to perform the four fundamental processes of arithmetic—addition, subtraction, multiplication, and division—all calculations being a combination of these processes. The field surveyor uses a calculating machine for only a small part of his working life, and often when tired and in poor lighting conditions. The operation of the machine should therefore be simple and not require concentration, many modern machines carry devices so infrequently used by the surveyor as to be unnecessary refinements. Essential features are clarity in figuring and ease of setting.

Semi-automatic and automatic electric machines reduce the manual labour in operation but for learning the techniques of operation the hand machine is the best. In this section the reader is assumed to be using a hand machine. Although calculating machines will always give the correct answer if properly used, the user would be well advised to check his results by a second calculation using a different machine process. As well a rough check on the results should always be made.

2.1.2. ELEMENTS OF DESIGN

The first calculating machine in any way similar to those of today utilized a stepped wheel termed a Leibniz wheel after its inventor. The Leibniz wheel, illustrated by figure 2.1, consists of a cylindrical gear wheel with nine teeth of decreasing length cut in its circumference. Rotation of the Leibniz wheel drives a secondary ten toothed gear wheel which is mounted on a spindle parallel to the axis of the Leibniz wheel. The amount of rotation depends upon the number of teeth of the Leibniz wheel that it comes into contact with and this partial rotation is indicated by the rotation of a counter wheel numbered from 0 to 9. The secondary ten toothed gear wheel is moved along its spindle by means of a setting lever.

An alternative mechanism is the Odhner wheel, figure 2.2 which consists of a flat disc with nine retractable teeth. As a setting lever is moved through ten positions a slot moves across a portion of the disc retracting or releasing the teeth so that the number standing proud from the disc correspond to the numbered position of the setting lever. The teeth standing proud from the disc come into contact with a secondary ten toothed gear wheel which with complete rotation of the Odhner wheel is partially rotated and the amount of rotation is indicated by an attached or directly driven numbered drum.

In the modern machine a bank of Leibniz or Odhner wheels is rotated by a handle or electric motor, 'tens transmission' ensures that for each complete

38   *Practical Field Surveying and Computations*

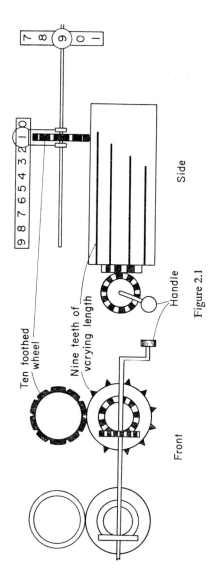

Figure 2.1

Introduction to Computation

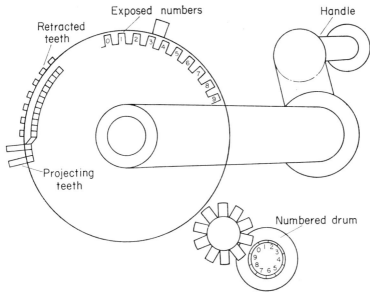

Figure 2.2

rotation of the secondary gear wheel a projection is brought into contact with the next secondary gear wheel causing it to rotate through one unit, i.e. a tenth of a rotation.

With components as described above the calculating machine has a setting mechanism which positions the secondary gear wheels, a rotating mechanism and a series of drums indicating the amount of rotation of the secondary gear wheels. The setting is indicated in a setting register S.R., the rotation of the Leibniz or Odhner wheels by a multiplier register M.R., and the resultant product appears in a product register P.R. These basic features are illustrated by figure 2.3.

Any multiplicand can now be multiplied by a small number, a large number would require excessive turning of the handle. To eliminate this the setting register can be moved sideways with respect to the product register so that one rotation of the bank of Leibniz or Odhner wheels can be made to correspond to 10^n turns of the handle. A position indicator P.I. indicates in the multiplier register the position of the setting register relative to the product register and thereby the value of n. Sometimes the position indicator is not clearly marked,

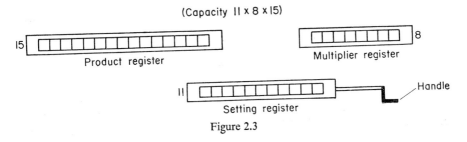

Figure 2.3

on an unfamiliar machine of this type rotation of the handle forward (clockwise) will bring a 1 in the window of the multiplier register where the position indicator should be. Usually a mark will then be found to lie near to or under this window, this mark moving with movement of the setting register.

2.1.3. THE SETTING REGISTER

There are three usual ways of entering a number in the setting register.

2.1.3.1. *Lever setting*

A series of levers carrying the ten toothed gear wheels can be moved individually along a bank of Odhner or Leibniz wheels to occupy one of ten positions marked on the casing (see figure 2.4). A number entered can be changed at will

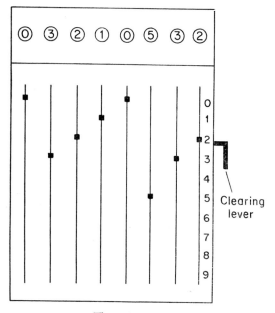

Figure 2.4

by moving individual levers. A single clearing is usually provided, essentially a simple bar, which will bring all the setting levers back to zero.

2.1.3.2. *Keyboard setting*

Instead of moving manually the ten toothed gear wheel its position is controlled by a column of ten keys numbered 0 to 9 (see figure 2.5). Each wheel has its own column of keys, depression of a key releases any other key already depressed, thus a number can be changed at will as in the lever setting machine. A release key for all keys is always provided.

2.1.3.3. *Ten key setting*

In this the bank of Odhner wheels can be moved across the machine and as each wheel passes from right to left through a gate the ten toothed gear wheel

Introduction to Computation 41

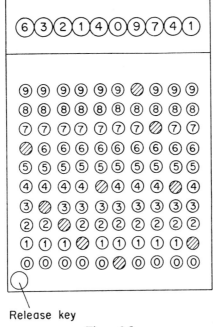

Release key

Figure 2.5

is positioned by depression of a key, its position varying according to which of ten numbered keys is depressed (see figure 2.6). A clearing lever is provided to draw the Odhner wheels back through the gate returning each to a zero setting. The disadvantage of this system lies in that the setting, once made, cannot be changed, the register is however very compact and setting can be by touch as in typewriting.

2.1.4. THE PRODUCT REGISTER AND DECIMAL POINT INDICATORS

By rotation of the handle the number set is transferred into the product register. The relative positions of the setting and product registers can be varied and a product obtained. Tens transmission ensures a carry-over into the next column of the product register when necessary. Each register is provided with a number of sliding markers which permit the decimal point to be indicated in any calculation. The positioning of the decimal indicator in the product

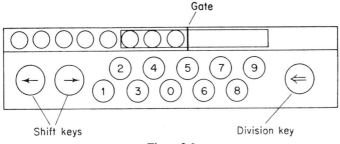

Figure 2.6

register depends upon the positioning of the other two decimal indicators, the normal rules of arithmetic apply, for instance a number with three digits after the decimal point multiplied by a number with two gives a product with five digits after the decimal point. This point is considered in each of the examples given below.

2.1.5. CLEARING LEVERS, TRANSFER, AND SHIFT KEYS

Levers are provided to clear all three registers, in some machines an extra lever will clear two registers together, usually the setting and multiplier registers. In all keyboard machines the setting register will automatically clear after a single turn of the machine unless a locking device, usually labelled 'Rep', is depressed.

The normal action of the machine is to transfer the number set to the product register. The reverse operation is a feature present in some machines, frequently this is accomplished by use of the clearing levers. The advantage of such a device is considered in the examples given below.

The movement of the setting register relative to the product register is controlled by shift keys and at the same time the position indicator moves in the multiplier register or that register moves bodily relative to a fixed mark. Tabulation devices to control this movement exist on some machines, usually these return the position indicator and setting register to a pre-selected point after each calculation.

The transfer device may be linked to a separate register not necessarily visible in which numbers can be stored and transferred to the setting register at will. Frequently such stores do not have tens transmission and clear before a new number is transferred in, so cannot accumulate totals.

2.1.6. MISCELLANEOUS FEATURES

There are many features particularly on electric machines which appeal to the experienced computor; frequently they require a different technique on different machines. A list of the more common would include:

Automatic squaring, double product registers, product counter, print-out of products, automatic square roots, addition in the multiplier register.

2.2. BASIC PROCESSES IN MACHINE CALCULATION

2.2.1. ADDITION

Example: 398·176 + 275·34 + 1 729·0438

(1) Clear all registers and position the P.I. as shown

                                                                                                                                                     P.I.

P.R. 0000 000 000 000 000   M.R. 000 000 00

S.R. 00 000 000

(2) Examine the calculation and note that the third number has four digits after the decimal point; all the rest must be regarded as having the same number of digits after the decimal point thus the example becomes

398·1760 + 275·3400 + 1 729·0438

Introduction to Computation 43

(3) Set 398·1760:

P.I.
P.R. 0000 000 000 000 000   M.R. 000 000 00
S.R. 03 98ỉ 760

N.B. The position indicator is indicated by the P. of P.I. and the decimal point by · over the first digit after the decimal point.

(4) Make one forward turn of the handle:

P.I.
P.R. 0000 000 003 98ỉ 760   M.R. 000 000 01
 03 98ỉ 760

Set the position of the decimal point marker in the P.R.

(5) Continue entering the remaining numbers in turn clearing only the setting register, the final position being:

P.I.
P.R. 0000 000 024 02ṡ 598   M.R. 000 000 03
S.R. 017 29ȯ 438

2.2.2. SUBTRACTION

Example: 729·043 8 − 264·374 13 − 36.176 12

(1) Clear all registers and position P.I. as shown:

P.I.
P.R. 0000 000 000 000 000   M.R. 000 000 00
S.R. 00 000 000

(2) Examine the calculation as in (ii) of addition, it becomes

729·043 80 − 264·374 13 − 36·176 12

(3) Set 729·043 80 and transfer to the P.R. by a forward turn of the handle giving:

P.I.
P.R. 0000 000 072 9ȯ4 380   M.R. 000 000 01
S.R. 72 9ȯ4 380

(4) Clear the S.R. and set 264·374 13 giving:

P.I.
P.R. 0000 000 072 9ȯ4 380   M.R. 000 000 01
S.R. 26 4ṡ7 413

(5) Make a backward turn of the handle giving:

P.I.
P.R. 0000 000 046 4ȯ6 967   M.R. 000 000 00
S.R. 26 4ṡ7 413

(6) Clear the S.R., set 36·176 12 and make a backward turn of the handle giving:

P.I.
P.R. 0000 000 042 8ȧ9 355   M.R. 999 999 99
S.R. 3 6ỉ7 612

N.B. (a) In the final operation of the subtraction of the complement in M.R. indicates that in that register the values is negative. Viz. 1 forward or positive turn and 2 backward or negative turns. This acts as a check on the number of operations performed. In some machines complements in the M.R. appear as red figures or by displaying a red indicating device. In the other registers negative amounts always appear as complements (see §2.2.3 below).

(b) The addition of zero to increase the number of digits after the decimal point is only a convenience for machine calculation. The final answer should always be brought to its correct value, that is to the same number of digits after the decimal point as in the figure with the smallest number of digits after the decimal point, viz, in addition 1 729.04 and in subtraction 428·4936.

2.2.3. COMPLEMENTS IN SUBTRACTION

Example: 37 614 − 38 414

(1) Setting 37 614 first and operating the machine as in normal subtraction described above gives:

P.I.
P.R. 9999 999 999 999 200   M.R. 000 000 00
S.R. 00 038 414

N.B. A series of nines has spread across the P.R.—these are called the 'bridge of nines' indicating that the amount there is negative. On most machines, but not all, a bell rings when the 'bridge of nines' appears or disappears.

(2) To find the correct negative value subtract each digit from 9 working from left to right until the last non-zero digit is reached which is deducted from 10. Zeros at the right hand end of the complement are written down as zero, In example: 0000 000 000 000 800, i.e. −800.

(3) The correct negative value is set in the S.R. so:

P.I.
P.R. 9999 999 999 999 200   M.R. 000 000 00
S.R. 00 000 800

(4) Make a forward turn of the handle giving:

P.I.
P.R. 0000 000 000 000 000   M.R. 000 000 01
S.R. 00 000 800

The zero value in the P.R. indicates that the value in the S.R. is correct.

2.2.4. ADDITION AND SUBTRACTION OF VERY LARGE NUMBERS

If numbers exceeding the capacity of the setting register have to be added or subtracted they can be entered into the product register in parts.

Example:
389 647·126 + 347 264 194·004

With an eight digit setting register

(1) Set the last seven digits of the first addendum after bringing all numbers to the same number of digits after the decimal point giving:

P.I.
P.R. 0000 000 000 000 000   M.R. 000 000 000
S.R. 09 647 126

Introduction to Computation

(2) Make one forward turn of the handle, clear the S.R., set the remaining part of the number and move the setting register along seven places giving:

P.I.
P.R. 0000 000 009 647 į26   M.R. 000 000 01
S.R. 000 000 38

(3) One forward turn of the handle sets the remaining part of the number. The second and remaining addendums are entered in the same way.

2.2.5. MULTIPLICATION

Example: 389·467 × 27·42.

(1) Clear all registers and enter the multiplicand (always the number with the greatest number of digits) in the S.R. if the multiplier consists of four digits move the S.R. along 3 places giving:

P.I.
P.R. 0000 000 000 000 000   M.R. 000 000 00
S.R. 00 389 4̇67

(2) Short cut the setting of 27 in the M.R. by making three forward turns of the handle in the position shown above followed by three backward turns after moving the S.R. back one place to the right.

N.B. Given 'tens transmission' in the M.R. 'short cutting' is the most efficient way of entering a number. The objective is to ensure that the handle is never turned more than five times in any direction. If a digit is 6 or greater, the number to the left is increased by 1, for instance if the multiplier is 3 827 426 there are 32 normal forward turns of handle but in short cutting the procedure is:

$$\overset{+}{0}\ \overset{}{4}\ \overset{-}{2}\ \overset{+}{3}\ \overset{-}{3}\ \overset{+}{4}\ \overset{+}{3}\ \overset{-}{4} \quad 23 \text{ turns}$$

Numbers should always be entered from left to right and the first turn of the machine should always be positive as some machines will record negative turns as a positive number if made from the start. Similarly some machines possess a device which when depressed causes negative turns to be counted as positive. This should in simple multiplication be set to the positive position.

(3) Continue by moving the S.R. one place to the left and making four forward turns of the handle; make a further move of the S.R. to the left and make two forward turns giving:

P.R. 0000 001 067 9į8 514   M.R. 000 027 4̇2
S.R. 00 389 4̇67

N.B. The position of · indicating the decimal point in the P.R. is given by adding the number of digits after the decimal points in the M.R. and the S.R. Indicatory numbers simplify the process as

Setting in S.R. + Setting in M.R. = Setting in P.R.

The indicatory numbers lie over the first digit after the decimal point or over the decimal point so:

$$\begin{array}{cccccccc} 8 & 7 & 6 & 5 & 4 & 3 & 2 & 1 \\ & & 3 & 8 & 9 \cdot & 4 & 6 & 7 \end{array}$$

or

$$\begin{array}{cccccccc} 8 & 7 & 6 & 5 & 4 & 3 & 2 & 1 \\ & & & 3 & 8 & 9 \cdot & 4 & 6 & 7 \end{array}$$

2.2.6. ADDITION AND SUBTRACTION OF PRODUCTS

Example: 389·467 × 27·42 + 14·264 × 16·18 − 381.341 × 12.81

(1) Carry out the first multiplication as described above.

(2) Clear the S.R. and the M.R., carry out the second multiplication giving:

P.R. 0000 001 090 9$\dot{9}$7 666   M.R. 000 016 $\dot{1}$8
S.R. 00 014 $\dot{2}$64

(3) Clear the S.R. and the M.R., if necessary set the M.R. to count negatively, set 381·341 in the S.R.

(4) Enter the multiplier negatively in the M.R. starting from the fourth position to the left, the procedure is:

Direction of turn	−	−	+	−
Number of turns	1	3	2	1
Number in M.R.	1	2	8	1

This gives:

P.I.
P.R. 0000 000 602 4$\dot{9}$9 845   M.R. 000 012 $\dot{8}$1
S.R. 00 381 $\dot{3}$41

2.2.7. DIVISION

Example: $\dfrac{1\,574\cdot763}{36\cdot43}$ the result being required to five places of decimals.

The result of a division appears in the M.R. so the setting of the dividend is given by

2 (Number of decimals in divisor 36·43) + 5(Number of decimals required in result) hence 7.

(1) Clear all registers and set 1574·763, move S.R. along five places giving:

P.I.
P.R. 0000 000 000 000 000   M.R. 000 000 00
S.R. 157 4$\dot{7}$6 3

(2) Enter the dividend in the P.R., position the decimal point marker and *clear the S.R. and M.R.*. This last step must not be overlooked.

(3) Set 36·43 in the S.R. and move S.R. along seven places giving:

P.I.
P.R. 0000 157 4$\dot{7}$6 300 000   M.R. 000 000 00
S.R. 000 036 $\dot{4}$3

(4) If necessary set the M.R. to count negatively, make backward turns of the handle until the 'bridge of nines' appears, this occurs after five turns.

(5) Make a single forward turn of the handle so that the 'bridge of nines' disappears. Move the S.R. one place to the right and repeat backward turning

Introduction to Computation

of the handle as in (4) and (5) in each position of the S.R. until the first position is reached giving:

P.I.
P.R. 9999 999 9̇9̇9 996 558   M.R. 43̇2̇ 270 94
S.R. 00 003 64̇3

N.B. It is worth cultivating the habit of estimating when the 'bridge of nines' will appear, so that excessive turning of the handle can be avoided.

It will be noted that the full capacity of the machine has been used to give an extra decimal place; if the result had been greater than 100 this could not have been done and it is more usual to position the leading figure of the divisor under the leading figure of the dividend. The extra figure after the decimal point is used to round off the fifth figure correctly, this can be done by estimation if the capacity of the machine will not permit the introduction of an extra figure.

2.2.8. DIVISION BY MULTIPLICATION

Taking the same example as in §2.2.7.

(1) Clear all registers and set 36·43 in the S.R.

(2) Move the S.R. to its extreme left position and make forward turns of the handle until the number in the P.R. exceeds the dividend, 1 574·763, then make one backward turn of the handle.

(3) Move the S.R. one place to the right and again make forward turns until the dividend is exceeded, make one backward turn and repeat in each position of the S.R. giving:

P.I.
P.R. 0000 157 47̇6 299 799   M.R. 43̇2̇ 270 93
S.R. 00 003 64̇3

2.2.9. RECIPROCALS

Example: $\frac{1}{441}$ to five significant figures.

(1) Clear all registers and set 441 in the S.R.

(2) Move the S.R. five places to the left and perform a division by multiplication to give 1·000 000 00 in the P.R. or nearly so. The final position is:

P.I.
P.R. 0000 000 1̇0̇ 000 278   M.R. 0̇02 267 58
S.R. 00 000 441

The result is 0·00 226 76.

2.2.10. SQUARE ROOTS

2.2.10.1. *Approximate Methods*

Given an approximate value for $N^{\frac{1}{2}}$ say a, then
$$N = (a + x)^2$$
This may be expressed as
$$x + \frac{x^2}{2a} = \frac{N - a^2}{2a}$$
or
$$a + x + \frac{x^2}{2a} = \frac{1}{2}\left(\frac{N}{a} + a\right)$$

Example: 726·43½ to three places of decimals.
Both above expressions are used and in both cases $x^2/2a$ is treated as negligible.
(1) Find an approximate value for 726·43½ such as 26·9.
(2) Divide 26·90 into 726·43 26·90 times. This leaves 2·82 being $N - a^2$.
(3) Divide 2·82 left in the P.R. by 53·8 (2a) giving 0·05241.
The required result is 26·952.
The error in the result is

$$\frac{(0.052)^2}{53 \cdot 8} = \text{approximately } \frac{1}{20{,}000} = 0 \cdot 000\ 05$$

This indicates an uncertainty in the fourth place, the result is correct to the third place. By this method a result can always be obtained to twice the number of digits in the approximate value with a slight inaccuracy possible in the last digit.

Alternative method from (1)
(2) Divide 26·90 into 726·43 giving to four decimal places 27·0048. This is N/a.
(3) $\frac{1}{2}(N/a + a)$ is 26·9524.
The required result is 26·952.
The error in the result and the requirement for the approximate value is as in the first method.

2.2.10.2. *An exact method*

This method is based on the relationship of the sum of the odd numbers to squares, thus:

$$1 + 3 = 4 = 2^2$$
$$1 + 3 + 5 = 9 = 3^2$$

This method can only be used on keyboard machines and those with lever setting

Example: 51 764·356.½
(1) Mark off the number in pairs working outward from the decimal point, thus: 5, 17, 64 · 35, 6. Clear all registers, set this number, and transfer it to the P.R. As there are eight significant figures in the number the root should not be taken to more than eight significant figures; the marking off of the number indicates that there will be three digits before the decimal point with a S.R. of eight digits this leaves five digits after the decimal point and consequently ten in the P.R. The transfer to the P.R. is made after movement of the S.R. to achieve this positioning as explained in division.

P.I.
P.R. 0517 643 560 000 000   M.R. 000 000 00
S.R. 517 643 56

(2) Clear S.R. and M.R. and if necessary set the M.R. to count negatively. Mark the decimal points working from the right-hand end of the M.R. and S.R. Move the S.R. so that where the first digit of the root will appear lies under the right-hand digit of the extreme left-hand pair in the P.R. The P.I. should

similarly lie so that the first digit lies in its correct position relative to the decimal point marker. Thus:

P.I.
P.R. 0517 64̇3 560 000 000   M.R. 000 0̇00 00
S.R. 000 0̇00 00

(3) Set and subtract in turn 1, 3, etc. until the 'bridge of nines' appears giving:

P.I.
P.R. 9617 64̇3 560 000 000   M.R. 300 0̇00 00
S.R. 500 0̇00 00

(4) Turn the handle forward and then reduce the 5 in the S.R. to 4. The value in the M.R. is now 2.

(5) Move the S.R. one position to the right and set 1 in the second position to the right in the S.R. giving:

P.I.
P.R. 0117 64̇3 560 000 000   M.R. 200 000 00
S.R. 41 0̇00 000

(6) Subtract 41, 43, etc. until the 'bridge of nines' appears giving:

P.I.
P.R. 9988 64̇3 560 000 000   M.R. 230 000 00
S.R. 45 0̇00 000

(7) Turn the handle forward and reduce the 5 in the S.R. to 4. The value in the M.R. is now 22.

(8) Repeat for the remaining 6 positions of the P.I. this procedure giving in the final position:

P.I.
P.R. 0000 00̇0 021 328 039   M.R. 227 5̇17 81
S.R. 45 5̇03 562

The required root is in the M.R. and as a check, the value in the S.R. is double this value.

2.2.11. MULTIPLICATION IN PARTS

It is sometimes required to multiply numbers larger than the capacity of the machine, the following method increases the capacity of an 8 digit machine to 11 digits.

Example: 3·264 578 761 4 × 2·117 068 895 5.

(1) Multiply the first eight digits of each number giving in the P.R. 06·911 337 710 914 56

(2) Clear M.R. and S.R. and without moving the decimal point markers multiplying 0·002 117 1 by 0·000 0614.

N.B. This is the same as multiplying 2·117 1 by 0·000 000 0614.

This gives in the P.R. 06·911 337 840 904 50.

(3) Clear M.R. and S.R. and without moving the decimal point markers multiply 0·003 2646 by 0·000 0965.

This gives the final result in the P.R. as:

6·911 338 155 938 40

2.2.12. MULTIPLE PRODUCTS

Example: 71·26 × 46·84 × 37·92

(1) Clear all registers and multiply 71·26 by 46·84 as described in multiplication. This gives in the P.R. 0000 000 033 378 184.

(2) If direct transfer from the P.R. to the S.R. can be carried out transfer 33 37·8 184 to the S.R.. Otherwise clear the M.R. and S.R. set 33 37·8 184 and subtract it from the value in the P.R., the full register of zeros indicates the correctness of the setting.

N.B. If the number of digits in the P.R. exceeds the capacity of the S.R. the product must be rounded off to the required number of digits. On some machines knobs and wheels adjacent to the P.R. enable this to be carried out before transfer; on lever and keyboard machines the number can be altered in the S.R.; on ten key machines the number in the P.R. may have to be increased or decreased by addition or subtraction before transfer. The usual rules for rounding off applies, viz.

Numbers greater than 5 are rounded off upward.
Numbers less than 5 are rounded off downward.
5 itself is rounded off to the even preceding figure
for example 7·65 to 1 place of decimals is 7·6
but   7·55 to 1 place of decimals is also 7·6.

(3) Clear the M.R. and multiply the first product by 37·92 as described in multiplication. The change in the position of the decimal point in the S.R. and the P.R. should be noted. This step gives:

P.I.

P.R. 0000 126 570 073 728   M.R. 000 037 92
S.R. 33 378 184

To conform with the data the example must be rounded off to two places after the decimal point.

2.2.13. COMBINED MULTIPLICATION AND DIVISION

Example: (a) (71·26 × 46·84)/12·87 to four places of decimals.

(1) Clear all registers and multiply 71·26 by 46·84 as described in multiplication.

(2) Transfer the product into the S.R. by one of the processes described in (2) of multiple products.

(3) As in division calculate the number of places of decimals required in the dividend viz. 4 + 2 = 6 (add one place for rounding-off if possible) and reset the product in this position in the P.R.

(4) Clear the M.R. and S.R. and set the divisor giving:

P.I.

P.R. 0000 033 378 184 000   M.R. 000 000 00
S.R. 000 012 87

N.B. Often the transfer and back transfer required to correctly position the product can be eliminated by increasing the number of digits after the decimal point in the top line multiplication; in this case it could be

$$71 \cdot 260 \times 46 \cdot 840$$

which would give the correct setting in the P.R.

The answer to this example is obtained from:

P.I.
P.R. 0000 000 000 001 162   M.R. 259 348 74
S.R. 00 001 287

The answer being 259·3487.

Example: (b)
$$\frac{12 \cdot 87}{71 \cdot 26 \times 46 \cdot 84}$$

This is best treated as the reciprocal of the previous example, viz. 1/259·3487 which is 0·0038 558.

N.B.

If $$\frac{a}{b \times c} = K \quad \text{and} \quad \frac{b \times c}{a} = N$$

then $$\frac{1}{N} = K$$

hence $$dN \cdot N^{-2} = -dK$$

This indicates that the number of significant figures required in N to obtain K correct to a specified number of figures depends upon the size of N. The larger the value of N the fewer significant figures are required.

2.2.14. CONCLUSION

The use of a calculating machine in survey computations is today almost essential. Its correct use is largely a matter of practice and there are many techniques not described above which may in particular problems be useful. Special machines such as the twin Brunsviga speed up many routine calculations. Adding listing machines are valuable in extraction of final coordinates and provide a print out of the results, operation is similar to addition and subtraction on a hand machine.

2.3. TYPES OF CALCULATING MACHINE

2.3.1. DESK MACHINES

The following five machines represent only a few of the many machines available. All these machines are in current production and are in the opinion of the author well suited to survey computations.

Diehl: Model V.S.R.; 9 × 9 × 18; electric, fully-automatic; keyboard setting; transfers and storage.

A most sophisticated machine.

Monroe: Model 88N-213 (Monroe-Matic); 10 × 11 × 21; electric, fully-automatic; keyboard setting; double product and multiplier registers which act as stores; transfer.

Comparable in sophistication to the Diehl.

Brunsviga: Model B20; 12 × 11 × 20; hand machine; lever-type setting; split product register; transfer.

A robust machine of proven reliability.

Facit: Model C.M. 2-16; 11 × 9 × 16; hand machine; ten key setting; transfer.

A reliable machine, particularly quiet to operate.

Curta Model II; 11 × 8 × 15; hand machine; lever setting; no transfer.

A pocket-sized machine 5 in by 3 in and weighing 18 oz.

2.3.2. DIGITAL COMPUTERS (ELECTRONIC COMPUTERS)

These machines work at speeds hundreds of thousands of times faster than desk calculating machines. Within the machines there are enormous storage facilities and unlimited storage of other information, such as tables of trigonometrical functions, is available to the machine on magnetic tapes.

Programmes have been prepared for many survey computations and these can be hired together with time on a machine. The speed of operation enables computations to be repeated and each of several doubtful observations can be used until the most acceptable result is obtained.

However the basic preparatory work for any survey computation still remains. Abstraction and punching data tapes to feed into the computer, and collecting and listing the results takes a comparatively long time. Data fed into a computer must be correct. The preparation of two sets of tapes from the same field data is generally advocated because mechanical comparision can be carried out easily. This does however increase the amount of preparatory work. Ideally the data tapes should be prepared in the field by, for instance, the theodolite. This has been achieved by Kern for the DKM3A but the weight of the additional equipment is prohibitive for all work except first order astronomical observations.

The solution of sets of simultaneous equations is a field where the computer has enormously reduced the repetitive work in survey computing. One of the largest solutions was achieved by the Oxford University Computer Laboratory with a Ferranti Mercury Computer, data being supplied by the Directorate of Overseas Surveys in the form of a tape containing 781 observation equations with 234 unknowns. The normal equations were solved by an iterative technique and such methods seem quicker than a direct solution, particularly as in survey adjustments there are many zero coefficients.

Both Geodimeter and Tellurometer observations can be reduced using existing programmes. With the Tellurometer observations two programmes are used, one to compute slant distances from the transit time, wet and dry bulb temperature readings, and pressure. Heights are then computed by hand. The heights are then used with the slant distances, azimuth, and latitude to compute spheroidal lengths.

Undoubtedly more and more survey work will be computed by electronic methods and many jobs which previously would not be considered feasible are now possible. Of great significance is the possibility of producing final coordinates shortly after completion of the field work so obviating the need for provisional coordinates for mapping purposes.

2.4. MATHEMATICAL TABLES

The following information is based on extracts from *An Index of Mathematical Tables*, A. Fletcher, J. C. P. Miller, L. Rosenhead, L. J. Comrie; 2nd edition London 1962, Scientific Computing Service Ltd.

2.4.1. TABLES OF TRIGONOMETRICAL FUNCTIONS

No of decimals	Interval of argument	Remarks	Author
	sin x (Degrees, Minutes, Seconds)		
33	1" 0–100" 100" 100"–1,000" 1,000" 1,000'–45°		Van Ostrande and Shoultes
17	9'	1st to 6th variations for 10"	Andoyer
15	10"	1st differences	
8	1"	Indication of 1st differences	Peters
7	10"	1st differences	Brandenburg
7	1'	1st differences	Pryde
6	10"	1st differences	Peters
6	1'	1st differences	Comrie
5	10"	No differences	N.A.O.
	tan x (Degrees, Minute, Seconds)		
15	10"	1st differences	Andoyer
8 8 figures 5–7 figures	1", 0° – 71° 30' 1", 71° 30' – 88° 12' 1", 88° 12' – 90°	Indication of 1st differences	Peters
7 7 figures	10", 0° – 84° 1", 84° – 90°	1st differences	Brandenburg
7 8 figures	1', 0 – 84°17' 1', 84°18' – 90°	No differences	Pryde
6 6 figures 4–6 figures	1', 0° – 60° 10", 60° – 90° 1", 88° 40 – 90°	1st differences	Peters
6 5–6 figures	1', 0° – 70° 1', 70° – 90°	1st differences	Comrie
5 5 figures 5 figures	10", 0° – 63° 10", 63° – 82° 30' 1", 82° 30' – 90°	No differences	N.A.O.

2.4.2. TABLES OF LOGARITHMS

No. of decimals	Interval of argument	Remarks	Author
\multicolumn{4}{c}{Logarithms of Natural Numbers}
7	10,000 —(1) — 100,000	10ths of all differences	Bruhnes
7	10,000 —(1) — 100,000	10ths of all differences	Vega—Bremiker
7	10,000 —(1) — 100,000	Proportional parts	Pryde
8	100,000 —(1) — 108,000		
6	10,000 —(1) —100,009	10ths of all differences	Bremiker
6	10,000 —(1) —100,009	10ths of all differences	Comrie
\multicolumn{4}{c}{Logarithms of sin x, tan x, etc}
14	10"	1st differences	Andoyer
7	1"	Proportional parts of Means differences	Shortrede
7	1", 0°— 6°	1st differences	
7	10", 6°— 84°	10ths of some differences	Bruhnes
7	1", 84°— 90°		
7	1", 0°— 5°	1st differences	Vega—
7	10", 5°— 90°	10ths of some differences	Bremiker
6	1", 0°— 5°	1st differences	
6	10", 5°— 90°	10ths of some differences	Bremiker
6	10", 1°20'— 10°	1st differences	
6	1', 10°— 80°		
6	10", 80°— 88°40'	10ths of some differences	Comrie
6	Critical, 88°40'— 90°		
5	10", 0°— 1°20'		

2.4.3. TABLES OF TRIGONOMETRICAL FUNCTIONS WITH ARGUMENT IN TIME

No. of decimals	Interval of argument	Remarks	Author
7	sin (1^S), 0—6h	Indication of	N.A.O.
7	tan (1^S), 0—4h	1st differences	(Comrie)
5—7 figure	tan (1^S), 4h—6h		

2.4.4. NUMBER OF FIGURES TO BE USED IN COMPUTATION

A useful guide to the number of decimals of trigonometrical functions normally required in computation, is that this number is the same as the number of figures used to express the angle concerned: thus

> five decimals are required for the angle 11° 11' 10"
> six for the angle 11° 11' 11", and seven for 11° 11' 11". 1.

Introduction to Computation

In any particular case, the exact precision can be derived by differentiation of the relevant function. For example if the tangent of an angle x is given from tables of a precision of one in the sixth place or 10^{-6} the angle x may be obtained to a precision given by $dx'' = dF \cos^2 x \times 206\,265$

where $\quad F = \tan x \therefore \dfrac{dF}{dx} = \sec^2 x \quad$ (dx in radians)

If $x = 1°$ this is a precision of $0''\!.2$.

2.4.5. AUXILIARY FUNCTIONS τ AND σ

The determination of a small angle x from the cotangent or cosecant of x is troublesome on account of the very large tabular differences, and the converse problem of calculating the cotangent or cosecant for small angles is similarly troublesome. Many tables therefore introduce the use of auxiliary functions τ and σ to assist in the calculations concerning the cotangent and cosecant respectively as follows.

$$\cot x = \frac{1}{x}(1 - \tfrac{1}{3}x^2 - \tfrac{1}{45}x^4 - \cdots)$$

$$= \frac{206\,265}{x''}(1 - \tfrac{1}{3}x^2 - \tfrac{1}{45}x^4 - \cdots) = \frac{\tau}{x''}$$

and $\quad \operatorname{cosec} x = \dfrac{1}{x}(1 + \tfrac{1}{6}x^2 + \tfrac{7}{360}x^4 + \cdots) = \dfrac{\sigma}{x''}$

whence $\cot x = \tau/x''$ and $\operatorname{cosec} x = \sigma/x''$ where x'' is in seconds of arc. For example if $x = 1°\,23'\,45''\!.67$; from tables $\tau = 206\,224$ and $x'' = 5\,025.67$, whence $\cot x = 206\,224/5\,025.67 = 41.0341$.

2.4.6. PROBABLE ERRORS OF TABULATED QUANTITIES.

When a table is constructed to a certain precision the last figure will have been rounded off from a figure which may range from zero to ± 0.5 in the last place, and no one particular end figure would occur more frequently than another. Any end figure is therefore just as likely to exceed 0.25 in the last place as it is to be less than 0.25, which means that the probable error of a tabular entry is one half of the maximum possible error, namely 0.25 in the last place, which for a six figure table is 0·000 000 25. The value of the function interpolated between two tabular entries T_1 and T_2 is given by

$$F = T_1 + n(T_2 - T_1)$$

where n is the fractional part of the argument for which the interpolate is required. The probable error σ_F in F is then given in terms of the probable errors in T_1 and T_2, σ_{T_1} and σ_{T_2} by

$$\sigma_F^2 = (1 - n)^2 \sigma_{T_1}^2 + n^2 \sigma_{T_2}^2$$

But $\sigma_{T_1} = \sigma_{T_2} = 0\cdot 25$ in the last place of tabulation therefore

$$\sigma_F{}^2 = (0\cdot 25)^2(1 - 2n + 2n^2)$$

in the last place. This interpolate has a minimum error when $n = \frac{1}{2}$ which is $0\cdot 25/2^{\frac{1}{2}}$, and a maximum when $n = 0$ or 1, which is $0\cdot 25$, in the last figure.

2.4.7. CRITICAL TABLES

Normally a table gives values of the function required, or the *respondent*, which correspond to values of the *argument* which take integral values at even intervals; for example the sines of angles are tabulated for every ten seconds of arc. In some cases the respondent changes so slowly within the range required that it may be tabulated for every value to be used. The values of the respondent where a change occurs of one unit in the last place when rounded-off are called the critical values. The values of the argument which correspond to the critical values of the respondent are calculated and tabulated, and in this case will not take regular integral values as in the ordinary table. Any value of the argument lying between successive tabular values will correspond to one value of the respondent which may be read from the table at a glance. The principle is best explained by an example. The correction for refraction r to be applied to an observed vertical angle H is given by $r = 58 \cot H$. Assuming that a precision of $1''$ is required for the range of H $30°$ to $31°$ the procedure is as follows:

(a) Find the value of r corresponding to $H = 30°$

i.e. r for $30° = 100\cdot 5$

(b) Calculate the values of H corresponding to the values of $r = 100\cdot 5$, $99\cdot 5$, etc. until H exceeds the required $31°$. i.e. we have

r	H
100·5	29° 59′
99·5	30° 14′
98·5	30° 29′
97·5	30° 45′
96·5	31° 01′

The calculated critical values of H are rounded *down* and not rounded *off*, and the convention is adopted that one ascends up the page when the entry value equals a critical value. The final critical table is then written

H	r
29° 59′	
	100″
30° 14′	
	99
30° 29′	
	98
30° 45′	
	97
31° 01′	

Introduction to Computation   57

The refraction for any value of H lying between tabular values may be read at once, e.g. r for $H = 30° 33'$ is $98''$.

The advantages of the critical table are that it is very quick to use, and it is slightly more accurate that the corresponding ordinary table. Its disadvantages are that it takes up more space, and inverse use is not convenient. Critical tables may be used to great advantage by the survey computer when many repetitive calculations are to be carried out; for example in field astronomy a critical table of the chronometer error can save much time spent in unnecessary interpolation.

2.5. LOGARITHMIC METHODS OF COMPUTATION

Little mention is made in this text of logarithmic methods of computation, the four basic processes of arithmetic are adequately described in any of the tables of logarithms given in §2.4.2, and any survey computation can be performed using the formulae given for machine methods. In the past, when most survey computations were performed using logarithms, special techniques and variations of basic formulae were evolved to facilitate easy computation. For instance:

$$X = (A^2 + B^2)^{\frac{1}{2}}$$

is unsuitable for logarithmic computation as it involves looking up 3 logarithms and 3 antilogarithms, but if

$$\tan \phi = \frac{B}{A}$$

then
$$X = A \sec \phi = \frac{A}{\cos \phi}$$

The stages being $\log B - \log A = \log \tan \phi$, find ϕ and hence $\log \cos \phi$, then $\log X = A - \log \cos \phi$.

Although logarithmic methods of computation are largely superseded by machine methods the value of logarithmic tables remains as illustrated by their use in problems of error propagation (see §§6.12.2 and 3).

2.6. THE SLIDE RULE

The use of a slide rule for many small computations is strongly recommended. Although special slide rules with scales suitable for reduction of tacheometry are available they are expensive. A simple 10 in rule is adequate for most small computations.

2.7. CONVERSION OF SMALL ANGLES FROM SEXAGESIMAL VALUES TO CIRCULAR VALUES

The sin of one second of arc differs from the circular measure of one second by a very small quantity. In all theoretical work circular measure is used and to convert an angle in seconds of arc into radians multiply by $\sin 1''$,

$$\text{viz arc } q = q'' \sin 1'' = \frac{q''}{206\,265}$$

Similarly but slightly more inaccurate

$$\text{arc } q = q' \sin 1' = \frac{q'}{3\,438}$$

2.8. SPHERICAL TRIGONOMETRY

2.8.1. DEFINITIONS AND CONCEPTS

(i) A *sphere* is a solid bounded by a surface equidistant from a single point, the centre.

(ii) The intersection of the surface of a sphere and a plane is a *circle*.

(iii) Any circle formed by the intersection of the surface of a sphere with a plane containing the centre is a *great circle*; while any circle formed by a plane not containing the centre is a *small circle*.

(iv) The line which passes through the centre of the sphere and is perpendicular to the plane of any small or great circle will pass through their centres and is termed the *axis* of those circles.

(v) The axis of any circle when extended meets the surface of the sphere in two points which are the *poles* of that circle.

(vi) In figure 2.7, where each of three planes pass through the centre of the sphere, the three sided figure PBA, the sides of which are all small arcs of great circles, is a *spherical triangle*.

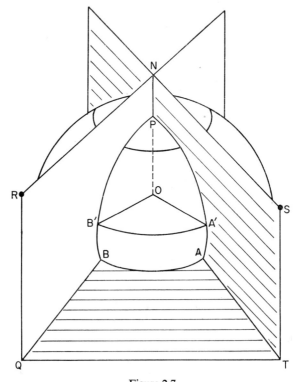

Figure 2.7

Introduction to Computation

(vii) When a great circle contains the pole of another great circle it is *secondary* to that circle. Meridional great circles are secondary to the equatorial great circle on any globe. The portion of a secondary intercepted between the pole and the primary great circle is a quadrant (90°).

(viii) The shortest distance between any two points on the surface of a sphere is along the great circle arc passing through them. In spherical trigonometry the angle subtended by the arc at the centre of the sphere is regarded as the 'length' of the side. In survey the spherical triangle is considered as having sides less than 180°.

(ix) The interior angles at the vertices of a spherical triangle can be defined in three different ways, viz:

(a) In figure 2.7, the spherical angle BPA is the dihedral angle between the planes PBQRN and PATSN. In magnitude it is the angle RNS in the plane parallel to the tangent plane at P.

(b) Alternatively the same spherical angle BPA is the arc intercepted on a great circle to which the two sides PA and PB are secondaries. If this arc is A'B' then in magnitude it is the plane angle A'OB', where O is the center of the sphere.

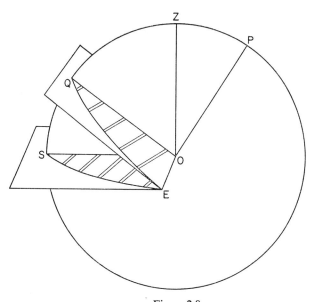

Figure 2.8

(c) In figure 2.8, the spherical angles QES between the spherical sides QE and SE is alternatively given by the angle between the poles of the great circles of which the adjacent sides form parts. In magnitude the plane angle ZOP.

It is convenient to remember all three definitions, each is useful for proving various expressions, both here and elsewhere.

(x) Unlike the plane triangle, which is a figure representing the distances between three points, the spherical triangle is a figure which represents the positions of three points on the surface of a sphere in terms of directions from the centre of the sphere. All dimensions are therefore angular.

2.8.2. PROPERTIES OF THE SPHERICAL TRIANGLE

Considering figure 2.9, which represents the solid formed by three inter-

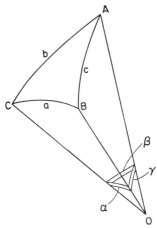

Figure 2.9

secting great circles.
  (i) The sum of any two plane surface angles at 0 is greater than the third. From this it follows that any two sides of a spherical triangle are together greater than the third.
  (ii) The sum of the three angles α, β, and γ i.e. the three sides, a, b, and c, will not exceed 360°, as the sum of the plane angles of a trihedral do not exceed 360°.
  (iii) The sum of the three angles between the plane surfaces A, B, and C, will exceed 180° and will not exceed 540°. (No dihedral angle can exceed 180°.) Consequently the sum of the angles in a spherical triangle lies between 180° and 540°. The amount by which this sum exceeds 180° is termed the spherical excess of the triangle.

2.8.3. SOLUTION OF SPHERICAL TRIANGLES

The seven interrelated elements of a spherical triangle are the three spherical angles usually denoted by upper case letters A, B, and C; the three sides usually denoted by lower case letters, a, b, and c; and the area.

The following formulae should be remembered.

The COSINE FORMULA

$$\cos a = \cos c \cos b + \sin c \sin b \cos A \qquad (1)$$

or
$$\cos b = \cos a \cos c + \sin a \sin c \cos B \qquad (2)$$

or
$$\cos c = \cos b \cos a + \sin b \sin a \cos C \qquad (3)$$

The SINE FORMULA
$$\frac{\sin A}{\sin a} = \frac{\sin B}{\sin b} = \frac{\sin C}{\sin c} \qquad (4)$$

The COT FORMULA
$$\cos a \cos B = \sin a \cot c - \sin B \cot C \qquad (5)$$

2.8.3.1. *The Cosine Formula*

In figure 2.9, the side lengths are in magnitude the angles α, β, and γ. The most important formula is that which gives the relationship between one side and the other two sides and their included angle.

In figure 2.10, angles PAO and QAO are right angles, hence QAP = Spherical angle A (see §2.8.1).

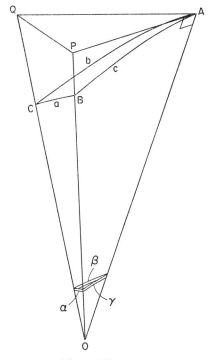

Figure 2.10

In triangle PQO by plane trigonometry
$$PQ^2 = PO^2 + QO^2 - 2PO \cdot QO \cos \alpha$$
in triangle PQA
$$PQ^2 = PA^2 + QA^2 - 2PA \cdot QA \cos A$$
in triangle QAO
$$QO^2 = QA^2 + AO^2$$
and in triangle PAO
$$PO^2 = PA^2 + AO^2$$

Equating and regrouping terms which equal AO^2 it is found that:

$$AO^2 = PO \cdot QO \cos \alpha - PA \cdot QA \cos A$$

or
$$\cos \alpha = \frac{AO}{PO} \cdot \frac{AO}{QO} + \frac{PA}{PO} \cdot \frac{QA}{QO} \cdot \cos A$$

which in terms of the angles at O is

$$\cos \alpha = \cos \gamma \cos \beta + \sin \gamma \sin \beta \cos A$$

By definition α is the Side a in angular measure, β Side b in angular measure, and γ Side c in angular measure

hence $\cos a = \cos c \cos b + \sin c \sin b \cos A$ (1)
similarly $\cos b = \cos a \cos c + \sin a \sin c \cos B$ (2)
and $\cos c = \cos b \cos a + \sin b \sin a \cos C.$ (3)

The symmetry of these expressions should be noted.

The Sine Formula

If any two of formula (1), (2), and (3) are regrouped in the form

$$-\cos A \sin b \sin c = \cos b \cos c - \cos a$$

then by squaring each side and substituting $1-\sin^2$ for $\cos^2$ it is established that

$$2 - \sin^2 a - \sin^2 b - \sin^2 c - 2 \cos a \cos b \cos c = -\sin^2 A \sin^2 b \sin^2 c$$
$$= -\sin^2 B \sin^2 a \sin^2 c$$
$$= -\sin^2 C \sin^2 b \sin^2 a$$

From which it follows that

$$\frac{\sin A}{\sin a} = \frac{\sin B}{\sin b} = \frac{\sin C}{\sin c} \quad (4)$$

N.B. Ambiguity exists as $\sin \alpha = \sin (180 - \alpha)$.

The Cot Formula

Taking any two of formulae (1), (2), and (3) and substituting for one in the other

$$\cos c = \cos^2 a \cos c + \sin a \cos a \sin c \cos B + \sin a \sin b \cos C$$

When reduced by $\sin a$ this gives

$$\cos c \sin a = \cos a \sin c \cos B + \sin b \cos C$$

As $\sin b \sin C = \sin c \sin B$

$$\frac{\cos c \sin a}{\sin c \sin B} = \frac{\cos a \sin c \cos B}{\sin c \sin B} + \frac{\sin b \cos C}{\sin b \sin C}$$

then $\cos a \cos B = \sin a \cot c - \sin B \cot C$ (5)

Introduction to Computation

This relates two sides and two angles, and the following mnemonic can be used to form five further formulae:

cos Inner side cos Inner angle = sIn Inner side cOt Outer side − sIn Inner angle cOt Outer angle.

2.8.4. THE POLAR TRIANGLE

In figure 2.11 the arcs BA′, CA′, AC′, BC′, AB′, CB′ are 90°. Then A′ is the pole of side BC, B′ is the pole of side AC, and C′ is the pole of AB. Also each

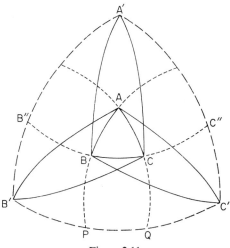

Figure 2.11

side of the triangle A′B′C′ has a pole at the vertices of the original triangle. Consequently

$$BA'C'' = 90 \text{ and } CA'B'' = 90$$

but

$$BA'C = a.$$

Hence $B''A'C'' = B'A'C' = 90 + 90 - a = 180 - a$.

It follows that $A'B'C' = 180 - b$ and $B'C'A' = 180 - c$.

Also $C'P = 90$ and $B'Q = 90$, but $B'C' = a'$ and $A = PQ$.
As $C'P + B'Q - B'C' = PQ$

$$A = 180 - a'$$

It follows that $B = 180 - b'$ and $C = 180 - c'$.

2.8.5. APPLICATIONS OF THE POLAR TRIANGLE

In triangle A′B′C′

$$\cos a' = \cos b' \cos c' + \sin b' \sin c' \cos A'$$

(from equation (1)) using the relationships of the polar triangle

$$\cos (180 - A) = \cos (180 - B) \cos (180 - C)$$
$$+ \sin (180 - B) \sin (180 - C) \cos (180 - a)$$

which is

$$-\cos A = \cos B \cos C - \sin B \sin C \cos a \qquad (6)$$

Similar formulae for $\cos B$ and $\cos C$ may be obtained.

2.8.6. THE RIGHT ANGLED SPHERICAL TRIANGLE

The general formulae are considerably simplified if one of the angles is a right angle as $\sin 90° = 1$, $\cos 90° = 0$, and $\cot 90° = 0$.

If angle C is 90° the following ten relationships exist:

$$\left.\begin{aligned}\sin a &= \sin A \sin c = \tan b \cot B \\ \sin b &= \sin B \sin c = \tan a \cot A\end{aligned}\right\} \qquad (7)$$

$$\cos c = \cos b \cos a = \cot A \cot B \qquad (8)$$

$$\left.\begin{aligned}\cos A &= \cos a \sin B = \tan b \cot c \\ \cos B &= \cos b \sin A = \tan a \cot c\end{aligned}\right\} \qquad (9)$$

2.8.7. NAPIERS RULE OF CIRCULAR PARTS

To remember the above formulae use may be made of Napiers rule of circular parts.

If angle C is 90° then the remaining five elements form the circular parts being the two sides adjacent to the right angle, the complement of the hypotenuse, and the complement of the two angles; viz, a, b, $90° - c$, $90° - A$, $90° - B$.

These are recorded in five divisions of a circle in their order of occurrence (see figure 2.12).

The rule specifies that:

$$\begin{aligned}\text{SIne of mIddle part} &= \text{product of tAngents of Adjacent parts} \\ &= \text{product of cOsines of Opposite parts}\end{aligned}$$

Any part may be taken as a middle part and the above ten formulae can be formed.

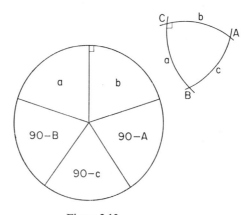

Figure 2.12

2.8.8. HALF ANGLE FORMULAE

Using formulae (1), (2), and (3) and the relationship $1 - \cos A = 2 \sin^2 \tfrac{1}{2} A$ it can be established that given sides a, b, and c:

$$\sin \tfrac{1}{2} A = \left[\frac{\sin (s - b) \sin (s - c)}{\sin b \sin C} \right]^{\tfrac{1}{2}} \tag{10}$$

where $2s = a + b + c$.

Formulae for $\sin \tfrac{1}{2} B$ and $\sin \tfrac{1}{2} C$ take a similar form.
Similarly from the relationship $1 + \cos A = 2 \cos^2 \tfrac{1}{2} A$ it can be established that:

$$\cos \tfrac{1}{2} A = \left[\frac{\sin s \sin (s - a)}{\sin b \sin c} \right]^{\tfrac{1}{2}} \tag{11}$$

Formulae for $\cos \tfrac{1}{2} B$ and $\cos \tfrac{1}{2} C$ take a similar form.
Combining (10) and (11)

$$\tan \tfrac{1}{2} A = \left[\frac{\sin (s - b) \sin (s - c)}{\sin s \sin (s - a)} \right]^{\tfrac{1}{2}} \tag{12}$$

If the required angle A be less than 90° formula (10) is the most suitable. If A be greater than 90° formula (11) is the most suitable. When all angles of a spherical triangle are required, use (12) the $\tan \tfrac{1}{2} A$ formula which gives a result that is always sufficiently accurate.

2.8.9. DELAMBRES FORMULAE

Using the half angle formulae and the processes of plane trigonometry it can be shown that:

$$\frac{\sin \tfrac{1}{2}(A + B)}{\cos \tfrac{1}{2} C} = \frac{\cos \tfrac{1}{2}(a - b)}{\cos \tfrac{1}{2} c} \tag{13}$$

$$\frac{\sin \tfrac{1}{2}(A - B)}{\cos \tfrac{1}{2} C} = \frac{\sin \tfrac{1}{2}(a - b)}{\sin \tfrac{1}{2} c} \tag{14}$$

$$\frac{\cos \tfrac{1}{2}(A + B)}{\sin \tfrac{1}{2} C} = \frac{\cos \tfrac{1}{2}(a + b)}{\cos \tfrac{1}{2} c} \tag{15}$$

$$\frac{\cos \tfrac{1}{2}(A - B)}{\sin \tfrac{1}{2} C} = \frac{\sin \tfrac{1}{2}(a + b)}{\sin \tfrac{1}{2} c} \tag{16}$$

$$\tan \tfrac{1}{2}(A + B) \cot \tfrac{1}{2} C = \frac{\cos \tfrac{1}{2}(a - b)}{\cos \tfrac{1}{2}(a + b)} \tag{17}$$

$$\tan (A - B) \cot \tfrac{1}{2} C = \frac{\sin \tfrac{1}{2}(a - b)}{\sin \tfrac{1}{2}(a + b)} \tag{18}$$

2.8.10. SPHERICAL EXCESS

This is the amount by which the sum of the angles in a spherical triangle exceeds 180°.

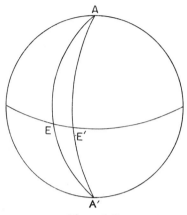

Figure 2.13

Considering figure 2.13, in which AEA'E' is a lune; that is the figure formed by the intersection of two great circles. The area of this figure is

$$(A/360)4\pi R^2 \text{ (the area of the sphere)}$$

where A is the angle between the two great circles.

Each pair of adjacent sides in a spherical triangle make up a lune. If the angles at the vertices are as before A, B, and C then the area of the lunes are:

$$(A/360)4\pi R^2 = (A/2\pi)\, 4\pi R^2 = 2AR^2$$
$$= 2BR^2,$$
$$= 2CR^2$$

where A, B, and C are in radians.

From figure 2.14 it is found that

$$2R^2(A + B + C) = 3 \times \text{Area ABC} + \text{Area A'BC}$$
$$+ \text{Area AB'C} + \text{Area ABC'}$$

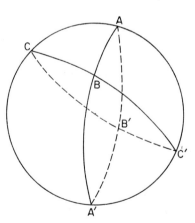

Figure 2.14

Introduction to Computation

But Triangle AB'C = Triangle A'BC'

and Area ABC + Area A'BC + Area A'BC' + Area ABC'
$$= 2\pi R^2 \text{ (a hemisphere)}$$
$$\therefore 2R^2(A + B + C) = 2 \times \text{Area ABC} + 2\pi R^2$$

If E is the spherical excess as defined above, then $A + B + C = \pi + E$

$$\therefore \pi + E = \text{Area ABC}/R^2 + \pi$$

Hence $E = \text{Area ABC}/R^2$ or

$$E'' = \frac{\text{Area ABC}}{R^2 \sin 1''} \tag{19}$$

N.B. It is usually sufficient to calculate ABC from the plane trigonometrical values for a triangle, strictly speaking it is however the area of the spherical triangle that is used in this formula.

2.9 INTERPOLATION

2.9.1. INTERPOLATION OF INTERMEDIATE VALUES IN A TABULATED FUNCTION

If a function F is tabulated for regular and equal changes of its argument intermediate values of the function are derived using an interpolation formula, the most convenient and accurate of which is Bessel's. For example, the table 2.1 shows values of function F tabulated against the latitude ϕ, *the argument*,

Table 2.1

1	2	3	4	5	6
$\phi°$	F	Δ^{I}	Δ^{II}	Δ^{III}	Δ^{IV}
12	2 079·117				
		+170·394			
13	2 249·511		−0·686		
		+169·708		−0·051	
14	2 419·219		−0·737		0
		+168·971		−0·051	
15	2 588·190		−0·788		0
		+168·183		−0·051	
16	2 756·373		−0·839		−0·001
		+167·344		−0·052	
17	2 923·717		−0·891		+0·002
		+166·453		−0·050	
18	3 090·170		−0·941		
		+165·512			
19	3 255·682				

at intervals of 1°. The value of F corresponding to a latitude of $\phi = 15°\ 52'\ 29''.9096$ is required. Obviously F_ϕ will lie between the values for 15° and 16°. Interpolation is always carried out with the ascending order of the argument positive. In column (3), table 2.1 the differences between successive values of F in column (2) are listed in the sense of $F_{16} - F_{15}$, etc., in this case these *first differences* Δ' are all positive. In column (4) the *second differences* Δ'', i.e. the differences between the first differences in column (3), are listed with the same sign convention, in this case they are all negative. Columns (5) and (6) give the *third* and *fourth differences* Δ''' and Δ^{iv} respectively. It will be seen that successive differences become rapidly smaller, that is they form a convergent sequence. These differences are used to find the intermediate value of the function F. Denoting $_{15}\Delta_{16}'$ and $_{16}\Delta_{17}'$ as the first differences between the values of the function with arguments ϕ_{15} and ϕ_{16} and, ϕ_{16} and ϕ_{17}, respectively, Δ_{16}'' as the second difference between these, and which lies in line with F_{16} and so on, the value of F is given by Bessel's Formula

$$F_\phi = F_{15} + n\,_{15}\Delta_{16}' + \frac{n(n-1)}{1.2}\tfrac{1}{2}(\Delta_{15}'' + \Delta_{16}'') + \frac{n(n-1)(n-\tfrac{1}{2})}{1.2.3}\,_{15}\Delta_{16}'''$$

$$+ \frac{(n+1)(n)(n-1)(n-2)}{1.2.3.4}\tfrac{1}{2}(\Delta_{15}^{iv} + \Delta_{16}^{iv})$$

$$+ \frac{(n+1)(n)(n-1)(n-2)(n-\tfrac{1}{2})}{1.2.3.4.5}\,_{15}\Delta_{16}^{v} \tag{1}$$

This may be written

$$F_\phi = F_{15} + n\,_{15}\Delta_{16}' + B''(\Delta_{15}'' + \Delta_{16}'')$$

$$+ B'''\,_{15}\Delta_{16}''' + B^{iv}(\Delta_{15}^{iv} + \Delta_{16}^{iv}) + B_{15}^{v}\Delta_{16}^{v}$$

The quantities $B'' = \dfrac{n(n-1)}{2.2!}$, etc. are the *Besselian coefficients* which are not identical to those of the binomial expansion after the third term. These Besselian coefficients may be computed for each particular value of n, or obtained from tables to be found in such books as *Chamber's Shorter Six Figure Tables*, or the *Apparent Places of Fundamental Stars*. The fractional part n of the interval of tabulation of the argument is expressed as a decimal of that interval which cannot exceed unity. In the example, $n = 1/60$ of $(52'\ 29''.9096)$

i.e. $n = 0{\cdot}874\,975\ _{15}\Delta_{16}'$ $= +\,168{\cdot}183$

$n - 1 = -0{\cdot}125\,025\ \tfrac{1}{2}(\Delta_{15}'' + \Delta_{16}'') = \tfrac{1}{2}(-0{\cdot}788 - 0{\cdot}839) = -0{\cdot}8135$

$n - 2 = -1{\cdot}125\,025\ _{15}\Delta_{16}'''$ $= -\,0{\cdot}051$

$n - \tfrac{1}{2} = +\,0{\cdot}374\,975\ \tfrac{1}{2}(\Delta_{15}^{iv} + \Delta_{16}^{iv}) = \tfrac{1}{2}(+\,0{\cdot}001) = +0{\cdot}005$

$n + 1 = 1{\cdot}874\,975$

Hence
$$F_\phi = F_{15} = 2\ 588\cdot 190$$
$$\therefore\quad n_{15}\Delta_{16}' \qquad\quad + 147\cdot 156$$
$$\tfrac{1}{2}n(n-1)\tfrac{1}{2}(\Delta_{15}'' + \Delta_{16}'') \quad + 0\cdot 044$$
Other terms negligible
$$F_\phi = 2\ 735\cdot 390$$

It will be apparent that many terms will be negligible and that it will be convenient if this information can be ascertained without calculating those terms. The significance of the interpolation terms may be found from a consideration of the maximum values that the Besselian coefficients may take, and the numerical values of the respective differences to be multiplied by these coefficients. We assume a computational precision of 0·5 in the last place of decimals throughout. In the above example we would work to $0\cdot 5 \times 0\cdot 001 = 0\cdot 0005$. Using the notation above:

F_{15} must be known to 0·5 in the last place of decimals $n\Delta'$ likewise; since the maximum value of n is 1, the first difference Δ' must not exceed 0·5 in the last place, if it is to be neglected in the interpolation. Consider the term $B''(\Delta'' + \Delta'')$, the maximum value of B'' is $\tfrac{1}{16}$ when $n = \tfrac{1}{2}$, hence the term will be significant if the double second difference $\Delta'' + \Delta''$ exceeds 8 in the last place, or the single second difference Δ'' exceeds 4. In a similar manner, single third and single fourth differences of 60 and 20 respectively are significant. These various maxima are found by differentiating the Besselian coefficients, finding the turning values of n, and evaluating the coefficients for these turning values. The fourth differences in table 2.1 are all really zero, their variation derives from the rounding off of the tabular values of F.

2.9.2. ERRORS IN A TABLE

Because of the cumulative effect of a tabular error on the various differences, taking out these differences enables a table to be checked. For example an error of +1 in F increases according to the scheme below

```
                    +1
                +1
            +1      −3
    +1          −2      etc.
        −1          +3
            +1
                −1
```

i.e. according to the binomial coefficients. Hence if other than smooth changes in the various differences arise, an error can be located and corrected.

2.9.3. INVERSE INTERPOLATION

If the intermediate value of the function F_ϕ is known and the corresponding argument ϕ is required, a process of successive approximation in formula (1) is

carried out. For example if $F_\phi = 2735 \cdot 390$ the first approximation using two terms is
$$F_\phi = F_{15} + n_{15}\Delta_{16}'$$
whence
$$n = (F_\phi - F_{15})/_{15}\Delta_{16}'$$
i.e.
$$n = +0 \cdot 875\ 237 \text{ in the first approximation.}$$

This value of n is substituted in the third term giving its approximate value, i.e. $0 \cdot 406\ 75 \times 0 \cdot 875\ 237 \times 0 \cdot 124\ 763 = +0 \cdot 044$. Whence in the second approximation
$$n = (F_\phi - F_{15} - 0 \cdot 044)/+ 168 \cdot 183 = +0 \cdot 874\ 975.$$

This value is substituted again in the third term to see if there is a change from $0 \cdot 044$. Since there is none, the approximation is sufficient and the desired result obtained.

2.9.4. INTERPOLATION AT THE EDGE OF A TABLE

Sometimes interpolation is required at the edge of a table when the necessary differences required by the Besselian formula are unobtainable. In this instance *Newton's formula* is used which is, in the notation of the example,

$$F_\phi = F_{12} + n_{12}\Delta_{13}' + \frac{n(n-1)}{2!}\Delta_{13}'' + \frac{n(n-1)(n-2)}{3!}{}_{13}\Delta_{14}'''$$

In practice it is seldom necessary to proceed further than the second difference.

Table 2.2. Transverse Mercator Eastings in Metres

Latitude ϕ	2° 10'	15'	20'	25'	30'	35'
16° 00'		740 789·1				
15° 55'	731 963·2	740 888·7	749 814·7	758 741·1	767 668·0	
50'	732 058·7	740 987·9	749 917·5	758 847·6	767 778·1	
45'	732 153·6	741 086·5	750 019·8	758 953·5	767 887·8	Clarke 1880 spheroid U.T.M.
40'	732 248·1	741 184·6	750 121·5	759 058·9	767 996·8	776 935·1
35'	732 342·1	741 282·2	750 228·8	759 163·8	768 105·3	
15° 30'	732 435·6	741 379·3	750 323·5	759 286·1	768 213·2	

λ = Longitude from central meridian 33°E

Introduction to Computation 71

If much computation is required using differences higher than this, the interval of tabulation is too large and it will be worth while reducing this by interpolation and printing a new table. For example table 2.1 could be halved by putting $n = \frac{1}{2}$ for all intervals, and subsequent work would be reduced.

2.9.5. INTERPOLATION IN TWO DIRECTIONS

If a function F is dependent upon two variables E, and N, i.e. $F = f(E, N)$, intermediate values may be interpolated from a *double entry table*, again assumed to have regular fixed arguments. Table 2.2 gives the values of the transverse Mercator eastings at the intersections of a five minute graticule in the vicinity of three points A, B, and C whose geographical coordinates are

	Longitude λ	Latitude ϕ
A	35° 18′ 16″.7559	15° 52′ 29″.9096
B	35° 23′ 37″.4540	15° 39′ 22″.5964
C	35° 28′ 22″.7070	15° 44′ 56″.7567

Since the central meridian of the projection is at longitude 33° the longitudes are reduced by this amount. The various differences are taken out as for the above interpolation, only in this case, differences are obtained both vertically and horizontally. Taking the example of point A, it is required to obtain its easting. The square in which A lies is treated as follows:

$$\begin{array}{ccc} \Delta N'' & F_2 & F_3 \\ & \Delta N' \; \Delta EN'' & \\ & F_1 \; \Delta E' \; F_4 & \\ & \Delta E'' & \end{array}$$

The functions F_1, F_2, F_3, and F_4 are the appropriate values for the corners of the graticule in which point A lies, and F_1 is the tabular entry to which interpolated values are to be applied.

$\Delta E'$ is the first difference in the table in the easting direction.
$\Delta N'$ is the first difference in the table in the northing direction.
$\Delta E''$ is the second difference in the easting direction.
$\Delta N''$ is the second difference in the northing direction.
$\Delta EN''$ is the second difference in the northings direction between the first differences in the easting direction for the square in which A lies, and this is equal to the second difference in the eastings direction between the first differences in the northings direction for that square. The numerical example should make this notation clear. It must be noted that the tabular entires are the easting values only, and that the interpolation for the northings of A will require to be carried out entirely separately. The nearest tabular entry to and numerically less than the coordinates of A is at the graticule intersection 15° 50′ ϕ and 2° 15′ λ. Let n be the fractional difference between the latitude of A and this tabular entry, and m be the fractional difference between the longitude of A and this tabular entry, then

$$n = (2' \; 29''.9096)/5 = 0.499\;700$$

72   Practical Field Surveying and Computations

and
$$m = (3'\ 16\overset{''}{.}7559)/5 = 0 \cdot 655\ 852$$

Denoting the nearest tabular entry to A by E_T the required easting of the point A is given by

$$E_A = E_T + n\ \Delta N' + m\ \Delta E' + nm\ \Delta EN'' + \tfrac{1}{2}n(n-1)\ \Delta N''$$
$$+ \tfrac{1}{2}m(m-1)\ \Delta E''$$

In this example:

$$n = 0 \cdot 499\ 700, \quad m = 0 \cdot 655\ 852, \quad mn = 0 \cdot 3277$$
$$n - 1 = -0 \cdot 500\ 300, \quad m - 1 = -0 \cdot 344\ 148$$
$$\tfrac{1}{2}n(n-1) = -0 \cdot 1250, \quad \tfrac{1}{2}m(m-1) = -0 \cdot 1128$$

	E_T	740 987·9
$\Delta N' - 99\cdot 2$	$n\ \Delta N'$	$-49\cdot 57$
$\Delta E' + 8\ 929\cdot 6$	$m\ \Delta E'$	$+5\ 856\cdot 50$
$\Delta EN'' - 3\cdot 6$	$nm\ \Delta EN''$	$-1\cdot 18$
$\Delta N'' - 0\cdot 4$	$\tfrac{1}{2}n(n-1)\ \Delta N''$	$+0\cdot 04$
$\Delta E'' + 0\cdot 5$	$\tfrac{1}{2}m(m-1)\ \Delta E''$	$-0\cdot 06$
	E_A	746 793·6

The northings of point A may be evaluated from the data given in table 2.3 using the same values of m and n as above. As further examples, the reader

Table 2.3. Transverse Mercator Northings in Metres

	2° 10'	15'	20'	25'	30'	35'	
16° 00'		1770 083·2					
15° 55'	1760 765·6	1760 860·0	1760 958·0	1761 059·5	1761 164·6		
50'	1751 543·0	1751 636·9	1751 734·4	1751 835·5	1751 940·1		Clarke 1880 spheroid U.T.M.
45'	1742 320·4	1742 413·9	1742 510·9	1742 611·5	1742 715·6		
40'	1733 097·8	1733 190·9	1733 287·5	1733 387·6	1733 491·2	1733 598·3	
35'	1723 875·3	1723 967·9	1724 064·1	1724 163·7	1724 266·8		
15° 30'	1714 652·9	1714 745·1	1714 840·7	1714 939·9	1715 042·5		

λ = Longitude from central meridian 33° E

may derive the grid coordinates of points B and C from the data given in these same tables 2.2 and 2.3. The required coordinates† are

	Easting	Northing
A	746 793·6	1 756 309·4
B	756 612·7	1 732 209·7
C	764 991·4	1 742 581·7

Interpolation of coordinates in this manner is very useful in cartographic work where the required precision is not too high and many hundreds of points may have to be computed.

2.10. COORDINATE TRANSFORMATION

2.10.1. INTRODUCTION

A very common survey problem is to convert a system of coordinated points into some other system based on a new origin, with perhaps a different scale or a different unit of length, and referred to a different bearing. The problem takes two forms: (a) when these datum, scale and bearing changes are known at the outset, and (b) when these changes are not known directly, but coordinates on the two systems exist for a few common points. A simple example will be worked to illustrate how the transformation is effected in both cases.

The following coordinates of a survey are based on an origin at point O

	Easting (ft)	Northing (ft)
A	33 400·2	83 672·3
B	31 886·9	81 771·6
C	29 770·0	91 633·8
D	7 693·3	9 700·5

The original bearing through O of 96° 27′ 00″ was later found to be incorrect, and should have been 93° 56′ 00″. The coordinates of the origin (0, 0) were tied into a national system in metres and were found to be

$$351\ 620·00\ E \quad 1\ 836\ 766·52\ N.$$

The coordinates of A, B, C, and D are required on the national system and referred to the correct bearing.

2.10.2. SCALE CHANGE FROM FEET TO METRES

Using the foot/metre conversion of 1 ft = 0·304 799 472 metres the co-ordinates become

	E	N
A	10 180·36	25 503·27
B	9 719·11	24 923·94
C	9 073·88	27 929·93
D	2 344·91	2 956·71

† The reader should refer to the example in §4.2.3.2 where the coordinates of these same points are derived by more rigorous formulae.

These metric coordinates are precise to 0·03 m only (0·1 ft), and although the second decimal is retained, a statement of this precision should be clearly made on the final coordinate list. Since there are many different foot/metre conversion factors, great care should be exercised when selecting the correct relationship for scale change.

2.10.3. CORRECTION FOR BEARING

The error in the original bearing is $+2° 31' 00''$, i.e. the correction to the bearing is equal to this and of opposite sign. This correction is in effect a forward rotation of the coordinate axis system of $+2° 31' 00'' = \theta$ say. In figure 2.15

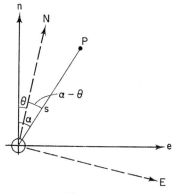

Figure 2.15

the original coordinates are denoted by (e, n) and the original bearings of points with respect to the origin 0 are α. With respect to the new axes, the coordinates are (E, N) and all bearings are $(\alpha - \theta)$. Consider a point P at a distance S from the origin and at bearing α. Then the original coordinates of P are

$$e_P = S \sin \alpha, \quad n_P = S \cos \alpha$$

And on the new system are

$$E_P = S \sin (\alpha - \theta)$$
$$= S \sin \alpha \cos \theta - S \cos \alpha \sin \theta;$$
$$= e_P \cos \theta - n_P \sin \theta \qquad (1)$$

$$N_P = S \cos (\alpha - \theta)$$
$$= S \cos \alpha \cos \theta + S \sin \alpha \sin \theta$$
$$= n_P \cos \theta + e_P \sin \theta \qquad (2)$$

Formulae (1) and (2) are the conversion formulae for the rotation of axes. In the example $\cos \theta = +0.999\ 03549$ and $\sin \theta = +0.043\ 91000$. Thus applying formulae (1) and (2) to the metric coordinates of A, B, C, and D the new values are

	Eastings	Northings
A	9 050·69	25 925·69
B	8 615·33	25 326·67
C	7 838·72	28 301·43
D	2 212·82	3 056·82

2.10.4. CHANGE OF ORIGIN

Finally the coordinates are referred to the national system by addition of the coordinates of the origin on that system to those of A, B, C, and D, giving:

	Easting	Northing
A	360 670·69	1 862 692·21
B	360 235·33	1 862 093·19
C	359 458·72	1 865 067·95
D	353 832·82	1 839 823·34

2.10.5. COMBINED CHANGE OF SCALE, BEARING, AND ORIGIN

The three changes considered separately above may be combined together as follows. For a scale factor k the new coordinates become

$$e_1 = ke \quad \text{and} \quad n_1 = kn$$

For a rotation of axis θ of the new coordinates are given by

$$e_2 = e_1 \cos \theta - n_1 \sin \theta \qquad n_2 = e_1 \sin \theta + n_1 \cos \theta$$

For a change of origin by factors E_0, N_0

$$E = E_0 + e_2 \quad \text{and} \quad N = N_0 + n_2$$

The final combined formulae are then

$$E = E_0 + ke \cos \theta - kn \sin \theta \qquad (3)$$
$$N = N_0 + ke \sin \theta + kn \cos \theta \qquad (4)$$

2.10.6. EMPIRICAL TRANSFORMATION METHOD

It often happens that the three variation factors for scale, rotation of axes, and origin are not directly known, but the coordinates of some points are known on both systems. Suppose that points A, B, C, and D are a few points of an old survey which were tied into a national framework, and it is required to derive formulae to convert all the coordinates of the old work into the new system. It is possible to find the values of the scale, bearing, and origin changes from a consideration of various common points, and then use formulae (3) and (4). However it is more convenient to obtain the values of E_0, N_0, $k \cos \theta$, and $k \sin \theta$ directly. Equations (3) and (4) may be written

$$E = \alpha + \beta e - \gamma n \qquad (5)$$
$$N = \delta + \gamma e + \beta n \qquad (6)$$

where $\alpha = E_0$, $\delta = N_0$, $\beta = k \cos \theta$, and $\gamma = k \sin \theta$. From the coordinates (e, n) and (E, N) of two points common to both systems, the four unknowns α, β, γ, and δ are obtained by solution of four simultaneous equations, two of the form of (5), and two of the form of (6). Usually more than the necessary two common points are available and the transformation formulae can be checked for points other than those used to derive the formulae. Suppose that the coordinates of points A, B, C, and D of § 2.10.1, and those of the same points in §2.10.4 are known, and it is required to derive the transformation formulae directly. To obtain the greatest precision, the two most distant points are used, i.e. A and D in this case. From point A the equations for eastings and northings are respectively

$$360\ 670 \cdot 69 = \alpha + 33\ 400 \cdot 2\beta - 83\ 672 \cdot 3\gamma$$
$$1\ 862\ 692 \cdot 21 = \delta + 33\ 400 \cdot 2\gamma + 83\ 672 \cdot 3\beta$$

From point D the respective equations are

$$353\ 832 \cdot 82 = \alpha + 7\ 693 \cdot 3\beta - 9\ 700 \cdot 5\gamma$$
$$1\ 839\ 823 \cdot 34 = \delta + 7\ 693 \cdot 3\gamma + 9\ 700 \cdot 5\beta$$

Solution of these equations gives:

$\alpha = 351\ 620 \cdot 00$
$\beta = 0 \cdot 304\ 5054$ (The foot/metre conversion $\times \cos \theta$)
$\gamma = 0 \cdot 013\ 3837$ (The foot/metre conversion $\times \sin \theta$)
$\delta = 1\ 836\ 766 \cdot 52$

Hence the transformation formulae become

$$E = 351\ 620 \cdot 00 + 0 \cdot 304\ 5054e - 0 \cdot 013\ 3837n$$
$$N = 1\ 836\ 766 \cdot 52 + 0 \cdot 013\ 3837e + 0 \cdot 304\ 5054n$$

These formulae are then tested out on the other points B, and C to ensure that they are correct.

2.10.7. DIFFERENTIAL SCALE CHANGE

In some cases it is necessary to take into account the effects of a differential scale change between one coordinate system and another, for example when the transformation involves two different projections, or a change of datum on the same projection. The procedure is similar to that of §2.10.6 except that six variables are required. We assume that the coordinate transformation formulae are of the form

$$E = \alpha + \beta e + \gamma n$$
$$N = \delta + \omega e + \phi n$$

At least three common points are required to evaluate the six unknowns $\alpha, \beta, \gamma, \delta, \omega,$ and ϕ.

2.10.8. TRANSFORMATIONS WITH LEAST SQUARES

In practice exact values of the coefficients will not be obtainable for all points common to both surveys unless the bare minimum of data is available. For each pair of points a slightly different value will be obtained for each coefficient because of inherent errors in the two surveys that produced the two sets of coordinates. The procedure adopted is to form equations for all data and, by applying the principle of least squares to these observation equations (see Chapter 5), obtain the most probable values of the coefficients.

2.10.9. SECOND ORDER TRANSFORMATIONS:

In some cases it is not sufficient to assume a linear relationship between two sets of coordinates, particularly if they are on different types of projection. In this case a second order transformation formula is required, i.e. one of the form

$$E = \alpha + \beta e + \gamma n + \delta e^2 + \epsilon n^2 + \phi en$$

which requires six common points for solution. The reader may use the data of tables 2.2 and 2.3 to form transformation formulae for geographical into transverse Mercator coordinates.

2.11. COMPUTATION IN THE PLANE RECTANGULAR SYSTEM

2.11.1. FUNDAMENTAL CONCEPTS

In surveying, a system of plane rectangular coordinates† is used in which (*a*) the scale along both axes is the same, (*b*) the origin is taken at the extreme south and west of the area so that all coordinates are positive, (*c*) coordinates are reckoned positive from west to east, and from south to north, (*d*) the direction of a point from the origin, or its *bearing* is reckoned clockwise from north, even in the southern hemisphere, (*e*) since there is no internationally accepted convention concerning the use of X and Y, it is less ambiguous to name the axes east and west, i.e. coordinates are (E, N). The convention is to quote eastings before northings. (*f*) A difference in eastings from point A to B denoted by $_A\Delta E_B$ is defined to be $E_B - E_A$. Similarly $_A\Delta N_B = N_B - N_A$. Thus $E_B = E_A + {_A\Delta E_B}$, and $N_B = N_A + {_A\Delta N_B}$. Thus if E_A is greater than E_B, $_A\Delta E_B$ is negative. The difference in northings between the two ends of a survey line whose length is L is often called the *latitude* of the line; the corresponding term for a difference in easting is the *departure*. These terms are normally used only in traverse work.

2.11.2. TRIGONOMETRICAL FUNCTIONS

Since the bearing of a line is reckoned clockwise from North in surveying, the signs of the trigonometrical functions are positive as follows

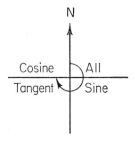

It will be remembered that the angle $-A$ is in the fourth quadrant if $A < 90°$, hence $\cos(-A) = +\cos A$. Since the signs are of great importance they should be given separate attention before the actual values of the functions are

† See Chapter 4 for qualifications to this simplification.

found. When evaluating functions in quadrants other than the first or third, the following procedure is best:

For sin t, look up cos $(t - 90)$ or cos $(t - 270)$
for cos t, look up sin $(t - 90)$ or sin $(t - 270)$
for tan t, look up cot $(t - 90)$ or cot $(t - 270)$

This avoids making mistakes when subtracting minutes and seconds. A convenient alternative to the subtraction of 90° or 270°, or of 180° for angles in the third quadrant, is to accept the last number of degrees, and add any previous numbers until the sum is less than or equal to 9. This will give the angle to be referred to in tables, counting 9 itself as zero. For example

sin 112 = sin (1 1)(2) = cos (1 + 1)(2) = cos 22
sin 196 = sin (1 9)(6) = sin (1 + 9)(6) = sin (1 + 0)(6) = sin 16
sin 318 = cos (31)(8) = cos (3 + 1)(8) = cos 48, etc.

Forward and back bearings

The *forward bearing* from A to B is denoted by Bg AB or by t_{AB}, that from B to A, the *back bearing*, is denoted by Bg BA or t_{BA}, etc. Thus $t_{AB} = t_{BA} + 180°$ in all cases. If the resultant bearing exceeds 360° or a whole revolution, 360° are subtracted from the result. In practice, if the original bearing t_{AB} is greater than 180°, the same numerical result is produced if 180° is immediately subtracted from t_{AB}. In mathematical text however t_{AB} is always $t_{AB} + 180°$ to avoid confusion. Bearings reckoned round all four quadrants are called *whole circle bearings* to distinguish them from other less convenient ways of considering bearings. Only whole circle bearings are considered in this book.

2.11.3. COMPUTATION OF COORDINATES FROM BEARING AND DISTANCE

A line AB of length L runs at a bearing of t. Then

$$_A\Delta E_B = L \sin t \quad \text{and} \quad _A\Delta N_B = L \cos t$$

Thus
$$E_B = E_A + L \sin t$$
and
$$N_B = N_A + L \cos t$$

Normally the coordinates of the second point, in this case B, are obtained on the machine without writing down the values of ΔE and ΔN.

2.11.4. BEARING FROM COORDINATES

The bearing t is obtained from $\tan t = \Delta E/\Delta N$ or from $\cot t = \Delta N/\Delta E$. Most people preferring that which gives the trigonometrical function as a decimal fraction.

2.11.5. DISTANCE FROM COORDINATES

The distance L is obtained from one of three expressions:
(a) Pythagoras, i.e. $L = (\Delta E^2 + \Delta N^2)^{\frac{1}{2}}$. To obtain sufficient precision, the calculating machine must have sufficient capacity to retain all figures of the squares, e.g. if ΔE is a five digit number, its square will be a 10 digit number at most, and the final square root will give a five digit number once again.

(b) From the bearing t the distance L may be calculated in two ways. Either from $L = \Delta N \cos t$ or $L = \Delta E \sin t$. If the bearing t is small, the former expression gives the better result, and if it is close to 90° the latter gives the better result. This is seen from figure 2.16 which shows the effect of small errors dE in ΔE and dN in ΔN in the corresponding length L, computed via t which is assumed free of error. The error dL_E arising from dE, is very much greater than that from dN, dL_N, when t is small. A corresponding figure will

Figure 2.16

show that the reverse is true for t close to 90°. For intermediate angles it is usual to accept the cosine under 30° and the sine over 60°, with no preference between 30° and 60°. In important work, both formulae will be used as a check on gross error.

Examples of the above: The above computations are now illustrated by an example in which $t = 22°\ 10'\ 03''\!.42$ and $L = 26\ 023 \cdot 250$ metres. The forward bearing is $22°\ 10'\ 03''\!.42$, therefore the back bearing is $202°\ 10'\ 03''\!.42$. Since t is in the first quadrant, both $\sin t$ and $\cos t$ are positive, and therefore so are ΔE and ΔN. The values of $\sin t$ and $\cos t$ are $+0 \cdot 377\ 31743$ and $+0 \cdot 926\ 08399$ respectively, giving $\Delta E = +9\ 819 \cdot 026$, and $\Delta N = +24\ 099 \cdot 715$. If the coordinates of the first point A are $E_A = 746\ 793 \cdot 673$, and $N_A = 8\ 243\ 690 \cdot 628$ the coordinates of the second point B are $E_B = 756\ 612 \cdot 699$, and $N_B = 8\ 267 \cdot 343$. From the values of ΔE and ΔN we have

$$\tan t = 9\ 819 \cdot 026 / 24\ 099 \cdot 715$$
$$= +0 \cdot 407\ 43328$$

whence $\quad t = 22°\ 10'\ 03''\!.42$

To obtain the length of AB $= L$, by Pythgoras

$$L = (\Delta E^2 + \Delta N^2)^{\frac{1}{2}} = 6\ 7720\ 9534 \cdot 670001^{\frac{1}{2}} = 26\ 023 \cdot 250$$

From the bearing t, $L = \Delta E/\sin t$ and $L = \Delta N/\cos t$, which give 26 023·2505 and 26 023·2500 respectively, the latter being the more precise because $t < 30°$.

2.11.6. CHECK COMPUTATION BY THE AUXILIARY BEARING METHOD

To obtain an independent check using different figures the following device is often employed. All bearings are increased by an angle of 45°. In this example the bearing becomes 67° 10′ 03″·42 whose sin and cos are respectively +0·921 64398 and +0·388 03656. From these new bearings a new $\Delta e = L \sin(t + 45)$ and $\Delta n = L \cos(t + 45)$ are obtained. This addition of 45° to the bearing has in effect rotated the axes back through -45°, and to bring the coordinates back to the original system, the formula for rotation of axis forward through +45° is applied, i.e.

$$\Delta E = \Delta e \cos 45 - \Delta n \sin 45 \quad \text{and} \quad \Delta N = \Delta e \sin 45 + \Delta n \cos 45$$

In this case $\sin 45° = \cos 45° = (1/2)^{\frac{1}{2}} = 0·707\ 10678$. Denoting $L \cos(t + 45)/(2)^{\frac{1}{2}} = C$, and $L \sin(t + 45)/(2)^{\frac{1}{2}} = S$, $\Delta E = S - C$ and $\Delta N = S + C$. In the example $S = 16\ 959·3704$ and $C = 7\ 140·3448$, whence $\Delta E = 9\ 819·0256$ and $\Delta N = +24\ 099·7152$, which checks the direct computation above. The reader will observe that this example is part of that computed by different means in §2.11.5.

2.11.7. SOLUTION OF A TRIANGLE (See also §§6.9.2 and 6.9.3)

In triangle ABC, the adjusted angles A, B, and C are given below, together with the length of the side $AB = c$. It is required to compute the length of the other two sides $AC = b$ and $BC = a$. The formulae used are

$$a = \sin A (c/\sin C) \quad \text{and} \quad b = \sin B (c/\sin C)$$

$A = 30° 48′ 03″·70, \quad B = 61° 05′ 48″·81, \quad C = 88° 06′ 07″·49$

whose sum is 180° 00′ 00″·00, and $AB = c = 26\ 013·282$ metres. Then $\sin A = 0·512\ 05827$, $\sin B = 0·875\ 43831$, $\sin C = 0·999\ 45142$, whence $c/\sin C = 26\ 027·560$ and $a = 13\ 327·627$, $b = 22\ 785·523$. The common factor $c/\sin C$ need not be written down, but is merely retained on the machine for multiplication in turn by $\sin A$ and $\sin B$. If logarithms are used a simple layout involving a minimum of writing is:

a	13 327·63
log a	4·124 7528
log sin A	9·709 3194
log c	4·415 1951
log cosec C	0·000 2383
log sin B	9·942 2255
log b	4·357 6589
b	22 785·52

The reader will notice that this triangle ABC is that used to illustrate Legendre's theorem (see example in §4.1.10).

3 *The Theodolite and Level*

3.1. THE THEODOLITE

The theodolite and the level will probably remain for many years standard surveying equipment. The development of the automatic level has simplified in part, the process of levelling, but has not changed the fundamental procedures used in the field. In this chapter the construction of these basic instruments and their modern accessories are examined in detail. It is proposed to consider basic concepts, adjustments, and field usage.

3.2. COMPONENT PARTS OF THE THEODOLITE

Figure 3.1 illustrates some of the essential components of a theodolite. The use and function of these and other essential components will now be considered in detail.

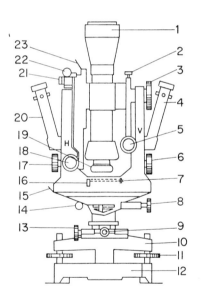

1. Telescope
2. Clamp for 5
3. Focusing screw
4. Vertical circle reading telescope
5. Vertical circle slow motion screw
6. Vertical circle micrometer screw
7. Plate bubble adjusting scews
8. Upper plate slow motion screw
9. Clamp for 13
10. Tribrach
11. Foot screw
12. Trivet stage
13. Lower plate slow motion screw
14. Clamp for 8
15. Horizontal circle mirror
16. Plate bubble pivot
17. Horizontal circle micrometer screw
18. Altitude bubble setting screw
19. Eyepiece
20. Horizontal circle reading telescope
21. Split bubble reader
22. Altitude bubble
23. Vertical circle mirror

Figure 3.1

3.2.1. THE TRIPOD

The purpose of the tripod is to provide support for the instrument. It may be rigid or telescopic. Before use it must be examined for any slackness between the legs and shoes at one end and the tripod head at the other; means are always provided to tighten these connections. When in use the tripod must be well pressed into the ground and wing nuts, if present at the tripod head, tightened. Wood is usually used for the manufacture of the tripod legs as it has a low coefficient of expansion. Aluminium alloy tripods are equally suitable but are more liable to serious damage.

3.2.2. THE TRIVET STAGE AND TRIBRACH

The trivet stage receives the shoes of the levelling screws which should be maintained in good contact. Movement of the trivet stage relative to the tripod head is frequently possible to permit accurate centering of the instrument over the ground mark. This is carried out by means of a plumb bob or an optical plummet which will give the line of the main axis of rotation of the instrument.

The threads of the levelling screws should be a tight fit in the tribrach, any looseness must be taken up. The tribrach receives the levelling screws and provides a bearing surface for the axes (see figure 3.2).

3.2.3. THE ALIDADE

The term alidade is applied to the whole of that part of the instrument which carries the telescope and rotates on a bearing surface provided by the tribrach. It is usually cast in one piece and includes the inner main axis of rotation. Figure 3.2 illustrates the simplified design of two principal types of main axes

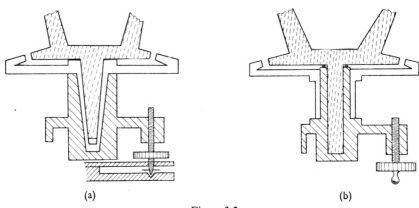

Figure 3.2

(*a*) conical and (*b*) cylindrical. It is seen that in (*a*) the outer axis is adjacent to the inner axis and in (*b*) the outer axis is outside of the tribrach. Conical axes have a long life and are found on all older instruments as well as on some modern types. Equal wear on all bearing surfaces allows the axis to move downwards whilst maintaining a tight fit. Conical gun metal axes are readily annealed to free them from internal stress and are as a result very reliable. Axes of this type require careful fitting and the machining necessary does not readily allow mass production. Cylindrical axes can more easily be produced with high speed grinding machines to a high degree of precision but, due to the slow release of internal stresses in the hardened stainless steel used, axis strain may be developed and is now recognized as an important potential source of error. The magnitude of this error is however only significant in work of a very high precision.

The alidade carries the plate level set in its adjustable mount. By means of the levelling screws the inner axes can be set to within a few seconds of the vertical; vertical being the direction taken up by a suspended plumb line. The level bubble usually has a sensitivity of 30 or 40 seconds per 2 mm run. Difficulty in maintaining the bubble in one position may be attributed to a poor fit of one of the axes and is more readily revealed by levelling with the more sensitive altitude level. This condition, frequently termed obliquity of the axes, is rare and difficulty in levelling with the plate level is more likely to be due to an unstable tripod, slackness in the footscrews, or use of an unsystematic procedure in levelling the instrument.

The movement of the alidade is read against a fixed horizontal, graduated, brass or glass circle. In all modern instruments this circle can either be rotated with the alidade, or by means of a directly connected knob without movement of the alidade.

3.2.4. THE TRANSIT AXIS AND TELESCOPE

The telescope is carried on an axis variously termed the 'transit', 'trunnion' or 'horizontal' axis. (Section 3.3 which deals with the telescope in detail.) The transit axis has its bearings in the standards of the alidade and at right angles to it should lie the line of sight or line of collimation of the telescope, defined as the straight line passing through the optical centre of the object glass and the intersection of the hairs in the reticule.

Besides carrying the telescope the transit axis carries either the vertical circle or the indices against which the vertical circle is read. In modern instruments it is usual to find that the vertical circle is attached to the transit axis and, like the horizontal circle, it is made of glass.

In modern instruments a sensitive altitude level is used to make the indices horizontal, mutual movement being effected by means of an index setting screw which is similar to a slow motion screw. In such cases the relationship between the index and the altitude bubble remains fixed unless deliberately adjusted. On very old instruments an index setting screw is not present and the instrument is levelled using the altitude level, the clip screws provided and the footscrews. The instrument is not relevelled for each vertical angle determination but the displacement of the bubble is recorded and a correction applied.

3.2.5. CLAMPS AND SLOW MOTION SCREWS

The clamps and slow motion screws enable the telescope to be pointed accurately at some distant object. All clamps should be definite in action and should always be lightly applied. Many instrument manufacturers suggest that the slow motion screws should always be turned in a clockwise direction, that is against the action of the spring plungers. Many surveyors do not follow this practice, but in all circumstances there should be no change in the direction of rotation of the slow motion screw in completing the pointing onto the object.

Pointing before use of the slow motion screw should be such that one turn of the slow motion screw brings the cross hairs onto the object.

In instruments with clamps for both the alidade or upper plate and the circle or lower plate, the function of the upper plate clamp is to bind together the alidade and circle and the function of the lower plate clamp is to bind the circle to the tribrach. As stated earlier some instruments have no lower plate clamp, the circle being rotated by a knob directly connected to the circle which otherwise is firmly clamped to the tribrach.

3.2.6. THE TRAVELLING CASE

The packing of an accurate instrument to withstand the shocks of transportation presents difficulties. Figure 3.3 illustrates various forms of stowage. The instrument must be securely held to the bottom of the case so that no damage results if the lid is not secured. Also illustrated is a well proven form of travelling case.

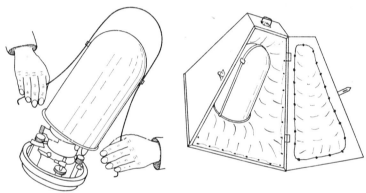

Figure 3.3

3.2.7. GENERAL

A theodolite should be handled lightly and not subjected to undue stress by being carelessly carried when out of its case. All operations should be carried out systematically. When observing a round of angles it is normal to swing in one direction and point to the various stations in turn. If a station is overshot rotation is continued in the same direction through a full circle and the station reapproached. All instruments should be protected from the direct rays of the sun as unequal expansion of both instrument and tripod may occur.

Small instrumental errors can often be cancelled out by use of a suitable procedure. The standard method of measuring all horizontal and vertical angles is with the vertical circle first to the left (face left) and then to the right (face right) of the line of sight. This is shown later to eliminate many of these errors.

3.3. THE TELESCOPE

An open sight involves aligning three objects in different planes, namely the target, back sight and fore sight. The telescope, regardless of its magnification, permits a higher degree of accuracy than an open sight. The aiming point is placed in the common focal plane of the eyepiece and object lens; now the aiming point and the image of the target are in the same plane.

As a result the 'surveying telescope' is fundamental to both the theodolite and level and is considered in detail. In its simplest form it consists of a cylindrical tube carrying a convex lens at each end. On viewing an object an image is formed by one lens on a fixed reticule. This image is magnified by being viewed with the other lens. The reticule must lie in the common focal plane of the two lenses. Figure 3.4 illustrates this simplified concept.

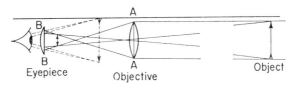

Figure 3.4

The essential conditions to be fulfilled in focusing the telescope are that the real image shall be formed by the object lens on the plane of the reticule and that the eye lens or eyepiece shall be focused on the reticule. Failure to achieve this state will result in the presence of parallax (see §3.4.1).

3.3.1. MAGNIFICATION

Magnification is the ratio of the angles subtended at the eye by the virtual image and the object. In terms of the focal lengths of the lenses in the telescope

$$M = F_o/F_o$$

where M is the magnification, F_o the focal length of the objective, and F_i the focal length of the eyepiece. This expression remains true regardless of the number of lenses forming the eyepiece or the objective.

3.3.2. BRIGHTNESS OF IMAGE

The brightness of the image depends upon the size of the aperture of the telescope at the objective end and upon the magnification. The relationship is expressed by

$$I = \frac{CA^2}{M^2}$$

where I is the brightness of the image, C a constant depending upon the number of lenses in the telescope, their quality, and thickness and A the diameter of the bundle of rays received and passed by the objective.

The aperture may be increased up to a maximum expressed by

$$A = Md$$

where d is the diameter of the iris of the human eye. To increase the aperture beyond this maximum will gain nothing in perceptible brightness and in fact the image will become blurred. Due to difficulties in exact positioning of the eye at the eyepiece the diameter of the objective never approaches this maximum value. Furthermore, as the resolving power of the eye is a maximum when $d = 0''08$, it is inadvisable to increase magnification beyond a value of $16A$ (when $d = 0''08$, $M = 12 \cdot 5A$), A being in inches.

For optimum conditions of brightness of the image A/M should be a maximum, the aperture size is fixed but, by changing the eyepiece, magnification can be reduced and hence the brightness increased. When observing under bad lighting conditions it is possible by changing the eyepiece for one of a greater focal length to reduce the magnification of the telescope and so produce a brighter image. In high quality instruments a second eyepiece is provided for this purpose.

3.3.3. RELATIVE RESOLVING POWER OR RESOLUTION

As far as a telescope is concerned relative resolving power is the relationship between detail that can be seen with the unassisted eye and detail that can be seen with the telescope. The unassisted eye can distinguish between two marks subtending an angle at the eye of one or two minutes of arc. For instance the two sides of a ranging rod at under 300 feet. If under certain light conditions the eye can distinguish between the graduations on a staff (0·01 feet) at 30 feet and under identical light conditions these same graduations can be distinguished with a telescope at 750 feet, it would be correct to say that the resolving power of that telescope was twenty five times that of the human eye.

The resolving power depends upon

(i) The magnification of the telescope.
(ii) The brightness of the image.
(iii) The quality of the lenses and their combination in the telescope.

Of these, magnification is the most important although the provision of an excessive magnification is a disadvantage as it reduces the brightness of the image, time is wasted in focusing the telescope, and the size of the field of view is reduced.

3.3.4. FIELD OF VIEW

This is the proportion of the horizon that can be seen on looking through the telescope; and, in terms of the optics of the telescope, it is the angle subtended by this visible field at the optical centre of the objective. This visible field is formed by the cone of rays refracted to enter the eye and figure 3.5 shows that this angle is effectively that subtended at the centre of the objective by the eyepiece. The important point that arises is that the size of the field of view

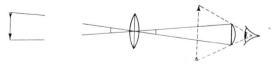

Figure 3.5

depends upon the diameter of the eyepiece and the distance between it and the objective but is independent of the size of the objective. It is stated in §3.3.3. that an increase in magnification will cause a decrease in size of the field of view. In terms of the angular definition of size of the field of view

$$M = \alpha'/\alpha$$

where, as before, M is the magnification, α the field of view, and α' the apparent field of view, that is the angle subtended at the eye by the image of the visible field of view. As α' remains constant any increase in magnification will be accompanied by a decrease in α, the actual field of view.

In telescopes fitted to theodolites a magnification of up to ×30 is usual with a field of view of 1°5 to 2°0. In the level the field of view can conveniently be smaller and telescopes with a 1° field are common.

To determine the size of the field of view, the image is brought first to one side of the field of view and then to the other. The difference in the horizontal circle readings will give the value required. Alternatively, the two extreme readings on a level staff held a known distance from the instrument will give the required value in circular measure. This is expressed by

$$\alpha = s/D$$

where s is the length of staff intercepted and D is the distance to the staff.

3.3.5. DEFECTS OF THE IMAGE

Observations made with a surveying telescope are almost always made to the centre of the field of view and as a consequence at or close to the optical centres of the various lenses. If precautions were not taken during manufacture the image so formed would be effected by axial spherical aberration and chromatic aberration.

Figure 3.6 illustrates axial aberration, where various rays from a point source of light do not come to a common focus due to differential refraction across the lens.

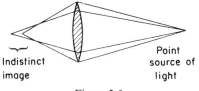

Figure 3.6

If a thin convex lens is used to form the image of a white object or source of light a series of images of different colours are formed at different distances from the lens. The violet image being the nearer and the red image the further. As a rule in focusing the yellow green image is made sharp as this is the brightest part of the spectrum. Superimposed upon this image are the other images out of focus, creating the impression of a blurred image which is incapable of an exact focus (see figure 3.7).

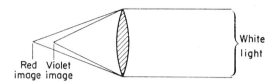

Figure 3.7

3.3.6. ELIMINATION OF DEFECTS OF THE IMAGE

By using a combination of two lenses of different glasses the defects referred to above can be reduced considerably. A doublet of flint glass and crown glass will bring the focal point of the red and blue images close to the focal point of the yellow green image. The violet image is barely perceptible. In this achromatic doublet suitable radii of curvature for the outside faces of the lens can be chosen to produce minimum axial spherical aberration. The two lenses are cemented together and mounted so that the light is incident on the crown glass lens first.

There are certain residual defects remaining in the image but as all careful scrutiny is carried out in the central portion of the field of view none of these defects should be perceptible.

3.3.7. CONSTRUCTION OF THE TELESCOPE

Figure 3.4 illustrates the simplified telescope. The previous paragraphs show some of the factors that are considered in modifying this concept. The objective is not a single lens but usually a doublet consisting of a convex lens of crown glass with a back concavo-convex lens of flint glass. The single lens eyepiece is replaced by two identical plano-convex lenses, separated by $\frac{2}{3}$ the focal length of either.

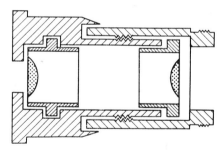

Figure 3.8

The eyepiece so formed is the 'Ramsden eyepiece' which is illustrated by figure 3.8. It does not completely eliminate chromatic aberration of the image but has advantages over other combinations that fulfil achromatic requirements. The important feature of the Ramsden eyepiece which makes it particularly suitable for surveying instrument telescopes is that the focal point falls outside of the lens combination. Thereby the eyepiece is moved bodily to focus on the fixed reticule. Furthermore the image of the cross hairs, seen through the whole of the eyepiece will be as effectively corrected for aberration as will be the image of the distant object. In the Huygens eyepiece used in the Galilean telescope the focus falls between the lenses and as a consequence the reticule must be positioned between the lenses. The effect of this is for the image of the cross hairs to be uncorrected for aberration. This would be a very serious defect particularly in tacheometric work. The Huygens eyepiece does however completely correct for aberration in respect of light passing right through the eye piece. It also produces an upright image, as it is placed to receive the rays from the objective before a real image is formed.

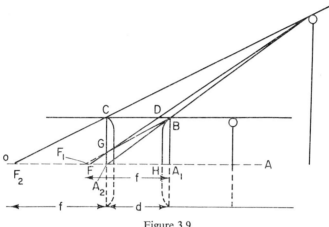

Figure 3.9

The geometrical optics of the Ramsden eyepiece are illustrated by figure 3.9 from which it follows that:

as CA_2, DF, BF_1 meet at G and F_2C is parallel to F_1B

$$\frac{CD}{FA_2} = \frac{CB}{F_1A_2} = \frac{d}{f-d} \tag{1}$$

and

$$\frac{DB}{FA_2} = \frac{CB}{F_2A_2} = \frac{d}{f} \tag{2}$$

by addition of (1) and (2)

$$\frac{CB}{FA_2} = \frac{d(2f-d)}{f(f-d)} = \frac{d}{FA_2}$$

$$\therefore FA_2 = \frac{f(f-d)}{(2f-d)} \tag{3}$$

hence substituting for FA_2 in (1)

$$CD = \frac{fd}{2f-d}$$

but by definition d, the separation of the lenses is $\tfrac{2}{3}f$

$$FA_2 = \tfrac{1}{4}f/4 \quad \text{and} \quad CD = \tfrac{2}{3}f/2$$

The focal length of the combination is HF

$$HF = CD + FA_2 = \tfrac{3}{4}f$$

Modified Ramsden eyepieces are sometimes used in surveying telescopes. In the Steinheil eyepiece both lenses are doublets of flint and crown glass. The Kellner eyepiece has an eye lens formed by a doublet of flint and crown glass.

Reference has been made earlier to the reticule (or graticule) where the image should be brought to a focus. It will usually be a circle of plane glass upon which a regular pattern of lines is etched. These etched lines are termed the cross hairs and various forms are used by different manufacturers. The term cross hairs arose from the use in older instruments of spiders web to produce a form as illustrated by figure 3.10. The spiders web had at that time certain advantages,

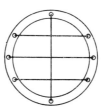

Figure 3.10

not least of which was the case of repair in the field. It was also preferred to lines etched on glass as the latter absorbed some of the light and was subject to a troublesome deposition of moisture with a falling temperature. With the introduction of effectively sealed telescopes (the internal focusing telescope) this dewing does not occur; also the modern telescope with modern lenses produces a brighter image and so loss of brilliance by absorbtion of light by a glass plate is no longer a serious problem. In all modern instruments, etched lines on glass have replaced the spiders web. Their various forms are illustrated by figure 3.11. The glass circle is carried in a metal cell which can be withdrawn

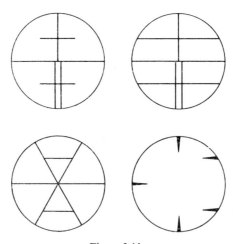

Figure 3.11

without disturbing the diaphragm. The diaphragm being a flanged brass ring held in place in the telescope by 3 or 4 screws (usually 4 as shown in figure 3.12).

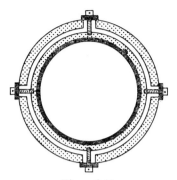

Figure 3.12

The screws pass through slots in the telescope tube permitting movement of the diaphragm in two directions and a small degree of rotation about the telescope axis. The initial perpendicularity of the hairs in an etched glass reticule can be relied upon, unlike the old spiders web cross hairs.

To move the diaphragm, for instance to the left, the top and bottom screws are released, the right-hand screw is slackened and the diaphragm is drawn over to the desired position by tightening the remaining screw. On achieving the desired degree of displacement the slackened screws are tightened and the verticality of the cross hairs checked. Exact verticality can be achieved by a slight rotation of the diaphragm with all screws loosened.

3.3.8. THE EXTERNAL FOCUSING TELESCOPE

The external focusing telescope (figure 3.13) is the simplified telescope (figure 3.4) modified in the manner described in §3.3.7. It consists of two tubes one within the other, the tube carrying the objective is moved by means of a rack and pinion so that a focus is achieved at the reticule. It is essential that the movement along the axis of the optical system shall be free from shake.

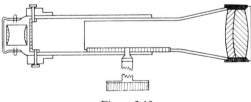

Figure 3.13

In the anallactic telescope of Porro an additional lens is introduced between the eyepiece and the objective (see §9.5.5). Focusing of this telescope is by movement of the eyepiece.

The external focusing telescope is not of constant length, cannot be sealed against moisture, and after some time has a tendency to droop, i.e. non-axial movement of the objective. This type of surveying telescope is now outdated and is replaced by the internal focusing telescope.

3.3.9. THE INTERNAL FOCUSING TELESCOPE

In this telescope the objective and eyepiece are fixed, focusing being achieved by means of a double convex lens set in a tube within the telescope. This internal tube is moved axially by means of a rack and pinion. Figure 3.14 shows

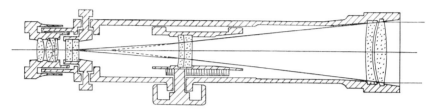

Figure 3.14

a section through a typical modern telescope. The simplified concept of the internal focusing telescope is illustrated by figure 3.15. Considering the geometrical optics of this lens arrangement:

Let f_0 be the focal length of the object lens AA and f_i be the focal length of the internal lens BB. DD is the reticule where the image is brought to a focus.

Then
$$\frac{1}{f_0} = \frac{1}{v} + \frac{1}{u} \tag{1}$$

where v is the distance of the image from the optical centre of lens AA, and u the distance of the object from the optical centre of lens AA.

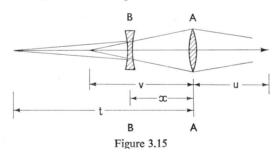

Figure 3.15

If a concave lens is inserted
$$\frac{1}{f_i} = \frac{1}{t-x} - \frac{1}{v-x} \tag{2}$$

where t is the distance between the optical centre of lens AA and the reticule, and x is the distance between the optical centre of lenses AA and BB.

It is usual to keep the distance t constant but the value of v will depend upon the value of u and can be determined from equation (1) above. Consequently only x can change for different values of v and this is achieved by moving the lens BB by means of the milled headed focusing screw. When the telescope is focused, i.e. the image falls on the reticule, equation (2) is satisfied.

When the object is distant, u is large compared to v, u may then be regarded as being infinite; consequently v will approximate to f_0 and it is taken that $v = f_0$.

Then
$$\frac{1}{f_i} = \frac{1}{t-x} - \frac{1}{f_0 - x} \tag{3}$$

from which the quadratic equation

$$x^2 - (f_0 + t)x + t(f_0 + f_i) - f_i f_0 = 0 \tag{4}$$

is formed.

It should be noted that the function of the eyepiece in both the external and internal focusing telescope is identical. This is to magnify the image formed on the reticule, and the cross hairs.

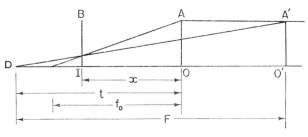

Figure 3.16

Figure 3.16 shows that the focal plane of the lens combination lies at $A'O'$ beyond the convex lens. The focal length F of the combination is represented in the figure by the distance $O'D$.

In the general case

$$F = \frac{v(t-x)}{v-x} \tag{5}$$

or when v is taken to be f_0

$$F = \frac{f_0(t-x)}{f_0 - x} \tag{6}$$

As $t - x$ is always greater than $f_0 - x$, owing to the use of a concave lens at BI, the internal focusing telescope is always shorter than an external focusing telescope of the same focal length.

As magnification in the surveying telescope varies directly with the focal length of the objective lens system (see §3.3.1) the magnification in an internal focusing telescope corresponds to that of a considerably longer external focusing telescope.

The internal focusing telescope is of constant length and balance, can be sealed against the entry of moisture and can be constructed to be virtually anallactic (see §9.5.4). Of little real significance as most observations are taken without change of focus, is the small effect of a lateral movement of the internal lens. A lateral movement which would in the external focusing telescope produce a displacement of the image of y would in the internal focusing telescope produce only an image displacement of $\tfrac{1}{4}y$.

There is some loss of brilliance arising from scattering of about 5% of the incident light by the additional glass air surfaces.

3.4. USE OF THE TELESCOPE

Focusing of the eyepiece is achieved by use of the encircling milled ring. In focusing the eyepiece the cross hairs are made to appear to the eye in a sharp and well defined form. The focusing ring carries a graduated scale and the reading of this should be noted as an aid to focusing the eyepiece at night, when exact focusing should be carried out by sighting onto some bright object such as the moon.

Once the eyepiece is correctly positioned the telescope is directed at some distant object and by use of the telescope focusing ring a clear image is formed on the reticule. This image is viewed through the eyepiece and should fall exactly on the plane of the cross hairs in the reticule, if it does not, parallax is present; the effect of which is considered below. It should be appreciated that the distant object referred to above must be at such a distance that the principal rays from it to the telescope are sensibly parallel.

Failure to ensure that the image of the distant object falls on the plane of the cross hairs, termed parallax, can be detected by moving the eye to different parts of the eye lens which will reveal apparent movement of the cross hairs relative to the image of the distant object (see figure 3.17). Under such circumstances

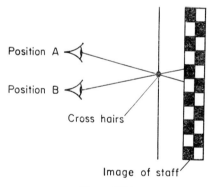

Figure 3.17

accurate sighting is impossible, as the line of sight can be made to intersect different points according to the position of the eye. The eye will frequently accommodate a small but inconvenient amount of parallax, but momentary closing of the eye will remove this capacity and the presence of parallax will be revealed.

To remove parallax the telescope is directed at the sky and the eyepiece adjusted until the cross hairs appear in sharp focus. The eye is kept on the cross hairs whilst the telescope is lowered until a well-defined object on the horizon is sighted and can be brought to a focus on the reticule.

It has been stated earlier that the function of the eyepiece is to magnify the cross hairs and the image formed by the objective. The distance of this magnified image from the eye will vary with individuals, according to variations in the optimum viewing distance. Consequently the position of the eyepiece will vary with individuals using the telescope. Only one eye is used but the experienced observer will frequently keep the other eye open. To keep one eye closed for a long period leads to eye strain and the habit of observing with both eyes open is worth cultivating. If difficulty is experienced in observing with both eyes open the non-viewing eye should be closed for the first viewing through the telescope and only opened after both eyepiece and objective have been focused. It may be necessary to close the non-viewing eye momentarily whenever the eye is replaced at the telescope. Exact pointing should be made with both eyes open.

3.4.1. THE OPEN SIGHT

It is common to find that the experienced surveyor makes little use of the open sight on surveying telescopes. Very frequently this is due to failure on the part of manufacturers to make sufficiently precise marks on the sights. A sight should be capable of adjustment and can be tested by picking up an object, first with the sight, and then seeing that the reversed image is approximately in the centre of the field of view of the telescope. Ideally there should be two pairs of sights so that on either face there will always be one pair above the telescope.

3.4.2. THE DIAGONAL EYEPIECE

For observation of angles of elevation in excess of 45° it is difficult to position the eye at the eyepiece. The diagonal eyepiece turns the line of sight through 90°. The simplest type consists of a simple prism which clips over the end of the eyepiece. Figures 3.18 and 3.19 show that as a consequence the reflection of the image no longer appears inverted but is still reversed, the left-hand side appearing on the right. A more complex arrangement, which is inserted to replace the usual eyepiece, consists of a normal Ramsden eyepiece with a

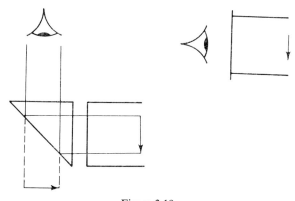

Figure 3.18

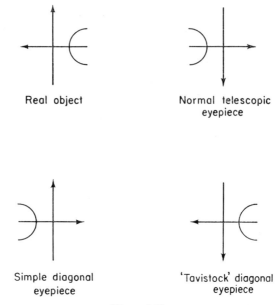

Figure 3.19

reflecting surface interposed between the lenses. In this case the image is no longer inverted but is still reversed. When the reading microscope adjoins the eyepiece an additional diagonal eyepiece is required which must not disturb the image normally viewed through the reading microscope. In such cases it is usual to find that the observing telescope has a diagonal eyepiece which produces an inverted but not reversed image. It is advisable to check the nature of the image formed by a diagonal eyepiece of an unfamiliar instrument. Figure 3.19 illustrates various images formed.

3.5. SPIRIT LEVEL

When measuring angles with a theodolite the verticality of the main axis is achieved by use of a spirit level, commonly termed the plate bubble. Thereby the measuring surface, the graduated arc, is made horizontal and consequently truly horizontal angles are measured. This surface is an element of a plane which is normal to the direction of gravitational attraction. Vertical angles are measured by reference to another local horizontal plane established with the altitude level. A third horizontal plane is established with the striding level, this plane is the reference surface for all corrections arising from dislevelment of the transit axis and non-verticality of the vertical axis.

3.5.1. THE GLASS VIAL

The spirit level consists of a glass vial partially filled with liquid, see figure 3.20. The remaining space contains air which finds its way to the highest point in the vial. The vial is so ground that its internal form is barrel shaped. In longitudinal section its radius is constant; in cross section its radius increases

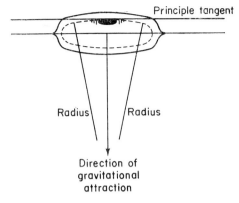

Figure 3.20

to a maximum value at the centre. The liquid used to fill the vial is usually alcohol, ether, or chloroform, these liquids being less viscous than water and freezing only at excessively low temperatures. The central tangent to the longitudinal curve of the vial is here termed the principal tangent of the level. In practice it is this tangent which is normal to the vertical axis of rotation when the instrument is in adjustment.

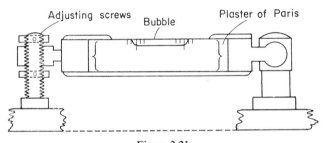

Figure 3.21

The glass vial is set in a case mounted where required on the instrument (see figure 3.21). Permanence of this setting is achieved by surrounding the ends of the vial with plaster of Paris, this must be completely removed before a new vial is fitted.

The glass vial is frequently graduated at 2mm intervals as can be seen from figure 3.21.

3.5.2. THE PLATE LEVEL

The purpose of this level is to secure verticality of the main axis of rotation. When the axis of rotation coincides with the direction of gravitational attraction, the air space, commonly termed the bubble, will not be displaced on rotation of the instrument. If coincidence is not present the bubble will move along the vial and this run is approximately, but not exactly, divided by the position taken up by the bubble in the vial when the main axis of rotation is vertical. See figure 3.22.

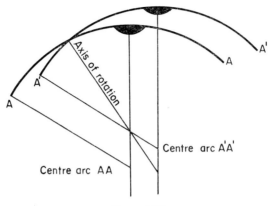

Figure 3.22

The setting of the main axis of rotation vertical is termed levelling the instrument. The procedure, a process of successive approximations, is as follows (see figures 3.23 and 3.24).

(a) Set the instrument so that the main axis is approximately vertical and ensure that the footscrews are in the middle of their run.

(b) Turn the alidade so that the longitudinal axis of the bubble is parallel to a line joining any two footscrews (A and B).

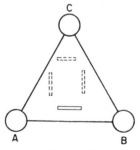

Figure 3.23

(c) Bring the bubble to the central position in the vial by rotating the two footscrews A and B in opposite directions. The bubble will move in the direction of the left-hand thumb. The central position in the vial is established by the graduations thereon.

(d) Turn the alidade through 90° and using the third footscrew C bring the bubble to the central position.

Theoretically if the instrument is in adjustment its vertical axis is now vertical. It is unlikely that a perfect setting will be achieved even if it is in adjustment.

(e) Return to the first position and repeat (c).

(f) Turn the alidade through 180° and bring the bubble by means of the footscrews A and B to a position midway between that which it takes up and the central position. Note this position.

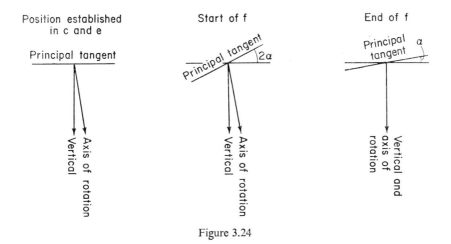

Figure 3.24

(g) Turn the alidade through 90° and using footscrew C bring the bubble to the mean position noted in (f).

(h) Turn the alidade to various positions completing a full rotation; any perceptible movement of the bubble is indicative of non-verticality of the main axis. If the movement is perceptible the procedure outlined above is repeated using the mean position established in (f) as if it were the central position for the bubble.

When an instrument is in constant use the mean position noted in (f) is remembered and in levelling this position is used from the start, rather than the central position indicated by the graduations. This point is considered later in the adjustments of the plate level in §3.8.2.

The possible poor fit of the axes (obliquity) referred to in §3.2.3 is revealed by levelling the instrument with rotation taking place first about the inner and then the outer axis. With obliquity the bubble assumes different positions in each case. Some modern instruments have only one main axis, the graduated horizontal axis 'floating' around it. As a result the same axis is used whether upper or lower plate clamps are fastened and so obliquity cannot exist.

3.5.3. THE ALTITUDE LEVEL

The function of this level is to establish a horizontal line for the indices against which the vertical circle can be read. When the instrument is in adjustment the principal tangent of the bubble will be approximately parallel to the line of the indices, the exact relationship also depends upon the relationship of the line of the indices to the line of sight. This point is considered in more detail in §3.8.1.

When the bubble in the altitude level is brought to the centre of its run with its setting screw the indices will at the same time be displaced by the same amount. The altitude level is usually viewed through a coincidence reader in which the two halves of each end of the bubble are made to coincide, when viewed in the prism (figure 3.25). The coincidence method of viewing permits

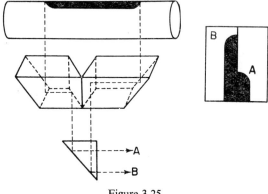

Figure 3.25

setting of the bubble to a higher degree of precision than is possible with a graduated vial viewed directly or through a mirror. It permits the use of a less sensitive vial which will come to rest more rapidly whilst at the same time accurately indicating the true horizontal. It is important that the mount for the altitude level be secure, in many modern instruments the mount can not be easily inspected but periodic checks should be made, failure to detect a loose mounting will give rise to non-systematic errors in the observations.

3.5.4. THE STRIDING LEVEL

The striding level is a sensitive graduated spirit level whose function is to establish the degree of dislevelment of the transit axis. The principal tangent of this level is, when this accessory is in adjustment, parallel to the line passing through the V notches at the foot of the legs (see figure 3.26).

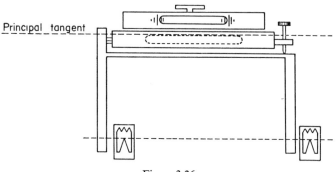

Figure 3.26

The two V notches are placed upon exposed portions of the transit axis facilitating accurate levelling of the instrument and determination of any dislevelment arising from either the main or transit axes in the direction normal to the line of sight. The treatment of errors arising from this source is considered in §§3.10.1 and 3.10.2, and 3.11.3.

3.5.5. GRADUATIONS AND SENSITIVITY OF THE VIAL

The plate level and the striding level are usually graduated in 2mm divisions and these graduations are designed to have a particular angular value, for example 20″. This means that for a 2mm movement of the centre (or one end) of the bubble the principal tangent of the bubble will have been tilted through 20″ of arc. The greater the radius of the vial the smaller will be its value for one division.

If x is the value of one division

$$x = \frac{d}{R} \qquad (1)$$

or

$$x'' = \frac{d \times 206\,265}{R} \qquad (2)$$

where d is the distance between the graduations and R the radius of the inner surface of the longitudinal section of the vial, in the same units.

To determine the value of one division proceed as follows:

(i) Set the theodolite (or level) a measured distance of around 300 feet from a vertical levelling staff. Level the instrument.
(ii) With the longitudinal axis of the spirit level vial along the line to the staff one end of the bubble to an exact coincidence with a graduation. Where necessary rotate the instrument and read the intersection of the centre hair on the staff.
(iii) Return the instrument to the position in (ii) and displace the bubble through a number of graduations. Where necessary rotate the instrument and read the intersection of the centre hair on the staff.

The following information is now available:

The distance D from instrument to staff,
the difference s between the two staff readings, and
the number y of graduations the bubble is moved through.

Then by reference to figure 3.27

$$\alpha = \frac{s}{D} \text{ radians} \qquad (3)$$

$$x = \frac{s}{dy} \text{ radians} \qquad (4)$$

From (1) and (4) above

$$\frac{s}{Dy} = \frac{d}{R}$$

$$R = \frac{dDy}{s} \qquad (5)$$

and

$$x'' = \frac{s \times 206\,265}{Dy} \qquad (6)$$

This method of determining the graduation value of a level is very suitable for the tilting or dumpy level where tilting is effected without rotation of the instrument.

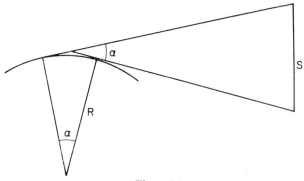

Figure 3.27

In the theodolite, where this procedure can be used to determine the value of the graduations on the plate level vial, the tilting of the level is effected with the footscrews and it is advisable to set the instrument so that one footscrew lies on the line from the instrument to the staff. In the case of the striding level it is not sufficiently accurate, for instance if the level vial had a sensitivity of 4″ and was run over 5 divisions with a staff at 300 feet the staff intercept would be about 0·03 feet.

Considering the general expression above (6), this may be written as

$$\log x'' = \log s + \log 206\,265 - \log D - \log y$$

Differentiating

$$dx''/x'' = ds/s - dD/D - dy/y$$

let ds be 0·005 feet, dD 1 foot, and dy 0·1 divisions, then the error arising from ds is 0″69, from dD 0″01, and from dy 0″08.

It will be seen that the error arising from determination of the staff intercept can be very serious, particularly as there is a very real possible error in a change of the degree of non-verticality of the staff. This method is generally regarded as unsuitable for the striding level where high precision is required.

A first alternative method suitable for the striding level is to set the striding level on a level trier, which is a beam 18–24 inches long supported at one end by a pivot and at the other by a micrometer screw. The degree of inclination of the beam can be read off the micrometer and thereby changes of inclination of an attached striding level directly determined for various movements of its bubble. This is an extremely accurate laboratory method.

A second method is to place the striding level longitudinally on the telescope. If the telescope is then tilted through a run of the bubble of several divisions the angular value of this tilt will be given by the change in the vertical circle readings. An objection to this method is the difficulty in securing the striding level to the telescope.

A third method described in Clark's *Plane and Geodetic Surveying for Engineers* (London, Constable 1944) Vol II is suitable for finding the value of each individual graduation on the striding level vial. The method is as follows:

(i) Set up and level the theodolite with the striding level in position. Align the telescope over one footscrew and clamp the upper and lower plates and the vertical circle. Level the altitude level and read the vertical and horizontal circles.

(ii) Rotate the footscrew over which the telescope is aligned to throw the main axis of rotation off by an amount i. The altitude bubble is relevelled and the vertical circle read. The difference between this reading and that taken in (i) will give the value of i, the component of the inclination of the main axis of rotation to the vertical in the plane at right angles to the transit axis.

(iii) With the lower plate clamped, turn the alidade in azimuth. The bubble in the striding level will move off by an amount d, where

$$\tan d = \sin a \tan i$$

In practice the alidade is turned until the bubble is displaced by one, two, three, or more divisions in turn, the value of a being read for each division in turn. By this method it is possible to find a value for each graduation on the level vial.

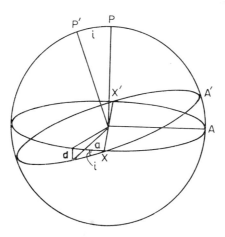

Figure 3.28

The expression given above for d is proved by use of spherical trigonometry. Referring to figure 3.28 the telescope is initially directed at a point on the great circle through AA'PP'. The striding level, on the transit axis, will lie along the line XX'. With rotation of the alidade by an amount a the line XX' will move through an equal amount but be displaced from the horizontal by an amount d.

Hence
$$\sin a = \tan d \tan (90 - i)$$

or
$$\tan d = \sin a \tan i$$

This method may also be used to find the value for the graduations on the altitude level vial. It is doubtful if on modern instruments this is required. The bubble is always centered before reading and in many cases the vial is not graduated.

3.5.6. THE CIRCULAR LEVEL

This consists of a circular box capped by a convex spherical lens of low power. This produces a circular bubble, which when central shows that the axial plane is horizontal. The circular bubble is usually of low sensitivity but is useful for approximate levelling of tripods. Many level staves are fitted with circular bubbles, if rigidly fixed to the staff they maintain good adjustment, but if hinged as is frequently the case with metal staves, they easily cease to be reliable. To check the level attached to a staff the verticality of the staff when the bubble is central is compared with a plumb line. Three small screws beneath the level casing permit its adjustment.

3.6. THE CIRCLES

With a linear scale, such as a steel rule, comparison is made with known standards. Linear units of length are arbitrary and are frequently defined by legal statute. Circular scales are only arbitrary in so far as subdivision of the principal unit, the whole circle, is concerned. With a circular scale any progressive error in each division is accumulated in the last space. The whole circle may be divided into 360 degrees (the sexagesimal system) 400 grades (the centesimal system) or 64 units (the millieme system).

The division of the circle into the required number of parts has exercised the ingenuity of instrument makers over the last two centuries. The maker is faced with the problem of spreading any small residual error evenly around the circle, so that if at any point a space between two graduations is too small there will be another point where the space between two graduations will be too great.

In 1793 Edward Troughton produced a machine for the accurate division of circles for surveying instruments. This machine can be seen at the Science Museum, London. The circle is mounted on a table which rotates about a vertical axis. The rotation is intermittent and whilst the table is at rest a line is engraved on the circle. This principle is still in use today although the machinery is more refined.

It follows that the accuracy of the divided circle must be less than the accuracy of the tooling of the dividing engine. By continual use the dividing engine will become progressively less accurate and the whole process is slow and expensive. In 1921 Heinrich Wild produced a new type of theodolite in which the graduations were etched on optical glass, formerly all graduations being engraved on silver or platinum strips set in brass circles. Optical glass is remarkably uniform and permits the passage of light. It is possible with an accurately divided master circle to produce by photographic processes many further circles of very nearly the same degree of accuracy. For the most precise instruments a circle produced by a dividing engine is still required but photographic methods can be used for circles for instruments of a lower quality and price. In both cases the

The Theodolite and Level 105

circles are made of optical glass varying in detailed specification from manufacturer to manufacturer.

As the circle permits the passage of light, an image of the graduations would be of a higher degree of brilliance than is found on a silvered brass of gunmetal circle. This enables the image to be magnified considerably and most modern theodolites incorporate a powerful compound microscope for reading the circle.

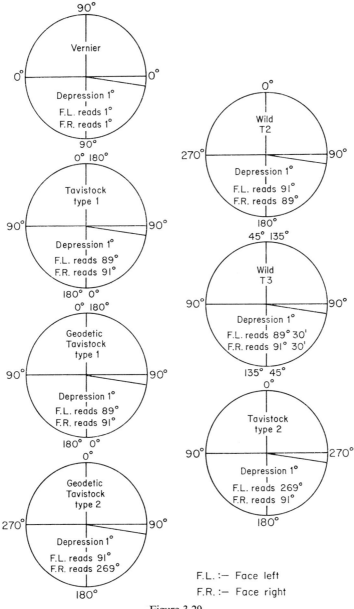

Figure 3.29

In all modern instruments the graduations on the horizontal circle increase numerically as the alidade is rotated in a clockwise direction. When the sexagesimal system is used the circles are figured from 0–359. Regrettably no convention exists for vertical circle graduation. Instruments fitted with micrometers must be figured so that the numbers increase continuously from zenith to nadir. Vernier instruments are customarily figured so that horizontal readings are 0 and the zenith and nadir 90. Some typical figurings are given in figure 3.29.

3.7. THE READING MECHANISM

All theodolites are fitted with a device to facilitate the reading of the graduated circle. In all cases this device permits the reading to be taken to a degree of precision greater than the smallest unit of the graduated circle. A typical case is the vernier which permits a reading of the circle to be taken to 20 seconds, whereas the circle is graduated in units of 20 minutes.

3.7.1. THE VERNIER

To achieve a reading to 20 seconds with a circle graduated in units of 20 minutes requires that each coarse division be divided into 60 parts. This is accomplished by dividing the vernier so that 60 divisions subtend exactly the same angle at the centre of the circle as 59 divisions of the circle, that is 19° 40′. Each division of the vernier is 59/60 of a circle division. In effect each vernier division is 1/60 of a circle division short, that is 20″ short. Coincidence of a vernier graduation with a circle graduation will now indicate the proportion of a circle graduation between the zero of the vernier and the previous circle graduation (this is illustrated by figure 3.30).

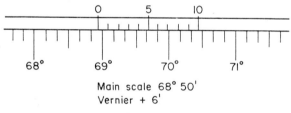

Main scale 68° 50′
Vernier + 6′

Figure 3.30

The vernier suffers from the disadvantage that a large proportion of the main scale is required to effect the necessary coincidence. In the case of the theodolite the main scale is the graduated circle and in the case above the vernier must be 19° 40′ long and must in consequence be markedly curved. It is also essential that the two scales be coplanar, so that the readings are unaffected by the position of the eye. The vernier was very common on instruments with metallic circles and although many are still in use the numbers being manufactured are declining.

Usually two verniers are fitted at opposite ends of a diameter of the graduated circle, these are frequently labelled Vernier A and Vernier B. The readings are booked for Vernier A in degrees, minutes, and seconds but for Vernier B in minutes, and seconds only. The mean of the two readings is taken for minutes and seconds only the degree value for Vernier A being taken. For example:

Vernier A	Vernier B	Mean
227° 38′ 00″	38′ 40″	227° 38′ 20″
	(47° not booked)	

3.7.2. THE OPTICAL SCALE

The glass circle permits magnification and light can be concentrated on that part of the circle being read. It is possible to insert between two graduations a

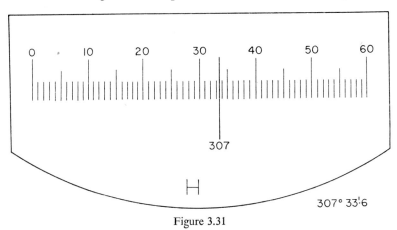

Figure 3.31

one minute scale, the simplest type being that fixed in the Wild T 16. This is illustrated by figure 3.31, the circle being graduated in units of 1 degree, and the inserted scale can be estimated to one place of decimals of a unit, viz. 0′.1. With less magnification the obsolete C.T.S. T 63 utilizes a master scale in units of one minute and two auxiliary scales offset by 20″ and 40″. Figure 3.32 illustrates

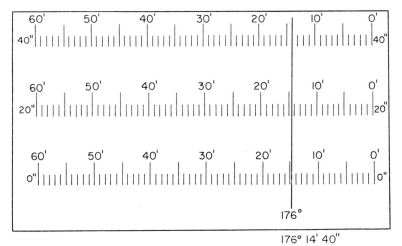

Figure 3.32

this method of reading in which the minutes and seconds are given by the point where the circle graduation falls upon or close to a graduation on one or other scales. In both cases the essential condition is that the length of the scale should conform exactly to the distance between the images of the circle graduations. There are no moving parts, reading is quick and easy but is taken to only one part of the circle.

Kern have produced optical scale theodolites in which a reading is taken of both sides of the circle and a mean value obtained. The purpose of obtaining a mean value from both sides of the circle is considered later. The method of reading the Kern DK2 theodolite is illustrated in figure 3.33. In this instrument use is made of a coincidence method of reading, but there are no moving

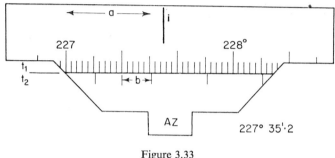

Figure 3.33

parts. Each circle carries two sets of graduations one at two minute intervals t_1 and the other at 20 minute intervals t_2. Points at opposite ends of the diameter of the graduated circle are brought together by an optical train into the field of the reading microscope. The index i visible in the middle of the field of view indicates whole degrees and tens of minutes (a). (It also gives an approximate value for the angle but this is ignored in reading.) The separation (b) of the twenty minute graduations on t_1 and t_2 will give in terms of the finely graduated scale twice the number of single minutes and decimals of a minute required. The required value can easily be obtained by treating each 2-minute graduation as a single minute. The reading being 227 35'·2.

An important principle is illustrated by this instrument (see figure 3.34);

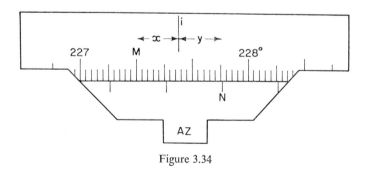

Figure 3.34

the fine and coarse scales appear in the field of view running in opposite directions. The reading on the fine scale with reference to the index could be regarded as $227° 20' + x$, $(M + x)$; the reading on the coarse scale could similarly be regarded as $N + y$, where N is some value in units of 20'. As the coarse graduations come from the other end of the diameter of the graduated circle N will be $47° 20'$ The mean value from the two ends of the diameter will be $227° 20' + \frac{1}{2}(x + y)$. The quantity $\frac{1}{2}(x + y)$ is easily obtained by counting the number of graduations from M to N treating each division as 1'.

In the particular case illustrated the coarse reading is taken directly from the index and is $227° 30'$ the fine reading being obtained by counting the number of graduations and parts of a graduation from $227° 40'$ to N; the ten graduations from M to $227° 40'$ being accounted for by the additional ten minutes in the coarse reading. This principle, of measuring the distance between the image of two corresponding graduations from either end of the diameter, is extensively used in the mean reading optical micrometer.

3.7.3. THE MICROMETER MICROSCOPE

The construction of the micrometer microscope is illustrated by figure 3.35. A magnified image of the graduated circle is viewed and a pair of travelling lines are arranged to move across the image at right angles to the graduations of

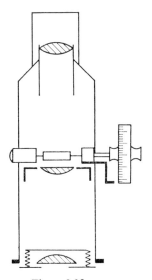

Figure 3.35

the circle. Movement of the travelling lines is effected by means of a micrometer head. The magnification of the circle graduations is usually such that a 10' division on the circle is traversed by one rotation of the micrometer head.

The micrometer head carries a drum divided into 60 parts each corresponding to 10". Two similar micrometer microscopes are mounted at either end of a diameter of the graduated circle, as a result graduations 180° different are viewed. To obtain a reading each micrometer head is turned until its travelling lines evenly span a circle graduation (see figure 3.36). Now the V notch indicates the coarse value, i.e. the reading to the nearest 10' and the fine value is obtained from the micrometer drum. In the case illustrated the coarse value is 276° 40' and the micrometer drum reading is 3' 35". This will be booked at 276° 43' 35",

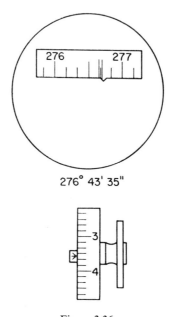

Figure 3.36

for the other micrometer only minutes and seconds are booked. The micrometer drum reading represents the angular displacement, along the graduated circle, moved by the travelling hairs from the position they would take up if the drum read zero. This position will correspond to the position of the V notch.

The magnification of the circle graduations must be correctly related to the pitch of the screw which controls the movement of the travelling lines. If the relationship is not correct 'an error in run' is said to be present. To detect 'an error in run' the following procedure is adopted for each microscope.

(i) The travelling lines are placed astride a circle graduation in the middle of the field of view and the drum read.
(ii) The micrometer head is rotated until the adjacent graduation falls midway between the travelling lines. The drum reading is compared with that obtained in (i).
(iii) The procedure is repeated at least four times using different parts of the graduated circle so that the error will not be dependent upon an error in the graduation of the circle.

Example

Circle reading	Micrometer A	Micrometer A
00° 10′	7′ 24″	7′ 21″
46° 40′	5′ 12″	5′ 08″
89° 30′	3′ 54″	3′ 48″
134° 00′	0′ 06″	0′ 05″
Sum	16′ 36″	16′ 22″

Error in run over 4 runs 14″ short
Run being 9′ 56″.5
Mean error in run = 3″.5

If the discrepancies are small some surveyors are prepared to tolerate individual errors in run in each micrometer provided the mean run of the pair is correct. For instance if in the case above the micrometer B had an error in run over 4 runs of about 14″ great, i.e. a run of 10′ 03″.5, then the instrument would be regarded as having no error of run, particularly if the circle values used for micrometer B were 180°, 225°, 270°, 315° or approximately so. It is however undesirable that there should be any error in run on either micrometer when precise work is being undertaken.

To correct for the error in run it is necessary to change the magnification of the microscope. This is achieved by movement of the object glass, which necessitates in turn a movement of the whole microscope to focus perfectly the image of the circle graduations in the plane of the travelling lines. Failure to do so will leave parallax in the microscope. By moving the object glass towards the graduated circle magnification is increased and thereby the run. It is found easiest to make a complete half turn of the object glass, refocus the whole microscope, and test the new run.

The two V notches should be at opposite ends of the diameter of the graduated circle. With the slow motion screw one V notch is set over a circle graduation, the other V notch should now lie over a circle graduation differing by 180°. If it does not, a screw on the micrometer box will move the V notch to the desired position. This screw is spring-loaded and its final movement should be clockwise.

The position of the V notch provides an approximate value for the reading and the drum should read zero when the travelling lines are astride the apex of the V notch. If the reading is not approximately zero the drum can be rotated against the friction drive, whilst holding the travelling lines fixed with the knurled head. The micrometer will now be in adjustment.

3.7.4. THE OPTICAL MICROMETER

The principle of the optical micrometer is that the image of the portion of the graduated circle presented in the field of view of a reading telescope is displaced by means of a wedge or parallel plate. One graduation is then brought into a set position with respect to a fixed index and the portion of a division on the circle measured. The image is produced by passing light through, or reflecting from, the glass graduated circle. The displacement of the image, effected by the wedge or parallel plate of optical glass, is obtained by turning a micrometer head which in turn causes a movement of a glass scale viewed in the same field as the image. The displacement of this glass scale with respect to a second index records the displacement of the image of the circle graduation necessary to effect a setting.

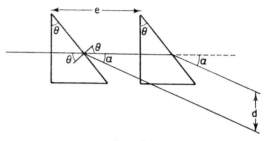

Figure 3.37

The travelling wedge produces a displacement of the image of the circle graduation which is directly proportional to the travel of the wedge; provided that the movement of the wedge is along the line of the incoming light ray See figure 3.37 in which

$$\tan \alpha = d/e$$

where d is the displacement of the image and e the movement of the travelling wedge. The angle α is a constant and by use of Snell's law of refraction the expression

$$\mu = \frac{\sin(\theta + \alpha)}{\sin \theta}$$

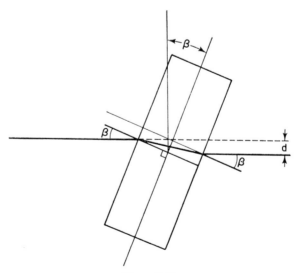

Figure 3.38

is obtained. In this expression μ is the refractive index of the glass in the wedge and θ the apex angle of the wedge.

The parallel plate consists of a plate of glass in which the opposite faces have been carefully ground exactly parallel. See figure 3.38 in which the plate is turned through an angle β with respect to the normal to the incoming ray of light.

By Snell's law of refraction

$$\mu = \frac{\sin \beta}{\sin r}$$

From the geometry of the figure

$$\sin (\beta - r) = \frac{d}{PQ} \quad \text{and} \quad \cos r = \frac{t}{PQ}$$

Hence

$$d = t \sec r \sin (\beta - r)$$
$$= t \sec r (\sin \beta \cos r - \cos \beta \sin r)$$
$$= t\left(\sin \beta - \frac{\cos \beta \sin \beta}{\mu \cos r}\right)$$
$$= t \sin \beta \left(1 - \frac{\cos \beta}{\mu \cos r}\right)$$

But

$$\cos^2 r = 1 - \sin^2 r = 1 - \frac{\sin^2 \beta}{\mu^2}$$

$$d = t \sin \beta [1 - (1 - \sin^2 \beta)^{\frac{1}{2}} (\mu^2 - \sin^2 \beta)^{-\frac{1}{2}}]$$

It will be seen that the relationship is complex and the linear displacement of the image depends upon the change of $\sin \beta$ and $[1 - (1 - \sin^2 \beta)^{\frac{1}{2}} (\mu^2 - \sin^2 \beta)^{-\frac{1}{2}}]$ with β. For a regular movement of the micrometer head, it is possible, by means of an accurately cut cam to produce a regular displacement of the image of the graduated circle.

It should be noted that if β is small, where it can be assumed that

$$\sin \beta = \beta$$

and

$$1 - (1 - \sin^2 \beta)^{\frac{1}{2}} (\mu^2 - \sin^2 \beta)^{-\frac{1}{2}} = 1 - 1/\mu$$

then the displacement of the ray of light is directly proportional to the thickness of the plate.

Then

$$d = t\beta(1 - 1/\mu)$$

This relationship is of significance in considering the parallel plate micrometer used in the geodetic level (see §3.14.1).

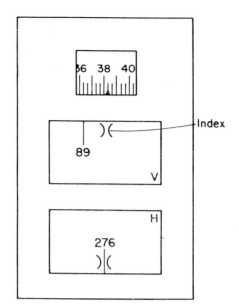

Horizontal reading
276° 38' 20"

Figure 3.39

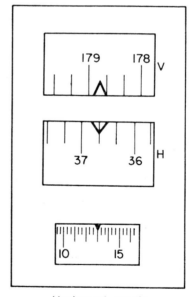

Horizontal reading
36° 53' 00"

Figure 3.40

As with optical scale and micrometer microscope instruments magnification of the portion of the graduated circle viewed must be exactly related to the travel of the fine reading scale. Instruments reading against a fixed index use only one side of the graduated circle and are not mean reading in the sense of

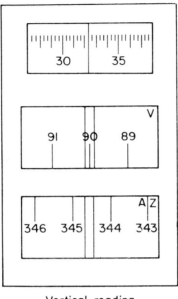

Vertical reading
90° 32' 20"

Figure 3.41

obtaining a mean value from both ends of its diameter. Typical instruments in this class are the C.T.S. V301, the Watts Microptic No. 1 and the Wild T1. The reading of these instruments is illustrated in figures 3.39, 3.40, and 3.41.

3.7.5. THE MEAN READING OPTICAL MICROMETER

Due to the effects of a possible eccentricity of the circle (see §3.9.2) it is essential for precise observations to read the circle at either end of its diameter. A mean of the two readings is required and the characteristic feature of the mean reading optical micrometer is that this is obtained with a single setting. Graduations from opposite ends of the diameter are brought into the same field of view and are read in a single telescope. This concept is illustrated by figure 3.42. In figure 3.43 one portion of the graduated circle is viewed directly and as shown the 66° graduation falls short of an index by an amount x, this reading may be taken to be 66° + x. At the other end of the diameter the 246° graduation falls short of a point diametrically opposite the index by an amount y. If these two readings could be booked and meaned in the usual way the value would be 66° + $\frac{1}{2}(x + y)$.

If by means of an optical train the 246° graduation is brought into the same

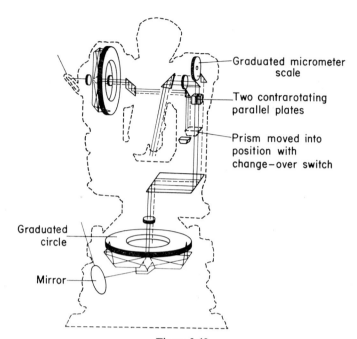

Figure 3.42

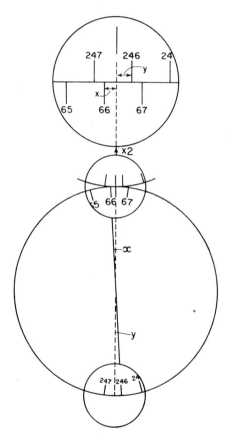

Figure 3.43

field of view as the directly viewed graduations it is possible to determine the quantity $x + y$ which will correspond to twice the required fine reading. Later with an optical micrometer the quantity $x + y$ is measured by bringing graduations from either end of the diameter into a exact relationship to each other. The index in the mean reading instrument is generally superfluous, but it is usually retained to give an approximate value of the reading. In the Zeiss and Wild instruments the images of the circle graduations from opposite ends of the diameter are made to coincide by means of an equal and opposite displacement, thereby measuring $x + y$. This is effected by movement of two glass parallel plates. The rotation of these plates, equally in opposite directions is effected by a single rotation of the micrometer head which is also connected, by a suitable cam, to a moving scale which will give the fine reading. In the Watts Microptic No. 2 there is only one parallel plate which displaces the image through the total amount $x + y$ to effect coincidence with a stationary image from the other side of the circle. In the C.T.S. or Vickers Tavistock theodolites a pair of moving wedges effect the necessary coincidence. An equal movement of the images again takes place. The methods of reading these instruments are illustrated by figures 3.44–3.47.

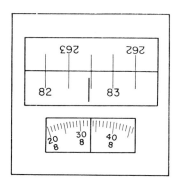

Figure 3.44   82° 48′ 33″

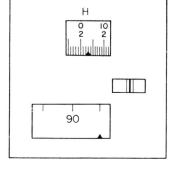

Figure 3.45   90° 22′ 03″

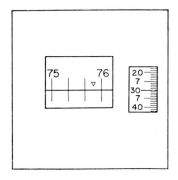

Figure 3.46   75° 57′ 30″

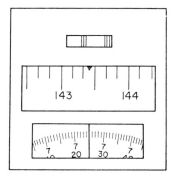

Figure 3.47   143° 27′ 25″

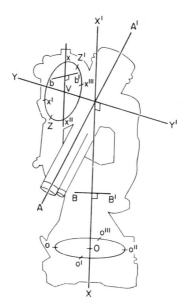

Figure 3.48

3.8. ADJUSTMENTS OF THE THEODOLITE

Due to wear, change in structure of the materials used in construction, and sometimes rough usage, the relationships between the various parts of the theodolite undergo slight changes in the course of time. For this reason theodolites are provided with means of adjustment. It is the responsibility of a surveyor using a theodolite for the first time to check that its maladjustments are small. Large instrumental discrepancies arising from a disturbance of the relationships between the various parts are inconvenient and will not be automatically cancelled out by a systematic observing procedure with the same certainty as small instrumental discrepancies. The methods of testing and adjusting a theodolite vary according to the particular design of the instrument and where possible the manufacturers handbook should be referred to before commencing the tests and adjustments. The methods given below are applicable to most instruments and can in general be followed in the absence of the manufacturers handbook.

3.8.1.

To understand the principles which underlie the design and adjustment of a theodolite refer to figure 3.48. It will be seen that the following conditions apply:

(i) Plane $o\, o'\, o''\, o'''$, the horizontal circle, should be normal to the main axis of rotation XX'.
(ii) Plane $x\, x'\, x''\, x'''$, the vertical circle, should be normal to the 'horizontal' transit axis YY'.

The Theodolite and Level 119

(iii) The main axis of rotation XX' should pass through the point from which the graduations of the horizontal circle radiate O.

(iv) The transit axis of rotation YY' should pass through the point from which the graduations of the vertical circle radiate V.

These four conditions are very nearly achieved during construction, it is unlikely that with normal use they will be disturbed and no provision is made for their adjustment. It should be noted that in the case of conditions (iii) and (iv) the design of the reading mechanism may almost completely eliminate any residual defects of construction. Any residual defects arising from (i) or (ii) will be negligible provided reasonable care has been taken during construction.
The following further conditions apply:

(v) The plate level BB', which is used to set the main axis of rotation vertical, should be positioned such that its principal tangent (see §3.5.1) be normal to the main axis of rotation XX'. Adjustment 1 (§3.8.2) describes how this condition may be achieved.

(vi) The line of sight AA' should be normal to the transit axis YY' regardless of the direction in which it is pointed. Adjustment 2 (§3.8.3) describes how this condition may be achieved.

(vii) In modern theodolites the graduated vertical circle rotates with the vertical rotation of the telescope about the transit axis. Then the condition is that the line joining the zero or zeros (or equivalent circle graduation) ZZ' to V the centre of the graduated circle should be parallel to the line of sight AA' and at the same time the line of the indices II' should be horizontal when the principal tangent of the altitude level bb' is horizontal (see figure 3.49). From an observational point of view the results will not be in error if non-fulfilment of one condition is counteracted by an equal and opposite non-fulfilment of the other (see figure 3.50). (See §3.8.4.) In instruments in which the indices move with the telescope the circle remaining fixed, II' must be parallel to AA', and ZZ' with bb'.

(viii) The transit YY' axis should be normal to the main axis of rotation XX'. In many modern instruments no provision is made for this adjustment,

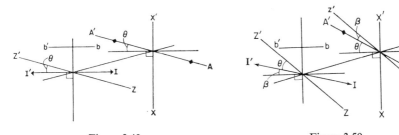

Figure 3.49   Figure 3.50

hardened steels and alloys wear less in the standards and the smaller weight will cause less wear. Adjustment 4 (§3.8.5) describes how this condition may be achieved.

(ix) The main axis of rotation XX′ should meet the transit axis YY′ at the same point as the line of sight AA′ meets this axis. For this to be achieved exactly in any theodolite is rare and no attempt is made during adjustment to achieve this condition. The manufacturer will usually ensure that the axis of the telescope tube (not necessarily the same as the line of sight) meets the other two axes at a point. The amount by which the line of sight is offset is always very small and on distant objects will have no appreciable angular value. On near objects an appreciable although small error may be found, even though adjustment 2 has been correctly carried out. The direction of the displacement is however reversed with change of face and the mean of both faces is error free.

(x) The line of sight should maintain the same position with a change of focus. If this condition is not fulfilled it will only give rise to errors when the angle between objects at different distances is measured and refocusing is found necessary. By observing on both faces the direction of displacement is reversed and the mean value is error free.

Besides the conditions considered above there are others peculiar to particular types of instruments usually relating to the reading mechanism. Some of these have been considered in §§3.7.1–5, reference should also be made where possible to the manufacturers handbook, although often no provision is made for adjustment by the surveyor.

3.8.2. ADJUSTMENT OF THE PLATE LEVEL. ADJUSTMENT 1

(i) Level the instrument as described in §3.5.2, performing the operation with only the lower plate clamped. This is standard levelling procedure. If the bubble takes up the same though not central position for all directions of the alidade adjustment is required.

(ii) Bring the bubble to a central position by means of the capstan headed screws, or nuts, at the end of the tube. The bubble should now remain in the same central position for all directions of the alidade.

(iii) Unclamp the lower plate and clamp the alidade, the bubble should remain central for all directions of the alidade. If it does not then the axis about which the lower plate rotates is not vertical and therefore not parallel to that about which the alidade rotates. (The amount by which the bubble moves off centre represents the angle between the two axes), and the instrument may require workshop attention. It should be noted that a small defect of this type will only produce exceedingly small errors in the measured angle which in most work can be tolerated.

Perfectly precise work can be done with a maladjusted bubble, provided it is not so far off centre as to be against the end of its run. Nevertheless most surveyors prefer a central bubble.

3.8.3. ADJUSTMENT OF THE LINE OF SIGHT (HORIZONTAL COLLIMATION). ADJUSTMENT 2.

Explanation: Figure 3.51 represents a plan of a horizontal theodolite telescope, if the line of sight is correctly positioned then on transiting the telescope the line of sight will describe a plane normal to the transit axis. If the line of sight

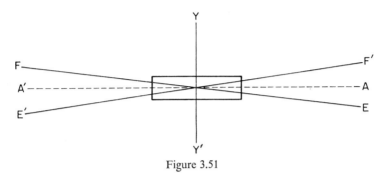

Figure 3.51

is in a direction such as EF, then when the telescope is transitted E will move to E' and F to F'. If now the alidade is rotated through 180° the line of sight will be F'E'.

(i) Choose a level stretch of ground 600 to 800 feet long and set up the theodolite midway. Focus and level in the usual way.
(ii) Clamp the upper and lower plates and intersect an object (chaining arrow) about 300 feet away, point A of figure 3.52.

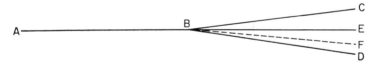

Figure 3.52

(iii) Transit the telescope and on the line of the vertical hair mark point C, the same distance away as point A. Point C is marked with an arrow or as a pencil line on a strip of white card, alternatively a levelling staff may be laid horizontal normal to the direction of observation.
(iv) Unclamp the upper plate and rotate the alidade to reintersect A. Leave plates clamped.
(v) Transit the telescope and mark, on the line of the vertical hair, point D adjacent to C.
(vi) Mark point E such that CE = DE, and F such that DF = $\frac{1}{4}$CD.
(vii) Move the cross hairs of the graticule horizontally, by means of the adjusting screws, until the vertical hair exactly covers point F. The upper and lower diaphragm screws must be eased before any movement of the graticule is attempted (see §3.3.7).
(viii) Transit the telescope and using either horizontal plate slow motion screw reintersect A.

(ix) Transit the telescope, the vertical hair should exactly intersect E. If it does not the whole operation should be repeated until this condition is achieved.

An alternative method not independent of error in the horizontal circle graduations is as follows:

(i) Set up the theodolite, focus and level in the usual way, clamp the lower plate.
(ii) Observe onto a distant point that can be accurately intersected, read the horizontal circle.
(iii) Change face and repeat (ii).
(iv) The reading should differ by exactly 180°; if they do not the difference being say $180 + a$ the second reading should be diminished by $\frac{1}{2}a$. Calculate and set the circle to this reading using the upper plate slow motion screw thereby throwing the vertical hair off the distant point.
(v) Move the cross hairs of the graticule horizontally by means of the adjusting screws until the vertical hair exactly covers the distant point (see §3.3.7).
(vi) Repeat the test until the two readings agree, (to within 10" for a 1" theodolite).

Example: First reading on distant point

$$\begin{aligned} \text{Face left} & \quad 176° \ 24' \ 38" \\ \text{Face right} & \quad 356° \ 25' \ 16" \\ a = & \quad 38" \\ \text{Required face right reading} & \quad 356° \ 24' \ 57" \end{aligned}$$

Second reading on distant point

$$\begin{aligned} \text{Face left} & \quad 176° \ 24' \ 54" \\ \text{Face right} & \quad 356° \ 24' \ 58" \end{aligned}$$

Instrument now regarded as in adjustment.

As the diaphragm is moved during this adjustment is must always be followed by adjustment 3 below and it is essential that a check by made on the verticality of the vertical hair. Strictly speaking what is required is that the vertical hair be set in a plane perpendicular to the transit axis; provided that the manufacturer has not erred this will also ensure that the horizontal hair is parallel to the transit axis and at right angles to the main axis of rotation. The test and adjustment is as follows:

(i) Focus the instrument and level with great care.
(ii) Set the vertical hair on some well-defined object and move the telescope in the vertical plane. If the object remains on the hair no adjustment is required.
(iii) If the hair moves off, ease all the diaphragm screws and rotate the diaphragm until the vertical hair remains on the well-defined object with elevation of the telescope.
(iv) Tighten the diaphragm screws and repeat the test.

If the diaphragm has been rotated the tests for the line of sight and adjustment of the indices must be repeated. Usually no perceptible maladjustment is introduced if care is taken to slacken and tighten the screws by equal amounts.

Perfectly precise observations can be made with a theodolite in which the line of sight is not perpendicular to the transit axis provided observations are made on both faces.

3.8.4. ADJUSTMENT OF THE INDICES (INDEX ERROR OR VERTICAL COLLIMATION). ADJUSTMENT 3

Explanation: This adjustment ensures that a vertical angle read on one face is correct.

(i) Set the instrument 200 feet from a white wall, post, or vertically held staff. Focus and level in the usual way and sight on the wall, post, or staff.
(ii) Bring the altitude level bubble to the centre of its run and set the vertical circle to read exactly zero or the equivalent horizontal reading.
(iii) Make a fine pin point A where the line of the horizontal hair falls on the white surface or alternatively read the staff.
(iv) Change face and repeat (ii) above. If the horizontal hair intersects the same position as in (iii) no adjustment is required. If it does not make a second mark B or take a second reading.
(v) Measure the distance between A and B and put in a third mark C midway between alternatively note the midway staff reading.

The procedure now depends upon the instrument, for all modern theodolites except those of the C.T.S. Tavistock range the procedure is as follows:

(vi) With the vertical slow motion screw bring the horizontal hair onto mark C, this will throw the reading off zero.
(vii) With the altitude level setting screw bring the circle reading back to zero and by means of the altitude level adjusting screws bring the bubble back to the centre of its run.

For the C.T.S. Tavistock range the procedure is as follows:

(vi) With the vertical slow motion screw bring the horizontal hair onto mark C, this will throw the reading off the horizontal value, usually 90°.
(vii) With the vertical circle adjusting screw adjacent to the telescope bring the circle reading to the horizontal value. This will usually amount to recentring the light gap or re-effecting coincidence.

For older instruments where a clip screw alters the pointing the procedure is as follows:

(vi) With the clip screw bring the horizontal hair onto mark C, this will displace the bubble but not affect the circle reading.
(vii) The bubble is now brought to the centre of its run by means of the altitude level adjusting screws.

In all cases the test is repeated to ensure that the adjustment has been successfully carried out.

An alternative method not independent of error in the vertical circle graduation is as follows:

(i) Set up the instrument, focus and level with care.
(ii) Observe onto a distant point that can be accurately positioned on the horizontal hair.
(iii) Carefully centre the bubble of the altitude level and read the vertical circle.
(iv) Change face and repeat (ii) above, the reading should when reduced be the same as obtained on the other face. If it is not the instrument requires adjustment, for modern instruments (except C.T.S. Tavistock range) the procedure is as follows:
(v) The reduced vertical angles are meaned and the mean value used to determine the reading required on the vertical circle for the instrument when in adjustment. Calculate and set the vertical circle to this reading with the altitude level setting screw thereby displacing the bubble from its central position.
(vi) With the altitude level adjusting screws bring the bubble back to the centre of its run.

Example: Instrument Wild T2.
First reading on distant point

	Face left	88° 26' 16"
	Face right	271° 32' 54"
	Sum	359° 59' 10"
2 × maladjustment		50"

Reduced vertical angles:

	Face left	01° 33' 44" elevation
	Face right	01° 32' 54" elevation
	Mean	01° 33' 19"
Required face right reading		271° 33' 19"

N.B. In this case the value may also be obtained and the maladjustment derived by adding the face left and face right readings together and subtracting from 360°. It should be appreciated that this difference must always be divided by 2 before being applied to the observed value.

For the C.T.S. Tavistock range the procedure is as follows:

(v) The reduced vertical angles are meaned and the mean value used to determine the required reading on the vertical circle when the instrument is in adjustment. Calculate and set the vertical circle to this reading using the vertical circle adjusting screw adjacent to the telescope. The bubble is not displaced.

For older instruments where a clip screw alters the pointing procedure is as follows:

(v) The reduced vertical angles are meaned and the mean value used to determine the required reading on the vertical circle when the instrument is in adjustment. Calculate and set the vertical circle to this reading using the altitude slow motion screw.
(vi) With the clip screw reposition the horizontal hair on the distant point thereby displacing the bubble from its central position.
(vii) With the altitude level adjusting screws bring the bubble back to the centre of its run.

Perfectly precise work can be carried out with the indices incorrectly placed provided that observations are made on both faces and the mean value taken. This is inconvenient when the error is large.

3.8.5. ADJUSTMENT OF THE TRANSIT AXIS. ADJUSTMENT 4

The transit axis is usually carefully set by the manufacturers, rough usage may make the adjustment necessary and on the older instruments in which the bearing surfaces were soft wear took place and an occasional adjustment was necessary. In many modern instruments no provision is made for this adjustment, for instance in the Wild T2. The Geodetic Tavistock has the transit axis deliberately set 6 seconds off horizontal to ensure greater stability in the instrument.

Explanation: This adjustment is carried out to set the transit axis perpendicular to the main axis of rotation.

(i) Set the instrument as near to a high building as possible, an object at an elevation of 30° is required. The distance to the foot of the building should be at least 100 feet and this point should be accessible.
(ii) Level the instrument with great care completing the levelling with the altitude level which is usually more sensitive than the plate level. The procedure is as follows:
Complete normal levelling with the plate level, place the altitude level parallel to the line of two footscrews and bring the bubble to the centre of its run with the altitude level setting screw. Turn the alidade through 180° and if the altitude bubble moves off recentre it half with the footscrews and half with the setting screw. Turn the alidade through 90° and if the bubble is not central recentre it with the third footscrew. Repeat until the altitude level bubble remains central in all positions of the alidade.
(iii) Place a strip of white card or, lay a level staff horizontal, at the foot of the building.
(iv) With the upper and lower plates clamped sight to an elevated point on the building and depress the telescope to sight onto the card or staff. Make a mark on the card or note the reading on the staff.

(v) Unclamp the upper plate, change face and with the upper and lower plates clamped again sight to the elevated point on the building. Depress the telescope and again sight onto the card or staff. Make a new mark or take a new reading. If both marks coincide or the readings are identical no adjustment is required.

(vi) If adjustment is required, point the telescope at the mean reading or exactly between the two marks. Elevate the telescope to the altitude of the point on the building and lower or raise one end of the transit axis with the adjusting screws until the elevated point is intersected.

3.8.6. ADJUSTMENT OF THE TRANSIT AXIS WITH STRIDING LEVEL

An alternative method by use of a striding level (see figure 3.26) may be used but it must be appreciated that this first involves its adjustment.

(i) Set the striding level in position upon the transit axis, the theodolite having previously been levelled with the plate level in the usual way.

(ii) Incline the striding level by a small amount to each side of the vertical plane containing the transit axis. (*If movement takes place the lateral adjusting screw of the level tube mounting is moved until the bubble remains steady.*) When the bubble in the striding level remains in the same, though not central, position its axis is coplanar with the line joining points midway between each V shaped foot of the striding level.

(iii) Using the footscrews bring the striding level bubble to its central position. Replace the striding level end for end. If the bubble remains in the same central position it is in adjustment. If movement takes place bring the bubble back half way with the vertical adjusting screw of the level tube mounting and complete the centring of the bubble with the footscrews.

(iv) Replace the striding level end for end and check that no further movement of the bubble takes place. Repeat the adjustment if any movement occurs. By means of this adjustment the bubble axis is made parallel to the line joining points midway between each V shaped foot of the striding level.

Procedure:

(i) With the striding level in position upon the transit axis make slight corrections to the footscrews to centre its bubble.

(ii) Turn the alidade through 180° in azimuth. If the bubble moves away from its central position adjustment is required. If the bubble is $2n$ divisions from its central position, half of this displacement is caused by the main axis of rotation being not truly vertical and half by the transit axis being not perpendicular to the main axis of rotation. To adjust, bring the bubble back n divisions towards its central position by means of the adjusting screws of the transit axis. Exactly centre the bubble with the footscrews and repeat the test until the bubble remains in a central position.

3.9. EFFECT OF MALCONSTRUCTION IN THE THEODOLITE

In §3.8.1 reference is made to the four conditions in theodolite construction which are nearly achieved during manufacture. In §3.6 reference is made to graduation of the circles and the possibility of small errors. In §3.2.3 reference is made to errors arising from axial strain in cylindrical axes. These points will now be considered.

3.9.1. EFFECT OF THE GRADUATED CIRCLES BEING INCLINED TO THEIR RESPECTIVE AXES OF ROTATION

This applies to both horizontal and vertical circles, the magnitude of the malconstruction is always small and the effect can be regarded as negligible. To find an expression for the magnitude (see figure 3.53): Let ABO be the true

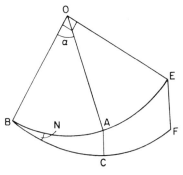

Figure 3.53

normal surface and CBO be the plane of the graduated circle, the inclination between these planes being N.

If an angle α' be measured on the plane CBO it will in error by $d\alpha$ from the required value α.

Where
$$d\alpha = \alpha - \alpha'$$

From the figure $\tan \alpha = AB/r$ and $\tan \alpha' = CB/r$

hence $\qquad \tan \alpha / \tan \alpha' = AB/CB = \sec N$

or $\qquad \tan \alpha' = \tan \alpha \cos N$

also $\qquad \tan(\alpha - \alpha') = \dfrac{\tan \alpha - \tan \alpha'}{1 + \tan \alpha \tan \alpha'}$

hence $\qquad \tan d\alpha = \dfrac{\tan \alpha (1 - \cos N)}{1 + \tan^2 \alpha \cos N}$

As N is small it can be taken that
$$\cos N = 1 - \frac{N^2}{2!}$$

hence
$$\tan d\alpha \simeq \frac{\tan \alpha N^2}{2 \sec^2 \alpha} = \tfrac{1}{2} \sin \alpha \cos \alpha N^2$$

From this it can be seen that $d\alpha$ is small

and
$$d\alpha'' \sin 1'' = \tfrac{1}{4} \sin 2\alpha N''^2 \sin^2 1''$$

or
$$d\alpha'' = \tfrac{1}{4} \sin 2\alpha N''^2 \sin 1''$$

Considering this expression, if N is $1'$, an exceedingly large malconstruction for any modern instrument, then the maximum value of $d\alpha$ is given when α is $45°$.

Then
$$d\alpha'' = 3\ 600''/4 \times 206\ 265 = 0\overset{\prime\prime}{.}004$$

3.9.2. EFFECT OF THE AXIS OF ROTATION NOT PASSING THROUGH THE POINT FROM WHICH THE GRADUATIONS ON A CIRCLE RADIATE

Eccentricity is the term usually applied to this condition.

See figure 3.54 where O is the centre of the graduations O' is the position of the axis of rotation about which the alidade, telescope and indices rotate.

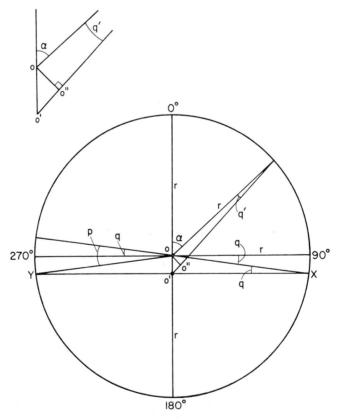

Figure 3.54

In the illustrated case the error will be zero when the indices lie along the line OO'. When the alidade is swung through an angle of 90° the readings of the indices will be:

index X 90° 00' 00" + q which is too great by q

and   index Y 270° 00' 00" + $q - p$

but $p = 2q$

Hence by taking the mean in the usual way the reading is 90° 00' 00" which is the required value.

It will be seen that by taking readings from two indices 180° apart the errors arising from eccentricity of the graduated circle are eliminated.

When the pointing is made in a direction not normal to the line OO' the reading α is in error by an amount q' less than q. Taking values for q and q' in radians it is found that:

as OO' is small $rq = $ OO'

and $rq' = $ OO"

But $\sin \alpha = \dfrac{OO''}{OO'} = \dfrac{rq'}{rq}$

$q' = q \sin \alpha$

Thus the error is variable from zero on the line OO' to a maximum of OO'/r on the direction normal to OO'. The error varying as $q \sin \alpha$. If the indices are not at 180° to each other the mean of the two readings will still give very nearly the correct value.

Consider the case when the indices are at 179° to each other and q is 30".

If one index lies on the line OO' and its reading be 00° 00' 00", the reading of the other index will be 179° 00' 00" − 30" sin 1°, viz. 178 59' 59".5.

When the alidade is turned through 90° exactly the first index will read 90° 00' 00" + 30" sin 90°, viz. 90° 00' 30", and the second index will read 269° 00' 00" − 30" sin 89°, viz. 268° 59' 30".

It will be seen that in this case the mean angle is in error by 0".25. But it must be realized that it is most improbable that an index be mispositioned by 1°, or that q be 30".

The effects of eccentricity are eliminated if the mean of two equally spaced indices are taken, similarly for any number of equally spaced indices. Whilst it is most unusual for more than two indices to be fitted it is relevant to note that the probable error of the mean reading of the graduated circle varies inversely as the square root of the number of indices.

The importance of errors of eccentricity has greatly increased with modern instruments. Many modern instruments have only one index and most of these instruments have very small circles. If the circle is small the value of r in $q = $ OO'/r is small and consequently q will be relatively large. It is essential with a fixed index instrument that every pointing be balanced by a second pointing made on the opposite face without change of zero, so creating conditions identical to those when a reading is made with two indices 180° apart.

3.9.3. EFFECT OF SMALL ERRORS OF GRADUATION OF THE CIRCLE AND READING MECHANISMS

The accuracy of graduation of modern glass circles is very high and for all but the most precise instruments is in excess of the precision of the reading mechanism.

Errors of graduation of the circle fall into two classes:

(i) Periodic errors, which occur at regular intervals and appear to follow some regular law.
(ii) Random errors, which do not appear to follow any regular law.

To minimize the effects of these errors different parts of the circle are used to make a number of separate determinations of the angle. Thereby those spaces between graduations which are too large will be included in some of the determinations and the spaces which are too small will be included in others. It is for this reason that different zeros are used (see §6.5.2.4).

Of similar origin to errors in graduation of the circle are errors in graduation of the micrometer or optical scale. These can be reduced by using different parts of the micrometer or scale for each zero.

3.9.4. ERRORS DUE TO AXIS STRAIN IN LIGHT WEIGHT GEODETIC THEODOLITES

The Canadian Geodetic Survey first drew attention to the presence of errors in results obtained with early models of the Wild geodetic theodolite. An investigation made in the Physics Laboratories of the National Research Council of Canada established that these errors were created by imperfections in the design of the bearings and the instrument was slightly modified and now produces perfectly satisfactory results.

The error can in part be attributed to a varying torque of the axis probably due to distortion of the metal, which in turn leads to a temporary displacement of the optical system. This displacement remains unaltered until the relationship of the axis and its bearings is changed and as a result the readings show little variation but are inaccurate. An error of this type is difficult to detect unless the observations are compared with those obtained with a larger geodetic instrument. Messrs J. L. Rannie and W. M. Dennis of the Geodetic Survey of Canada give an account of this investigation and an ingenious method of revealing the presence of errors of this type in a paper entitled 'Improving the Performance of Primary Triangulation Theodolites as a Result of Laboratory Tests', *The Canadian Journal of Research*, Vol. 10, No. 3, March 1934. Reference should also be made to Vol. 14 pp. 93–114, 1936 of the same journal.

To eliminate possible effects of axis strain Rannie and Dennis suggest the following procedure for primary observations:

(i) Keep the footscrews tight.
(ii) Divide the programme of observing into three parts the direction of the footscrews being altered by 120° during each part.

(iii) To avoid error due to the transit axis climbing in the bearings in the standards the top centre of the telescope should be tapped whenever the elevation of the pointings is altered.

3.10. THE EFFECT OF MALADJUSTMENTS IN THE THEODOLITE

In the following explanation it is assumed that I is a small angle (not greater than 5′), so that the relationship $I = \sin I = \tan I$ can be used without sensible error. It is also assumed that observations will not be made within a few degrees of the zenith.

3.10.1. THE EFFECT OF AN INCLINED MAIN AXIS OF ROTATION ON ANGLES OF AZIMUTH

Through imperfect levelling a theodolite main axis is inclined to the vertical by a small angle I_L and α is the angle between the plane of maximum inclination (i.e. the plane containing I_L) and a direction observed, say to a star, at elevation H (figure 3.55 illustrates this situation).

The angle recorded on the plane ABB′C is in error by an amount β, represented on the figure by the arc BB′. The elevation of the telescope along the plane B′S having consumed that excess portion of the horizontal angle.

From the spherical triangle ZZ′S and the right angled spherical triangle B′BS

$$\frac{\sin S}{\sin I_L} = \frac{\sin (180 - \alpha)}{\sin (90 - H)} = \frac{\sin \alpha}{\cos H}$$

and
$$\sin H = \tan \beta \cot S$$

Hence
$$\sin S = \sin I_L \sin \alpha \sec H \simeq S - \frac{S^3}{3!}$$

and
$$\tan S = \tan \beta \operatorname{cosec} H \simeq S + \frac{S^3}{3!}$$

As long as S remains small, which it will do provided I_L is less than 5′ and H less than 88°,

$$\sin S = \tan S$$

hence
$$\tan \beta = \sin I_L \tan H \sin \alpha$$

or
$$\beta = I_L \tan H \sin \alpha \qquad (1)$$

There is a second order error which has no effect on angles measured with a conventional theodolite, it arises from the measurement of the 'horizontal' angle on a graduated circle which is inclined by an amount I_L. It has been shown in §3.9.1. that the error in an angle measured on an inclined plane is given by $-\frac{1}{2} \sin 2\alpha I_L^2$.

If I_L is 5′, an excessively large amount, and α takes its maximum value of 45° this error is 0″.01.

3.10.2. THE EFFECT OF AN INCLINED TRANSIT AXIS ON ANGLES OF AZIMUTH

If through failure to adjust the theodolite correctly the transit axis is inclined to the horizontal by an amount I_T, the instrument having been correctly levelled, then the line of sight will again sweep out a plane represented by B'SZ' in figure 3.55. This plane will be inclined to the vertical by an amount I_T.

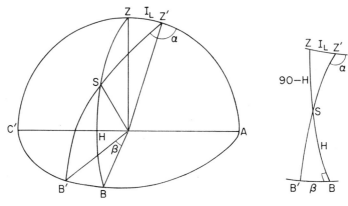

Figure 3.55

This condition is identical to that described in §3.10.1 *except* that the displacement I_T, is always at right angles to the line of sight. With change of face the inclination is reversed and as a consequence changes in sign. Consequently the effect of this maladjustment is cancelled out on observing to a point on both faces. The error in a single pointing arising from this source is given by t, where

$$t = I_T \tan H \tag{2}$$

The magnitude of the combination of I_L and I_T at right angles to the line of sight can be measured with a striding level. If this is not available I_L only can be measured with the plate level.

3.10.3. THE EFFECT OF AN INCLINED TRANSIT AXIS ON VERTICAL ANGLES

Provided the vertical circle is normal to the transit axis only a small error is introduced in any measurement of a vertical angle when the transit axis is inclined.

Referring to figure 3.56 the recorded vertical angle will be the arc B'S whereas the correct value is the arc BS. Designating these quantities H' and H respectively it can be established that:

$$\sin H' = \sin H \sec I_T \simeq \sin H (1 + \tfrac{1}{2} I_T^2)$$

hence
$$\sin H' - \sin H \simeq \tfrac{1}{2} \sin H \, I_T^2$$

and
$$2 \sin \tfrac{1}{2}(H' - H) \cos \tfrac{1}{2}(H' + H) \simeq \tfrac{1}{2} \sin H \, I_T^2$$

As the difference between H' and H is small we may be

$$(H' - H) = \tfrac{1}{2} \tan H \, I_T^2 \tag{3}$$

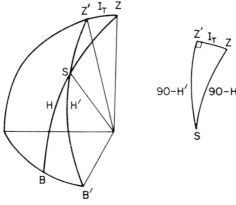

Figure 3.56

3.10.4. THE EFFECT ON ANGLES OF AZIMUTH OF THE LINE OF SIGHT BEING NOT NORMAL TO THE TRANSIT AXIS

If the line of sight makes an angle of $90° \pm I_C$ to the transit axis, then each pointing will be in error.

Referring to figure 3.57 illustrating observations made to a star S at an elevation H, the pointing is in error by arc PQ. Arc PQ is the angle c at Z in the

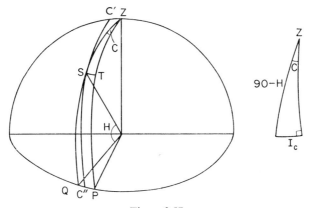

Figure 3.57

spherical triangle ZST, ST can be regarded as of magnitude I_C as C'SC" is a portion of a small circle swept out by the line of sight.

It follows that:
$$\sin I_C = \cos H \cos (90 - c)$$
hence
$$\sin c = \sin I_C \sec H$$
which approximates to
$$c = I_C \sec H \qquad (4)$$

Where H is unchanged between two pointings on opposite faces to the same object the mean horizontal circle reading is not in error from this effect. Similarly if two objects at the same elevation are observed on one face only the determination of the included angle is not in error. If the two points are at different elevations the determination of the included angle on one face only is is error by an amount e, where:

$$e = I_C(\sec H_1 - \sec H_2)$$

It should be noted that when H is an angle of depression sec H is still positive.

3.10.5. THE EFFECT OF AN INDEX ERROR ON ANGLES OF ELEVATION

When vertical angles are measured on one face an error will be present in the result arising from maladjustment of the indices (see §3.8.1). The error results from a dislevelment of the indices, the zeros, or the combined effect of both. The difference in vertical angles between faces will represent twice the error. Change of face completely eliminates the error.

3.11. PROCEDURE TO MINIMISE THE EFFECTS OF MALCONSTRUCTION AND MALADJUSTMENT OF THE THEODOLITE

It has been indicated in §§3.9 and 3.10 that by adoption of a suitable procedure many of the effects of malconstruction and maladjustment can be minimized and in some cases eliminated.

3.11.1. COMPONENTS OF PROCEDURE (see Table 3.1)

3.11.2. PRACTICAL HINTS IN HORIZONTAL ANGLE MEASUREMENT

(1) For precise work, rigidly emplace the instrument on a concrete pillar usually with the aid of cords or straps. If no pillar is available the tripod feet should be cemented to bed rock or to specially positioned concrete blocks.

For less precise work care is taken to ensure the rigidity of the tripod by digging the feet well into the ground and tightening the screws at the top of the tripod.

(2) For precise work shelter the instrument from the sun with a suitable umbrella. This avoids uneven heating of the instrument and its legs.

(3) After centring over the ground mark focus the cross hairs and telescope and eliminate parallax (see §3.4). These settings are left undisturbed throughout a set up at a station. Set the slow motion screws to the middle of their run.

(4) On completion of (3) level the instrument (see §3.5.2). Levelling is checked after every round.

(5) Illuminate the circles taking care to ensure that, in mean reading instruments, both sides of the circle are equally illuminated.

(6) In instruments with a direct drive to the graduated circle the zero is set after directing the line of sight onto the reference object (R.O.). The method is to set the micrometer to the required minutes and seconds and then rotate the graduated circle until a setting or coincidence is effected in the requisite part of the circle.

In instruments with a lower plate and slow motion screw setting or coincidence is effected by movement of the alidade with the lower plate clamped. A pointing is then made to the R.O. with this reading held, the lower plate clamp being

Table 3.1

Description	Figure	Remedy	Effect	Magnitude and remarks
Malconstruction				
(1) Circles not normal to axes	3·53	None	$d\alpha'' = \frac{1}{4}\sin 2\alpha \, N^2 \sin 1''$	The effect is small, less than the precision of the instrument
(2) Axis and centre of graduations not coinciding, eccentricity	3·54	Two indices 180° apart	$q = 00'$ r $q' = q \sin \alpha$	The smaller the circle the more significant is the effect.
(3) Small errors in graduation		Change of zero		This malconstruction is rare in modern instruments except for minute errors.
(4) Axis strain		Alteration in position of footscrews		Significant only in precise work.
Maladjustments				
(5) Inclined main axis	3·55	None. Careful levelling. If H large measure dislevelment and calculate effect (3-11-3)	$\beta = I_L \tan H \sin \alpha$	Significant in all astronomical observations when H is large.
(6) Inclined transit axis	3·55	Change of face	$t = I_T \tan H$	I_T is usually small, the remarks under (5) also apply.
(7) Line of sight not normal to transit axis	3·57	Change of face	$c = I_C \sec H$	I_C may be quite large.

loosened until the exact bisection is made with the lower plate slow motion screw.

(7) With the required first zero set and the instrument in the face left position rotate the alidade several times in a clockwise direction to take up slack in the axes. The instrument is rotated by gripping the lower part of the standard with the fingers extended around the plate covers. The instrument should not be gripped near the transit axis, any rough handling may cause strain which is released during observation. In clamping the plate a very light pressure is used.

(8) Make a careful pointing on the R.O. approaching it from the left. The cross hair will lie to the right in the telescope and will appear to move from right to left as the approach is made.

On face left the alidade is usually rotated in a right-hand or clockwise direction and on face right in a left-hand or anticlockwise direction. This rule also applies to movement with the slow motion screws even though this may involve movement with the retaining spring. If the R.O. or any station is overshot the alidade is completely rotated and a new approach made. Intense concentration is required in making the final intersection, the object is examined after each movement of the slow motion screw and the approach is made in a series of diminishing steps.

(9) Once the setting has been made on the R.O. the reading is taken. Intense concentration is required and the fingers are removed from the micrometer head while the setting is examined. The final movement of the micrometer is made in the same direction during each round but from a different direction in different rounds, say from the left on face left and from the right on face right. The pointing is checked after the reading has been taken.

(10) The alidade is rotated (swung) right making a pointing to the next station as described in (8). After intersecting the last station face is changed and the procedure repeated in the opposite direction as explained in (8) above.

3.11.3. DISLEVELMENT CORRECTION

This may be necessary when a horizontal angle is measured to an object at an appreciable elevation or depression. It is usually of considerable significance in the determination of azimuth by astronomical means.

The magnitude of the correction is deduced from observations taken with either a striding level or from the plate level.

The correction is given by

$$I'' \tan H = \text{Correction in seconds}$$

where I'' is the dislevelment of the transit axis in seconds and H the vertical angle to the object observed. This correction follows from §§3.10.1 and 2.

To determine I the positions of the ends of the bubble in the level are recorded for each pointing. The level vial is considered to be graduated outwards from the centre and the left-hand reading L is recorded first.

With a single pointing

$$I'' = \frac{L-R}{2} v''$$

where v'' is the value of one division of the vial in seconds of arc (see §3.5.5). If $L = 4$, $R = 2$ and $v'' = 20''$, $I'' = +20''$.

It should be noted that I is positive when $L > R$ as then the transit axis is high on the left-hand side and a portion of the clockwise angle is consumed on elevation of the telescope. Since theodolite horizontal circles are graduated clockwise the reading is less than it should be.

It is not necessary to correct each pointing, although in some cases it may be desirable to do so. The correction to the mean observed direction is given by

$$+ \frac{\Sigma L - \Sigma R}{2n} v'' \tan H = \text{Correction in seconds}$$

where n is the number of pointings made to the object. For angles of depression the usual convention is adopted and $\tan H$ is negative.

3.12. TYPES OF THEODOLITE

No detailed description is given here of any particular model, reference should be made to the relevant manufacturers' literature which is usually most comprehensive. Table 3.2 over leaf gives technical details of most modern theodolites in common use.

3.13. COMPONENTS OF THE LEVEL

The function of the surveyors level, more commonly termed simply the level, is to establish a horizontal line of sight or nearly so. In its simplest form it consists of a telescope with a defined line of sight and a spirit level tube to enable this line of sight to be set horizontal.

The level is supported on a tripod similar to those used for theodolites (see §3.2.1). The difference being an increase in weight for greater stability and the absence of a movable centring device.

The base of the level is similar to the theodolite tribrach and trivet stage described in §3.2 or may be a ball and socket arrangement. Either device carries a vertical spindle similar to the main axis of rotation of a theodolite. This spindle carries either the telescope to which it is rigidly connected at 90° or a cradle to which the telescope is hinged.

The telescope is similar in design to the description given in §§3.3 and 3.4. The magnification of the level telescope controls the accuracy with which readings on the staff may be made. As stated in §3.3.3 the size of the field of view and the brightness of the image are reduced by an increase in magnification. For levelling a large field of view is not important. It is usual to have a large aperture to improve the brightness of the image. The relationship

$$M = 16A$$

where M is the magnification and A the effective aperture size, will usually be slightly exceeded.

The features of the spirit levels attached to the telescope are identical to those described in §3.5.1. Coincidence reading is commonly used with a split bubble as described in §3.5.3 and illustrated by figure 3.2.5. Mirror bubble readers are used on low priced instruments, the mirror hinging over the vial housing which

Table 3.2

Instrument	T3	T2	T1	T16	DKM 3	DKM 2	DKM 1	Microptic No. 3	Microptic No. 2	Microptic No. 1
Manufacturer	Wild	Wild	Wild	Wild	Kern	Kern	Kern	Watts	Watts	Watts
Telescope power	x 24, 30 & 40	x 28	x 27	x 28	x 45	x 30	x 20	x 40	x 28	x 25
Field of view at 1,000 ft	28 ft	29 ft	29 ft	29 ft	28 ft	28 ft	28 ft	17 ft	28 ft	28 ft
Horizontal circle										
Diameter	360°	360°	360°	360°	360°	360°	360°	360°	360°	360°
	5.9 in	3.5 in	3.1 in	3.1 in	4.0 in	3.0 in	2.0 in	3.8 in	3.85 in	3.0 in
Graduation	4'	20'	1°	1°	10'	10'	20'	5'	10'	20'
Vertical circle										
Diameter	180°	360°	360°	360°	360°	360°	360°	360°	360°	360°
	3.7 in	2.8 in	2.8 in	3.1 in	4.0 in	2.8 in	2.0 in	3.0 in	3.0 in	2.5 in
Graduation	8'	20'	1°	1°	10'	10'	20'	5'	10'	20'
Micrometer interval	0"2 ①	1"0	20"	Optical scale 1'	0"5	1"0	10"0	0"2	1"0	20"0
Plate level	7"/2mm	20"/2mm	30"/2mm	30"/2mm	10"/2mm	20"/2mm	30"/2mm	10"/2mm	20"/2mm	45"/2mm
Altitude level	12"/2mm	30"/2mm	30"/2mm	30"/2mm	10"/2mm	20"/2mm	30"/2mm	20"/2mm	20"/2mm	30"/2mm
Weight instrument/case	24.2/8.3 lb	12.3/4.8 lb	11.2/3.7 lb	9.8/3.9 lb	26.9 lb	8 lb	4 lb	16/14	13.8/9 lb	21 lb
Reading system	Optical micrometer coincidence	Optical micrometer coincidence	Optical micrometer fixed index	Optical scale by estimation to 0'2	Optical micrometer light gap	Optical micrometer light gap	Optical micrometer light gap	Optical micrometer straddling coincidence	Optical micrometer straddling coincidence	Optical micrometer fixed index
Remarks	Mean reading of opposite graduations	Mean reading of opposite graduations	Reading to one side of circle only	Reading to one side of circle only	Light gap formed by opposite graduations. Mean reading	Light gap formed by opposite graduations. Mean reading	Light gap formed by opposite graduations. Mean reading	Mean reading of opposite graduations	Mean reading of opposite graduations	Reading to one side of circle only

Instrument	Geodetic Tavistock	Tavistock Type 1	Tavistock Type 2	V 308	V 22	010	030
Manufacturer	C.T.S. (Vickers)	C.T.S. (Vickers)	C.T.S. (Vickers)	C.T.S. (Vickers)	Vickers	Zeiss / Jena	Zeiss / Jena
Telescope power	x 20-30	x 25	x 25	x 25	x 25	x 31	x 25
Field of view at 1,000 ft	24 ft	33 ft	30 ft	35 ft	35 ft	21 ft	27 ft
Horizontal circle	360°	360°	360°	360°	360°	360°	360°
Diameter	5·0 in	3·5 in	3·35 in	3 in	3·1 in	3·5 in	3·8 in
Graduation	20'	20'	20'	20'	1°	20'	1°
Vertical circle	360°	360°	360°	360°	360°	360°	360°
Diameter	2·75 in	2·75 in	2·75 in	2 in	2·5 in	2·5 in	3·0 in
Graduation	20"	20"	20"	20"	1°	20"	1°
Micrometer interval	0"·5	1"·0	1"·0	20"·0	Optical scale 1'	1"·0	Optical scale
Plate level	40"/2mm	40"/2mm	20"/2mm	30"/2mm	45"/2mm	20"/2mm	30"/2mm
Altitude level	10"/2mm	20"/2mm	20"/2mm	40"/2mm	90"/2mm	20"/2mm	Stabilizer [2]
Weight instrument/case	24/18 lb	13/14 lb	11/9 lb	9/5 lb	9/7 lb	11·5/10·5 lb	9·5/10·5 lb
Reading system	Optical micrometer coincidence	Optical micrometer light gap	Optical micrometer coincidence	Optical micrometer fixed index	Optical scale by estimation to 0'·2	Optical micrometer coincidence	Optical scale by estimation to 0'·2
Remarks	Mean reading of opposite graduations	Mean reading of opposite graduation	Mean reading of opposite graduations	Reading to one side of circle only		Mean reading of opposite graduations	Reading to one side of circle only

Footnote:-
① Marked as 0"·1 and read twice
② A pendulum makes zeros automatically horizontal.

protects it when not in use. Provision is made for the movement of the level vial within its housing by means of adjusting screws similar to those of a theodolite plate level. A small circular bubble is frequently provided for preliminary levelling of the instrument.

Figure 3.58 illustrates a simplified typical level in which the telescope body is hinged to the top of the vertical spindle and is tilted by means of a sensitive elevating screw. This screw being used to bring the rigidly attached level

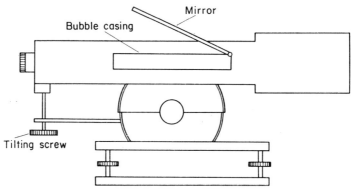

Figure 3.58

bubble to a central position. Such a level is termed a *tilting level*. All modern levels are of this type except automatic levels details of which are given in §3.15. Some tilting levels are provided with a tilting screw graduated in gradients.

An obsolecent level is the *dumpy level* where the telescope body is not hinged at the top of the spindle, being fixed at 90°.

In modern instruments a clamp and slow motion screw is usually provided enabling the telescope to be clamped to the base and given a fine relative motion.

3.14. THE GEODETIC LEVEL

The geodetic level differs little from the normal tilting level except that it is larger and thereby more stable. It is usually made of nickel steel which has a low coefficient of expansion.

The distinctive features are:

(i) A coincidence bubble viewed in the field of view of the telescope (see figure 3.59). The level usually has a sensitivity between 4" and 10". A scale may be provided to measure small dislevelment, this saves time in the field but requires correction later.
(ii) A special reticule as shown in figure 3.59.
(iii) A parallel sided plate of glass mounted on a horizontal axis and fitted in front of the object glass. This is connected by a rod to a graduated drum which records the amount of tilt and thereby the displacement of the staff image in the vertical plane. This is termed the parallel plate micrometer.

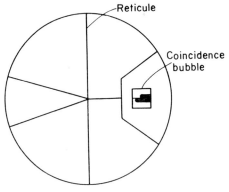

Figure 3.59

3.14.1. THE PARALLEL PLATE MICROMETER

In taking an observation with the geodetic level the plate is tilted until the image of the nearest staff graduation is brought onto the horizontal hair. It has been shown earlier in §3.7.4 that the displacement d of a parallel sided plate of glass can be sensibly regarded as proportional to the thickness t

$$d = t\beta(1 - 1/\mu)$$

Therefore suitable regular graduation of the drum will enable an exact measurement to be made of a fraction of a staff graduation (see figures 3.60 and 3.61).

The Watts Precise Level has a drum with 20 divisions each of which corresponds to a vertical displacement of the line of sight of 0·001 ft. The Watts staff is graduated in units of 0·02 ft and hence the fractions of each staff graduation can be determined. The drum is designed so that with the plate vertical the

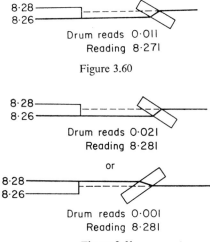

reading is 10, viz. 0·010 ft, thereby each and every reading is 0·010 ft too great, i.e. as if the instrument was set 0·010 ft higher. With forward tilting of the plate, drum readings increase.

3.15. THE AUTOMATIC LEVEL

For work of all but the highest precision automatic levels can be used. Various types are manufactured and in all cases the basic concept is of a stabilizer which will influence the horizontal line of a sight so that it passes from the objective to the graticule. Thereby delicate levelling of the instrument with a sensitive spirit level is no longer necessary. A stabilizer is built into the body of the instrument which once the instrument is approximately levelled with a circular level is free to swing. Swinging freely under the influence of gravity the stabilizer aligns the line of sight in the horizontal plane. With an ordinary tilting level the line of sight can be set to within about 3″ of the horizontal; with a geodetic level the line of sight can be set with certainty to 1″ or better; the stabilizer is required to set the line of sight to a comparable accuracy.

3.15.1. THEORETICAL CONCEPTS

Following largely the treatment of F. B. R. Hogg, and J. A. Armstrong in their paper *Two New Self-Aligning Levels* in *Empire Survey Review*, No. 111, Vol. XV, January 1959 a number of points emerge.

(1) Consider a telescope, with two reflecting surfaces, tilted through an angle α to the horizontal (see figure 3.62). The divergence of the horizontal

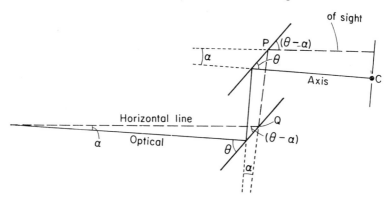

Figure 3.62

ray from the optical axis remains constant. The function of the stabilizer is to deflect this ray so that it intersects the graticule on the horizontal hair.

(2) If the mirror Q (figure 3.63) is allowed to take up a constant inclination to the vertical $(90° - \theta)$ whilst the mirror P remains fixed, the horizontal ray is turned through an angle 2α and will converge onto the horizontal hair. Similarly, although not illustrated, mirror P could be rotated in the opposite direction to give the same effect.

(3) A mechanism similar in principle to figure 3.63 requires that the stabilizer be midway in the optical train. If the movement of the mirror could be amplified

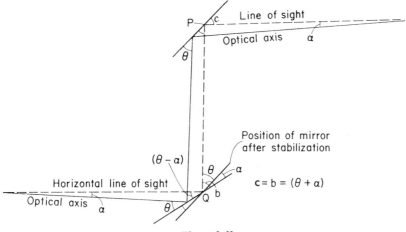

Figure 3.63

the stabilizer could be nearer to the graticule with a consequent reduction in the size of the converging cone of rays. Alternatively the change in direction of the principal ray might be increased for the same mirror rotation. It is this latter feature which C.T.S. and Hilger and Watts have adopted in their stabilizer; from here on only the theoretical concepts behind that stabilizer will be described.

(4) If three reflecting surfaces are used (figure 3.64) and rigidly connected to a tilted telescope the divergence of the horizontal ray will remain constant at α. If the two outer mirrors P and R swing into a position where they have a constant inclination to the vertical the horizontal ray will be deflected by 4α towards the graticule (see figure 3.65).

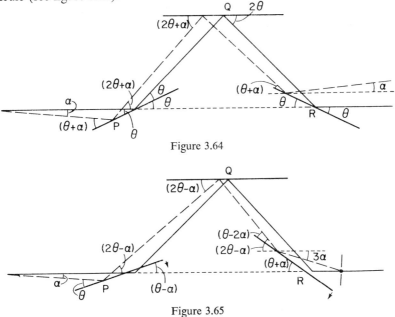

Figure 3.64

Figure 3.65

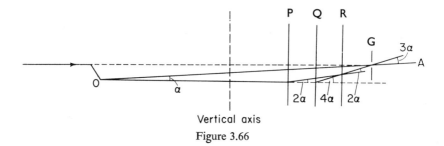

Figure 3.66

(5) This arrangement is illustrated by the principal ray diagram in figure 3.66, where if $QG = \tfrac{1}{4}OG$ the ray is brought onto the graticule, QG being the focal length of the lens system. It should be noted that the horizontal ray shown is that which intersects the point where the vertical axis meets the optical axis, this ray having been effectively refracted at the objective.

(6) In the C.T.S./Hilger Watts stabilizer the theoretical mirrors are replaced by prisms. The result is that the telescope forms an erect image, also the traditional negative internal focusing lens is split into two components so increasing the distance between the focusing lens and the graticule. In this increased length is placed the stabilizer.

(7) The stabilizer incorporates two moving and one fixed prism. The suspension system for the two moving prisms must be frictionless, robust, completely reliable, compact, and accurate to one second of arc. Knife edges and jewelled pivots are not sufficiently accurate but a flexure pivot will give these requirements although it is markedly limited in the amount of permissible rotation. This makes the setting of the telescope to within 20 minutes of arc essential, this is achieved with the circular level.

(8) Figure 3.67 illustrates the essential features of the British made stabilizers. A mount for the two swinging prisms is supported on four steel alloy straps

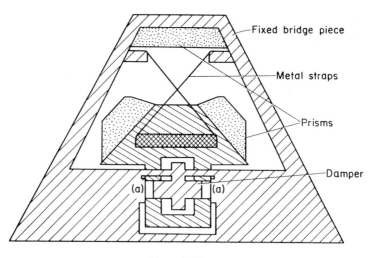

Figure 3.67

mounted in crossed pairs on either side. The intersection of these straps is made the centre of rotation of the mount. The upper ends of the straps are secured to blocks keyed into the bridge piece which carries the roof prism. An air damper is provided in the form of a cylindrical chamber in the swinging prism mount into which pistons from the bridge piece project.

(9) The whole unit is carefully designed to overcome errors which may arise from temperature changes and the metal straps are hardly stressed so have a constant modulus of elasticity. When the instrument is carried the bridge piece supports the prism mount and for this reason a vertical position for the telescope is desirable when the instrument is encased.

3.15.2. USE OF THE AUTOMATIC LEVEL

There is no significant difference between using the automatic level and using any other level. The following points should be noted:

(i) The preliminary levelling of the instrument with the circular level must always be carried out. This requires that the circular level be always in adjustment or nearly so.
(ii) The stabilizer will vibrate in a strong wind and the image will appear to shimmer, some uncertainty will exist in the readings. In these conditions several readings should be taken and the mean value adopted.
(iii) The low sensitivity of the level to solar radiation adds to its advantages over the conventional level.
(iv) The ray brought to the horizontal hair will only be the horizontal ray if the horizontal hair is correctly positioned and the instrument must be tested, as any other level is tested, for the horizontality of its line of sight (see §3.16 below).
(v) A parallel plate micrometer can be used with the level and although there are theoretical objections due to the tilt of the telescope tube no perceptible error is introduced.

3.16. ADJUSTMENT OF THE LEVEL

3.16.1. ADJUSTMENT OF THE TILTING LEVEL

Only one adjustment is made this establishes the line of sight parallel to the principal tangent of the spirit level (see figure 3.68).

(i) Two points A and B are selected 200 feet apart and firm ground marks established.
(ii) Set the level midway between the marks, centre the bubble and take readings to a staff held at A and B in turn. Let these readings be R_A and R_B then the difference in height $\Delta h = R_A - R_B$.
   This is the correct difference in height regardless of the adjustment of the instrument.
(iii) Set the instrument close to A and measure its height above the ground mark at A with either a simple rule or by viewing the staff using the objective as the ocular (the centre of the field can be estimated on the staff). Let this be i.

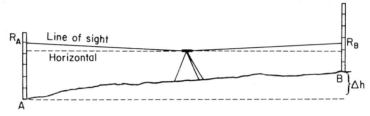

Figure 3.68a

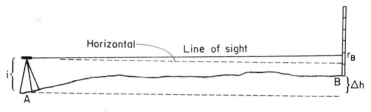

Figure 3.68b

(iv) Take a staff reading on B, let this be r_B.

If $i - r_B$ does not equal Δh the instrument needs adjustment.

(v) Calculate a value r_B' such that $i - r_B' = \Delta h$ and direct the horizontal hair onto that reading by means of the tilting screw; this will displace the bubble from its central position.

(vi) Bring the bubble back to its central position by means of the level adjusting screws and repeat the tests until $i - r_B = \Delta h$.

If $r_B' - r_B$ does not exceed 5 mm (0·02 ft) this adjustment is not usually carried out.

Example

R_A	8·27	i	4·78		4·78
R_B	6·34	r_B	2·80		1·93
Δh	1·93		1·98 $\neq \Delta h$	r_B'	2·85

3.16.2. DETERMINATION OF THE LEVEL ERROR

See figure 3.69. This adjustment is usually only carried out as part of the procedure in geodetic levelling (see §8.3.6). Metric units will be used throughout this explanation.

(i) Two points A and B are selected 100 metres apart and firm ground marks established.

(ii) Set the level at A' such that AA' is about 10 metres, read the staves at A and B. Let the middle hair readings be N_A and D_B and the stadia intercepts be A'A and A'B.

(iii) Set the level at B' such that BB' is about 10 metres and repeat the readings made in (ii). Let these be N_B, D_A, B'B, and B'A. If the ratio of the displacement of the line of sight from the horizontal to the stadia

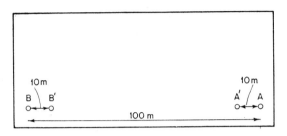

Figure 3.69

intercept be C. Then:

$$\Delta h = (D_B - N_A) + (A'B - A'A)C$$

and

$$\Delta h = (N_B - D_A) - (B'A - B'B)C$$

Thus

$$C = \frac{(N_A + N_B) - (D_A + D_B)}{(A'B + B'A) - (A'A + B'B)}$$

C should be calculated to the nearest digit in the third decimal place but no further.

(iv) If C exceeds 0·010 the instrument must be adjusted as follows:

Move the telescope by means of the tilting screw so that the reading of the middle hair is changed by an amount equal to C times the total stadia intercept. Move the hair upward on the staff if C is positive or downward if C is negative.

(v) The effect of (iv) is to displace the bubble from its central position to which it is returned by use of the level vial adjusting screws. Redetermine C.

3.16.3. ADJUSTMENT OF THE DUMPY LEVEL

Two adjustments are involved:

(i) To make the principal tangent of the level vial perpendicular to the vertical axis.
(ii) To make the line of sight parallel to the principal tangent of the level vial.

Adjustment (i) is carried out in the same manner as the adjustment of the theodolite plate level (see §3.8.2).

Adjustment (ii) is carried out as with the tilting level except that the horizontal hair is brought onto the required staff reading by movement of the reticule, see §3.3.7, thereby leaving the level vial as set by adjustment (i).

3.16.4. ADJUSTMENT OF THE AUTOMATIC LEVEL

As indicated in §3.15.2 this instrument can be out of adjustment due to mis-positioning of the reticule. If after carrying out the test described in §3.16.1 for the tilting level the instrument is found to be in need of adjustment the horizontal hair is brought to the correct reading by movement of the reticule (see §3.3.7).

On some models various other adjustments relating to the positioning of the stabilizer are required. The manufacturers handbook should be consulted and followed.

4 *The Spheroid and Projections*

4.1. THE SPHEROID

4.1.1. THE GEOMETRY OF THE SPHEROID

For many purposes in land surveys it is possible to neglect the Earth's curvature, and to treat its surface as plane over a limited area. This greatly simplifies computation. Sometimes it is sufficient to consider the Earth as a sphere and work in terms of spherical trigonometry. In the most refined work a *spheroid of reference* has to be used. A spheroid is the figure described by the rotation of an ellipse about one of its axes; if rotation is about its minor axis, an *oblate spheroid* is produced, and if about its major axis, a *prolate spheroid* results.

If the land masses were covered by a network of canals through which the waters of the oceans were permitted to flow freely under gravity, neglecting tidal effects, the surface of water on the canals and the oceans would form an equipotential surface called the *geoid*, which simply means 'Earth-shaped'. For a further consideration of this topic the reader should refer to Chapter 8 and to Bomford's *Geodesy*.† The simplest mathematically regular surface which best fits this geoid is an oblate spheroid. As a result of various measurements carried out since the 18th century in different parts of the world a number of geodesists have produced estimates of the size of spheroid which best fits the geoid for the part of the Earth considered in their calculations. Today spheroids have been adopted in various countries, e.g. the United Kingdom is computed on Airy's spheroid, much of Africa is computed on the Clarke 1880 spheroid. It is often sufficiently accurate to consider the surface of a sphere which closely approximates to the spheroid at the particular place in question, and whose radius equals the radius of curvature of the spheroid at that point. Such an approximate treatment enables one to calculate the results of surveys covering lines of up to say 25 miles long. The permissible length varies with latitude and precision required. In this book we shall confine out attention to this semi-rigorous approach since it enables the various principles of computation to be presented in a limited space, and with sufficient accuracy for most surveys. More refined methods may be found in textbooks on geodesy.†

4.1.2. THE MERIDIAN ELLIPSE

The ellipse which defines a spheroid is called the *meridian ellipse*. The parallels of latitude are small circles in planes parallel to the equator, which is a great circle.

An ellipse is defined in many ways and has a multiplicity of geometrical properties. We shall consider it defined with respect to its semi-major axis a

† G. Bomford, *Geodesy*, (Oxford: Clarendon Press, 1952).

150   Practical Field Surveying and Computations

and semi-minor axis b by the equation

$$\frac{x^2}{a^2} + \frac{y^2}{b^2} = 1 \tag{1}$$

where the coordinates of a point on the ellipse with respect to the origin at its centre are (x, y). The following properties will also be used:

(i)
$$b^2 = a^2(1 - e^2) \tag{2}$$

where e is the eccentricity of the ellipse. For a terrestrial spheroid $e \simeq \frac{1}{12}$.

(ii) The flattening $f = (a - b)/a$. For the terrestrial spheroid $f \simeq \frac{1}{298}$.

(iii) The area of the ellipse is πab.

Given any two of the parameters $a, b, e,$ or f the others may be derived. It is usual to define a meridian ellipse, and therefore a spheroid, in terms of a and f. Table 4.1 shows the values of these parameters for some of the major spheroids.

Table 4.1. Spheriodal Parameters

Spheroid	Semi–major axis – a	Flattening – f
Everest's 1830	6 377 304 metres	1: 300·8
Bessel 1841	6 377 397 "	1: 299·2
Clarke 1858	6 378 293 "	1: 294·3
Clarke 1866	6 378 206 "	1: 295·0
Clarke 1880	6 378 249 "	1: 293·5
Helmert 1906	6 378 200 "	1: 298·3
Hayford 1910	6 378 388 "	1: 297·0

N.B. The I.U.G.G. (International Union of Geodesy and Geophysics) recommended Hayfords spheroid for international use, hence it is called the *International* or *Madrid Spheroid*, the latter after the place at which the conference of 1924 was held.

4.1.3. TYPES OF LATITUDE

Since the reference spheroid will not fit the geoid exactly at all points, two main types of latitude may be distinguished:

(i) *Geodetic latitude* ϕ_G is the angle between the normal to the spheroid at a point and the equator (see figure 4.1).

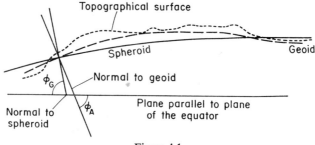

Figure 4.1

The Spheroid and Projections

(ii) *Astronomical latitude* ϕ_A is the angle between the equator and the meridian component of the normal to the geoid at that point. The normal to the geoid is the direction of the plumb line and will not usually lie in the plane of the meridian ellipse. The angle between the direction of the plumb line and the normal to the spheroid is *the deviation of the vertical*. This depends on the spheroid chosen and the point or points at which the spheroid and geoid are related to each other.

Latitude is positive North of the Equator.
Others definitions of latitude will not concern us here.

4.1.4. TYPES OF LONGITUDE

In a like manner to latitude we distinguish two types of longitude which arise out of the lack of coincidence between the direction of the plumb line and the normal to the spheroid.

(i) *Geodetic longitude* λ_G is the angle measured along the equator between the meridian of Greenwich and the meridian ellipse of the place.

(ii) *Astronomical longitude* λ_A is the angle measured along the equator between the astronomical meridian of Greenwich and the astronomical meridian of the place; the astronomical meridian of a place being defined as the plane containing the Earth's axis and the astronomical zenith of the place.

Longitude is positive East of Greenwich.

4.1.5. TYPES OF AZIMUTH

Since there are two meridians at any place, the astronomical and the geodetic, directions referred to these meridians, or azimuths, will be different, i.e. we have *astronomical azimuth* α_A and *geodetic azimuth* α_G. Since computation is in terms of geodetic azimuth, and only astronomical azimuth may be observed to control a survey, the geodetic value corresponding to an observed astronomical azimuth must be obtained. Azimuth is positive clockwise from North.

Let the geodetic and astronomical longitudes and azimuths be respectively λ_G, λ_A, and α_G, α_A. Then

$$\alpha_A - \alpha_G = (\lambda_A - \lambda_G) \sin \phi \tag{3}$$

(ϕ is the latitude). This is the *Laplace azimuth equation*. This may be proved by reference to figure 4.2 in which the following are shown: P the pole, O the centre of the Earth; Z_A and Z_G the respective astronomical and geodetic zeniths, N_A and N_G the astronomical and geodetic north points on the horizon; and astronomical and geodetic azimuth and longitude differences $\Delta\alpha$ and $\Delta\lambda$.

In the triangle $PN_A N_G$, arc $N_A N_G = \Delta\alpha$, and angle $PN_G N_A$ is a right angle if $N_A N_G O$ is the geodetic horizon. Then by Napier's rules

$$\sin \Delta\alpha = \sin \Delta\lambda \sin \phi$$

since both $\Delta\alpha$ and $\Delta\lambda$ are small angles

$$\Delta\alpha \simeq \Delta\lambda \sin \phi \tag{4}$$

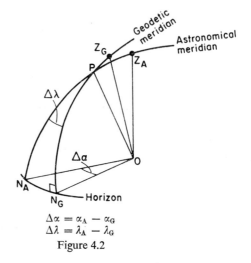

$$\Delta\alpha = \alpha_A - \alpha_G$$
$$\Delta\lambda = \lambda_A - \lambda_G$$

Figure 4.2

4.1.6. RADII OF CURVATURE OF THE SPHEROID

The surface of the spheroid is one of double curvature which is usually resolved in the plane of the meridian ellipse, and at right angles to this plane. The respective radii of curvature in these two directions are denoted by the Greek letters ρ and ν, and are given by

$$\rho = \frac{a(1-e^2)}{(1-e^2\sin^2\phi)^{\frac{3}{2}}} \qquad (5)$$

$$\nu = \frac{a}{(1-e^2\sin^2\phi)^{\frac{1}{2}}} \qquad (6)$$

where ϕ is the latitude of the place. Thus it is seen that both radii are functions of latitude alone, for any given spheroid for which the values of a and e are defined.

The radius of curvature R at latitude ϕ at any azimuth α is given by

$$\frac{\cos^2\alpha}{\rho} + \frac{\sin^2\alpha}{\nu} = \frac{1}{R} \qquad (7)$$

where ρ and ν are evaluated for ϕ. Equation (7) is *Euler's theorem*, which is used to calculate R for computations involving survey *lines*.

The mean value of the radius of curvature for a point is given by

$$(\rho\nu)^{\frac{1}{2}} \qquad (8)$$

This expression is used when *areas* are considered, e.g. in the computation of spherical excess. The following sections outline the proofs of expressions (5), (6), (7), and (8).

4.1.6.1. *Principal Radii of Curvature of the Spheroid; ρ and ν*

Consider figure 4.3 showing the meridian ellipse, whose equation is

$$\frac{x^2}{a^2} + \frac{y^2}{b^2} = 1 \qquad (1)$$

The Spheroid and Projections

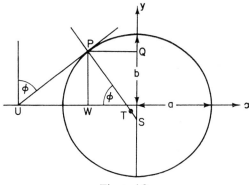

Figure 4.3

The point T is the centre of curvature of the meridian ellipse, and S is the centre of curvature of the spheroid at right angles to the plane of the meridian ellipse. The radius of curvature of a plane curve is given in terms of the first and second differentials by

$$\rho = \left\{1 + \left(\frac{dy}{dx}\right)^2\right\}^{\frac{3}{2}} \div \frac{d^2y}{dx^2} \qquad (9)$$

Also

$$\frac{dy}{dx} = -\cot\phi \qquad (10)$$

since PU is a tangent to the ellipse at P. Differentiating (1) with respect to x

whence from (10) $\qquad \dfrac{dy}{dx} = -\dfrac{b^2 x}{a^2 y}$

$$b^2 x \sin\phi = a^2 y \cos\phi$$

From (1) $\qquad b^2 x^2 + a^2 y^2 = a^2 b^2$

eliminating y gives

$$x^2 = \frac{a^4 \cos^2\phi}{a^2 \cos^2\phi + b^2 \sin^2\phi} \qquad (11)$$

And eliminating x gives

$$y^2 = \frac{b^4 \sin^2\phi}{a^2 \cos^2\phi + b^2 \sin^2\phi} \qquad (12)$$

Now

$$1 + \left(\frac{dy}{dx}\right)^2 = \frac{a^4 y^2 + b^4 x^2}{a^4 y^2}$$

and

$$\frac{d^2y}{dx^2} = -\frac{a^2 b^2 y^2 + b^4 x^2}{a^4 y^3}$$

whence substituting in (9) after some re-arrangement and remembering (2)

$$\rho = \frac{a(1 - e^2)}{(1 - e^2 \sin^2\phi)^{\frac{3}{2}}}$$

Again PS $= \nu = x \sec \phi$ whence substituting for x from (11)

$$\nu = \frac{a}{(1 - e^2 \sin^2 \phi)^{\frac{1}{2}}}$$

When $\phi = 90°$, $\rho = \nu$; when $\phi = 0°$, $\nu = a$ and $\rho = a(1 - e^2)$ and in general ν is always greater than or equal to ρ.

4.1.6.2. Euler's Theorem

Consider a point P on the surface of the spheroid touched by a tangent plane AA' (see figures 4.4(a), (b), and (c)). If BB' is a plane parallel to AA' at a small distance z below it, it will cut the spheroid whose surface will describe an ellipse

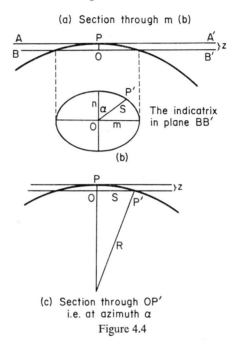

Figure 4.4

in plane BB'. This ellipse is called the indicatrix. Now consider the normal section through P at azimuth α and let P' be a point on the indicatrix at this azimuth a distance s from P. Let the semi-major and semi-minor axes of the indicatrix be m and n respectively. Then from figure 4.4(c) $z \simeq s^2/2R$ and similarly when $s = m$ and $s = n$ $z \simeq m^2/2\nu$ and $z \simeq n^2/2\rho$. And the equation of the indicatrix is

$$\frac{s^2 \sin^2 \alpha}{m^2} + \frac{s^2 \cos^2 \alpha}{n^2} = 1$$

therefore

$$\frac{R \sin^2 \alpha}{\nu} + \frac{R \cos^2 \alpha}{\rho} = 1$$

$$\frac{\cos^2 \alpha}{\rho} + \frac{\sin^2 \alpha}{\nu} = \frac{1}{R}$$

which is Euler's theorem.

4.1.6.3. The Mean Radius of Curvature

The mean value of R is given by

$$\frac{1}{2\pi}\int_0^{2\pi} R \, d\alpha$$

and from the indicatrix this is

$$\frac{1}{2\pi}\int_0^{2\pi} \frac{s^2}{n^2} \rho \, d\alpha$$

But

$$\frac{1}{2}\int_0^{2\pi} s^2 \, d\alpha$$

is the area of the ellipse which is πmn whence the mean radius $= m\rho/n = \rho(\nu/\rho)^{\frac{1}{2}}$, i.e. the mean radius of curvature is $(\rho\nu)^{\frac{1}{2}}$.

4.1.7. DISTANCES ALONG A MERIDIAN

The length of a small arc of the meridian in the vicinity of the point P whose latitude is ϕ is given by

$$ds = \rho \, d\phi \tag{13}$$

where ds is the arc length subtended by $d\phi$. This expression (13) is often used in practice for approximate computations when tables of meridional distances are not available. For longer arcs, e.g. between latitudes ϕ_1 and ϕ_2 the length of arc is given by

$$s = \int_{\phi_1}^{\phi_2} \rho \, d\phi$$
$$= a(1-e^2)\int_{\phi_1}^{\phi_2} \frac{d\phi}{(1-e^2\sin^2\phi)^{\frac{3}{2}}} \tag{14}$$

To evaluate this elliptical integral (14) we expand the elliptical part $1/(1-e^2\sin^2\phi)^{\frac{3}{2}}$ as a power series and integrate term by term to the accuracy required, obtaining a result of the form

$$s = a(1-e^2)[A(\phi_2 - \phi_1) - \tfrac{1}{2}B(\sin 2\phi_2 - \sin 2\phi_1) + \tfrac{1}{4}C(\sin 4\phi_2 - \sin 4\phi_1)\cdots$$

where A, B, and C are constants for the particular spheroid, being power series of the eccentricity of the meridian ellipse. For the Clarke 1866 spheroid these coefficients have the following values $A = 1\cdot005\,1093$; $B = 0\cdot005\,1202$; $C = 0\cdot000\,0108$.

4.1.8. LENGTH ALONG A PARALLEL

The length of a parallel is simply found, because the parallel is a circular arc of radius $\nu \cos \phi$. Hence for a difference in longitude of $\Delta\lambda''$ the corresponding arc length Δs is given by $\Delta s = \nu \cos \phi \, \Delta\lambda'' \sin 1''$.

4.1.9. GEODETIC TABLES

Any survey department which carries out computations in spheroidal terms has tables available for the factors ρ, ν and of meridional arcs. Also of assistance are tabulated values of $1/\rho \sin 1''$, $1/\nu \sin 1''$ and $1/2\rho\nu \sin 1''$ the significance

4.1.10. LEGENDRE'S THEOREM

To compute the geographical coordinates of the points of a triangulation network the lengths of the sides must be known. Most of these lengths will be derived via the adjusted angles from a few measured sides. The sides of a large triangle may be computed by spherical trigonometry, but with normal terrestrial triangles whose sides are of the order of 25 miles, the small angles subtended at the centre of curvature are troublesome in computation. A simpler solution is obtained using *Legendre's theorem*. One third of the spherical excess is deducted from each of the adjusted spherical angles of the triangle to give the *Legendre angles*. It is then sufficiently accurate to compute the side lengths using these Legendre angles and plane trigonometry. It will be noted that the angles are first adjusted to 180° plus the spherical excess ϵ, and that the various adjustments will not as a rule be equal. The application of $-\frac{1}{3}\epsilon$ to these angles is merely a computational dodge to reduce the work.

Example: We wish to compute the sides BC and AC of triangle ABC given the following data:

Adjusted spherical angles:
A 30° 48′ 03″96
B 61° 05′ 49″07
C 88° 06′ 07″74
Sum 180° 00′ 00″77

Since the angles are adjusted, the spherical excess ϵ is 0″77. However, to illustrate the method, the reader may recompute ϵ from the formula

$$\epsilon = \frac{c^2 \sin A \sin B}{2\rho\nu \sin 1'' \sin C} = 0\cdot 77$$

where $1/\rho\nu \sin 1'' = 0\cdot 2550 \times 10^{-8}$ (see §2.8.10). Subtracting 0″26 from A and B, and 0″25 from C, we obtain the Legendre angles

A′ 30° 48′ 03″70
B′ 61° 05′ 48″81
C′ 88° 06′ 07″49
Sum 180° 00′ 00″00

Whence the sides BC and AC are respectively 13 327·628 m, and 22 785·523 m, given that AB is 26 013·282 m (see §2.11.7 for computation).

4.1.10.1. Proof of Legendre's Theorem

Consider the spherical triangle ABC whose sides in linear units are a, b, and c. With ample accuracy the surface of the spheroid may be taken to approximate to a sphere of radius $r = (\rho\nu)^{\frac{1}{2}}$ where ρ and ν are the principal radii of curvature of the spheroid at the mid latitude of triangle ABC. The angles subtended by the sides are then a/r, b/r, and c/r.

Now
$$\cos\frac{a}{r} = \cos\frac{b}{r}\cos\frac{c}{r} + \sin\frac{b}{r}\sin\frac{c}{r}\cos A$$

i.e. $\cos A = \dfrac{\cos a/r - \cos b/r \cos c/r}{\sin b/r \sin c/r}$

Expanding and neglecting powers above the fourth

$$\cos A = \frac{[1 - (a^2/2r^2) + (a^4/24r^4)\cdots] - [1 - (b^2/2r^2) + (b^4/24r^4)][1 - (c^2/2r^2) + c^4/24r^4)\cdots]}{(b/r - b^3/6r^3 + \cdots)(c/r - c^3/6r^3)}$$

$$= \frac{N}{D}$$

$$N = 1 - \frac{a^2}{2r^2} + \frac{a^4}{24r^4} - \left\{1 - \frac{c^2}{2r^2} + \frac{c^4}{24r^4} - \frac{b^2}{2r^2} + \frac{b^2c^2}{4r^4} + \frac{b^4}{24r^4}\right\}$$

$$= \frac{1}{2r^2}(b^2 + c^2 - a^2) + \frac{1}{24r^4}(a^4 - b^4 - c^4) - \frac{b^2c^2}{4r^4}$$

$$D = \frac{bc}{r^2} - \frac{bc^3}{6r^4} - \frac{b^3c}{6r^4} = \frac{bc}{r^2}\left(1 - \frac{c^2 + b^2}{6r^2}\right)$$

$$\therefore \frac{1}{D} = \frac{r^2}{bc}\left(1 - \frac{b^2 + c^2}{6r^2}\right)^{-1} \simeq \frac{r^2}{bc}\left(1 + \frac{b^2 + c^2}{6r^2}\right), \left\{\frac{1}{6r^2}(b^2 + c^2) \text{ is small}\right\}$$

$$\therefore \cos A = \frac{N}{D} = \frac{r^2}{bc}\left(1 + \frac{b^2 + c^2}{6r^2}\right)\left(\frac{b^2 + c^2 - a^2}{2r^2} + \frac{a^4 - b^4 - c^4}{24r^4} - \frac{b^2c^2}{4r^4}\right)$$

which reduces to the following on rejecting powers greater than the fourth:

$$\cos A = \frac{b^2 + c^2 - a^2}{2bc} - \frac{1}{6}\frac{2a^2b^2 + 2a^2c^2 + 2b^2c^2 - a^4 - b^4 - c^4}{4r^2bc}$$

Now consider the plane triangle whose side lengths in linear units are also a, b, and c, and whose angles are A', B', and C'. Then

$$\cos^2 A' = \left\{\frac{b^2 + c^2 - a^2}{2bc}\right\}^2$$

$$\sin^2 A' = 1 - \cos^2 A' = \frac{4b^2c^2 - \{(b^2 + c^2)^2 - 2(b^2 + c^2)a^2 + a^4\}}{4b^2c^2}$$

$$= \frac{2b^2c^2 + 2a^2b^2 + 2a^2c^2 - a^4 - b^4 - c^4}{4b^2c^2}$$

$$\therefore \cos A = \cos A' - \frac{1}{6}\frac{bc}{r^2}\sin^2 A'$$

Now let $A = A' + x$, where x is a small angle, then

$$\cos A = \cos(A' + x) = \cos A' - \sin A' \, x \cdots \quad \text{(Taylor's Theorem)}$$

i.e.
$$\cos A' - \sin A' x = \cos A' - \frac{1}{6}\frac{bc}{r^2}\sin^2 A'$$

$$\therefore \quad x = \frac{1}{6}\frac{bc}{r^2}\sin^2 A' = \tfrac{1}{3}\epsilon \quad \text{since} \quad \epsilon = \left(\frac{1}{2r^2}\right)bc \sin A'$$

Hence $\quad A = A' + \tfrac{1}{3}\epsilon \quad \text{or} \quad A' = A - \tfrac{1}{3}\epsilon$

This result may also be proved in an identical manner for angles B and C, i.e.
$$B' = B - \tfrac{1}{3}\epsilon \quad \text{and} \quad C' = C - \tfrac{1}{3}\epsilon.$$

Thus the plane triangle whose sides are equal to the spherical triangle has Legendre angles A', B', and C'. Sometimes in very refined geodetic work, a further spheroidal term has to be considered, but this falls outside the scope of this book.

4.1.11. COMPUTATION OF GEODETIC POSITIONS

The main factors deciding the choice of formulae to be used for the computation of surveys on the surface of a reference spheroid are:

(i) The latitude of the survey.
(ii) The precision required.
(iii) The lengths of the lines involved.
(iv) The computational aids available to the computer.

In this book it is impossible to consider even briefly some of the many different formulae which may be used for spheroidal computations but a useful resumé is given in Bomford's *Geodesy*† in which the author considers the most important formulae, and gives useful references for further study.

In this book, consideration is given only to a simple version of the mid-latitude formulae, the application of which should be restricted to lines under 25 miles long and latitudes of less than 80°. These formulae however illustrate most of the basic principles involved in spheroidal computation.

Consider figure 4.5, showing the following:

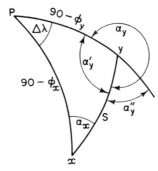

Figure 4.5

† G. Bomford, *Geodesy* (Oxford: Clarendon Press, 1952).

The Spheroid and Projections

Terrestrial pole is at P; x and y are points on the surface of a sphere approximating to the spheroid, whose respective latitudes and longitudes are ϕ_x, ϕ_y, λ_x, and λ_y.

α_x is the azimuth of y from x

α_y is the azimuth of x from y; α_y'' is $\alpha_y - 180$.

Because of the convergence of the meridians α_x will not equal $\alpha_y - 180°$ as a rule. The forward and back azimuths will differ by 180° only if x and y are on the same meridian, or if both are on the equator.

The length of xy is s, which is not greater than 25 miles. If the position of x, the forward azimuth α_x, and the distance s are given, we require to compute the position of y and the reverse azimuth from y.

The convergence of the meridians is zero at the equator, and increases towards the poles until it equals the difference in longitude between the two points considered. At any intermediate latitudes the convergence is given by Dalby's theorem, viz.

$$\tan \tfrac{1}{2}\Delta\alpha = \tan \tfrac{1}{2}\Delta\lambda \sin \tfrac{1}{2}(\phi_x + \phi_y) \sec \tfrac{1}{2}\Delta\phi \qquad (15)$$

where $\Delta\alpha$ is $\alpha_y'' - \alpha_x$. This expression holds for all observable distances on the surface of the spheroid, and is not limited to 25 miles. Equation (15) is proved as follows

$$\tan \tfrac{1}{2}\Delta\alpha = \cos \tfrac{1}{2}(\alpha_y' + \alpha_x) = \tan \tfrac{1}{2}\Delta\lambda \frac{\cos \tfrac{1}{2}(90 - \phi_y + 90 - \phi_x)}{\cos \tfrac{1}{2}(90 - \phi_y - 90 + \phi_x)}$$

(see §2.8.9 equation (17)).

$$\tan \tfrac{1}{2}\Delta\alpha = \tan \tfrac{1}{2}\Delta\lambda \sin \tfrac{1}{2}(\phi_x + \phi_y) \sec \tfrac{1}{2}\Delta\phi$$

4.1.11.1. *Approximate version of* (15)

If $\Delta\phi$ is 25 miles, i.e. xy is a North to South line, then $\Delta\phi$ is 22' of arc, and $\sec \tfrac{1}{2}\Delta\phi$ differs from unity by only 5 in the fifth place. Therefore we may use the expression

$$\tan \tfrac{1}{2}\Delta\alpha = \tan \tfrac{1}{2}\Delta\lambda \sin \tfrac{1}{2}(\phi_x + \phi_y)$$

Again, for small values of $\Delta\alpha$ and $\Delta\lambda$ we may write

$$\Delta\alpha = \Delta\lambda \sin \tfrac{1}{2}(\phi_x + \phi_y) \qquad (16)$$

with ample accuracy for most work. Equation (16) is often written

$$\Delta\alpha = \Delta\lambda \sin \phi_m \qquad (17)$$

where ϕ_m is the mean latitude.

4.1.12. COMPUTATION OF POSITION BY THE MID-LATITUDE FORMULAE

These formulae usually accredited to Gauss were first published in English by the great computor James O'Farrell of the Ordnance Survey, in *Philosophical*

Magazine of 1861. The formulae are:

$$\Delta\alpha'' = \Delta\lambda'' \sin\phi_m \tag{17}$$

$$\Delta\phi'' = s \cos\alpha_m/\rho_m \sin 1'' \tag{18}$$

$$\Delta\lambda'' = s \sin\alpha_m/\nu_m \cos\phi_m \sin 1'' \tag{19}$$

Where α_m is the mean azimuth omitting the 180°, $\Delta\phi''$ and $\Delta\lambda''$ are differences in latitude and longitude as before, and ρ_m and ν_m are the values of the principal radii of curvature of the spheroid taken out for the mean latitude. Formula (17) is the abridged version of Dalby's theorem. Formulae (18) and (19) are proved as follows: Let the angular value of the length s be S, i.e. $S = s/R$. Then in figure 4.5

$$\frac{\sin(90-\phi_x)}{\sin\alpha_y'} = \frac{\sin S}{\sin\Delta\lambda} \tag{20}$$

$$\frac{\sin(90-\phi_y)}{\sin\alpha_x} = \frac{\sin S}{\sin\Delta\lambda} \tag{21}$$

whence
$$\cos\phi_y - \cos\phi_x = \frac{\sin S (\sin\alpha_x - \sin\alpha_y')}{\sin\Delta\lambda}$$

$$\therefore 2\sin\tfrac{1}{2}(\phi_x+\phi_y)\sin\tfrac{1}{2}(\phi_y-\phi_x) = \frac{\sin S\, 2\cos\tfrac{1}{2}(\alpha_x+\alpha_y'')\sin\tfrac{1}{2}(\alpha_x-\alpha_y'')}{\sin\Delta\lambda}$$

$$\sin\tfrac{1}{2}\Delta\phi = \tfrac{1}{2}\sin S \cos\alpha_m \cos\tfrac{1}{2}\Delta\alpha \sec^2\tfrac{1}{2}\Delta\lambda$$

But $\cos\tfrac{1}{2}\Delta\alpha \simeq \sec^2\tfrac{1}{2}\Delta\lambda \simeq 1 \quad \therefore \quad \Delta\phi \simeq S\cos\alpha_m$

Since $\Delta\phi$ is in radians and S is an angle, converting to seconds of arc and linear s gives

$$\Delta\phi'' = \frac{s\cos\alpha_m}{\rho_m \sin 1''}$$

which is equation (18). In a similar manner considering from (20) and (21)

$$\cos\phi_x + \cos\phi_y = \frac{\sin S(\sin\alpha_x + \sin\alpha_y')}{\sin\Delta\lambda}$$

we obtain equation (19) after elimination and rearrangement.

Since both equations (18) and (19) require a knowledge of the mid-latitude ϕ_m both for itself and to obtain the values of ρ_m and ν_m from geodetic tables, successive approximation is required. The following slight adaptation of the above formulae gives simple computational versions. Since α_m the mean azimuth is given by $\alpha_x + \tfrac{1}{2}\Delta\alpha$ (18) may be written

$$\Delta\phi'' = (s/\rho_m \sin 1'') \cos(\alpha_x + \tfrac{1}{2}\Delta\alpha)$$

Expanding by Taylor's theorem

$$\Delta\phi'' = \frac{s}{\rho_m \sin 1''}(\cos\alpha_x - \sin\alpha_x \tfrac{1}{2}\Delta\alpha \cdots)$$
$$= \frac{s \cos\alpha_x}{\rho_m \sin 1''}(1 - \tan\alpha_x \tfrac{1}{2}\Delta\alpha'' \sin 1'') \tag{22}$$

Similarly the expanded version of (19) is

$$\Delta\lambda'' = \left(\frac{s \sin\alpha_x}{\nu_m \cos\phi_m \sin 1''}\right)(1 + \cot\alpha_x \tfrac{1}{2}\Delta\alpha'' \sin 1'') \tag{23}$$

and from (17)

$$d\alpha'' = d\lambda'' \sin\phi_m \tag{24}$$

where $d\alpha''$ is the correction to $\Delta\alpha''$ resulting from the correction $d\lambda''$ to $\Delta\lambda''$. In computing with formulae (22), (23), and (24) first obtain an approximate value of ϕ_m the mid-latitude from a diagram, or if none is available, from a slide rule computation.

4.1.13. EXAMPLE

Table 4.2 gives the data to be used in this and other examples in this section. We shall compute the latitude and longitude of B, and the reverse azimuth from B to A. The computation is set out in table 4.3. Part (2) gives the provisional computation of ϕ_m. The provisional value of ϕ_m was $-15°\,46'$ which was used to give $1/\rho \sin 1''$ and $1/\nu \sin 1''$ from the table of geodetic factors, table 4.4. The corrective terms are worked with four figure tables only.

Table 4.2. Data for Geodetic Triangle ABC.

ADJUSTED ANGLES	SIDE LENGTHS (metres)
A 30° 48' 03".96	a = BC = 13 327·628
B 61° 05 49·07	b = AC = 22 785·523
C 88° 06 07·74	c = AB = 26 013·282
Spherical excess ϵ = 0".77	

Longitude λ	Latitude ϕ
A + 35° 18' 16".755 9	− 15° 52' 29".909 6
B + 35 23 37·454 0	− 15 39 22·596 4
C + 35 28 22·707 0	− 15 44 56·756 7

Azimuths	
AB 21° 32' 28".16	BA 201° 31' 01".02
AC 52 20 32·11	CA 232 17 47·00
BC 140 25 11·95	CB 320 23 54·74

Table 4.3. Computation of Position and Reverse Azimuth by Mid-Latitude Formulae.

		LINE AB	Notes
(1) STARTING DATA	ϕ_A	$-15°\ 52'\ 29''.909\ 6$	$\Delta\phi'' = s\cos\alpha/\rho_m \sin 1''$
	λ_A	$+35°\ 18'\ 16''.755\ 9$	$\Delta\lambda' = s\sin\alpha/\nu_m \cos\phi_m \sin 1''$
	α_{AB}	$21°\ 32'\ 28''.16$	$\Delta\alpha' = \Delta\lambda' \sin\phi_m$
	S_{AB}	$26\ 013\cdot 282\ m$	
(2) $\cos\alpha_{AB}$		$0\cdot 930\ 15406$	Computation of approx. mid-latitude ϕ_m
$1/\rho_1 \sin 1''$		$325\cdot 354 \times 10^{-4}$	
$\Delta\phi'$		$+787\cdot 2$	
$\frac{1}{2}\Delta\phi'$		$+393''.6$	
		$+6'\ 33''.6$	
ϕ_A		$-15\ 52\ 29\cdot 9$	
ϕ_m		$-15\ 45\ 56\cdot 3$	
(3) $1/\rho_m \sin 1''$		$325\cdot 357\ 79 \times 10^{-4}$	$\Delta\phi = \Delta\phi'' + d\phi$
$\Delta\phi''$		$+787\cdot 247\ 4$	$\tan\alpha + 0\cdot 394\ 7$
$d\phi$		$+0\cdot 065\ 7$	$d\phi = \Delta\phi'' \tan\alpha \frac{1}{2}\Delta\alpha'' \sin 1''$
$\Delta\phi$		$+787\cdot 313\ 1$	use $\Delta\alpha'$ from part (5)
		$13'\ 07''.313\ 1$	
ϕ_A		$-15\ 52\ 29\cdot 909\ 6$	
ϕ_B		$-15\ 39\ 22\cdot 596\ 5$	
(4) $\sin\alpha_{AB}$		$0\cdot 367\ 169\ 45$	$\Delta\lambda = \Delta\lambda' + d\lambda$
$1/\nu_m \sin 1''$		$323\cdot 306\ 60 \times 10^{-4}$	$\cot\alpha + 2\cdot 533$
$\cos\phi_m$		$0\cdot 962\ 381$	$d\lambda = \Delta\lambda' \cot\alpha \frac{1}{2}\Delta\alpha'' \sin 1''$
$\Delta\lambda'$		$+320\cdot 870\ 1$	use $\Delta\alpha'$ from part (5)
$d\lambda$		$-0\cdot 171\ 8$	
$\Delta\lambda$		$+320\cdot 698\ 3$	
		$+5'\ 20''.698\ 3$	
λ_A		$+35\ 18\ 16\cdot 755\ 9$	
λ_B		$+35\ 23\ 37\cdot 454\ 2$	
(5) $\sin\phi_m$		$-0\cdot 271\ 704$	$\Delta\alpha = \Delta\alpha' + d\alpha$
$\Delta\alpha'$		$-87\cdot 181\ 7$	$d\alpha = d\lambda \sin\phi_m$
$d\alpha$		$+0\cdot 046\ 7$	
$\Delta\alpha$		$-87\cdot 135\ 0$	
		$-01'\ 27''.14$	
α_{AB}		$21°\ 32'\ 28''.16$	
α_{BA}		$201°\ 31'\ 01''.02$	

The Spheroid and Projections

Table 4.4. Geodetic Factors [Clarke 1880 Spheroid]

Latitude	$10^4/\nu \sin 1''$ (m)	$10^4/\rho \sin 1''$ (m)
15° 42'	323·307 253	325·359 775
43	086	271
44	·306 919	·358 767
45	752	263
46	585	·357 757
47	417	252
48	249	·356 746
49	082	239
50	·305 914	·355 732
51	746	224
52	577	·354 716
15° 53	323·305 409	325·354 208

Latitude	$10^8/2\rho\nu \sin 1''$
14°	0·255 042
15°	0·255 012
16°	0·254 981

Reproduced by permission of the Department of the Army, United States of America.

The sign of $\Delta\alpha$ is best seen by inspection of a diagram such as figure 4.6 in which all lines are drawn straight. This diagram gives a useful method of remembering the mid-latitude formulae, since the linear value of 'ΔN' is $\rho_m \Delta\phi'' \sin 1''$ and of ΔE is $\nu_m \cos \phi_m \sin 1''$.

It is left to the reader to calculate the coordinates of C both from A and B as a check. The results of these computations are given in Table 4.2.

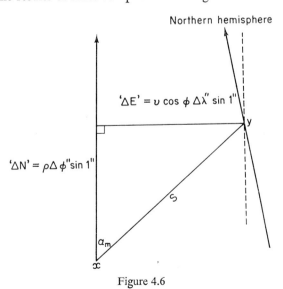

Figure 4.6

4.1.14. REVERSE COMPUTATION

As a check on the computation of the position of y and $\Delta\alpha$ the computation should be carried out in reverse order commencing with the coordinates of x and y. Since the mid-latitude is obtainable at once, no successive approximation is necessary. From (17), $\Delta\alpha$ is computed. Then from (19) divided by (18) we obtain $\tan \alpha_m$ and thence α_x, and α_y. Finally the length s is computed from both (18) and (19) to check. The values obtained should agree exactly with the starting data. In practice this check will always be carried out, particularly if a traverse is being computed on the spheroid.

4.1.15. ELECTRONIC COMPUTATION

It is customary practice today with large survey organizations to compute all spheroidal work on a digital electronic computer using more rigorous formulae such as Clarke's and for many small firms to contract out this work to specialists. Although these methods are much more rigorous and complicated, and produce more accurate results for long lines, the basic principles involved are the same as outlined here.

4.1.16. CHECK COMPUTATION ON A PROJECTION

As a completely independent check on spheroidal computation the survey may be computed on a map projection system and the results compared for each point. In §4.2.6 the triangle of table 4.2 is computed on the transverse Mercator projection and results compared for identity. If very long lines are involved, such computation on a projection becomes excessively tedious. Hence if a surveyor does not have access to an electronic computer it is usual to compute long lines on the spheroid only, and short lines on the projection only since plane trigonometry may be used. Organizations with an electronic computer are tending to compute all work on the spheroid and convert the results to grid coordinates for mapping purposes.

4.1.17. LOCATION OF GEOGRAPHICAL BOUNDARIES

Sometimes the surveyor has to set out or relocate a geographical boundary defined by a meridian, a great circle between two points, or a parallel of latitude. The first two cases are comparatively simple because the theodolite can be made to point along a great circle, for all practical purposes. The parallel of latitude however has to be set out from a great circle by offsets, as with a railway curve.

In all three operations one has first to locate a point on the required line. This may be carried out in many possible ways using any recognized survey method. The basic principle is to locate a point as close to the required position as possible by reconnaissance surveys or aerial photography, then establish its exact position and finally to compute an offset to the required boundary. Suppose that x is the provisional position whose coordinates can be calculated through the survey. Then assuming that the point y lies on the required meridian or parallel, the azimuth and distance of y from x are computed as in §4.1.13, and the required point is located on the ground, making due allowance for height above sea level, slope, etc.

Points are then set out along the required route, with checks applied from time to time by tying-in to control surveys or making astronomical determinations. Such surveys can often involve the greatest of ingenuity on the part of the surveyor to overcome obstacles, though the advent of electromagnetic distance measurement has made the work easier.

4.1.17.1. *Setting out a Parallel*

This operation may be carried out by offsets from a tangent to the parallel, or by offsets from a 'chord' of the parallel. Refer to figure 4.7. AC is the tangent to the parallel AEB at A, and AFB is the chord of this parallel. Both

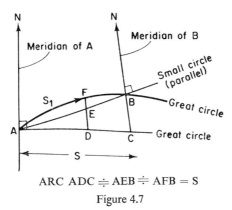

ARC ADC ≑ AEB ≑ AFB = S

Figure 4.7

arcs AFB and AC are great circles, and AEB is a small circle. The length CB is the offset from the tangent at a distance ADC equal to s. Since BC is small it is sufficiently accurate to take the lengths AFB, AEB, and ADC to be equal to s. The diagram is grossly exaggerated for clarity. Let $\Delta\alpha$ be the difference between the forward and back azimuth from A to C, neglecting 180°. Then angle ACB is $90 - \Delta\alpha$ and the offset BC is given by

$$BC = s^2 \tan \phi_m / 2\nu \qquad (25)$$

For
$$BC = \rho(\phi_B - \phi_C)'' \sin 1'' = s \cos(\alpha + \tfrac{1}{2}\Delta\alpha) = s \cos \alpha - \tfrac{1}{2}\Delta\alpha \sin \alpha s$$
But $\alpha = 90°$
$$\therefore \quad BC = -\tfrac{1}{2} s \Delta\alpha \sin \alpha = -\tfrac{1}{2} s \Delta\lambda \sin \phi_m \sin \alpha$$
$$= -\frac{\tfrac{1}{2}(s)(s \sin \alpha) \sin \phi_m \sin \alpha}{\nu \cos \phi_m} = -\frac{\tfrac{1}{2} s^2 \tan \phi_m}{\nu}$$

($\sin \alpha \simeq 1$)

To locate B directly along the chord from A sight along AFB at an azimuth of $90 - \tfrac{1}{2}\Delta\alpha$, and proceed a distance s. Intermediate points along the parallel are often required say at intervals of one mile. The offset at an intermediate point F, i.e. FE is given by

$$FE = \tfrac{1}{2} s_1 \tan \phi_m (s - s_1)/\nu \qquad (26)$$

where s_1 is the distance AF $\simeq$ AD. For

$$FE = FD - ED = \tfrac{1}{2}s_1 \Delta\alpha - \frac{\tfrac{1}{2}s_1^2 \tan\phi_m}{\nu}$$

$$= \frac{\tfrac{1}{2}s_1 BC}{s} - \frac{\tfrac{1}{2}s_1^2 \tan\phi_m}{\nu} = \frac{\tfrac{1}{2}s_1 \tan\phi_m(s - s_1)}{\nu}$$

Example: Suppose the distance s is 50 miles, $\nu = 3\,960$ and $\phi = 45$, the offset from the tangent BC is

$$\frac{2\,500 \times 5\,280}{7\,920} \simeq 1\,667 \text{ feet}$$

and

$$\Delta\alpha = \frac{s \tan\phi}{\nu} = \frac{50 \times 3\,438}{3\,960} = 43\cdot4$$

And if F is at the mid-point of AB, i.e. s_1 has its maximum value of $\tfrac{1}{2}s$ the offset $FE = \tfrac{1}{2}(25)(25)/7\,920$ miles $= 417$ ft. When point B is reached, the next stage of the setting out is repeated as from A, and so on. In thickly wooded country the method of offsetting from a chord is preferable since less cutting is required for the shorter offsets by this method.

4.2. SURVEY PROJECTIONS

Although the coordinated positions of points may be computed on the surface of a reference spheroid, the final end-product of surveying, viz the map, has to be drawn on a plane surface. Hence some system of map projection is required. Again, since computation by hand machines on the curved surface of a spheroid or a sphere is more tedious than on a plane surface it is preferable to compute on the latter if possible. As will be seen below, most minor survey work is computed on a map projection instead of on the spheroid. The exception to this is a large national survey organization which may compute on the spheroid with the aid of an electronic digital computer and convert to map projection terms for plotting purposes.

4.2.1.1. *A Map Projection is any Orderly System of Representing Points of the Sphere or Spheroid on a Plane.*

We shall confine our theoretical deliberations to the sphere since the main principles can be more easily understood by this approach. With the exception of the stereographic projection, the projections used by surveyors cannot easily be drawn by direct geometrical perspective projection, but are mathematical projections. The two most commonly used projections are the *transverse Mercator* and *Lambert's conical orthomorphic*, both of which are *conformal* or *orthomorphic* projections, i.e. on these projections the scale factor at a point is independent of azimuth. Attention is further confined in this book to these two surveyor's projections, with brief mention of the Cassini projection as an interim stage in the development of the theory. The formulae derived are restricted to closed spherical trigonometrical formulae, except in a very few

The Spheroid and Projections

cases, notably the *'arc to chord'* or $t - T$ formula, and the *scale factor* formula for the transverse Mercator projection. These latter form the basis of practical computation on the projection, whilst the spherical trigonometrical formulae are useful for cartographic purposes. Since the accurate conversion of geographical or spheroidal coordinates to projection coordinates is carried out with the aid of tables, this aspect has been given consideration.

4.2.1.2. Basic Concepts of Projection

The following basic concepts are essential for a grasp of the subject (see figure 4.8).

A *graticule* is the network of lines formed by the meridians and parallels either on the sphere or on the map, the latter being the projected positions of the former.

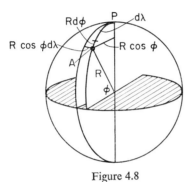

Figure 4.8

A *grid* is a system of squares drawn on a map, from which the map graticule and points of detail are plotted. It is unusual, though not impossible, to plot directly in terms of a graticule.

Let A be a point on the sphere whose pole is P The map projection system adopted relates the latitude and longitude (ϕ, λ) of A, to its grid coordinates (E, N) by some formulae, i.e. $N = f(\phi, \lambda)$, $E = F(\phi, \lambda)$, and $\phi = f'(E, N)$, $\lambda = F'(E, N)$.

The *scale factor* K is the ratio of the straight line on the map joining two points, to the corresponding distance on the sphere or spheroid. The scale factor is often considered under the headings: *nominal scale*, and *differential scale* (or *scale error*). The nominal scale of a map is its representative fraction, say 1:50 000, starting *in a general way* the magnitude of the scale factor. The scale factor over the whole map is not constant but varies from this nominal scale by small differential amounts, often called scale errors. Surveyors usually work implicitly with a nominal scale of unity (at life size); hence the differential scale or scale error is the amount by which the scale factor differs from 1. For example if the scale factor at a point on a projection is 0·999 6 the corresponding scale error is −0·000 4. The subsequent reduction of coordinates from life size to some convenient map scale merely alters the nominal scale of the survey. The use of the word 'error' is perhaps unfortunate, since it connotes the idea of a mistake, whereas the introduction of a scale factor is deliberate and its effects are predictable.

The concept of *map scale* is explained as follows: Let a small linear element of the meridian of A in the vicinity of A be $R\,d\phi$, and a small element of the parallel of A in the vicinity of A be $R\cos\phi\,d\lambda$. Then if dN and dE are the elements on the map corresponding to $R\,d\phi$ and $R\cos\phi\,d\lambda$, the *scale* in a North–South direction is $K_1 = dN/R\,d\phi$ and in an East–West direction is $K_2 = dE/R\cos\phi\,d\lambda$.

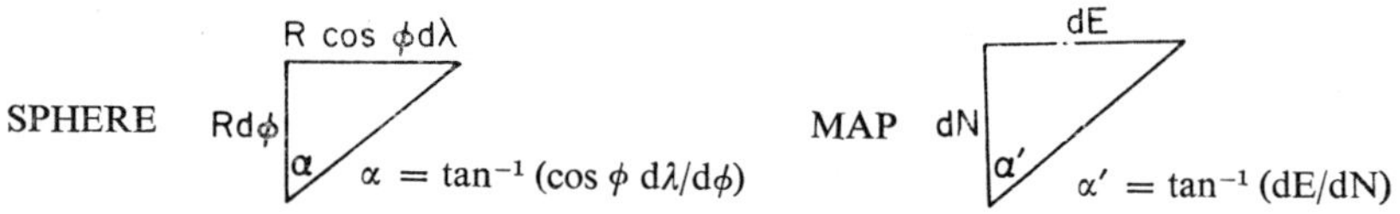

SPHERE $R\,d\phi$ $R\cos\phi\,d\lambda$ $\alpha = \tan^{-1}(\cos\phi\,d\lambda/d\phi)$

MAP dN dE $\alpha' = \tan^{-1}(dE/dN)$

Figure 4.9

The projection is *conformal* or *orthomorphic* if $K_1 = K_2$. A corollary to this is that *for a small element*, angles on the surface of the sphere are correctly represented on the map. For, in figure 4.9

$$\alpha = \tan^{-1}\left(\frac{R\cos\phi\,d\lambda}{R\,d\phi}\right)$$

$$= \tan^{-1}\left(\frac{dE}{K_2}\right)\left(\frac{K_1}{dN}\right)$$

$$= \tan^{-1}\left(\frac{dE}{dN}\right)$$

(since $K_1 = K_2$).

It is important to note that these properties of a conformal projection hold only for *small differential amounts*. Since the surface of the sphere cannot be correctly represented on a plane over large areas, the scale factor *for long lines* will not be independent of azimuth, neither will angles be correctly preserved in conversion from the sphere to the map. Just what is meant by 'long lines' depends on the ultimate accuracy desired. The angular distortion over a line of 5 miles is normally less than 1 second of arc on a conformal projection, hence the projection effect could be neglected for most work over this distance, and the survey computed in simple rectangular coordinates. Before computing a survey over a wide area it is usual to investigate the angular and scale distortion which will be accumulated as a result of neglecting these factors, and reach a decision on the method of computation on the grounds of economy, paying due regard to possible future requirements. In non-conformal projections, angular and differential scale distortions seriously restrict the lengths of lines that may be computed without applying corrections.

4.2.2. THE CASSINI PROJECTION

Figure 4.10(a) shows the terrestrial sphere. P is the pole; OGHP is the central meridian of the projection system; O is the origin of the projection; A and B are two points on the surface of the sphere; G and H are the feet of the perpendiculars from A and B respectively to the central meridian; arc AC is

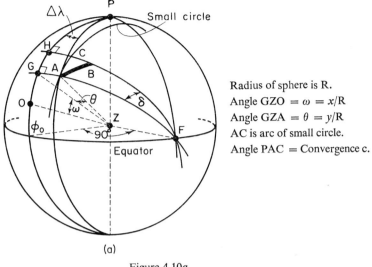

Radius of sphere is R.
Angle GZO = ω = x/R
Angle GZA = θ = y/R
AC is arc of small circle.
Angle PAC = Convergence c.

Figure 4.10a

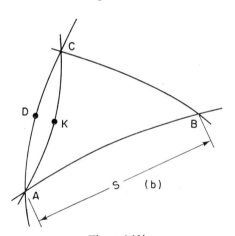

Figure 4.10b

part of a small circle passing through A lying in a plane which is parallel to that of the central meridian; F is the pole of the central meridian; the radius of the sphere is R; latitudes and longitudes of the various points are denoted by ϕ_A, λ_A, with the suffix referred to the point.

Let the angles subtended at Z, the centre of the sphere, by OG and GA be ω and θ respectively. Let the arc length of OG be x_A and that of GA by y_A; to construct a projection draw a straight line O'G'H', mark off O'G' = OG = x_A, and G'A' perpendicular to O'G' at G' equal to GA = y_A, the point A' so obtained is the position of A on the *Cassini* projection whose origin is at O and whose central meridian is OGH (O'G'H') (see figure 4.11). Figure 4.12 shows the spherical triangle formed by A, P, and G. G is a right angle. Since CA is perpendicular to GA and if the angle PAC = c, angle PAG is $90 - c$.

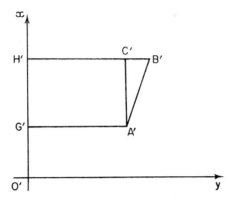

Figure 4.11

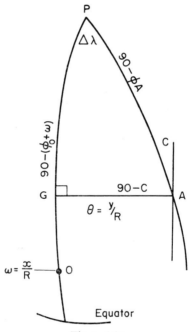

Figure 4.12

The line AC is projected as A'C' on the map (figure 4.11), which defines the direction of grid North. The angle c is therefore the difference between *true* and *grid* North, and is called the '*map convergence*' or simply '*the convergence*' at A.

By Napier's rules for circular parts (see figure 4.12)

$$\sin(y/R) = \sin\theta = \sin\Delta\lambda \cos\phi_A \qquad (1)$$
$$\cot(\phi_0 + x/R) = \cot(\phi_0 + \omega) = \cos\Delta\lambda \cot\phi_A \qquad (2)$$
$$\tan c = \tan\Delta\lambda \sin\phi_A \qquad (3)$$

or
$$c \simeq \Delta\lambda \sin\phi_A \qquad (4)$$

where

$$\Delta \lambda = \lambda_A - \lambda_0.$$

Hence given the value of R for the sphere the coordinates of A' on the map can be calculated from its given geographical coordinates (ϕ_A, λ_A). For the reverse computations the following expressions are used:

$$\sin \phi_A = \cos \theta \sin (\phi_0 + \omega) \tag{5}$$

and

$$\cot \Delta \lambda = \cos (\phi_0 + \omega) \cot \theta \tag{6}$$

4.2.2.1. Cutting Points on a Map Sheet

Although the details of a map will be plotted from the grid, it is convenient to have some parts of the graticule plotted. The boundary of the map sheet may even be a part of the graticule. At most scales it is sufficient to plot three points on each graticule line and draw in the line by fitting a very slightly curved spline to these points. However, when the grid and graticule intersect at a very small angle it is sometimes uncertain where the two cut, and in some cases it is uncertain if a cut lies on a particular map sheet at all, therefore these critical *cutting points* have to be computed.

Case (a)

Cutting point where an East–West line cuts a parallel. In this case it is required to compute the easting y of the cutting point, whose northing x and latitude ϕ are given, from the formula

$$\cos (y/R) = \cos \theta = \sin \phi \operatorname{cosec} (\phi_0 + \omega) \tag{7}$$

Case (b)

Cutting point where a North–South line cuts a meridian. In this case ω is derived from

$$\cos (\phi_0 + \omega) = \cot \Delta \lambda \tan \theta \tag{8}$$

whence x, since longitude and easting are known. Calculations using these formulae are usually sufficiently accurate for plotting purposes provided geodetic tables are used to obtain the values of the radii of curvature and the meridional distance from the origin.

4.2.2.2. Scale and Angular Distortions on the Cassini Projection

Figure 4.11 shows points A and B plotted in positions A' and B' on the map. C' lies on $H'B'$ such that $H'C' = G'A' = y$. Figure 4.13 shows an enlarged

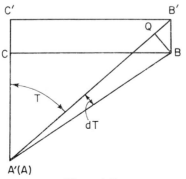

Figure 4.13

view of triangle A'B'C' with the original spherical triangle ABC superimposed upon it. The *grid bearing* of A'B' is denoted by T, and angle B'A'B' by dT. Since B'C' = BC the scale East–West is correct. But the line AC is plotted too long equal to A'C' = H'G' = HG = $R\delta$ where angle HFG = δ, or δ is the angle subtended by GH at the centre of the sphere (figure 4.10).

$$AC = \delta R \cos\theta = GH \cos\theta = A'C' \cos\theta$$

therefore the scale factor in a North–South direction at A is

$$A'C'/AC = \sec\theta = K \quad \text{say}$$

then

$$BB' = CC' = A'C' - AC = A'C'(1 - \cos\theta) \simeq A'C'\tfrac{1}{2}\theta^2 \simeq \tfrac{1}{2}s \cos T\theta^2 \quad (9)$$

4.2.2.3. Angular Distortion

The angular distortion dT is given approximately as follows:

$$dT = BQ/s = BB' \sin T/s = \tfrac{1}{2}s \sin T \cos T\theta^2/s$$
$$= \tfrac{1}{4}\theta^2 \sin 2T = (y^2/4R^2) \sin 2T \quad (10)$$

Hence dT is zero for a line North to South, and a maximum at azimuths of 45, 135, 225, and 315°.

4.2.2.4. Scale Distortion

The line AB becomes A'B' on the map, hence the scale is increased by

$$B'Q = BB' \cos T = \tfrac{1}{2}s \cos^2 T\theta^2$$

Scale error as a fraction of $s = \tfrac{1}{2} \cos^2 T\theta^2$

$$= \tfrac{1}{2} \cos^2 T(y/R)^2 \quad (11)$$

i.e.   when $T = 0°$ or $180°$,   scale error = $\tfrac{1}{2}(y/R)^2$
        and $T = 90°$ or $270°$,   scale error is zero.

Hence the scale factor is $\sec\theta$ in a North–South direction, and unity in an East–West direction.

Because of this excessive angular distortion and differential scale distortion the projection was not well suited to survey computation, though it was widely used in the past on account of its simplicity.

4.2.3. THE TRANSVERSE MERCATOR PROJECTION

The angular distortion at a point such as A will be eliminated by increasing the projection easting y of point B' until B' lies in position B″ (figure 4.14), i.e. until A'B B″ is a straight line. This is achieved if the scale factor for the element BC is made the same as for the element AC, i.e. BC is increased by $\sec\theta$. Thus each element of the arcs G'A' and H'B' will be increased by the value of $\sec(y/R)$ for all the points along these lines.

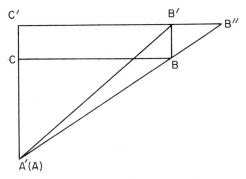

Figure 4.14

The easting thus obtained is the *Transverse Mercator*† easting E and is given by

$$E = \int_0^\theta R \sec\theta \, d\theta \tag{12}$$

$$= R \log_e \tan(\pi/4 + \theta/2) \tag{13}$$

Since (13) is awkward for computations, the following alternative approach is preferred. Sec θ is expanded as a power series of θ and integrated term by term

$$\begin{aligned}
E &= R\int_0^\theta \sec\theta \, d\theta = R\int_0^\theta (1 + \tfrac{1}{2}\theta^2 + \tfrac{5}{24}\theta^4 + \cdots)\, d\theta \\
&= R[\theta + \tfrac{1}{6}\theta^3 + \tfrac{1}{24}\theta^5 + \cdots]_0^\theta \\
&= R\theta(1 + \tfrac{1}{6}\theta^2 + \tfrac{1}{24}\theta^4 + \cdots)
\end{aligned} \tag{14}$$

The third term $(R/24)\theta^4$ has a value of 0·3 ft for $\theta = 3°$ which is normaly the maximum distance from the central meridian used. When larger extents East–West are projected, the area is divided into two or more projection 'belts' or 'zones' with a suitable overlap (see §4.2.5 on the Universal Transverse Mercator projection). Hence in practice the following is used

$$E = R\theta(1 + \tfrac{1}{6}\theta^2) = y(1 + y^2/6R^2) \tag{15}$$

The northings N on the T.M. projection are the same as for the Cassini, i.e.

$$N = x$$

Thus the coordinates on the T.M. projection are found from the formulae (1)–(8) inclusive, obtaining E from (15) above.

4.2.3.1. Reverse Case

To obtain θ from E, a method of successive approximation is required; First approximation

$$\theta' = E/R - \frac{1}{6R}(E/R)^3$$

† The transverse Mercator projection is also known as the Gauss–Krüger projection.

then, second approximation

$$\theta = E/R - \frac{1}{6R}(\theta')^3$$

which is usually sufficient.

4.2.3.2. Conversion of Coordinates by Projection Tables

For ease of computation, the closed formulae (1)–(6) and (14) are expanded as converging power series in terms of the various known quantities ϕ, $\Delta\lambda$, and E, whose coefficients are sufficiently closely tabulated to avoid the need for second difference interpolation. Most tables are of the following form:

$\Delta\lambda$ is the difference in longitude of the point from the central meridian positive eastwards and negative westwards; and $p = 0{\cdot}000\,1\,\Delta\lambda$. E' is the true easting of the point reckoned from the central meridian, positive eastwards and negative westwards. The easting $E = E' + \text{F.E.}$ where F.E. is the false easting introduced to avoid negative eastings. In the example below F.E. = 500 000 metres. The conversion from (λ, ϕ) to (E, N) is carried out by the following formulae:

Northings: $N = (\text{I}) + (\text{II})p^2 + (\text{III})p^4 + A_6$ (16)

Eastings: $E' = (\text{IV})p + (\text{V})p^3 + B_5$ (17)

Convergence: $C = (\text{XII})p + (\text{XIII})p^3 + C_5$ (18)

Example: Table 4.5 shows the necessary values of these coefficients to convert triangle ABC of table 4.2 into transverse Mercator projection coordinates. The working for A is as follows: The longitude of the central meridian λ_0 is 33° East. Then

λ_A 35° 18′ 16″.755 9
λ_0 33
$\Delta\lambda$ +2° 18′ 16·755 9 = +8 296″.755 9

whence $p = 0{\cdot}829\,675\,59$

Since all the powers of p up to the fourth are required they are conveniently evaluated together, i.e.

p 0·829 675 59
p^2 0·688 361 6
p^3 0·571 116 8
p^4 0·473 84

From table 4.5

(I)   1 754 951·147   (II)   1 971·918
(II)p^2 1 357·393   (III)   1·778
(III)p^4 0·842 $A_6 = 0$
∴ N 1 756 309·372

Whence the northing of A is 1 756 309·372 reckoned from the Equator southwards, since A is in the southern hemisphere. To avoid negative values in the

Table 4.5. Transverse Mercator Projection Factors to Convert Geographical Coordinates to Projection Coordinates.

(1) Northings factors:

Latitude	I	Δ for 1″	II	Δ for 1″	III	A_6
15° 39′	1 730 068·220		1 946·823		1·760	
		+30·72279		+0·03106		0
40′	1 731 911·587		1 948·687		1·762	
15°44′	1 739 285·087		1 956·134		1·767	
		+30·72302		+0·03100		0
45′	1 741 128·468		1 957·994		1·768	
15°52′	1 754 032·222		1 970·997		1·778	
		+30·72341		+0·03092		0
53′	1 755 875·627		1 972·852		1·779	

(2) Eastings factors:

Latitude	IV	Δ for 1″	V	Δ for 1″	B_5
15° 39′	297 717·065		100·341		
		−0·402 01		−0·000 73	+0·019
40′	297 692·944		100·297		
15° 44′	297 596·211		100·122		
		−0·404 10		−0·000 73	+0·024
45′	297 571·965		100·078		
15° 52′	297 401·541		99·771		
		−0·407 44		−0·000 73	+0·016
53′	297 377·095		99·727		

(3) Convergence factors

Latitude	XII	Δ for 1″	XIII	C_5
15° 39′	2 697·602		1·997	
		+0·046 68		0
40′	2 700·403		1·999	
15° 44′	2 711·605		2·006	
		+0·046 65		0
45′	2 714·404		2·008	
15° 52′	2 733·997		2·020	
		+0·046 62		0
53′	2 736·794		2·021	

Reproduced by permission of the Department of the Army, United States of America.

southern hemisphere, the origin is moved from the Equator to 10 million metres South of the equator. Thus the final Northing of A becomes

$$\begin{array}{r} 10\ 000\ 000 \\ -1\ 756\ 309 \cdot 372 \\ \hline 8\ 243\ 690 \cdot 628 \end{array}$$

Also from table 4.5

(IV)	297 389·3546	(IV)p	246 736·689
(V)	99·749	(V)p^3	56·968
B_5			+0·016
		E'	246 793·673

Whence the true easting is 246 793·673, and referred to the false origin, $E = 500\ 000 + E' = 746\ 793 \cdot 673$. Again from table 4.5

(XII)	2 735·391	(XII)p	2 269".487
(XIII)	2·020	(XIII)p^3	1·154
C_5	0		2 270·641 = 37' 50".64

and the convergence at A is 37' 50".64. Applying this convergence to the azimuth of AB from table 4.2 the *grid azimuth* $T_{AB} = 21° 32' 28".16 + 37' 50".64 = 22° 10' 18".80$. The grid azimuth is the spheroidal azimuth referred to grid North instead of true North. Figure 4.15 shows the relationships between azimuth, convergence, and grid azimuth for various positions of a point with respect to the central meridian and the equator. It is left as an exercise for the reader to compute the coordinates of and convergences at B, and C. The results are

	Easting	Northing	Convergence
A	746 793·673	8 243 690·628	37' 50".64
B	756 612·703	8 267 790·343	38' 46".83
C	764 991·410	8 257 418·274	40' 17".84

4.2.3.3. *Reverse Case*

The conversion of coordinates on the projection into geographical coordinates, i.e. the reverse of the above process, is carried out by the following:

$$\text{Latitude:} \quad \phi = \phi' - (\text{VII})q^2 + (\text{VIII})q^4 - D_6 \qquad (19)$$
$$\text{Longitude:} \quad \lambda = \lambda_0 + (\text{IX})q - (\text{X})q^3 + E_5 \qquad (20)$$
$$\text{Convergence:} \quad C = (\text{XV})q - (\text{XVI})q^3 + F_5 \qquad (21)$$

Where ϕ' is the value of latitude obtained by inverse interpolation for northing in factor (I) of Table 4.5 and $q = E' \times 10^{-6}$.

The Spheroid and Projections

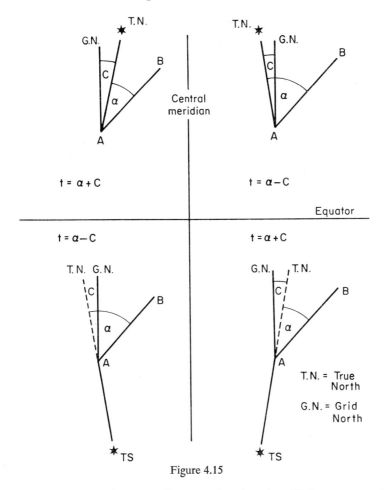

Figure 4.15

Example: The projection coordinates of point A will be converted into geographicals using the coefficients of table 4.6. Since A is South of the equator, the northings are converted to true northings with respect to the equator as follows

N_A	8 243 690·628	$q = E' \times 10^{-6} = (E - 500\,000)10^{-6}$
	10 000 000	$= 0·246\,793·673$
True N	1 756 309·372	q^2 0·060 907 12 q^3 0·015 031
		q^4 0·003 71

Using this value of the true northings ϕ' is found by inverse interpolation for the factor (I) in table 4.5, i.e. $\phi' = 15° 53' 14''·117\,7$, which is the argument for all the factors of table 4.6 From this table

ϕ' 15° 53' 14''·117 7

VII 726·310 47   VII q^2 −44·237 5
VIII 7·80   VIII q^4 +0·028 9
   D_6 0
   15° 52' 29''·909 4

Table 4.6. Transverse Mercator Projection Factors to Convert Grid to Geographical Coordinates.

(1) Latitude factors

ϕ'	I	Δ for 1"	VII	Δ for 1"	VIII	D_6
15° 40'	1 731 911·587		715·721		7·68	
		30·72284		0·01331		0
41'	1 733 754·958		716·520		7·69	
15° 45'	1 741 128·468		719·719		7·73	
		30·72309		0·01334		0
46'	1 742 971·853		720·519		7·74	
15° 53'	1 755 875·627		726·122		7·80	
		30·72347		0·01336		0
54'	1 757 719·035		726·923		7·81	

(2) Longitude factors

ϕ'	IX	Δ for 1"	X	Δ for 1"	E_5
15° 40'	33 591·659		160·191		
		0·045 41		0·001 02	0·001 3
41'	33 594·384		160·252		
15° 45'	33 605·316		160·498		
		0·045 69		0·001 03	0·001 3
46'	33 608·057		160·559		
15° 53'	33 627·338		160·993		
		0·046 12		0·001 04	0·001 3
54'	33 630·105		161·055		

(3) Convergence factors

ϕ'	XV	Δ for 1"	XVI	F_5
15° 40'	9 071·10		79·7	
		0·169 1		0·001
41'	9 081·25		79·8	
15° 45'	9 121·84		80·2	
		0·169 2		0·001
46'	9 132·00		80·3	
15° 53'	9 203·11		81·1	
		0·169 4		0·001
54'	9 213·28		81·2	

Reproduced by permission of the Department of the Army, United States of America

Whence the latitude of A is 15° 52' 29"·909 4 which checks the work to within the precision of the tables; which is 0"·001 in geographical from grid coordinates. Again from table 4.6

```
IX    33 627·989 1        IX q    8 299·174 8
X        161·007 7        X q³      −2·420 2
                          +E₅      +0·001 3
                          ────────────────
                          λ_A − λ_0  8 296·755 9    ∴ λ_A = 35° 18' 16"·755 9
```

Which checks the value of the longitude of A. Finally to check the convergence

$$\begin{array}{ll} \text{XV} & 9\ 205 \cdot 503 \\ \text{XVI} & 81 \cdot 1 \end{array} \qquad \begin{array}{ll} \text{XV}\ q & 2\ 271 \cdot 860 \\ \text{XVI}\ q^3 & -1 \cdot 219 \\ F_5 & +0 \cdot 001 \\ \hline & 2\ 270 \cdot 642 = 37'\ 50''\!\cdot\!642 \end{array}$$

which agrees with that found in example §4.2.3.2 above. The tables are designed to give the convergence to $0''\!\cdot\!01$ only.

It is left to the reader to check the coordinates of B and C. In practice, both of these conversions will always be done to check the work. In large national mapping organizations the conversion of coordinates is carried out with the aid of an electronic computer using the original formulae from which these tables were computed. The computation of the projection tables themselves was achieved with the aid of such a computer.

4.2.4. SCALE ERROR ON THE TRANSVERSE MERCATOR PROJECTION

It was seen above in §4.2.2.3 that the scale factor at a point (E, N) on the projection is given by $k = \sec \theta$, where $\theta = E/R$, E being the true easting from the central meridian. Since $\sec \theta$ increases rapidly with θ, the scale factor becomes inconveniently large away from the central meridian. For example if $\theta = 3°$, $\sec \theta = 1 \cdot 000\ 6$. Hence a length l' on the spheroid becomes $l = l' \times 1 \cdot 000\ 6 = l' + 0 \cdot 000\ 6l'$ i.e. the scale error is $0 \cdot 000\ 6l' \simeq l'/1\ 700$.

4.2.4.1. *Central Scale Factor* k_0

Since the scale is correct along the central meridian and increases away from it, the average scale error over the whole map will be reduced by the introduction of a negative scale factor to the whole map. This introduces a negative scale error along the central meridian, and the scale only becomes correct at some North–South line away from it.

Suppose θ_M is the maximum of E/R that will be encountered at the edge of the projection belt, then the maximum scale factor k_M is given by

$$k_M = \sec \theta_M = 1 + \tfrac{1}{2}\theta_M^2 + \tfrac{5}{24}\theta_M^4 + \cdots$$
$$\doteqdot 1 + \tfrac{1}{2}\theta_M^2$$

The maximum scale error is $+\tfrac{1}{2}\theta_M^2$.

The whole map is now reduced by a factor $k_0 = 1 - \tfrac{1}{4}\theta_M^2$, thus the scale factor at any point A now becomes

$$\begin{aligned} k_A = k_0 \sec \theta_A &\doteqdot (1 - \tfrac{1}{4}\theta_M^2)(1 + \tfrac{1}{2}\theta_A^2) \\ &= 1 + \tfrac{1}{2}\theta_A^2 - \tfrac{1}{4}\theta_M^2 - \tfrac{1}{8}\theta_M^2\theta_A^2 \\ &\doteqdot 1 + \tfrac{1}{2}\theta_A^2 - \tfrac{1}{4}\theta_M^2 \end{aligned} \qquad (22)$$

neglecting the fourth term which is very small. Thus the scale error along the central meridian is $-\tfrac{1}{4}\theta_M^2$ (putting $\theta_A = 0$ in (22)), and along the maximum extent East or West of the central meridian is $+\frac{1}{4}\theta_M{}^2$ (putting $\theta_A = \theta_M$). The reduced scale error at A will be zero when

$$\tfrac{1}{2}\theta_A{}^2 = \tfrac{1}{4}\theta_M{}^2, \text{ i.e. when } \theta_A \simeq 0{\cdot}70_M \text{ or } E_A \simeq 0{\cdot}7E_M$$

Thus at a point whose easting $E_A = 0{\cdot}7E_M$ the scale is correct. Figure 4.16 shows the scale factor plotted against easting for a value of $k_0 = 0{\cdot}999\ 6$.

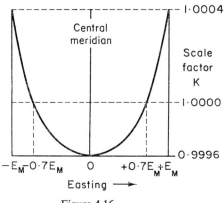

Figure 4.16

4.2.4.2. Scale Factor from Projection Tables

The scale factor k_A at any point A on the transverse Mercator projection may be written

$$k_A = k_0(1 + \text{XVIII}\, q_A{}^2 + 0{\cdot}000\ 03 q_A{}^4) \tag{23}$$

where $q_A = E_A \times 10^{-6}$ and E_A is the true easting from the central meridian. The factor XVIII is tabulated as shown in table 4.7.

Table 4.7. Scale Factor on Transverse Mercator Projection

Scale factor on transverse mercator projection	
Northing	XVIII
8 100 000	0·012 370
8 200 000	0·012 371
8 300 000	0·012 373

Reproduced by permission of the Department of the Army, United States of America

Example: To find the scale factor at A where the false easting = 500 000, $E_A = 746\ 794$, $N_A = 8\ 243\ 690$ proceeds as follows:

True easting = 246 794

$$q = 0{\cdot}246\ 794$$
$$q^2 = 0{\cdot}060\ 907$$
$$q^4 = 0{\cdot}003\ 71$$

Factor XVIII for A is 0·0123 72 whence $k_A = 1\cdot000\ 353\ 4$ and a short line in the vicinity of A whose spheroidal length is L becomes $1\cdot000\ 353\ 4L$ on the map.

It will be noted that formula (23) merely gives the scale factor *at a point* on the projection, and that this factor may only be used for *short lines* in the vicinity of this point. For a consideration of *line* scale factors, the reader should refer to §4.2.7.1 below.

4.2.5. THE UNIVERSAL TRANSVERSE MERCATOR PROJECTION SYSTEM (U.T.M.)

Since the scale error becomes excessive, even with a central scale factor, the practical width of a transverse Mercator projection is limited to about 3° on either side of the central meridian. This limitation also permits relatively simple formulae to be used for the calculation of scale and angular distortions with ample accuracy for most work. To map areas of greater East–West coverage than 6°, two or more *belts* or *zones* consisting of identical transverse Mercator projections are used. When a survey crosses a belt junction its coordinates are referred to the new central meridian; and the coordinates of points within about 25 km of the belt junction are often given with respect to both belts. Conversion from one belt to another is best carried out with the aid of special tables, rather than by the longer process of converting to geographical coordinates as an intermediate stage. The *Universal Transverse Mercator* projection system consists of sixty belts each 6° wide beginning with zone one at 180° West longitude as its western edge, i.e. its central meridian is 177° West longitude. The zones are then numbered eastwards, e.g. zone 31 has the Greenwich meridian as its western edge. Each zone consists of an identical transverse Mercator projection whose characteristics are as follows:

(i) The North–South extent is from 80°N to 80°S latitude.
(ii) In the northern hemisphere the origin is on the equator at a point 500 000 metres to the West of the central meridian. i.e. the false easting (F.E.) = 500 000 m.
(iii) In the southern hemisphere the origin is 10×10^6 metres to the South of the equator and 500 000 metres to the West of the central meridian.
(iv) The central scale factor $k_0 = 0\cdot999\ 6$ exactly.
(v) Various spheroids are recommended for different parts of the world. e.g. the Clarke 1880 spheroid is suggested for Africa.
(vi) Projection tables are available for the conversion of geographical coordinates on the recommended spheroids to grid coordinates and *vice versa*.
(vii) Tables are also available for zone-to-zone transformations.
(viii) Tables are available giving the grid coordinates to 0·1m of the intersections of the graticule at intervals of 5 minutes of arc, and at intervals of $7\frac{1}{2}$ minutes of arc. These are of great use to the cartographer since the edges of various map sheets may be plotted directly from the tables. See the data for the example of interpolation in Chapter 2. for a specimen of these tables (tables 2.2 and 2.3).

182   Practical Field Surveying and Computations

The various factors given in this book for the transverse Mercator projection are for the U.T.M. system referred to the Clarke 1880 spheroid obtained from the tables of the United States Army Map Service.

4.2.6. THE COMPUTATION OF A SURVEY ON A MAP PROJECTION

In §4.1.13 the computation of geographical coordinates, latitude, and longitude, on the surface of a reference spheroid by Legendre's theorem and the mid-latitude formulae was described, and in §4.2 etc. it was shown that these geographical coordinates can be projected on to a plane surface by the formulae for a particular projection system. Thus a set of plane rectangular coordinates representing the points of a survey is obtained, which will be used as the basis for mapping, for the calculation of areas, etc. As an alternative to computation on the surface of a spheroid, it is possible to compute directly in terms of plane trigonometry, provided the spheroidal lengths and the measured angles are distorted to suit the particular projection used. This method of computation is invariably the more convenient for most work unless an electronic computer is available. The computation of surveys on the transverse Mercator projection is considered at length to show the general principles of projection computation.

4.2.7. COMPUTATION OF SURVEYS ON THE TRANSVERSE MERCATOR PROJECTION

The data for triangle ABC used in the spheroidal computation will be computed on the Universal Transverse Mercator system Zone 36, central meridian 33° East longitude, origin at 10×10^6 m South, 500 000 m West. The spheroidal angles of the triangle are denoted by A', B', and C' whilst their corresponding projection or grid counterparts are denoted by A, B, and C. The spheroidal lengths are denoted by a', b', and c' whilst the corresponding grid lengths are denoted by a, b, and c. Let the angular distortions be dA, dB, and dC and the side distortions be da, db, and dc such that

$$\begin{aligned} A = A' + dA; & \quad B = B' + dB; \quad C = C' + dC \\ a = a' + da; & \quad b = b' + db; \quad c = c' + dc \end{aligned} \quad (24)$$

If the spherical excess of triangle $A'B'C'$ is ϵ, then since $A + B + C = 180°$, $dA + dB + dC = {}^-\epsilon$ paying due regard to signs. The grid angles and sides have been calculated from the grid coordinates of A, B, and C obtained in Example in §4.2.3.2 from the geographical coordinates of A', B', and C'. The corresponding spheroidal and grid data are:

Angles	Spheroidal △ $A'B'C'$ (1)	Plane △ ABC (2)	Difference (2) − (1)
A'	30° 48′ 03″96	A 30° 48′ 10″48	dA″ +06·52
B'	61 05 49·07	B 61 05 57·90	dB +08·72
C'	88 05 07·74	C 88 05 51·55	dC −16·01

Sides	(metres)			
a'	13 327·628	a	13 333·512	da +5·884
b'	22 785·523	b	22 794·865	db +9·342
c'	26 013·282	c	26 023·251	dc +9·969

The Spheroid and Projections

It is obvious that if the various distortions dA, da, etc. can be obtained from the spheroidal data, the grid values can be computed from equations (24), thence coordinates can be computed in plane trigonometry by the formulae of Chapters 6 and 7, etc. The formulae for the angular and scale distortions involve the known coordinates of the starting points and also those of the points required. Thus a process of successive approximation is required, though usually only one approximation is necessary in most work. The method is to compute by plane trigonometry approximate coordinates to a metre precision using the spheroidal data, or a mixture of spheroidal and grid data, and repeat the process with spheroidal lengths altered to grid lengths and spheroidal angles to grid angles. Whichever combination of fixation methods is used, a unique result is obtained. For example, an intersected point computed from grid angles will give the same result as the same fixation computed by a linear grid distance and a grid bearing using the same original data.

4.2.7.1. Line Scale Factors

Theoretically the scale factor $k = \sec \theta$ holds only for a point on the projection, but in practice it may be applied to short lines in the vicinity of this point. The validity of this practice depends on the precision required, the position of the line on the projection, the length of the line, and its direction. For example if a line runs North–South, θ does not vary sensibly over the whole line and k for any point of the line will be constant and therefore typical of the whole line.

Consider the worst possible case of a line L running East–West at the very edge of the projection, i.e. at $\theta = 3$. The scale error e is given by $e = +\frac{1}{4}\theta^2$. Hence a change de in e for a change $d\theta$ in θ is $\frac{1}{2}\theta\, d\theta$. Thus the corresponding change dL in a projected length L is given by $dL = \frac{1}{2}L\theta\, d\theta$. Since the line runs E–W, $d\theta = L/R$

$$\therefore\ dL = \frac{1}{2}\theta L^2/R$$
$$= \frac{1}{2}\frac{3}{57\cdot 3}\frac{L^2}{R} \qquad (25)$$

Putting $R = 6\cdot 37 \times 10^6$ metres, and supposing that an error of 0·01 m is acceptable in computation, i.e. $dL \leq 0\cdot 01$, $L = 1\cdot 5$ km (from (25)). Hence it is safe to assume that a line under 1·5 km long may be treated as a point for the purpose of evaluating its scale factor, without introducing a computational error in excess of 1 cm. For longer lines, the line scale factor is obtained from the point scale factors evaluated for its terminals and its mid-point by Simpson's rule. Consider the line AB whose mid-eastings and northings are $E_M = \frac{1}{2}(E_A + E_B)$ and $N_M = \frac{1}{2}(N_A + N_B)$, and whose point scale factors are k_A, k_M, and k_B. The scale factor for AB, i.e. k_{AB} is given by

$$k_{AB} = \tfrac{1}{6}(k_A + 4k_M + k_B) \qquad (26)$$

Example: Using the data of table 4.7 the scale factors for AB at A, its mid-point, and B are found to be respectively $k_A = 1\cdot 000\ 353\ 4$, $k_M = 1\cdot 000\ 383\ 6$, and $k_B = 1.000\ 414\ 5$ whence $k_{AB} = 1.000\ 383\ 7$. Since the spheroidal length of A'B' is 26 013·282, this gives the grid length AB = 26 023·263 which agrees with that obtained from coordinates to the precision expected. The student may work out the scale factors for lines b and c in a similar manner.

4.2.7.2. Alternative Formula for the Line Scale Factor k_{AB}

An alternative formula for k_{AB} is

$$k_{AB} = 1 + (E_A^2 + E_A E_B + E_B^2)/6R^2 \qquad (27)$$

This may be derived from (26) by substituting for

$$k_A = 1 + \frac{1}{2}\left(\frac{E_A}{R}\right)^2, \quad k_M = 1 + \frac{1}{2}\left(\frac{E_A + E_B}{2R}\right)^2, \quad k_B = 1 + \left(\frac{E_B}{R}\right)^2$$

4.2.7.3. Angular Distortion on the Transverse Mercator Projection

The grid angle A is derived from the grid *bearings* of AC and AB denoted by t_{AC} and t_{AB} respectively, i.e.

$$A = t_{AC} - t_{AB} \qquad (28)$$

Let T_{AB} and T_{AC} be the grid *azimuths* of AB and AC respectively, and let

$$\begin{aligned} dt_{AC} &= t_{AC} - T_{AC} \quad \text{or} \quad (t - T)_{AC} \\ dt_{AB} &= t_{AB} - T_{AB} \quad \text{or} \quad (t - T)_{AB} \end{aligned} \qquad (29)$$

Whence from (29) and (28) we have

$$\begin{aligned} A &= (T_{AC} + dt_{AC}) - (T_{AB} + dt_{AB}) \\ &= (T_{AC} - T_{AB}) + (dt_{AC} - dt_{AB}) \end{aligned}$$

Since the projection is orthomorphic the angle at A between the projected spheroidal arcs, or which is the same thing, the angle between the projected tangents to these arcs is the same as the spheroidal angle A', i.e. $A' = T_{AC} - T_{AB}$ whence

$$dA = A - A' = (t - T)_{AC} - (t - T)_{AB} \qquad (30)$$

In practice the values of $t - T$ for each direction are found first by the formula given below, and these are subtracted in pairs to give the distortion to each angle.

The sign of $t - T$ or 'arc to chord correction' to a direction is obtainable from a strict application of the formulae with due regard to signs, but in practice it is quicker to draw a diagram which gives the signs at once. In figure 4.17 A and B are the projected positions of A' and B', G.N. is grid North, T.N. true North, c = convergence at A. If P' is a point on the spheroidal line A'B', on the projection it plots as P on the dotted line which *curves away from the line of least scale*, i.e. away from the central meridian of the projection. Thus the sign of $t - T_{AB}$ is at once evident from the diagram. Figure 4.18 shows the projected spheroidal arcs for the triangle ABC lying to the East of the central meridian, and the projected spheroidal arcs for the triangle XYZ lying to the West of the

The Spheroid and Projections

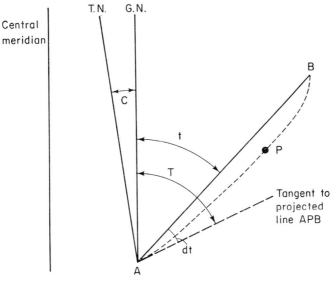

Figure 4.17

central meridian. Remembering that azimuths and bearings are reckoned clockwise from North, even in the southern hemisphere

Grid bearing = Grid azimuth + $(t - T)$ with due regard to signs.

e.g. the $(t - T)_{ZY}$ is negative and $(t - T)_{YZ}$ is positive, etc. The formula for the arc to chord correction for the bearing AB is

$$(t - T)_{AB} = - \frac{(2E_A + E_B)(N_B - N_A)}{6\rho\nu k_0^2 \sin 1''} \quad (31)$$

and that for the bearing BA is

$$(t - T)_{BA} = - \frac{(2E_B + E_A)(N_A - N_B)}{6\rho\nu k_0^2 \sin 1''} \quad (32)$$

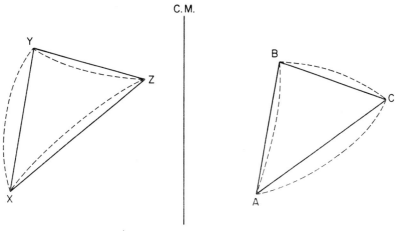

Figure 4.18

Table 4.8

Table of F_1 factor for $(t-T)$ on U.T.M.

$$(t-T_{AB}) = +F_1(N_A-N_B)$$

Tabular difference Δ varies from 0·008 52 to 0·008 47

Central scale factor 0·9996

$2E_A + E_B$ km	00	10	20	30	40	50	60	70	80	90
0	0·000 00	0·008 52	0·017 03	0·025 54	0·034 06	0·042 58	0·051 09	0·059 60	0·068 12	0·076 64
100	0·085 15	0·093 66	0·102 18	0·110 69	0·119 21	0·127 72	0·136 23	0·144 75	0·153 26	0·161 78
200	0·170 29	0·178 80	0·187 32	0·195 83	0·204 35	0·212 86	0·221 37	0·229 89	0·238 40	0·246 92
300	0·255 43	0·263 94	0·272 45	0·280 97	0·289 48	0·297 99	0·306 50	0·315 01	0·323 53	0·332 04
400	0·340 55	0·349 06	0·357 57	0·366 08	0·374 59	0·383 10	0·391 61	0·400 12	0·408 63	0·417 14
500	0·425 65	0·434 16	0·442 67	0·451 17	0·459 68	0·468 19	0·476 70	0·485 21	0·493 71	0·502 22
600	0·510 73	0·519 24	0·527 74	0·536 24	0·544 75	0·553 26	0·561 76	0·570 26	0·578 77	0·587 28
700	0·595 78	0·604 28	0·612 78	0·621 29	0·629 79	0·638 29	0·646 79	0·655 29	0·663 80	0·672 30
800	0·680 80	0·689 30	0·697 80	0·706 29	0·714 79	0·723 29	0·731 79	0·740 29	0·748 78	0·757 28
900	0·765 78	0·774 27	0·782 77	0·791 26	0·799 76	0·808 25	0·816 74	0·825 24	0·833 73	0·842 24
1000	0·850 72	0·859 21	0·867 70	0·876 19	0·884 68	0·893 16	0·901 65	0·910 14	0·918 63	0·927 12
1100	0·935 61	0·944 09	0·952 58	0·961 06	0·969 55	0·978 03	0·986 51	0·995 00	1·003 48	1·011 97
1200	1·020 45	1·028 93	1·037 41	1·045 88	1·054 36	1·062 84	1·071 32	1·079 80	1·088 27	1·096 75
1300	1·105 23	1·113 70	1·122 17	1·130 65	1·139 12	1·147 59	1·156 06	1·164 53	1·173 01	1·181 48

Tables 4.8 and 4.9 published by permission of the Director of Overseas Surveys, Tolworth, England.

It will be evident that the $(t - T)$'s at either end of the same line are not identically equal in magnitude, though they will be of the same order. The size of the correction varies with easting, i.e. increases away from the central meridian, and it also increases with the difference in northing. A line running East–West will have no correction.

4.2.7.4. Complete Formula for $(t - T)$

The formula (31) is the first and by far the largest term of a series from which the theoretically exact value of the $t - T$ correction is derived. In practice the short formula will normally be quite sufficient, though in geodetic work involving very long lines the full formula should be used. For a discussion of this question see A. G. Bomford, Transverse Mercator Arc to Chord etc. *Empire Survey Review* No. 125, Vol. XVI, July 1962.

4.2.7.5. Computation of $(t - T)$ Corrections

Since the denominator of formula (31) changes very slowly with latitude it may be tabulated together with values of $2E_A + E_B$ for particular spheroids and central scale factors k_0. Table 4.8 gives values of $(2E_A + E_B)/6\rho\nu k_0^2 \sin 1''$ with the argument $2E_A + E_B$ in km and for the Clarke 1880 spheroid. Thus formula (31) reduces to

$$(t - T)_{AB} = -F_1(N_B - N_A) \qquad (33)$$

where $F_1 = (2E_A + E_B)/6\rho\nu k_0^2 \sin 1''$†, $k_0 = 0.9996$ and $N_B - N_A$ is in km.

Since the algebraic sum of the $t - T$ corrections for the angles of a closed figure must equal the spherical excess of that figure, the spherical excess should always be calculated to check the arithmetic. Table 4.9 gives the value of $F_2 = 1/(2\rho\nu k_0^2 \sin 1'')$ at latitude $\phi = 0°$ for the Clarke 1880 spheroid, together with correction factors F to enable the values at other latitudes up to 20° to be evaluated. Thence the spherical excess of triangle ABC is $\epsilon = ab \sin C/2\rho\nu k_0^2 \sin 1''$

$$\therefore \quad \epsilon = ab \sin C \times F_2 \qquad (34)$$

or $$\epsilon = c^2 \sin A \sin B \operatorname{cosec} C \times F_2 \qquad (35)$$

4.2.7.6. Example of Solution of Triangle ABC

Given the projection coordinates of A and B and the adjusted angles of triangle ABC, the coordinates of C will be derived using the cotangent formula [(7) and (8) §6.9.2]. Refer to table 4.10 giving the layout of the complete computation. Provisional coordinates of C, E_C' and N_C' are obtained from the adjusted angles and their cotangents. From these provisional coordinates and the coordinates of A and B, the six $t - T$ corrections for the directions are derived, remembering to subtract the false easting of 500 000 m. The $t - T$ correction for each angle is obtained by subtraction of corrections for pairs of

---

† The quantity F_1 actually tabulated in table 4.8 also makes allowance for the higher order terms described in the article referred to in §4.2.7.4.

Table 4.9. Spherical Excess Factor F_2.

$$F_2 = \tfrac{1}{2}\rho v k_0^2 \sin 1'' = F \times 0.255\,245 \times 10^{-8}$$

$\phi =$	1°	2°	3°	4°	5°	6°	7°	8°	9°	10°
$F =$	1.000 00	0.999 98	0.999 96	0.999 93	0.999 89	0.999 85	0.999 80	0.999 74	0.999 67	0.999 59
$\phi =$	11°	12°	13°	14°	15°	16°	17°	18°	19°	20°
$F =$	0.999 50	0.999 41	0.999 31	0.999 20	0.999 09	0.998 97	0.998 84	0.998 70	0.998 56	0.998 41

The Spheroid and Projections 189

Table 4.10

Computation of Intersection on Transverse Mercator Projection by Cotangent Formula for Angles [clockwise convention]
Values of F_1 and F_2 from Tables 4.8 and 4.9
False Easting 500 000 m.

Point	Easting − F.E.	Northing	Adjusted angle	$t-T$	Grid angle	Provisional cotangents	Final cotangents
A	246 793·673	8 243 690·628	30° 48′ 03″·96	+6″·52	10″·48	+1·677 4427	3221
B	256 612·703	8 267 790·343	61 05 49·07	+8·72	57·79	+0·522 0989	0440
C′	264 990·5	8 257 418·5	88 05 67·74	−16·01	51·73	+0·033 1358	2126
C	264 991·40	8 257 418·28	180 00 00·77		+ Cot A / Cot B	+2·229 5416	+2·229 3661

All provisional values cancelled

Line	$2E_1+E_2$ km	F_1	N_2-N_1 km	$t-T$ directions	$t-T$ angles
AC	758·578	0·64558	−13·727	−8″·86	+6·52
AB	750·201	0·63846	−24·099	−15·38	
BA	760·020	0·64681	+24·099	+15·59	+8·72
BC	778·216	0·66228	+10·372	+6·87	
CB	786·593	0·66940	−10·372	−6·94	−16·01
CA	776·774	0·66105	+13·727	+9·07	
			Sum	−0·77	−0·77 ✓

$F_2 = 0·2550 \times 10^{-8}$, $c^2 = 6·767 \times 10^8$, sin A 0·5120, sin B 0·8755, sin C 0·9995
∴ ε = 0·774

directions and their algebraic sum is determined. The spherical excess of the triangle is found to check the sum of the $t - T$'s. The grid angles are then derived by application of the $t - T$ corrections to the adjusted angles, and the cotangents of the grid angles written down. The cotangent of the third angle C is also found so that the cotangent check may be applied (see §6.9.6). The final coordinates of C are computed using the cotangents of the grid angles. The provisional cotangents and coordinates should always be crossed out immediately after they have been used so that confusion is avoided in using the calculating machine, and provisional coordinates not mistaken for the final values.

In the semigraphic methods of fixation, the $t - T$ corrections are incorporated into the C–O terms at the approximation stage at which the coordinates are known to within a metre of their correct values.

4.2.7.7. *Other Fixations on the Projection*

This procedure of converting observed angles and measured sides into their grid counterparts is used in all methods of fixation, i.e. in traverse, trilateration, etc. (see Chapters 7 and 13). The process is similar to that given for the triangle, viz. scale factors and, or $t - T$ corrections are derived from provisional coordinates, thereafter the plane trigonometrical formulae are applied to grid lengths and grid bearings or grid angles as the case may be.

4.2.7.8. Consistency between Spheroidal and Grid Computations

The triangle A'B'C' on the spheroid has been solved in example §4.1.13 and its grid counterpart in §4.2.7.6. Within the precision of computation expected, the results of these two methods of solution should be identical when compared in similar terms. For example, the known azimuth of BA should agree with its valve obtained from the azimuth of BA by calculation through the projection. In figure 4.19 the azimuths of AB and BA are α_{AB} and α_{BA}, the convergencies at

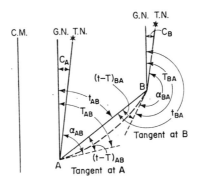

Figure 4.19 Southern hemisphere

A and B are C_A and C_B, the grid azimuths at AB and BA are T_{AB} and T_{BA}, and the grid bearings AB and BA are t_{AB} and t_{BA} respectively. Then paying due regard to signs:

$$t_{AB} = \alpha_{AB} + C_A + (t - T)_{AB}$$

$$t_{BA} = \alpha_{BA} + C_B + (t - T)_{BA}$$

$$t_{AB} + 180 = t_{BA}$$

Using the data of the chapter given in table 4.11, for the line AB:

$$\alpha_{AB} = 21° 32' 28''16, \quad C_A = +37' 50''64, \quad (t - T)_{AB} = -15''38$$

$$\therefore \quad t_{AB} + 180° = 202° 10' 03''42$$

$$\alpha_{BA} = 201° 31' 01''02, \quad C_A = +38' 46''83, \quad (t - T)_{BA} = +15''59$$

$$\therefore \quad t_{BA} = 202° 10' 03''44$$

i.e. the forward and back grid bearings agree to within the expected precision. The results of the other lines BC and AC are given in table 4.11(v). In very important geodetic work a survey might be computed both on the spheroid and on the projection and the results compared, though in practice this is a luxury that most surveyors can ill afford. In any case, most surveys are self-checking, for example a point is normally fixed by more observations than are absolutely necessary.

The Spheroid and Projections 191

Table 4.11

Computation on the Transverse Mercator Projection Table of Data for Triangle ABC

(i) Coordinates (m)

Point	Easting	Northing
A	746 793·673	8 243 690·628
B	756 612·703	8 267 790·343
C	764 991·410	8 257 418·274

(ii) Grid bearings and angles

Bearings		Angles	
AB	22° 10′ 03″·45	A	30° 48′ 10″·55
BC	141° 04′ 05″·60	B	61° 05′ 57·85
AC	52° 58′ 14″·00	C	88° 05′ 51·60

(iii) $(t-T)$ corrections

Bearings		Angles	
AB	−15″·38	A	+06″·52
BA	+15·59	B	+08·72
BC	+6·87	C	−16·01
CB	−6·94		
CA	+9·07		
AC	−8·86		

(iv) Grid lengths and scale factors

	Grid length	Scale factor
AB	1 26 023·251	1·000 383 2
BC	13 333·512	1·000 441 5
CA	22 794·865	1·000 410 0

(v) Check spheroidal and grid data

Line	AB	BA	BC	CB
Azimuth	21° 32′ 28″·16	201° 31′ 01″·02	140° 25′ 11″·95	320° 23′ 54″·74
Convergence	+37 50·64	+38 46·83	+38 46·83	+40 17·84
$t-T$	−15·38	+15·59	+06·87	−06·94
Grid bearing	22° 10′ 03·42	202 10 03·44	141 04 05·65	321 04 05·64

Line	AC	CA
Azimuth	52° 20′ 32″·11	232° 17′ 47″·00
Convergence	+37 50·64	+40 17·84
$t-T$	−08·86	+09·07
Grid bearing	52 58 13·89	232 58 13·91

4.2.8. PROOF OF THE FORMULA FOR THE ARC TO CHORD CORRECTION ON THE TRANSVERSE MERCATOR PROJECTION

At each stage of this proof of the $t - T$ formula for the transverse Mercator projection only those terms necessary to reach the simple formula (31) are retained, since a completely rigorous treatment would be too long for this book, and indeed, would be unnecessary since the full formula is seldom used in everyday computation. Consider figures 4.10(a) and 4.10(b). The figure 4.10(b) is an enlarged view of the triangle ABC.

The Cassini coordinates of A and B are respectively (y_A, x_A) and (y_B, x_B). and the transverse Mercator coordinates (E_A, N_A) and (E_B, N_B).

The length of AB is s. The great circle OGHP is the central meridian of the projection. The arc ADC is part of the *small circle* through A, in a plane parallel to that of the central meridian. Thus $AG = CH$. The arc AKC is part of the *great circle* passing through A and C. $HG = x_B - x_A = N_B - N_A$; $AG = y_A$ and $BH = y_B$. Let the spherical excess of the figure GAKCH, which is bounded by great circles, $= e$. In this figure, angle G = angle $H = 90°$, and angle A = angle C by symmetry. But

$$A + C + H + G = 360 + e$$
$$\therefore \quad A = C = 90 + \tfrac{1}{2}e$$

Also $\angle DAG = 90°$

$$\therefore \quad \angle DAK = \tfrac{1}{2}e = \angle DCK$$

Let the spherical excess of triangle ABC $= \epsilon$, and let the spherical bearing of $AB = T$, i.e. angle $DAB = T$. In triangle ABC $A = T - \tfrac{1}{2}e$, $C = 90 - \tfrac{1}{2}e$, and since $A + B + C = 180 + \epsilon$,

$$B = 180 + \epsilon - A - C$$
$$= 90 - T + \epsilon + e$$

Let the Legendre angles of triangle ABC be A_1, B_1, and C_1, where $A_1 = A - \tfrac{1}{3}\epsilon$, $B_1 = B - \tfrac{1}{3}\epsilon$, and $C_1 = C - \tfrac{1}{3}\epsilon$. Then

$$A_1 = T - (\tfrac{1}{2}e + \tfrac{1}{3}\epsilon), \quad B_1 = 90 - (T - e - \tfrac{2}{3}\epsilon)$$
$$C_1 = 90 - (\tfrac{1}{2}e + \tfrac{1}{3}\epsilon), \quad \therefore \sin C_1 \simeq 1 \text{ since } (\tfrac{1}{2}e + \tfrac{1}{3}\epsilon) \text{ is small}$$

$$AC = \frac{s \sin B_1}{\sin C_1} \simeq s \cos [T - (e + \tfrac{2}{3}\epsilon)] \tag{1}$$
$$\simeq s \cos T + (e + \tfrac{2}{3}\epsilon)s \sin T \cdots \quad \text{(Taylor)}$$

$$BC = \frac{s \sin A_1}{\sin C_1} \simeq s \sin [T - (\tfrac{1}{2}e + \tfrac{1}{3}\epsilon)] \tag{2}$$
$$\simeq s \sin T - (\tfrac{1}{2}e + \tfrac{1}{3}\epsilon)s \cos T \cdots \quad \text{(Taylor)}$$

Converting Cassini eastings into transverse Mercator eastings

$$E_A = y_A + y_A^3/6R^2 \quad \text{and} \quad E_B = y_B + y_B^3/6R^2$$
$$\therefore \quad \Delta E = E_B - E_A = y_B - y_A + (y_B^3 - y_A^3)/6R^2$$
$$= BC + (y_B^3 - y_A^3)/6R^2$$
$$= s \sin T - (\tfrac{1}{2}e + \tfrac{1}{3}\epsilon) \, s \cos T + (y_B^3 - y_A^3)/6R^2 \tag{3}$$

Now
$$e = \text{Area of GACH}/R^2 \simeq y_A s \cos T/R^2$$
$$\epsilon = \text{Area of } \triangle ABC/R^2 \simeq s^2 \sin T \cos T/2R^2$$

The Spheroid and Projections

Whence substituting for e and ϵ in (3)

$$\Delta E = s \sin T - \frac{s^3 \sin T \cos^2 T}{6R^2} - \frac{y_A s^2 \cos^2 T}{2R^2} + \frac{(y_B{}^3 - y_A{}^3)}{6R^2} \quad (4)$$

Now, $\quad \Delta N = N_B - N_A = x_B - x_A = GH = AC \sec \theta$

where $\theta = y_A/R$

$$\simeq AC(1 + \tfrac{1}{2}\theta^2) = AC + \tfrac{1}{2}AC\theta^2 \quad (5)$$

Putting $AC \simeq s \cos T$ and $\theta = y_A/R$ in the second term of (5) and remembering (1)

$$\Delta N = s \cos T + (e + \tfrac{2}{3}\epsilon)s \sin T + \frac{s \cos T y_A{}^2}{2R^2}$$

which on substitution for e and ϵ becomes

$$\Delta N = s \cos T + \frac{y_A s^2 \sin T \cos T}{R^2} + \frac{s^3 \sin^2 T \cos T}{3R^2} + \frac{y_A{}^2 s \cos T}{2R^2} \quad (6)$$

and since $y_A = y_B - s \sin T$, after substitution in (6)

$$\Delta N = s \cos T \left(1 - \frac{s^2 \sin^2 T}{6R^2} + \frac{y_B{}^2}{2R^2}\right) \quad (7)$$

If the *grid bearing* of $AB = t$, $\tan t = \Delta E \, \Delta N^{-1}$; and substituting for ΔE and ΔN from (4) and (7), expanding and neglecting powers of $1/R$ greater than the second,

$$\tan t = \frac{s \sin T}{s \cos T} - \frac{s^3 \sin T \cos^2 T}{6R^2 s \cos T} - \frac{y_A s^2 \cos^2 T}{2R^2 s \cos T}$$

$$+ \frac{(y_B{}^3 - y_A{}^3)}{6R^2 s \cos T} + \frac{s^2 \sin^3 T}{6R^2 \cos T} - \frac{y_B{}^2 s \sin T}{2R^2 s \cos T} \quad (8)$$

$$= \tan T + \alpha \quad \text{(say)}$$

Let $t = T + dT$, where dT is small

$$\therefore \quad \tan t = \tan (T + dT) = \tan T + \sec^2 T \, dT \cdots \quad \text{(Taylor)}$$
$$\therefore \quad dT = \alpha \cos^2 T \quad \text{(from (8))}$$

Whence on substitution for α, and remembering that $\Delta x \simeq s \cos T$, after some reduction

$$6s^2 R^2 \, dT/\Delta x = -\Delta y \, \Delta x^2 - 3y_A \, \Delta x^2 + y_B{}^3 - y_A{}^3 + \Delta y^3 - 3y_B{}^2 \Delta y$$
$$= -(2y_A + y_B)(\Delta x^2 + \Delta y^2)$$

But $\quad s^2 = \Delta x^2 + \Delta y^2$

$$\therefore \quad dT = -(2y_A + y_B)(x_B - x_A)/6R^2 \quad (9)$$

Since dT is small it is sufficiently accurate to put $E_A = y_A$ and $N_A = x_A$ in (9), whence

$$dT'' = (t - T_{AB})'' = -\frac{(2E_A + E_B)(N_B - N_A)}{6\rho\nu \sin 1''} \quad R = (\rho\nu)^{\frac{1}{2}}$$

4.2.9. THE LAMBERT CONICAL ORTHOMORPHIC PROJECTION

The Lambert conical orthomorphic projection, or simply Lambert's Projection, is much used to map areas which have a great extent East–West and a small extent (about 6°) from North to South. For example, about one half of the states of the U.S.A. are on this projection. It is easier to construct than the transverse Mercator projection, since its graticule is formed by meridians which project as straight lines, and parallels which plot as circular arcs. Interpolation of coordinates between tabulated portions of the graticule is relatively uncomplicated. One drawback is that the convergence becomes excessive away from the centre of the projection.

4.2.9.1. *Simple Conical Projection*

The Lambert projection is formed by making a simple conical projection conformal, in a similar way that the transverse Mercator projection is the Cassini projection made conformal. Simple conical coordinates are produced by developing the surface of a cone which touches the sphere along a parallel of latitude passing through the middle of the North–South dimension of the area to be mapped. This parallel is the *standard parallel* defined by the standard latitude ϕ_0. The apex of the cone lies on the prolongation of the axis of the sphere. In figure 4.20 the cone is shown touching the sphere along the standard parallel. A point at the centre of the East–West dimension of the area to be mapped and lying on the standard parallel is chosen to be the origin of coordinates. The meridian through this origin O is the *central meridian*. As will be seen later, the character of the projection depends much more on the standard parallel than on the central meridian. The latitudes of points to be mapped are reckoned from the standard parallel, and the longitudes from the central

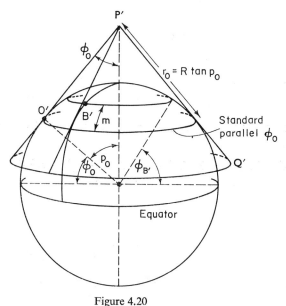

Figure 4.20

meridian. If the radius of the sphere is R, the slant length of the cone from the standard parallel, i.e.

$$P'O' = R \cot \phi_0 = R \tan p_0 \tag{1}$$

where p_0 is the co-latitude of the standard parallel, i.e.

$$p_0 = 90 - \phi_0$$

The cone is *developed* by cutting it along a line such as P'Q' producing a shape such as that shown in figure 4.21. The standard parallel on the map is plotted

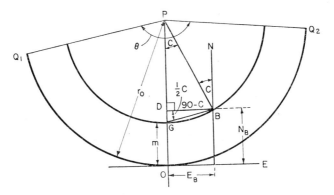

Figure 4.21

as a circular arc of radius $r_0 = $ P'O' centered on the projected apex on the cone, i.e. at P, and whose length equals the length of the standard parallel on the sphere. The length of the standard parallel on the sphere $= 2\pi R \cos \phi_0$ hence the arc $Q_1 O Q_2 = 2\pi R \cos \phi_0$. The angle $Q_1 P Q_2$ denoted by

$$\theta = \frac{2\pi R \cos \phi_0}{R \tan p_0} = 2\pi \sin \phi_0$$

Hence 2π radians of longitude on the sphere are represented by $2\pi \sin \phi_0$ radians on the map; in general, any longitude difference $d\lambda$ on the sphere will be represented by an angle $d\theta$ where

$$d\theta/d\lambda = \sin \phi_0 = n \tag{2}$$

The quantity n is called the *constant of the cone*, and *defines the spacing of the meridians on the map.*

The other parallels are represented by circular arcs centred on P such that their spacing equals the arc distance along the meridian separating these parallels on the sphere, e.g. consider a point B' on the sphere at latitude ϕ_B. The spherical arc distance $m = R(\phi_B - \phi_0)$ separates the projected parallel of B from the standard parallel. The radius of the projected parallel of B, i.e.

$$r_B = r_0 - m \tag{3}$$

Again, the projected point B lies on a projected meridian which makes an angle c with the central meridian such that

$$n = \frac{c}{\lambda_B - \lambda_0} = \frac{c}{\Delta\lambda}$$

i.e.
$$c = \Delta\lambda \sin \phi_0 \tag{4}$$

c is the map *convergence* at B. The map coordinates of B are given by (figure 4.21):

$$E_B = DB = r_B \sin c = PG \sin c = (r_0 - m) \sin c \tag{5}$$
$$N_B = OD = PO - PD = r_0 - r_B \cos c = r_0 - (r_0 - m) \cos c \tag{6}$$

4.2.9.2. Lambert's Projection

Since the point scale along a meridian does not equal the point scale along a parallel, other than the standard parallel, the projection is not orthomorphic, or it is not conformal. To make it conformal, the spacing of the parallels is altered to make the point scale factors along the projected meridian and parallel in the vicinity of a point equal.

Let K_1 be the scale factor along a parallel, and K_2 be the scale factor along a meridian, then

$$K_1 = \frac{r\, d\theta}{R \cos \phi\, d\lambda} = \frac{r\, d\theta}{R \sin p\, d\lambda} \quad (p = 90 - \phi)$$

but
$$d\theta/d\lambda = \sin \phi_0 = n$$

$$\therefore K_1 = \frac{nr}{R \sin p} \tag{7}$$

Also
$$K_2 = \frac{dr}{R\, dp} \tag{8}$$

For orthomorphism $K_1 = K_2$, i.e. $dr/R\, dp = nr/R \sin p$

$$\therefore \frac{dr}{r} = n \frac{dp}{\sin p} \tag{9}$$

Integrating
$$\int_{r_0}^{r} \frac{dr}{r} = n \int_{p_0}^{p} \frac{dp}{\sin p}$$

$$[\log r]_{r_0}^{r} = n[\log \tan \tfrac{1}{2}p]_{p_0}^{p}$$

$$\therefore \log\left(\frac{r}{r_0}\right) = n \log \left(\frac{\tan \tfrac{1}{2}p}{\tan \tfrac{1}{2}p_0}\right) = \log \left(\frac{\tan \tfrac{1}{2}p}{\tan \tfrac{1}{2}p_0}\right)^n$$

$$\therefore \frac{r}{r_0} = \frac{(\tan \tfrac{1}{2}p)^n}{(\tan \tfrac{1}{2}p_0)^n}$$

$$\therefore r = \frac{r_0}{(\tan \tfrac{1}{2}p_0)^n} (\tan \tfrac{1}{2}p)^n$$

$$= F (\tan \tfrac{1}{2}p)^n \tag{10}$$

The Spheroid and Projections

where
$$F = r_0/(\tan \tfrac{1}{2}p_0)^n$$

Since $n = \sin \phi_0$, F is a scale constant defined by the latitude of the standard parallel.

The computation of the grid coordinates (E, N) is identical to that for the simple conical projection with the exception that the radius of the parallel for latitude ϕ_A, i.e. r_A is given by equation (10), i.e.

$$r_A = F(\tan \tfrac{1}{2}p_A)^n \tag{11}$$

Because the radii of the projected parallels are very large, equation (11) proves troublesome in computation, and it is simpler to work in terms of corrections to the standard radius r_0. Let $r = r_0 + m'$, then it can be shown (see §4.2.9.5) that

$$m' = m + \frac{m^3}{6R^2} - \frac{m^4 \tan \phi_0}{24R^3} \cdots \tag{12}$$

where $m = R(\phi_0 - \phi)$, i.e. the meridional arc distance reckoned from the standard parallel, positive South, i.e. with increasing r and negative northwards, i.e. with decreasing r. The values of m' are tabulated in projection tables.

Since $m' \doteq m + m^3/6R^2$, which compares with the easting on the transverse Mercator $E \simeq y + y^3/6R^2$, where y is the Cassini easting (true arc distance from the central meridian), the character of the Lambert projection is very similar to that of the transverse Mercator projection, except that the scale variation on Lambert depends on the distance North or South of the standard parallel.

4.2.9.3. Scale Factor

The scale factor k is given by

$$\frac{\text{Map distance}}{\text{True distance}} = k$$

$$k = \frac{dm'}{dm} \simeq 1 + \frac{m^2}{2R^2} \tag{13}$$

An overall scale factor is introduced in the same way as for the transverse Mercator projection, which produces a negative net scale along the standard parallel and reduces the scale error at the maximum extent North or South to a similar numerical amount. If m_M is the maximum value of m, then the central scale factor $k_0 = 1 - m_M^2/4R^2$ and the scale factor at point A becomes

$$k_A = k_0(1 + m_A^2/2R^2) \tag{14}$$

4.2.9.4. The Arc to Chord Correction on the Lambert projection

By analogy with the transverse Mercator projection, the first term of the formula for the $t - T$ correction is

$$(t - T)_{AB} = -\frac{(2N_A + N_B)(E_B - E_A)}{6R^2 \sin 1''} \tag{15}$$

i.e. northings are interchanged with eastings.

4.2.9.5. *Proof of Equation* (12) *For a Sphere*

Let dr and dp be corresponding small changes in r and p respectively from the standard valves r_0 and p_0; i.e. $r = r_0 + dr$ and $p = p_0 + dp$. Since $r = F(\tan \tfrac{1}{2} p)^n$ we may let

$$r = f(p) \qquad (16)$$

(F and n are constants.)

$$\therefore \quad r_0 + dr = f(p_0 + dp)$$

$$= f(p_0) + f'(p_0)\,dp + f''(p_0)\frac{dp^2}{2!} + f'''(p_0)\frac{dp^3}{3!} + f^{\mathrm{iv}}(p_0)\frac{dp^4}{4!} + \cdots$$

(Taylor's theorem) (17)

where $f'(p_0)$, $f''(p_0)$, $f'''(p_0)$, $f^{\mathrm{iv}}(p_0)$ are the first, second, third, and fourth differentials of p when $p = p_0$. e.g.

$$f(p) = F(\tan \tfrac{1}{2} p)^n$$

$$\therefore \quad f'(p) = nF(\tan \tfrac{1}{2} p)^{n-1} \sec^2 \tfrac{1}{2} p \cdot \tfrac{1}{2}$$

$$= \frac{nF(\tan \tfrac{1}{2} p)^n \sec^2 \tfrac{1}{2} p}{2 \tan \tfrac{1}{2} p} \qquad (18)$$

$$= nf(p)\,\mathrm{cosec}\,p$$

When $p = p_0$, $r = r_0$

$$\therefore \quad f'(p_0) = r_0 \cos p_0 / \sin p_0 = R \qquad (19)$$

Similarly it may be shown that $f''(p_0) = 0$; $f'''(p_0) = R$, and $f^{\mathrm{iv}}(p_0) = -R \cot p_0 = -R \tan \phi_0$ whence on substitution in (17)

$$r_0 + dr = r_0 + R\,dp + \tfrac{1}{6} R\,dp^3 - \tfrac{1}{24} R\,dp^4 \tan \phi_0 \cdots$$

and remembering that $dp = m/R$, and putting $m' = dr$

$$m' = m + \frac{m^3}{6R^2} - \frac{m^4 \tan \phi_0}{24 R^3}$$

which is equation (12) for a sphere. Consideration of corresponding spheroidal formula is beyond the scope of this book.

4.2.10. TRANSVERSE AND OBLIQUE PROJECTIONS

In figure 4.10 a projection system was derived from the point P and the great circle PHGO. In the cases of the Cassini and transverse Mercator projections, P is the Earth's geographical pole, and PHGO is a meridian. However, the same formulae could be used to obtain similar projection systems, if P is *any* point on the surface of the sphere, and PHGO is *any* great circle; and provided the coordinates of points to be projected analogous to 'latitude' and 'longitude' with reference to this 'pole' and this 'meridian' can be calculated. In figure 4.22 Q is the new 'pole' and the great circle QO is the new 'central meridian' of a projection whose origin is at O. A is any point to be projected. The geographical latitudes and longitudes of Q, O, and A are respectively

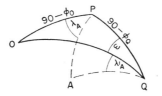

Figure 4.22

(ϕ_Q, λ_Q), (ϕ_0, λ_0), and (ϕ_A, λ_A), and the new 'latitudes' and 'longitudes' with respect to the pole at Q are (ϕ_Q', λ_Q'), (ϕ_0', λ_0'), and (ϕ_A', λ_A'). Since QO is the reference 'meridian' on the new system and Q the pole $\phi_Q' = 90°$, $\lambda_Q' = 0$ and $\lambda_0' = 0$. Let the spherical angle PQO $= \omega$. Then ϕ_0' and ω can be found by spherical trigonometry from spherical triangle OPQ. Also ϕ_A' and $\omega + \lambda_A'$ can be found on solution of the spherical triangle PQA. Thence a Cassini *type* or transverse Mercator *type* projection can be computed, based on pole Q and meridian QO and coordinates of points such as A, ϕ_A', λ_A'. When ϕ_Q and ϕ_0 are not zero, not 90°, nor equal, the general case of projection is produced. This general case is an *oblique* or *skew* form of projection. The skew projection is suited to a country which is orientated at some azimuth intermediate between 0° and 90°, e.g. the countries of Malaya and Borneo are on oblique Mercator projections. Although the projection character and formulae are based on a skew 'meridian' the coordinates obtained are finally rotated on to a North–South grid axis to accord with convention and reduce the convergence between the final grid and graticule. When $\phi_Q = 0°$ we have the *transverse* case.

In a similar manner a small circle through an origin O could be used as a 'standard parallel' for a conic projection and a skew conic projection derived.

The name transverse Mercator derives from the fact that this projection is the transverse case of the Mercator projection in the sense described here. The Mercator projection is based on the equator as the 'central meridian' and was developed historically before the transverse case and before the *Gauss–Kruger* projection as the latter is also called. Alternatively the Mercator projection could be called the 'transverse Gauss–Kruger' projection.

All map projections can be considered as conical projections. For example, if the standard parallel of the Lambert projection is the equator, Mercator's projection is produced, and thus the Gauss–Kruger projection is a special transverse case of Lambert's conical orthomorphic projection. If the standard parallel is at latitude 90°, i.e. at the pole, an azimuthal projection is obtained. A full treatment of map projections is beyond the scope of the book. For more information the reader should refer to R. Sacks, *E.S.R.*, No. 78, Oct. 1950 and Brigadier M. Hotine, *E.S.R.* Nos. 62, 63, 64, 65, and 66 1947.

5 Theory of Errors and Survey Adjustment

5.1. INTRODUCTION

At school we were once asked to draw a line 18 inches long and then measure it, using a six inch rule for both processes; and we were not a little surprised to discover that the exercise was not as ridiculous as it appeared at first. Some of the discoveries made were:

(i) No two lengths drawn by pupils were identically equal.
(ii) If we measured a length with a rule graduated in inches only the same result was obtained each time, but if a ruler with small precise divisions was used the results varied within small limits.
(iii) Different answers were obtained with different rulers.
(iv) In the end we were uncertain as to the exact lengths of all the lines, which were each supposed to be 18 inches long.

Thus this apparently simple exercise was full of problems; problems which arise in all attempts to set out measurements according to a specification, and to measure the exact value of a quantity. In addition it raised the problem of what is meant by 'the exact value'. Since land surveying is almost wholly concerned with measurements and setting out to specifications, these problems are of paramount importance; indeed in many instances it is almost as important to know the accuracy of a result as to know the result itself. This section of the book describes methods of analysing the errors of measurement, and the subsequent adjustment of discrepancies on the basis of the error theory propounded.

5.1.1. THEORY OF ERRORS

The following definitions and symbols are essential for an understanding of subsequent sections.

5.1.1.1. *True Value*

Just as the concept of 'truth' is an abstract idea, so also is the concept of true value in most cases. In general, the true value of a quantity will never be found; or if it is found we will never know that we have found it. This argument applies to a single observed quantity only. We often know the true value of a combination of quantities that are observed singly, e.g. the sum of the three angles of a plane is known to add up to 180°.

5.1.1.2. *True Error*

Like true value, the true error Δ of a single observed quantity can never be found; hence it is merely an abstract idea defined to be the difference between the true value T and the observed value O, i.e.

$$T - O = \Delta$$

5.1.1.3. *Most Probable Value*

The most probable value (m.p.v.) is that value which is more likely than any other to be the true value, judged on the evidence available. We shall accept the axiom that the most probable value that can be derived from a series of observations of a quantity is the arithmetic mean of the set, *provided that the observations are independent of each other and that we think they are of equal reliability*.

5.1.1.4. *Residual*

A residual v is defined to be the difference between the most probable value V and the observed value O.

$$\text{Most probable value} - \text{Observed value} = \text{Residual}$$

i.e. $$V - O = v$$

5.1.1.5. *Weight*

The weight of an observation is a measure of its trustworthiness *relative to other observations*, usually expressed as a number (see §5.4.3 below for further exemplification).

5.1.1.6. *Symbols*

Instead of the sigma common in mathematics, the square bracket is used to denote a summation of like terms in error theory and statistics, e.g.

$$[a] = a_1 + a_2 + a_3 + a_4 + \cdots + a_n$$
$$[a^2] = a_1^2 + a_2^2 + a_3^2 + a_4^2 + \cdots + a_n^2, \text{ etc.}$$

It is usually understood that there are n terms unless otherwise stated.

5.1.1.7. *Accuracy and Precision*

If a quantity is measured several times, the degree of agreement between the measures is the *precision* of the set. Thus if the residuals are small the observations are precise, and *vice versa*. The *accuracy* of the set is the difference between the most probable value and the true value. Hence in most cases the accuracy can never be found.

A high degree of precision is no indication of great accuracy. For example an expensive watch may be precise to a second, but could easily be two hours slow, i.e. it is inaccurate by two hours.

5.1.2. TYPES OF ERROR

5.1.2.1. *Systematic and Random*

In the classification of errors two main types are distinguished according to the way they affect a result.

(i) Errors which have an entirely cumulative effect are *systematic errors*.
(ii) Errors which have a tendency to compensate one another are *random errors*, often called *accidental* errors.

It is often very dangerous to cite examples of these types of error, out of context. For example to say that 'failure to record temperatures in taping is an example of systematic error', may either be correct or incorrect according to the circumstances of the measurement. If the standardization temperature of the tape is say 50°F, and all the measurements are taken at a temperature of 80°F the errors introduced are systematic. If on the other hand the field temperatures vary between 48°F and 52°F in a random manner, the errors introduced will tend to compensate each other. For this and other reasons the temperature at which a tape is standardized should be close to the expected field temperatures. This principle of changing a systematic into a random error is applied widely in surveying where possible.

A third type of error is sometimes distinguished on the basis of mere size. This is the *gross error* or *mistake*. However since there may be systematic and random mistakes this is not a genuine third type of error. It is however true to say that they are treated in a different manner by most surveyors simply because they may often be detected by a self-checking procedure built into the particular survey process involved. Gross errors are very serious on account of their size and care must be taken to avoid them. The following list of common mistakes should serve to indicate how serious they can be.

(i) Reading a micrometer or altimeter in the wrong direction.
(ii) Misreading scale divisions.
(iii) Transposition of numbers, e.g. writing 3457 for 3475.
(iv) Mistake between observer and booker due to faulty procedure, e.g. saying 'oh' instead of 'zero' which is mistaken for 'four'.
(v) Looking up the wrong trigonometrical function, e.g. tan instead of sin.
(vi) Misidentification of control on aerial photography.
(vii) Alteration of a bearing through 180°, because the wrong face is booked, and two zeros are not observed.
(viii) Misnumbering a station in a field book so that the survey is swung through the angle between the station named and that actually observed.
(ix) Using a wrong foot/metre conversion factor, e.g. using the factor 1 metre = 3·047 99 feet. (1 metre is 3·280 8 ft but 1 foot is 0·304 799 metres.) The result looks about correct in either case.

Mistakes are the most serious of all the errors. This fact can easily be forgotten in the light of the attention paid to the adjustment of small errors by the method of least squares, and to the space taken up by the latter in survey literature, including this book. The importance of a topic is not proportional to the number of words written on the subject nor to the effort required to understand it.

5.2. SYSTEMATIC ERRORS

Systematic errors arise from some physical phenomenon or psychological tendency on the part of the observer, and may only be eradicated if the laws governing these contributory factors are known, albeit empirically. A source of systematic error in levelling is refraction, and an example of a personal tendency is *personal equation* in astronomy. The magnitude of systematic error is often difficult to assess and equally troublesome to remove. Processes most likely to be affected significantly are those involving long repetitive techniques such as levelling and traversing.

In some cases, although the magnitude of a systematic error may be assessed, its sign may be unknown. For example, the standardization of a tape is itself subject to error. If the length of the tape is quoted as 100·001 ft all field measurements will be corrected by the 0·001. The standardization itself will be in error by say ±0·0005, i.e. either plus or minus. Whatever its sign it will affect the measurement in a systematic manner. Generally this sign is unknown. If ±0·0005 is the 'probable error' (see §5.3.7) of the standardization, it becomes a *probable systematic error* when affecting the field measurement.

5.2.1. PROPAGATION OF SYSTEMATIC ERRORS

As an example of the propagation of systematic errors, consider a base line of n bays each with a systematic error $e_1, e_2, e_3 \cdots e_n$. Then the resultant systematic error of the whole base is given by

$$E = e_1 + e_2 + e_3 + \cdots + e_n$$

i.e.
$$E = [e]$$

And if $e_1 = e_2 = e_3 = \cdots = e_n = e$

$$E = ne$$

If these values of e_1, e_2, e_3, etc. are probable systematic errors, they are *either all positive or all negative*. Hence E_{PS} the probable systematic error of the base is either $+[e]$ or $-[e]$. And again if the e's are all numerically equal we have $E_{PS} = +ne$ or $-ne$. For convenience the magnitude of the probable systematic error is considered in terms of its square, i.e. $E_{PS}^2 = n^2 e^2$ thus eliminating the positive and negative signs.

If the bays are of length $l_1, l_2, l_3 \cdots l_n$ respectively, the proportional systematic errors are

$$\frac{e_1}{l_1}, \frac{e_2}{l_2}, \frac{e_3}{l_3} \cdots \frac{e_n}{l_n}$$

Let these proportional errors equal $1/K_1, 1/K_2, 1/K_3 \cdots 1/K_n$ respectively then $e_1 = l_1/K_1$; $e_2 = l_2/K_2 \cdots e_n = l_n/K_n$. Then the error of the whole base is

$$E = [e] = \frac{l_1}{K_1} + \frac{l_2}{K_2} + \frac{l_3}{K_3} + \cdots + \frac{l_n}{K_n}$$

If $l_1 = l_2 = l_3 = \cdots = l$ and $e_1 = e_2 = e_3 = \cdots = e_n = e$ and $e/l = 1/K$, $E = nl/K$. But nl is the total length of the base, hence the proportional error of the base as a whole is also $1/K$. Hence we have the obvious result that the proportional error of the whole base is not improved by increasing the length of the base, which is contrary to the result for random errors.

5.2.2. ERROR PER UNIT

In some cases both systematic and random errors are treated in terms of a common unit even if this unit was not actually measured. For example in base measurement the unit is the bay or tape length, and in levelling the 'error per mile', or 'error per kilometre' is used. This is justified on the grounds that on average there will be the same number of actual measures per mile or kilometre, and therefore the same error produced. Hence figures in terms of 'error per mile, etc.' are quoted. One should always ascertain if the assumptions behind this simplification remain valid in any one case.

5.3. RANDOM ERRORS

The theory of random errors is now considered at length. However, attention is drawn again to the importance of mistakes and systematic error, neither of which conform to exactly the same laws. They must be studied in each particular case, and the adjustment of discrepancies carried out accordingly. In theory random errors are left after mistakes and systematic errors have been removed. In practice some systematic errors will remain and, for want of an alternative, they are treated along with random errors.

5.3.1. STATISTICAL ANALYSIS OF RESULTS

In the study of random errors we are concerned with probability. If there are two balls in a hat, one white one red, the probability that we shall draw out a red ball is one in two, or $\frac{1}{2}$. This does not mean that we shall draw out a red ball in one of two chances. We might pick a white ball in the first six attempts. It means that if we make *a very large number of attempts* say 1 000, 50% of the times we shall draw out a red ball. If there are two white and ten red balls in the hat, the probability of drawing out a white ball is two in twelve or 2/12, and that of drawing out a red ball is ten in twelve, again understood that a very large number of chances have to be involved. Hence we obtain the rule that

$$\text{The probability of a result} = \frac{\text{The number of ways of getting the result}}{\text{The total number of possible results}}$$

In this way, the probability of the occurrence of an error is related to the number of times this particular error can occur, and the total possible number of all errors that can occur. Hence we can find the probability or likelihood of the occurrence of errors by analysing and counting errors of various magnitudes. For such an analysis to be valid, a very large number of errors should be 'sampled'.

5.3.2. THE HISTOGRAM

The analysis of results will be explained in terms of the game of shove halfpenny. Figure 5.1 shows the board on which a number of parallel lines are ruled at a distance apart of just over an inch, i.e. slightly larger than the diameter of the halfpenny. The player places his coin at position A just protruding over the edge of the board. The idea is to strike the coin with the wrist and get it

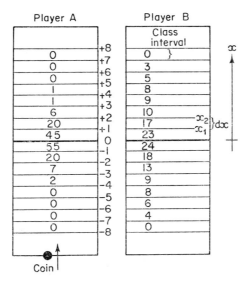

Figure 5.1

to land as close to the middle line as possible. The centre of the coin is taken as the reference point. If this lies within the range marked 0 to -1 the score is recorded for that group and so on. The figures show the results of two players A and B, both of whom took the same number of shots 157. It is obvious that A was a much better player than B. These bands between the lines are called the '*class intervals*' of the analysis. Their width has to be carefully chosen with respect to the skill of the players if a difference between them is to be detected. For instance, if the class interval had been chosen as the complete width of the board no difference in the respective skills would have been found. Again if the intervals had been very small, an abnormally large number of shots would have to be played to get more than one shot in each space. The number of times that a shot lands in a space is plotted on a block diagram called a *histogram* (figure 5.2). The histogram for A is bounded by the solid line and that for B by the broken line. The area of each histogram is the same since it represents the same number of shots; for a fair comparison to be made between two players this should always be the case.

Expressed mathematically, an error is denoted by x_1 and the class interval by dx. Hence the next error after x_1, i.e. x_2 is given by $x_2 = x_1 + dx$, and so on.

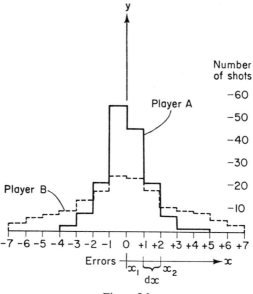

Figure 5.2

5.3.3. LAWS OF RANDOM ERROR

Continuing with the shove halfpenny example, if the number of shots is increased to a very great number, i.e. towards infinity, while at the same time the class interval is decreased, we can think of a continuous curve towards which the histograms will approach. The respective curves representing A and B's performances are given in figure 5.3. The area contained between each curve and the x-axis represents the total number of all possible errors in the system, an infinite number in theory. This is true for both curves, therefore neither curve will meet the x-axis but approach it asymptotically. For any point on the curve, the ordinate y can be considered as either the number of errors of abscissa x that occur, or the probability of the occurrence of an error of magnitude x (see §5.3.1). Hence the area under the curve represents either

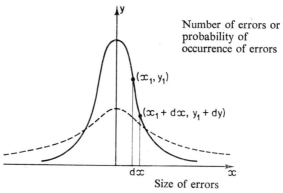

Figure 5.3

the total number of errors *n*, or the probability of all errors occurring, i.e. certainty, or a probability of 1.

From the shape of the curve the laws of random error are obtained as follows:

(i) Small errors are more frequent than large errors.
(ii) Positive and negative errors are equally likely to occur.
(iii) Very large errors seldom occur.

5.3.3.1. *Index of Precision*: h

It will be seen that the more precise an observation is the higher will be the curve, e.g. as shown in the histograms, A's performance was better than B's. Thus the height of the error curve indicates the precision of the set of observations. This index of precision is usually denoted by h. (The height of the curve is actually $h/\pi^{\frac{1}{2}}$; see §5.3.4.)

5.3.3.2. *Deviation Curve*

If player A in the shove halfpenny game had faulty eyesight he might always aim at the wrong position and thus produce a set of results with the same symmetrical distribution as above, but the whole set would spread about a position that was incorrect, i.e. he would obtain a curve as in figure 5.4 about

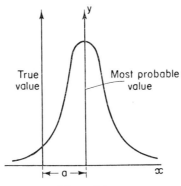

Figure 5.4

a value of $x = a$. This curve is a graph of the distribution of residuals about the most probable value. The amount by which its axis fails to coincide with $x = 0$ is the systematic error of the set of observations, in this case a. In theory it is often assumed that $a = 0$, by use of such phrases as 'assuming that there is no systematic error', and thus the graph of the distribution of residuals is taken to be coincident with the error curve. As a reminder that this is not always permissible it is better to talk in terms of 'deviations' about the most probable value, than about 'errors', unless we are actually dealing with errors.

This curve is called a *normal distribution* curve or *Gaussian curve*, after the great German mathematician who developed the subject. The equation of the curve is

$$y = \pi^{-\frac{1}{2}} h e^{-h^2 x^2} \tag{1}$$

where e is the base of natural logarithms. Since the derivation of this formula is too long to be given here, the reader should refer to H. F. Rainsford—*Survey Adjustments and Least Squares* (London: Constable, 1957).

5.3.4. PROPERTIES OF THE NORMAL CURVE

From equation (1) above when $x = 0$, $y = h\pi^{-\frac{1}{2}}$, i.e. the maximum height of the curve is $h\pi^{-\frac{1}{2}}$. Turning points occur when

$$\frac{dy}{dx} = -\pi^{-\frac{1}{2}} 2h^2 x h e^{-h^2 x^2} = -2h^2 xy$$

$$= 0 \quad \text{when} \quad x = 0;\ y = 0$$

Points of inflexion occur when

$$0 = \frac{d^2 y}{dx^2} = -2h^2 x \frac{dy}{dx} - 2h^2 y$$

$$\therefore \quad 0 = 4h^4 x^2 y - 2h^2 y = -2h^2 y(1 - 2h^2 x^2)$$

$$\therefore \quad -2h^2 y = 0 \quad \text{or} \quad 2h^2 x^2 = 1$$

i.e. $\quad x = \pm\infty \quad \text{or} \quad x = \pm 1/h2^{\frac{1}{2}}$.

Hence the curve has the shape shown in figure 5.5 and follows that of the histograms.

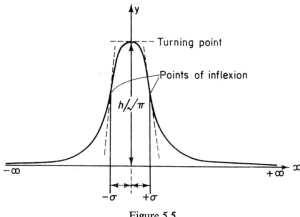

Figure 5.5

5.3.5. STANDARD ERROR, OR MEAN SQUARE ERROR (M.S.E.)

If the errors of a set of observations of a single quantity are $x_1, x_2, x_3, x_4 \cdots x_n$ we define the standard error of the set to be

$$\left(\frac{[x^2]}{n}\right)^{\frac{1}{2}} = \left(\frac{x_1^2 + x_2^2 + \cdots + x_n^2}{n}\right)^{\frac{1}{2}}$$

Standard error is usually denoted by σ. The square of the standard error is the *variance*.

Theory of Errors and Survey Adjustment

Then the variance is $\sigma^2 = \dfrac{\text{The sum of the squares of all the errors}}{\text{The total number of errors}}$

i.e. $$\sigma^2 = \int_{-\infty}^{+\infty} x^2 y \, dx \Big/ \int_{-\infty}^{+\infty} y \, dx = \int_{-\infty}^{+\infty} x^2 y \, dx \qquad (2)$$

Remembering that the area under the curve $\int_{-\infty}^{+\infty} y \, dx$ represents a probability of 1.

But $$\frac{d^2 y}{dx^2} = 4h^4 x^2 y - 2h^2 y \quad \text{from above}$$

$$\therefore \int_{-\infty}^{+\infty} \frac{d^2 y}{dx^2} dx = 4h^4 \int_{-\infty}^{+\infty} x^2 y \, dx - 2h^2 \int_{-\infty}^{+\infty} y \, dx$$

i.e. $$\left(\frac{dy}{dx}\right)_{-\infty}^{+\infty} = 0 = 4h^4 \sigma^2 - 2h^2 \quad \text{from (2)}$$

$$\therefore \sigma^2 = \frac{2h^2}{4h^4} = \frac{1}{2h^2};$$

$$\therefore \sigma = \pm 1/h(2)^{\frac{1}{2}} \qquad (3)$$

or $$h = \frac{1}{\sigma(2)^{\frac{1}{2}}}$$

The equation of the normal curve is sometimes written

$$y = \pi^{-\frac{1}{2}} h e^{-x^2/2\sigma^2}$$

5.3.6. AVERAGE ERROR η

The algebraic sum of all the errors is zero if an infinite number of errors is considered. Hence average error is defined to be the mean of all the errors taken without regard to sign, i.e. $[|x|] = \eta$. For example if the errors are $+4$, -4, $+2$, -2 the average error usually denoted by $\eta = 12/4 = 3$.

$$\eta = \frac{\int_{-\infty}^{+\infty} |x| y \, dx}{\int_{-\infty}^{+\infty} y \, dx} = \int_{-\infty}^{+\infty} |x| y \, dx, \quad \left(\text{since} \int_{-\infty}^{+\infty} y \, dx = 1\right)$$

$$= 2 \int_0^{\infty} xy \, dx = 2 \int_0^{\infty} x \frac{h}{\pi^{\frac{1}{2}}} e^{-h^2 x^2} dx$$

Let $u = -h^2 x^2$

$$\therefore du = -2h^2 x \, dx,$$

i.e. $$dx = -\frac{du}{2h^2 x}$$

when $x = 0$, $u = 0$; when $x = \infty$, $u = -\infty$.

$$\therefore \eta = -2 \int_0^{-\infty} \pi^{-\frac{1}{2}} x h e^u \frac{du}{2h^2 x} = -\frac{\pi^{-\frac{1}{2}}}{h} [e^u]_0^{-\infty} = \frac{1}{h \pi^{\frac{1}{2}}} \qquad (4)$$

5.3.7. PROBABLE ERROR

Consider the error curve again. There will be an error x that divides the area under the curve into two equal parts, each representing a probability of $\frac{1}{2}$. This error x is called the *probable error*. The term is unfortunate since there is nothing probable about it. The most likely error of a set, i.e. the most numerous is zero, i.e. when y is a maximum. This is common sense since we are more likely to get an observation right than to get it wrong. The term 50% error has been suggested instead of 'probable error' (p.e.) but has not been widely accepted.

The probable error will be given by the values of the limits $\pm\epsilon$ which satisfy the equation

$$h\pi^{-\frac{1}{2}} \int_{-\epsilon}^{+\epsilon} e^{-h^2 x^2} \, dx = \tfrac{1}{2}$$

This cannot be integrated exactly, but may be tabulated as follows. Let $hx = t$.
$\therefore$ when $x = \epsilon$, $t = h\epsilon$; when $x = 0$, $t = 0$, $dx = dt/h$ then

$$\tfrac{1}{2} = 2h\pi^{-\frac{1}{2}} \int_0^{+h\epsilon} e^{-t^2} \frac{dt}{h} = 2\pi^{-\frac{1}{2}} \int_0^{h\epsilon} e^{-t^2} \, dt$$

i.e.
$$\tfrac{1}{2} = 2\pi^{-\frac{1}{2}} \int_0^{h\epsilon} \left\{ 1 - t^2 + \frac{t^4}{2!} - \frac{t^6}{3!} + \cdots \right\} dt$$

$$= 2\pi^{-\frac{1}{2}} \left[t - \frac{t^3}{3} + \frac{t^5}{10} - \frac{t^7}{42} + \cdots \right]_0^{h\epsilon = t}$$

First approximation $2\pi^{-\frac{1}{2}} t = \tfrac{1}{2}$

$$\therefore \quad t \simeq 0{\cdot}45 \quad \text{or} \quad h\epsilon \simeq 0{\cdot}45$$

The expression $2\pi^{-\frac{1}{2}} \int_0^{h\epsilon} e^{-t^2} \, dt$ is called the 'probability integral' which is $\tfrac{1}{2}$ when $h\epsilon = 0{\cdot}4769$.

5.3.8. RELATIONSHIPS BETWEEN σ, η, AND ϵ

Using the values $h\epsilon = 0{\cdot}4769$, $\pi^{\frac{1}{2}} = 1{\cdot}7724$, $2^{\frac{1}{2}} = 1{\cdot}4142$ and remembering that $\sigma = 1/h2^{\frac{1}{2}}$; $\eta = 1/h\pi^{\frac{1}{2}}$; $\epsilon = 0{\cdot}4769/h$ it follows that:

$$\eta = 0{\cdot}797\,9\,\sigma \tag{5}$$
$$\epsilon = 0{\cdot}674\,5\,\sigma \simeq \tfrac{2}{3}\sigma \tag{6}$$

Thus we need only work in terms of any one of these statistical quantities. Although η is simpler to apply, σ is more commonly used since it gives a better estimate of precision for a relatively small number of observations (see example (1) §5.3.9).

5.3.9. STANDARD ERROR FROM RESIDUALS

Let $M_1, M_2 \cdots M_n$ be n measurements of a quantity whose true value is T, and whose most probable value is V. Assuming that there is no systematic error in the measurements, V will be close to T, and will approach closer and

Theory of Errors and Survey Adjustment 211

closer to T as the number of observations is increased to a very large number. Let the true errors Δ, and the residuals v, be given by, respectively,

$$\Delta_1 = T - M_1 \quad \text{and} \quad v_1 = V - M_1$$
$$\cdots \text{etc.} \quad\quad\quad\quad \cdots \text{etc.}$$
$$\Delta_n = T - M_n \quad\quad v_n = V - M_n$$

Then $\Delta_1 = (T - V) + v_1$ etc.

$$\therefore \quad [\Delta^2] = n(T - V)^2 + [v^2] + 2(T - V)[v]$$
$$= n(T - V)^2 + [v^2] \quad \left(\text{since } V = \frac{[M]}{n}, [v] = 0\right)$$

The standard error σ is given by

$$\sigma^2 = [\Delta^2]/n \tag{7}$$

Let the *estimate* of the standard error derived from *residuals* treated as errors be S, i.e. $S^2 = [v^2]/n$

Then
$$\sigma^2 = [\Delta^2]/n = (T - V)^2 + [v^2]/n$$
$$= (T - V)^2 + S^2$$

But $(T - V)$ is the error of the mean, the estimate of which is given by

$$(T - V)^2 = \sigma^2/n \quad \text{(see §5.4.2.2, equation (18))}$$

Whence
$$\sigma^2 = nS^2/(n - 1) = [v^2]/(n - 1)$$

For a very large number of observed quantities, $n - 1$ tends to n and the two formulae $[\Delta^2]/n$ and $[v^2]/(n - 1)$ using errors and residuals become numerically identical. For a limited number of observations, as in practice, the standard error derived from *residuals* is best obtained from

$$\sigma = \{[v^2]/(n - 1)\}^{\frac{1}{2}} \tag{8}$$

though if *errors*, such as triangle closures are used, the correct formula is (7).

Practical Examples:
(1) *Errors.* The triangular closures listed below were produced by two observers A and B using similar equipment and methods. By an analysis of the average, standard, and probable errors of each set of results, assess the relative quality of the observations produced by each surveyor. Sperical excess is negligible.

A's triangle closures were $+10''$, $-1''$, $+2''$, $-9''$, $+1''$, $-1''$, $+6''$, $-1''$, $+2''$, $-7''$; B's were $+5''$, $-3''$, $+4''$, $-4''$, $-6''$, $-2''$, $+1''$, $+7''$, $-3''$, $+5''$.

The average error in each case is the same, i.e. $\eta_A = 4''$ and $\eta_B = 4''$. Hence each observer's work seems of the same standard. Using (5) and (6) estimates of σ and ϵ are as follows:

$$\sigma = \eta/0{\cdot}797\,9 = 5''; \quad \epsilon = 0{\cdot}674\,5\,\sigma = 3{\cdot}''4$$

However, computing the respective values of σ from the formula

$$\sigma = \pm([\Delta^2]/n)^{\frac{1}{2}}$$

For A; $\sigma_A = (278/10)^{\frac{1}{2}} = \pm 5''\!.3$; $\epsilon_A = 0.674\ 5\sigma = \pm 3''\!.6$

which agrees closely with the above values. For B;

$$\sigma_B = (190/10)^{\frac{1}{2}} = \pm 4''\!.4; \qquad \epsilon_B = \pm 3''\!.0$$

The value of the probable error ϵ derived from $\epsilon = 0.674\ 5\ ([\Delta^2]/n)^{\frac{1}{2}}$ will only agree exactly with that derived from $\epsilon = 0.674\ 5\eta/0.797\ 9$ if a very large number of purely random errors is considered, which is not the case in most practical examples. Thus it is seen that B's observations are slightly better than A's, and that *σ gives a better estimate of accuracy than η.*

Example (2) *Residuals.* The following independent results were obtained for nine measures of an angle. Compute the standard and probable deviations for the set of results.

Observed angle (1)	Residuals (2)	(3)	(4)
109 25 06·3	1·4		1·96
07·2	0·5		0·25
10·4		2·7	7·29
04·3	3·4		11·56
09·6		1·9	3·61
05·8	1·9		3·61
07·9		0·2	0·04
08·3		0·6	0·36
09·1		1·4	1·96
9)68·9	+7·2	−6·8	30·64

M.P.V. 07·7
Remainder is 0·4

Column (2) gives the positive residuals, and column (3) the negative ones. The m.p.v. is the arithmetic mean taken to the nearest 0·1, thus leaving a remainder of 0·4. This is the amount by which the sum of the positive residuals will differ from the negative residuals, and therefore checks the calculation of the arithmetic mean. The residuals are calculated in the sense—m.p.v. minus observed, e.g.

$$07\cdot 7 - 06\cdot 3 = +1\cdot 4,\ \text{etc.}$$

When calculating $[v^2]$ there is normally no need to write down each v^2 since the total summation of squares may be accumulated on the calculating machine. They are written down here for clarity.

The standard deviation is then

$$\sigma = \pm(30\cdot64/8)^{\frac{1}{2}} = \pm1\cdot95$$
$$\epsilon = \pm0\cdot674\,5\sigma = \pm1\cdot32$$

A method of checking the computation of $[v^2]$ is given below in §5.4.6.

5.3.10. REJECTION OF OBSERVATIONS

The standard deviation provides the basis for deciding whether or not to reject an observation which differs from others. From the theory of random errors we would expect a large error now and again, but not very often, i.e. we can estimate the probability of the occurrence of errors outside certain limits, and if in practice errors occur more frequently than this, we will be justified in rejecting the observations which give rise to them. The criterion usually taken is three times the probable error, i.e. $\pm3\epsilon$. The problem is then to compute the value of the probability integral within the limits $+3\epsilon$ and -3ϵ, or twice the value of the integral from 0 to $+3\epsilon$. This gives the probability of the occurrence of errors within these limits, and subtracted from unity the chance of an error lying outside the limits.

When
$$2\pi^{-\frac{1}{2}} \int_0^{h\epsilon} e^{-t^2} dt = \tfrac{1}{2}$$

$h\epsilon$ is $0\cdot476\,9$, then we wish to find the value of this integral when $h\epsilon = t = 3 \times 0\cdot476\,9 = 1\cdot431\,0$.

Working out the series to the eleventh power gives

$$2\pi^{-\frac{1}{2}} \int_0^{1\cdot4310} e^{-t^2} dt = 0\cdot999$$

The probability that an error lies outside the limits -3ϵ and $+3\epsilon$ is 1:1 000. On this basis one may reject an observation whose residual exceeds five times the probable error. In the above example (2) a residual would have to be $6''\!.5$ before it could be rejected with safety. Generally speaking, the use of this rejection criterion causes more observations to be retained than one would probably do by inspection.

On a similar basis, the probable error or deviation is used as a guide to likely error. In this instance the limit of $\pm3\epsilon$ gives the probability that one error in 1000 will exceed 3ϵ. This is the basis on which work is planned, e.g. if coordinates are required to 0·1 m the survey would be designed to give a probable error of 0·03 m, with the expectation that only one coordinate in 1000 would be outside the required precision.

5.4. PRINCIPLE OF LEAST SQUARES

(A)

Let the arithmetic mean of n observed values $M_1, M_2, M_3, M_4 \cdots M_n$ be V, and let W be any estimate of the most probable value other than V. It is assumed that the observations are equally reliable. Then forming residuals for each of the values V and W we have

$$\left.\begin{array}{c} V - M_1 = v_1 \\ V - M_2 = v_2 \\ \cdots \text{etc.} \\ V - M_n = v_n \end{array}\right\}$$

$$\left.\begin{array}{c} W - M_1 = w_1 \\ W - M_2 = w_2 \\ \cdots \text{etc.} \\ W - M_n = w_n \end{array}\right\} \quad (9)$$

Taking the sum of the squares of each set of residuals we have

$$[v^2] = V^2 - 2VM_1 + M_1^2 + V^2 - 2VM_2 + M_2^2 + \cdots + M_n^2$$
$$= nV^2 - 2V[M] + [M^2]$$
$$[w^2] = W^2 - 2WM_1 + M_1^2 + W^2 - 2WM_2 + M_2^2 + \cdots + M_n^2$$
$$= nW^2 - 2W[M] + [M^2]$$
$$\therefore [w^2] - [v^2] = nW^2 - 2W[M] - nV^2 + 2V[M]$$

but $\quad V = [M]/n$

$$\therefore [W^2] - [v^2] = nW^2 - 2W[M] - \frac{n}{n^2}[M]^2 + \frac{2}{n}[M][M]$$

$$= n\left(W^2 - \frac{2W[M]}{n} + \frac{[M]^2}{n^2}\right)$$

$$= n\left(W - \frac{[M]}{n}\right)^2 \quad \text{which is positive}$$

$$\therefore [v^2] < [w^2] \quad (10)$$

Since W was *any* value other than V expression (10) holds for *all* values other than V. Hence we have the principle of least squares: *the sum of the squares of the residuals derived from the arithmetic mean is a minimum* (see §5.4B).

This principle is used in reverse to find the most probable value—the arithmetic mean—*sic*; the most probable value of a quantity is found from a series of observations of equal reliability if the sum of the squares of the residuals derived from it and the observed values is a minimum. For

$$[v^2] = (V - M_1)^2 + (V - M_2)^2 + \cdots + (V - M_n)^2 \quad (11)$$

This expression will have a turning point when $d[v^2]/dv = 0$, and the turning value corresponding to this turning point will be a minimum since the function (11) is formed of square terms, i.e.

$$\frac{d[v^2]}{dv} = \frac{d[v^2]}{dV} = 0 \quad \{M_1 \cdots M_n \text{ are constants}\}$$

$$0 = 2(V - M_1) + 2(V - M_2) + \cdots + 2(V - M_n)$$

i.e. $$nV = M_1 + M_2 + \cdots + M_n$$
i.e. $$V = [M]/n \quad \text{the arithmetic mean.}$$

(B)
A more general treatment is as follows.
If we have two white and ten red balls in one hat, and three white and seven red in another, the probabilities of selecting a white ball from each hat separately are 2/12 and 3/10 respectively. Suppose we now take one ball from each hat *simultaneously* the chances of taking a white ball from each hat are

$$\frac{\text{Number of ways of getting a result}}{\text{Total number of possible results}}$$

For each white ball in the first hat we have three chances of selecting a white ball from the second, therefore the total number of ways of getting two white balls is 2 × 3. For each of the 12 balls from the first hat there are ten possible selections from the second, therefore the total number of possible results is 12 × 10. Hence the chances of taking a white ball from each hat *simultaneously* are ($\frac{2}{12} \times \frac{3}{10}$). Hence the *probability of the simultaneous occurrence of two events which are independent of each other is the product of their respective probabilities of occurrence*. Consider now a series of errors which have a normal distribution. Let the probability of the occurrence of errors $x_1, x_2, x_3 \cdots x_n$ be respectively $y_1, y_2, y_3 \cdots y_n$ then

$$y_1 = \frac{h_1}{\pi^{\frac{1}{2}}} e^{-h_1^2 x_1^2} \cdots y_n = \frac{h_n}{\pi^{\frac{1}{2}}} e^{-h_n^2 x_n^2},$$

where $h_1, h_2 \cdots h_n$ are the respective indices of precision.
The probability of their occurrence *simultaneously* is

$$y_1 y_2 \cdots y_n = \left(\frac{h_1}{\pi^{\frac{1}{2}}}\right)\left(\frac{h_2}{\pi^{\frac{1}{2}}}\right) \cdots \left(\frac{h_n}{\pi^{\frac{1}{2}}}\right) e^{-(h_1^2 x_1^2 + \cdots + h_n^2 x_n^2)} \tag{12}$$

Now, we shall obtain the most probable value of the quantity if we can produce observations which sample all possible errors which will tend to cancel each other, i.e. the most probable value is obtained when the simultaneous occurrence of *all* errors is a *maximum*, i.e. when (12) is a maximum. This will occur when $[h^2 x^2]$ is a minimum. If all the observations are of equal reliability, as has been assumed to this stage, $h_1 = h_2 = \cdots = h_n = h$ say, then the most probable value is obtained when $h^2[x^2]$ is a minimum, or simply $[x^2]$ is a minimum: which is the principle of least squares. It will be seen below in §5.4.4 that the weights of observations of different reliabilities are proportional to the reciprocals of their respective standard errors, hence from (12), for observations of different weight, the most probable value is obtained when the sum of the squares of the weighted residuals (or errors) is a minimum.

5.4.1. THE COMBINATION OF RANDOM ERRORS

In many cases the value of a quantity which has not been directly observed is required, and also its standard error. For example if the sides a and b of a rectangle are measured with standard errors in each of σ_a and σ_b, we wish to know the standard error in the area A calculated from $A = ab$. Consider first actual errors da, db, and dA respectively in a, b, and A. Then

$$A + dA = (a + da)(b + db) = ab + a\,db + b\,da + da\,db$$

i.e. $\quad dA = a\,db + b\,da + da\,db$

Since da and db are small quantities, we neglect the term $da\,db$ of the second order of small quantities. It should however be noted that there are cases in which this assumption is invalid.

More generally if $A = f(a, b)$ then

$$dA = \frac{\partial f}{\partial a} da + \frac{\partial f}{\partial b} db = P\,da + Q\,db \tag{13}$$

where P and Q are the partial derivatives of A with respect to each variable a and b in turn.

Suppose the errors of a and b have a normal distribution with respective standard errors σ_a and σ_b, and we require to find the standard error in A. Suppose da can take any of m values independently, and db can take any of n values independently, both sets normally distributed. Then dA can take any of mn values, since each error da can be combined with any of the n values of db, and each error db can be combined with any of the m errors da. The standard error of A is then given by

$$\sigma_A{}^2 = \frac{[dA^2]_0^{mn}}{mn}$$

i.e. $\quad \sigma_A{}^2 = \dfrac{1}{mn}\{nP^2[da^2]_1^m + mQ^2[db^2]_1^n + 2PQ[da\,db]\}$

$$= P^2\frac{[da^2]_1^m}{m} + Q^2\frac{[db^2]_1^n}{n} + \frac{2PQ}{mn}[da\,db]$$

Now in the summation $[da\,db]$ each da is combined with all values of db, i.e. we have terms like $da_1[db]$. Since $[db]$ tends to zero the term $[da\,db]$ also tends to zero and may be neglected. This will occur *only if the errors da and db are independent*, i.e. if their observed values are also independent.

whence $\quad \sigma_A{}^2 = P^2\sigma_a{}^2 + Q^2\sigma_b{}^2$

or $\quad \sigma_A{}^2 = \left(\dfrac{\partial f}{\partial a}\right)^2 \sigma_a{}^2 + \left(\dfrac{\partial f}{\partial b}\right)^2 \sigma_b{}^2 \tag{14}$

In the above exposition we considered that A was a function of only two variables a and b. By a similar argument we obtain the general case for A a function of n independent variables $a_1, a_2, a_3 \cdots a_n$

$$\sigma_A{}^2 = \left(\frac{\partial f}{\partial a_1}\sigma_{a_1}\right)^2 + \left(\frac{\partial f}{\partial a_2}\sigma_{a_2}\right)^2 + \cdots + \left(\frac{\partial f}{\partial a_n}\sigma_{a_n}\right)^2 \tag{15}$$

Since the probable error ϵ is related to the standard error σ by $\epsilon = 0.6745\sigma$ expression (15) can be extended to probable errors as follows

$$\epsilon_A{}^2 = \left(\frac{\partial f}{\partial a_1}\epsilon_{a_1}\right)^2 + \left(\frac{\partial f}{\partial a_2}\epsilon_{a_2}\right)^2 + \cdots + \left(\frac{\partial f}{\partial a_n}\epsilon_{a_n}\right)^2 \qquad (16)$$

These results (15) and (16) are most important, and in conjunction with the definitions of standard error and standard deviation, form the basis for the solution of most problems of the propagation of random errors in surveying.

5.4.2. STANDARD ERRORS OF SOME COMMONLY USED FUNCTIONS

5.4.2.1. Standard Error of the Sum of n Quantities

If $A = a_1 + a_2 + a_3 + \cdots + a_n$ each with respective standard errors σ_A, $\sigma_1, \sigma_2, \sigma_3 \cdots \sigma_n$ then

$$\sigma_A{}^2 = 1 \cdot \sigma_1{}^2 + 1 \cdot \sigma_2{}^2 + \cdots + 1 \cdot \sigma_n{}^2$$
$$= \sigma_1{}^2 + \sigma_2{}^2 + \cdots + \sigma_n{}^2 = [\sigma^2]$$

If $\sigma_1 = \sigma_2 = \cdots = \sigma_n = \sigma$ this reduces to $\sigma_A{}^2 = n\sigma^2$, i.e.

$$\sigma_A = \pm \sigma n^{\frac{1}{2}} \qquad (17)$$

5.4.2.2. Standard Error of the Mean of n Quantities

If A is the arithmetic mean of n independent quantities $a_1, a_2, a_3 \cdots a_n$, i.e. $A = (a_1 + a_2 + a_3 + \cdots + a_n)/n$ and their respective standard errors are σ_A, $\sigma_1, \sigma_2 \cdots \sigma_n$

$$\sigma_A{}^2 = \frac{1}{n^2}(1 \cdot \sigma_1{}^2 + 1 \cdot \sigma_2{}^2 + \cdots + 1 \cdot \sigma_n{}^2)$$
$$= \frac{1}{n^2}(\sigma_1{}^2 + \sigma_2{}^2 + \cdots + \sigma_n{}^2) = [\sigma^2]/n^2$$

If $\sigma_1 = \sigma_2 = \cdots = \sigma_n = \sigma$ this reduces to

$$\sigma_A{}^2 = \frac{n\sigma^2}{n^2} = \sigma^2/n$$

i.e.
$$\sigma_A = \pm \sigma/n^{\frac{1}{2}} \qquad (18)$$

Deriving σ from residuals $\sigma = ([v^2]/n - 1)^{\frac{1}{2}}$, and the standard deviation of the mean is

$$\therefore \sigma_A = \pm \left\{\frac{[v^2]}{n(n-1)}\right\}^{\frac{1}{2}} \qquad (19)$$

Example: If $A = ka$ where k is a constant, say an integer, the standard error of A, σ_A is $\pm k\sigma_a$.

5.4.3. WEIGHTED OBSERVATIONS

So far we have only considered observations which are equally reliable, i.e. observations of equal weight. The theory is now extended to include observations of different weights.

Suppose an angle is observed six times giving values M_1, M_2, M_3, M_4, M_5, M_6 all of which are equally reliable, then the m.p.v. is the arithmetic mean

$$V_1 = \tfrac{1}{6}(M_1 + M_2 + M_3 + M_4 + M_5 + M_6)$$

and $\qquad 6V_1 = M_1 + M_2 + \cdots + M_6$

If the same angle is further observed three times to give values M_7, M_8, M_9, each with the same reliability as before, the m.p.v. for the last three is $V_2 = \tfrac{1}{3}(M_7 + M_8 + M_9)$

$$\therefore \quad 3V_2 = M_7 + M_8 + M_9 \tag{20}$$

and the m.p.v. for all nine observations is

$$V = \tfrac{1}{9}(M_1 + M_2 + \cdots + M_9)$$

Now if only V_1 and V_2 are known, and that they were derived from six and three observations respectively, we derive the m.p.v. as follows

$$V = \tfrac{1}{9}\{(M_1 + M_2 + \cdots + M_6) + (M_7 + M_8 + M_9)\}$$
$$= \frac{6V_1 + 3V_2}{9} \tag{21}$$

We have therefore weighted the values V_1 and V_2 according to the number of observations of equal weight from which they are derived. This principle is extended to apply to weighted observations even if they were not actually derived from a number of basic observations of equal weight; we merely consider that they have so been obtained, or derived from some *fictitious observations of unit weight*.

If an observed value x is considered to have weight p with respect to a value y of weight q, x can be considered as the mean of p observations w_1, w_2, $w_3 \cdots$ etc. of the unit weight, and y as the mean of q observations z_1, $z_2 \cdots$ etc. also of unit weight. Hence the m.p.v. is the arithmetic mean of $p + q$ observations of unit weight, i.e.

$$V = \frac{1}{p+q}\{(w_1 + w_2 + w_3 \text{ to } p \text{ terms}) + (z_1 + z_2 \text{ to } q \text{ terms})\}$$

$$V = \frac{1}{p+q}(px + qy) = \frac{px + qy}{p+q} \tag{22}$$

This is the *weighted mean*.

5.4.4. WEIGHTS AND STANDARD DEVIATIONS

Let the standard deviation of a fictitious observation of unit weight be σ, and the standard deviations of x and y be σ_x and σ_y. Then by (18) $\sigma_x = \sigma/p^{\frac{1}{2}}$ and $\sigma_y = \sigma/q^{\frac{1}{2}}$

$$\therefore \quad p/q = \sigma_y^2/\sigma_x^2$$

Hence the weights of the observations are inversely proportional to the squares of their respective standard deviations; (or probable deviations since $\epsilon = \sigma \times$ constant).

5.4.5. STANDARD DEVIATION OF THE WEIGHTED MEAN

The residuals v_x and v_y are given by

$$\left.\begin{array}{l} V - x = v_x \\ V - y = v_y \end{array}\right\} \tag{23}$$

Since x is the weight of p, v_x is also of weight p, and s.d. σ_x; and y is of weight q, v_y is also of weight q, and s.d. σ_y. Multiplying each residual by the square root of its weight, remembering the result of example (§5.4.2.2) above, we have

and
$$\begin{array}{l} \text{the s.d. of } v_x p^{\frac{1}{2}} \text{ is } \sigma_x p^{\frac{1}{2}} = \sigma \quad \text{(from §5.4.4)} \\ \text{the s.d. of } v_y q^{\frac{1}{2}} \text{ is } \sigma_y q^{\frac{1}{2}} = \sigma \end{array}$$

i.e. $v_x p^{\frac{1}{2}}$ and $v_y q^{\frac{1}{2}}$ are of equal weight, hence we apply (8) for the standard deviation σ_v of the set of observations now of equal weight. Thus

$$\sigma_v = \pm \left(\frac{pv_x^2 + qv_y^2}{n-1}\right)^{\frac{1}{2}}$$

And the s.d. of the weighted mean is then $\pm \sigma_v/(p+q)^{\frac{1}{2}}$ since the total number of fictitious observations of unit weight is $p+q$. In general, if we have observed values $x_1, x_2, x_3 \cdots x_n$ of respective weights $p_1, p_2, p_3 \cdots p_n$, the standard deviation is

$$\pm \left\{\frac{(p_1 v_1^2 + p_2 v_2^2 + p_3 v_3^2 + \cdots + p_n v_n^2)}{n-1}\right\}^{\frac{1}{2}}$$

i.e.
$$\sigma = \pm \left(\frac{[pv^2]}{n-1}\right)^{\frac{1}{2}} \tag{24}$$

The standard deviation σ_v of the weighted mean V (the m.p.v.) is

$$\sigma_v = \pm \left\{\frac{[pv^2]}{[p](n-1)}\right\}^{\frac{1}{2}} \tag{25}$$

and the standard deviation σ_r of an observation of weight p_r is

i.e.
$$\sigma_r = \pm \left\{\frac{[pv^2]}{p_r(n-1)}\right\}^{\frac{1}{2}} \tag{26}$$

5.4.6. CHECKS ON THE COMPUTATION OF THE MOST PROBABLE VALUE AND THE SUM OF THE SQUARES OF RESIDUALS

Consider n independent values $x_1, x_2, x_3 \cdots x_n$, whose respective weights are $p_1, p_2, p_3 \cdots p_n$.

The residuals $v_1, v_2, v_3 \cdots v_n$, are given by

$$v_1 = V - x_1, \quad v_2 = V - x_2; \quad \cdots \quad v_n = V - x_n \tag{27}$$

$$V = \frac{[px]}{[p]} \tag{28}$$

After slight reduction it follows that

$$[pv] = p_1 v_1 + p_2 v_2 + \cdots + p_n v_n = 0 \tag{29}$$

This check verifies the computation of V and the v's. For observations of equal weight this check reduces to $[v] = 0$. In practice neither of these expressions is satisfied exactly if the value of V is rounded off, though the remainder gives the amount by which these summations should disagree (see example (2) §5.3.9 above).

Again from (27) and (28) after slight reduction

$$[pv^2] = [px^2] - \frac{[px]^2}{[p]} \tag{30}$$

This formula is difficult to work with in practice on account of the large values involved. However it may be reduced to a manageable form. Let X be any convenient value of the observed quantity, not V, and from it form 'residuals' $l_1, l_2, l_3 \cdots l_n$, i.e.

$$l_1 = X - x_1; \quad l_2 = X - x_2; \cdots; \quad l_n = X - x_n \tag{31}$$

Substituting for the values of x in (30) gives

$$[pv^2] = p_1(X^2 - 2Xl_1 + l_1^2) + p_2(X^2 - 2Xl_2 + l_2^2) + \cdots$$
$$+ p_n(X^2 - 2Xl_n + l_n^2) - \{p_1(X - l_1)$$
$$+ p_2(X - l_2) + \cdots + p_n(X - l_n)\}^2/[p]$$

which reduces to

$$[pv^2] = [pl^2] - \frac{[pl]^2}{[p]} \tag{32}$$

Since the values of the l's are small (32) is a convenient formula.

For observations of equal weight (32) reduces to

$$[v^2] = [l^2] - \frac{[l]^2}{n} \tag{33}$$

5.4.7. EXAMPLE OF THE WEIGHTED MEAN AND ITS STANDARD DEVIATION

The following results together with their respective weights were obtained for an astronomical azimuth. Compute the most probable value of the azimuth, the standard deviation of the set, and the standard deviation of the m.p.v. The

x	l	p	pl	v	pv	pv²	pl²
273° 51' 16"	11	2	22	−3".4	6·8 +	23·1	242
18	13	1	13	−5·4	5·4	29·2	169
10	05	2	10	+2·6	5·2	13·5	50
05	00	1	00	+7·6	7·6	57·8	0
17	12	1	12	−4·4	4·4	19·4	144
19	14	2	28	−6·4	12·8	81·9	392
06	01	1	01	+6·6	6·6	43·5	1
07	02	1	02	+5·6	5·6	31·4	4
11	06	2	12	+1·6	3·2	05·1	72
12	07	2	14	+0·6	1·2	00·7	98
	[p] 15		[pl] 114		29·4 = 29·4	305·6	1172

$$V = x + \frac{[pl]}{[p]}$$

$= 273° 51' 05" + 07 \cdot 6$

$= 273° 51' 12" 6$

$[pl^2] = 1\ 172$

$-[pl]^2/[p] = -866 \cdot 4$

$\therefore [pv]^2 = 305 \cdot 6$ check

value of the base X is chosen to be $273° 51' 05"$. For the first observation $l = 273° 51' 16" - X = 11"$, etc. The weight of the observation is 2, hence $pl = 22$. The m.p.v. V is given by $V = X + [pl]/[p] = X + 114/15 = 07".6 + 273° 51' 05"$.

$v_1 = 273° 51' 12".6 - 273° 51' 16" = -3".4$, etc.

$p_1 v_1 = 2 \times -3".4 = -6".8$, $p_1 v_1^2 = 23 \cdot 1$, etc.

$\sigma = \pm \{[pv^2]/(n-1)\}^{\frac{1}{2}} = \pm (305 \cdot 6/9)^{\frac{1}{2}} = \pm 5 \cdot 83$

$\sigma_v = \pm \sigma/[p]^{\frac{1}{2}} = \pm 5 \cdot 83/15^{\frac{1}{2}} = \pm 1 \cdot 51$

5.5. THE ADJUSTMENT OF OBSERVATIONS BY THE METHOD OF LEAST SQUARES

The purpose of an adjustment is to render a series of observed quantities consistent within themselves and with geometrical or other data. Once consistency has been achieved, computations involving the adjusted values will give unique results. This is of great assistance to the computor since it permits him to check his work.

Since the observed values are the best available evidence concerning the true magnitude of the desired quantities, any adjustment process must achieve consistency with as little disturbance to the observations as possible.

A good adjustment method should be capable of giving due consideration to all relevant factors, with correct emphasis on the important ones, and permit the simultaneous interaction of these factors in the derivation of the adjusted results. In most semi-rigorous (i.e. non-least squares) adjustments arbitrary selections are required, and adjustment is carried out step by step, with the previous results affecting successive answers.

The least squares method of adjustment satisfies all of the above requirements, since it produces a set of consistent values—the most probable values—by simultaneous consideration of all factors, whilst at the same time causing 'least harm' to the observations.

5.5.1. OBSERVATION AND CONDITION EQUATIONS

In §5.4. it was seen that the most probable values—the m.p.v's—are obtained from the observed values, when the sum of the squares of the residuals derived from these m.p.v's and observed values, is a minimum. There are two slightly different methods of deriving the m.p.v's from the observations by the application of this principle of least squares, viz.

(i) The *Method of Observation Equations*.
(ii) The *Method of Condition Equations*.

To illustrate these methods, their differences and their strong similarities, we shall consider a simple adjustment example. In figure 5.6 the angles AOB,

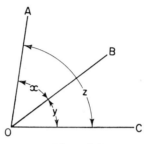

Figure 5.6

BOC, and AOC are independently observed; i.e. the value of one does not depend in any way on the value obtained for the others. The purpose of the adjustment is to obtain the most probable values of these angles.

Denoting the m.p.v's of angles AOB, BOC, and AOC by x, y, and z respectively, their corresponding observed values by M_x, M_y, and M_z and the residuals by v_x, v_y, and v_z, by definition of a residual

$$v_x = x - M_x$$
$$v_y = y - M_y \qquad (1)$$
$$v_z = z - M_z$$

These equations (1) are called '*observation equations*'. An *observation equation* is an equation expressing the result of the direct observation of a quantity. In some cases computed values of quantities are treated as though they were directly measured.

If $M_x = 23° 46' 20''$, $M_y = 17° 18' 23''$, and $M_z = 41° 04' 40''$ the observation equations are

$$v_x = x - 23° 46' 20''$$
$$v_y = y - 17° 18' 23'' \qquad (2)$$
$$y_z = z - 41° 04' 40''$$

The m.p.v.'s x, y, and z are given when $[v^2]$ is a minimum, i.e. $v_x^2 + v_y^2 + v_z^2$ is a minimum. There is however another factor to be considered; the m.p.v's

Theory of Errors and Survey Adjustment 223

must also be geometrically consistent, i.e. in this case $x + y$ must equal z. The m.p.v's are subject to a condition expressed as the '*condition equation*'

$$x + y = z \tag{3}$$

A condition equation is an equation expressing the relationship between most probable values, and or between most probable values and fixed quantities. Whichever method of adjustment is employed this condition equation must be satisfied exactly, and therefore the required unknowns are not independent of one another. Any two can be chosen as *independent* variables, and the third the *dependent* variable. x and y are chosen to be the independent variables, and z the dependent one. This choice will not effect the result in any way.

In the observations equations method of adjustment, the values of x and y are found directly in terms of the observed values M_x, M_y, and M_z; whereas, in the condition equations method, the values of the residuals v_x, v_y, and v_z are found from a consideration of the condition equation $x + y = z$, and thereafter the m.p.v's are derived by the application of these residuals to the observed values, equations (1). We shall work the algebraic and numerical examples above, to illustrate these points.

5.5.2. OBSERVATIONS EQUATIONS

Since $z = x + y$, the observation equations (1) and (2) are rewritten as

$$\begin{aligned} v_x &= x - M_x \\ v_y &= y - M_y \\ v_z &= x + y - M_z \end{aligned} \tag{4}$$

$$\begin{aligned} v_x &= x - 23° \, 46' \, 20'' \\ v_y &= y - 17° \, 17' \, 23'' \\ v_z &= x + y - 41° \, 04' \, 40'' \end{aligned} \tag{5}$$

Now the m.p.v's are given when $[v^2]$ is a minimum, i.e. when

$$v_x^2 + v_y^2 + v_z^2 = (x - M_x)^2 + (y - M_y)^2 + (x + y - M_z)^2 \tag{6}$$

is a minimum.

Since x and y are independent of each other the minimum value of (6) is given when the first derivatives with respect to x and y are both zero. A second order expression with positive coefficients has a minimum turning point when its first derivative is zero, i.e. when (differentiating with respect to x)

$$2(x - M_x) + 2(x + y - M_z) = 0$$

i.e. $$2x + y - M_x - M_z = 0 \tag{7}$$

and (differentiating with respect to y)

$$2(y - M_y) + 2(x + y - M_z) = 0$$

i.e. $$x + 2y - M_y - M_z = 0 \tag{8}$$

Eliminating y from (7) and (8)

$$x = \tfrac{1}{3}(2M_x + M_z - M_y)$$

Substituting the numerical values $x = 23° 46' 19''$.
   Again, eliminating x from (7) and (8)

$$y = \tfrac{1}{3}(2M_y + M_z - M_x)$$

and substituting values $y = 17° 18' 22''$, and since $z = x + y$, $z = 41° 04' 41''$. Substituting these m.p.v's in equations (5) gives $v_x = -1$, $v_y = -1$, and $v_z = +1$.

5.5.3. CONDITION EQUATIONS

In this method we concentrate on the condition equation $z = x + y$. From equations (1)

$$M_x + v_x + M_y + v_y = M_z + v_z \tag{9}$$

Putting $M_x + M_y - M_z = E$, (9) reduces to

$$v_x + v_y - v_z + E = 0 \tag{10}$$

or

$$v_z = v_x + v_y + E \tag{11}$$

This is the condition equation to be satisfied by the residuals. Now the m.p.v's are obtained when $[v^2]$ is a minimum, i.e. when

$$v_x^2 + v_y^2 + v_z^2 = v_x^2 + v_y^2 + (v_x + v_y + E)^2 \tag{12}$$

is a minimum. Since x and y are independent, v_x and v_y are also independent and (12) will be a minimum when the first derivatives with respect to v_x and v_y are both zero, i.e. when

$$2v_x + 2(v_x + v_y + E) = 0$$

i.e.

$$2v_x + v_y + E = 0 \tag{13}$$

and

$$v_x + 2v_y + E = 0 \tag{14}$$

Solving equations (13) and (14) gives:

$$v_x = -\tfrac{1}{3}E \qquad v_y = -\tfrac{1}{3}E$$

whence

$$v_z = -\tfrac{1}{3}E - \tfrac{1}{3}E + E = +\tfrac{1}{3}E$$

Substituting the numerical values:

$$E = 23° 46' 20'' + 17° 18' 23'' - 41° 04' 40'' = +3''$$

whence as in the observations equations method $v_x = -1''$, $v_y = -1''$, $v_z = +1''$, and the m.p.v's are

$$x = 23° 46' 19''$$
$$y = 17° 18' 22''$$
$$z = 41° 04' 41''$$

The reader should check by assigning values other than the most probable values obtained, that $[v^2]$ derived from any values other than the m.p.v's is greater than $(-1)^2 + (-1)^2 + (+1)^2 = 3$.

From the simple example considered above, it might appear that the method of condition equations is far superior to the observation equations method because it involves smaller numbers. As will be seen below, this objection does not apply in practice to the observation equations method. Both methods have wide applications in surveying, with one method preferred to the other according to circumstances which will be explained in the course of this book. Each method is now considered separately in more detail.

5.6. OBSERVATION EQUATIONS

When a quantity is observed or measured, the result may be written as an observation equation. If the most probable value of the observed quantity is denoted by x, the observed value denoted by M_x, then the residual v_x is by definition given by

$$v_x = x - M_x \tag{15}$$

For example, if a measured length is 230·6 feet, the observation equation is

$$v_x = x - 230 \cdot 6$$

5.6.1. REDUCTION OF THE NUMBER OF FIGURES

The labour of solving equations is shortened and simplified by the introduction of a device commonly used when finding the arithmetic mean of a number of quantities of similar size. An approximate value of the unknown is assumed to which will be added a small correction to give the most probable value, e.g.

let
$$x = P_x + dx \tag{16}$$

where the m.p.v. is x, the assumed or *provisional* value is P_x and the correction to this provisional value is dx. If dx is found, the m.p.v. can therefore be found from P_x. For example: Assume a value of 230·0 for P_x in the above example. Then $x = 230 \cdot 0 + dx$ and the observation equation may then be written $v_x = 230 \cdot 0 + dx - 230 \cdot 6$. In general the observation equation will be of the form:

Residual v = Correction to provisional value P
+ provisional minus observed values

i.e.
$$v_x = dx + (P_x - M_x) \tag{17}$$

from which dx is obtained, and finally the m.p.v. x is given by

$$x = P_x + dx$$

The purely numerical term $P - M$ is called the *absolute term* and is often denoted by $C - O$ ('computed minus observed' values).

Examples of the formation of observation equations: Lines of levelling were run over the routes shown in figure 5.7 with the results given below. The

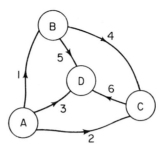

Figure 5.7

observation equations are to be formed in (a) direct form, (b) abbreviated form using provisional values.

Line	Number	Observed difference in height (feet)	
AB	1	$h_B - h_A$	+12·00
AC	2	$h_C - h_A$	+69·36
AD	3	$h_D - h_A$	−1·55
BC	4	$h_C - h_B$	+57·28
BD	5	$h_D - h_B$	−13·66
CD	6	$h_D - h_C$	−70·87

The reader should note the order in which the various quantities are listed, together with the signs of the differences in height. The procedure is to follow the points in correct alphabetical or numerical order, listing the height differences that are obtained from these points in turn to points lower down in the alphabet, or of higher number. In this way, no observed height difference is omitted nor is one listed twice. The lines are arrowed in the diagram to show the sense chosen for the height difference; e.g. going from B to D we go down 13·66 feet, etc.

Since heights are relative to one another, any point is chosen as datum and the other values are referred to it. In this case A is datum. Let the most probable values of the heights of B, C, and D be respectively x, y, and z.

(a) The direct observation equations for the differences in height are

Line 1: $v_1 = x - (+12·00)$
Line 2: $v_2 = y - (+69·36)$
Line 3: $v_3 = z - (-1·55)$
Line 4: $v_4 = y - x - (+57·28)$ (18)
Line 5: $v_5 = z - x - (-13·66)$
Line 6: $v_6 = z - y - (-70·87)$

(b) Since the above observation equations (18) involve large numbers the working is reduced by assuming values for x, y, and z and finding corrections dx, dy, and dz. Let these assumed values be $P_x = +12·00$, $P_y = +69·36$, and $P_z = -1·55$, i.e. we have chosen the values given by the first three observations for simplicity, but any values whatever could be chosen. Thus

$$P_y - P_x = +57\cdot36, \quad P_z - P_x = -13\cdot55, \quad \text{and} \quad P_z - P_y = -70\cdot91$$

$$x = +12\cdot00 + dx, \quad y = +69\cdot36 + dy, \quad \text{and} \quad z = -1\cdot55 + dz \quad (19)$$

And substituting for x, y, and z in equations (18) we obtain

$$\begin{aligned}
v_1 &= dx + (+12\cdot00 - 12\cdot00) = dx + 0 \\
v_2 &= dy + (+69\cdot36 - 69\cdot36) = dy + 0 \\
v_3 &= dz + (-1\cdot55 + 1\cdot55) = dz + 0 \\
v_4 &= dy - dx + (+57\cdot36 - 57\cdot28) = dy - dx + 0\cdot08 \\
v_5 &= dz - dx + (-13\cdot55 + 13\cdot66) = dz - dx + 0\cdot11 \\
v_6 &= dz - dy + (-70\cdot91 + 70\cdot87) = dz - dy - 0\cdot04
\end{aligned} \quad (20)$$

5.6.2. NON-LINEAR FUNCTIONS

If the quantity observed is not a linear function of the quantities whose most probable values are required, it has to be reduced to a linear form by Taylor's theorem. For example if the observed quantity is a length whose most probable value is s, and the most probable differences in easting and northing between the terminals are E and N respectively, s is a non-linear function of E and N, i.e. $s^2 = E^2 + N^2$. In general $s = f(E, N)$. If the observed value of s is M_s, provisional values are assumed as follows: P_s for s, P_E for E, and P_N for N, where the provisional values are consistent with each other, i.e. $P_s^2 = P_E^2 + P_N^2$ and the respective corrections are ds, dE, and dN, thus:

$$s = P_s + ds; \quad E = P_E + dE; \quad \text{and} \quad N = P_N + dN$$

the observation equation is $v_s = s - M_s$,

i.e.
$$v_s = P_s + ds - M_s \quad (21)$$

To obtain ds the procedure is as follows:
Consider the equation relating the respective m.p.v's, i.e. $s^2 = E^2 + N^2$ then

$$(P_s + ds)^2 = (P_E + dE)^2 + (P_N + dN)^2$$

i.e. $P_s^2 + 2P_s\,ds + (ds)^2 = P_E^2 + 2P_E\,dE + (dE)^2 + P_N^2 + 2P_N\,dN + (dN)^2$

$$\therefore \quad ds = \frac{P_E}{P_s} dE + \frac{P_N}{P_s} dN - (ds)^2 + (dE)^2 + (dN)^2$$

$$\therefore \quad ds \simeq \frac{P_E}{P_s} dE + \frac{P_N}{P_s} dN \quad (22)$$

Since ds, dE, and dN are small, we may neglect $(ds)^2$, $(dE)^2$, and $(dN)^2$ in most cases. This means that the provisional values used should be as close as possible to the m.p.v's.

Substituting for the value of ds from (22) in (21) the observation equation is

$$v_s = \left(\frac{P_E}{P_s}\right) dE + \left(\frac{P_N}{P_s}\right) dN + (P_s - M_s) \quad (23)$$

In general ds is obtained by partial differentiation of $f(E, N)$ neglecting powers of the differentials higher than the first,

i.e. $$ds = (\partial f/\partial E)\,dE + (\partial f/\partial N)\,dN$$

For examples of this procedure refer to §6.11.10.

5.6.3. OBSERVATION EQUATIONS IN TABULAR FORM

For convenience in handling and to reduce the amount of writing, the observation equations are written in a tabular form. Considering equations (20). The unknowns dx, dy, and dz are written at the head of the columns in table 5.1

Table 5.1 Observation Equations

dx	dy	dz	Abs
+1			0
	+1		0
		+1	0
−1	+1		+0·08
−1		+1	+0·11
	−1	+1	−0·04

with their respective coefficients placed in the table. The absolute terms are also tabulated in a column. Since the left-hand sides of the equations are the residuals, these are omitted together with equality sign.

5.6.3.1. NORMAL EQUATIONS AND THE PRINCIPLE OF LEAST SQUARES

Systems of linear equations in independent unknowns fall into three categories;

(i) When there are more unknowns than equations.
(ii) When there is an equal number of unknowns and equations.
(iii) When there are fewer unknowns than equations.

All three cases arise in survey adjustment. In case (i) an infinite number of solutions is possible, and in case (iii) no exact solution is possible, unless one invokes some principle to govern our result, such as the principle of least squares. In case (ii) there is a unique answer, and this is therefore considered to be the 'normal case', and the equations as *normal equations*.

The principle of least squares is invoked to reduce cases (i) and (iii) to the normal case (ii). Case (i) concerns 'condition equations', and case (iii), 'observation equations'.

By the principle of least squares, the most probable values of the unknowns are obtained when the sum of the squares of the residuals is a minimum (neglecting different weights meantime), i.e. the m.p.v's are derived when $[v^2]_1^n$ is a minimum; this minimum occurs when the various independent square functions v_1^2, v_2^2, etc. have turning values, i.e. when their first differentials are all zero. This differentiation produces a series of normal equations which are solved to give the m.p.v.'s. For example, the minimum function §5.5.2 (6) was differentiated to give the normal equations (7) and (8).

5.6.4. DIRECT FORMATION OF THE NORMAL EQUATIONS

It is unnecessary to derive the normals by partial differentiation of a minimum function such as (6) in each case. They may be formed directly from the observation equations according to a definite pattern which should be mastered before proceeding further.

To establish the rules for the direct formation of normals from observation equations consider three observation equations (24) in two unknowns x and y, whose coefficients given by theory are a, b, c, d, e, and f; and whose absolute terms given by observation are K, L, and M, i.e.

$$+ax + by + K = v_1$$
$$+cx + dy + L = v_2 \qquad (24)$$
$$+ex + fy + M = v_3$$

The m.p.v's, x and y, are given when $[v^2]$ is a minimum. This is the same as making $\frac{1}{2}[v^2]$ a minimum (The $\frac{1}{2}$ is introduced to simplify the differentiation.), i.e. when

$$\tfrac{1}{2}[v^2] = \tfrac{1}{2}(ax + by + K)^2 + \tfrac{1}{2}(cx + dy + L)^2 + \tfrac{1}{2}(ex + fy + M)^2 \qquad (25)$$

is a minimum. Differentiating (25) with respect to each unknown in turn and equating each result to zero gives; (differentiating with respect to x)

$$a(ax + by + K) + c(cx + dy + L) + e(ex + fy + M) = 0$$

i.e. $\qquad (aa + cc + ee)x + (ab + cd + ef)y + (aK + cL + eM) = 0 \qquad (26a)$

and differentiating (25) with respect to y

$$b(ax + by + K) + d(+cx + dy + L) + f(ex + fy + M) = 0$$

i.e. $\qquad (ab + cd + ef)x + (bb + dd + ff)y + (bK + dL + fM) = 0 \qquad (26b)$

Equations (26a) and (26b) are *the normal equations in x and y respectively*. Writing the observation equations (24), and the normals derived from them (26), in tabular form gives

Observation equations	x	y	Abs.
	a	b	K
	c	d	L
	e	f	M
Normal equations	x	y	Abs.
	$(aa + cc + ee)$	$(ab + cd + ef)$	$(aK + cL + eM)$
	$(ab + cd + ef)$	$(bb + dd + ff)$	$(bK + dL + fM)$

The reader should study the above patterns carefully to become fully conversant with the method of forming the normals, and should test that equations (7) and (8) §5.5.2 can be formed from observation equations (4) according to this pattern.

This procedure can be enlarged to the general case of n equations in m unknowns $x_1, x_2, \cdots x_m$, $(n > m)$. Let the observation equations be

5.6.4.1. *Observation Equations*

m unknowns $x_1, x_2 \cdots x_m$

Equation	x_1	x_2	x_3	$\cdots$	x_m	Abs.	
1	a_1	b_1	c_1	$\cdots$	m_1	K_1	
2	a_2	b_2	c_2	$\cdots$	m_2	K_2	
3	a_3	b_3	c_3	$\cdots$	m_3	K_3	(27)
.	.	.	.		.	.	
.	.	.	.		.	.	
.	.	.	.		.	.	
n	a_n	b_n	c_n	$\cdots$	m_n	K_n	

5.6.4.2. *Normal Equations*

Formed from the above equations (27)

Equation	x_1	x_2	x_3	$\cdots$	x_m	Abs.	
1	[aa]	[ab]	[ac]	$\cdots$	[am]	[aK]	
2	[ab]	[bb]	[bc]	$\cdots$	[bm]	[bK]	
3	[ac]	[bc]	[cc]	$\cdots$	[cm]	[cK]	(28)
.	.	.	.		.	.	
.	.	.	.		.	.	
.	.	.	.		.	.	
m	[am]	[bm]	[cm]	$\cdots$	[mm]	[mK]	

where $[aa] = a_1 a_1 + a_2 a_2 + a_3 a_3 + a_4 a_4 + \cdots + a_n a_n$, etc.

Since it is essential that this technique be mastered before proceeding further, a numerical example is given.

Example: Form the normals from the following observation equations in x, y, and z.

x	y	z	Abs.	
+1	−2	+1	−5	
+1	0	+2	−5	
+1	0	−1	+1	(29)
0	+1	−2	+5	
−1	0	−2	+5	

The normals then are

x	y	z	Abs.	
+4	−2	+4	−14	
−2	+5	−4	+15	(30)
+4	−4	+14	−36	

For:
$[aa] = (+1)(+1) + (+1)(+1) + (+1)(+1) + (0)(0) + (-1)(-1) = +4$
$[ab] = (+1)(-2) + (+1)(0) + (+1)(0) + (0)(+1) + (-1)(0) = -2$
$[ac] = (+1)(+1) + (+1)(+2) + (+1)(-1) + (0)(-2) + (-1)(-2) = +4$
$[aK] = (+1)(-5) + (+1)(-5) + (+1)(+1) + (0)(+5) + (-1)(+5) = -14$
etc.

When working with a calculating machine only the final summation of products is written down, the complete answer being accumulated on the machine. This has the advantage that any rounding off will be made at the end of each summation, rather than for every individual term.

5.6.5. SYMMETRY OF NORMAL EQUATIONS

It will be seen that the coefficients in the normal equations are symmetrical about the diagonal terms, i.e. about $[aa], [bb], [cc]$, etc. Hence we may omit from the table of normals, that part of the array of coefficients which lies to the left of and below the diagonal terms. We write the normals (30) and their algebraic equivalents in the form:

x	y	z	Abs.	
$[aa]$	$[ab]$	$[ac]$	$[aK]$	(31a)
	$[bb]$	$[bc]$	$[bK]$	
		$[cc]$	$[cK]$	

x	y	z	Abs.	
+4	−2	+4	−14	(31b)
	+5	−4	+15	
		+14	−36	

The normal equation in y is then obtained by reading *down the column headed y* to the diagonal term $[bb]$ or $+5$, then *along the line* to the absolute term. This arrangement is not only shorter, but it also enables one to adopt a special form of solution of the equations.

5.6.6. CHECK ON THE FORMATION OF NORMAL EQUATIONS

Some check on the formation of the normal is highly desirable since the summation of products may involve a large number of calculations. To check each stage, we introduce a summation term s to the observation equations. Each s term is the sum of all the coefficients along a horizontal row of an observation equation. For example, the arrays of coefficients or 'matrices' for equations (29) and their algebraic counterparts will have the check columns added as follows:

x	y	z	Abs.	s	x	y	z	Abs.	s
a_1	b_1	c_1	K_1	s_1	+1	−2	+1	−5	−5
a_2	b_2	c_2	K_2	s_2	+1	0	+2	−5	−2
a_3	b_3	c_3	K_3	s_3	+1	0	−1	+1	+1
a_4	b_4	c_4	K_4	s_4	0	+1	−2	+5	+4
a_5	b_5	c_5	K_5	s_5	−1	0	−2	+5	+2
Sum $[a]$	$[b]$	$[c]$	$[K]$	$[s]$	+2	−1	−2	+1	0

Each s term is obtained as follows

$$s_1 = a_1 + b_1 + c_1 + K_1, \text{ etc.} \qquad -5 = +1 \quad -2 \quad +1 \quad -5, \text{ etc.}$$

To check the formation of the s terms the total summations vertically and horizontally should be equal, i.e.

$$[a] + [b] + [c] + [K] \text{ should equal } s_1 + s_2 + s_3 + s_4 + s_5 = [s]$$

When forming the normals, the s terms are operated on in the same manner as the other coefficients, i.e. a further column of terms $[as]$, etc. is obtained where $[as] = a_1s_1 + a_2s_2 + a_3s_3 + a_4s_4 \cdots a_ms_m$. The equations (30) and their algebraic counterparts will be

x	y	z	Abs.	s	x	y	z	Abs.	s
[aa]	[ab]	[ac]	[aK]	[as]	+4	−2	+4	−14	−8
	[bb]	[bc]	[bK]	[bs]		+5	−4	+15	+14
		[cc]	[cK]	[cs]			+14	−36	−22

The check is that

$$[as] \text{ should equal } [aa] + [ab] + [ac] + [aK]$$
$$[bs] \text{ should equal } [ab] + [bb] + [bc] + [bK]$$

remembering that portion of the matrix which has been omitted.
In the numerical example

$$-8 = +4 - 2 + 4 - 14$$
$$+14 = -2 + 5 - 4 + 15$$
$$-22 = +4 - 4 + 14 - 36$$

This gives a check of the formation of the normals from the observation equations, but does not check the latter in any way. Where possible the observation equations should be checked in their final tabular form by a second computor working independently. Once the observation equations have been checked the rest of the computation is self-checking and may therefore be left to one individual. Where a second person is not available, the observation equations should be formed by an alternative method if one exists, and the two sets compared.

Example: Form the normal equations from the observation equations (20) applying the summation check. Equations (20) written in tabular form are

dx	dy	dz	Abs.	s
+1			0	+1
	+1		0	+1
		+1	0	+1
−1	+1		+0·08	+0·08
−1		+1	+0·11	+0·11
	−1	+1	−0·04	−0·04
−1	+1	+3	+0·15	+3·15

The normals are

	dx	dy	dz	Abs.	s
	+3	−1	−1	−0·19	+0·81
		+3	−1	+0·12	+1·12
			+3	+0·07	+1·07

Checks

$$+0\cdot81 = +3 - 1 - 1 - 0\cdot19$$
$$+1\cdot12 = -1 + 3 - 1 + 0\cdot12$$
$$+1\cdot07 = -1 - 1 + 3 + 0\cdot07$$

5.7. THE SOLUTION OF NORMAL EQUATIONS

Solution of the normal equations gives the values of the unknowns, i.e. the most probable values of the required quantities. For example, solution of equations (31b) gives the values $x = +1$, $y = -1$, and $z = +2$.

If there are only two normals, solution is best carried out by determinants. If the equations are

$$ax + by + c = 0$$
$$bx + dy + e = 0 \tag{32}$$

Then $\quad x = + \begin{vmatrix} b & c \\ d & e \end{vmatrix} \div \begin{vmatrix} a & b \\ b & d \end{vmatrix}$, and $y = - \begin{vmatrix} a & c \\ b & e \end{vmatrix} \div \begin{vmatrix} a & b \\ b & d \end{vmatrix}$

$$x = \frac{be - dc}{ad - bb}, \quad y = -\frac{ae - bc}{ad - bb}$$

With more than two equations, the determinants method of solution is cumbersome. Since the arithmetic involved in the solution of large systems of as many as several hundred equations is considerable, it is highly desirable that some tabular form of solution is adopted, which is capable of being checked at convenient intervals.

Two methods of solving normal equations are in common use today; the *Gauss–Doolittle* method, and the *Cholesky* method. Both lend themselves to simple tabular arrangement, both make special use of the symmetry of normals, and both are self-checking at convenient intervals.

At first sight these methods may seem to be very complicated, but on careful study, a simple pattern will emerge which one is able to memorize. The reader new to the subject should not attempt to master both methods at first, but concentrate on one only. The Gauss–Doolittle is easier for a beginner, though the Cholesky is theoretically the better method.

5.7.1. GENERAL DESCRIPTION OF THE METHODS

In both methods, the unknowns are eliminated in succession by a special routine procedure until one equation in one unknown is obtained This process is called *the forward solution*. The forward solution produces a triangular array of equations, which when solved give the same values for the unknowns as the original normals. For example, consider equations (31b) written in full

$$\begin{array}{cccc} x & y & z & \text{Abs.} \\ +4 & -2 & +4 & -14 \\ -2 & +5 & -4 & +15 \\ +4 & -4 & +14 & -36 \end{array} \quad (31b)$$

The values that satisfy these equations are

$$x = +1; \quad y = -1; \quad z = +2$$

The triangular array of equations given by the Gauss–Doolittle method is

$$\begin{array}{ccccc} x & y & z & \text{Abs.} & \\ -1\cdot 0 & +0\cdot 5 & -1\cdot 0 & +3\cdot 5 & (33a) \\ & -1\cdot 0 & +0\cdot 5 & -2\cdot 0 & (33b) \\ & & -1\cdot 0 & +2\cdot 0 & (33c) \end{array}$$

And that produced by the Cholesky method is

$$\begin{array}{ccccc} x & y & z & \text{Abs.} & \\ +2\cdot 0 & -1\cdot 0 & +2\cdot 0 & -7\cdot 0 & (34a) \\ & +2\cdot 0 & -1\cdot 0 & +4\cdot 0 & (34b) \\ & & +3\cdot 0 & -6\cdot 0 & (34c) \end{array}$$

Both of these sets of equations (33) and (34) are not abbreviated in any way. Equation (33c) in full is

$$-1\cdot 0 z + 2\cdot 0 = 0$$

whence $z = +2$, and equation (34c) in full is

$$+3\cdot 0 z - 6\cdot 0 = 0$$

whence $z = +2$ as before. This value of z is substituted back into equation (33b) or (34b) to give y. Thus

(33b) $-1\cdot 0 y + (+0\cdot 5)(+2\cdot 0) - 2\cdot 0 = 0,$ i.e. $y = -1$,
(34b) $+2\cdot 0 y + (-1\cdot 0)(+2\cdot 0) + 4\cdot 0 = 0,$ i.e. $y = -1$ as before.

Finally these values of x and y are substituted in equation (33a) or (34a) to give the value of $x = +1$.

This procedure of substituting back through the triangular array is called the *back solution*.

Hence it is seen that the general technique of solution is the same for both methods; a forward solution followed by a back solution. The following sections describe how the sets of equations (33) and (34) are obtained from the normals (31b). To assist the explanation, each space of tables 5.2 and 5.3 has been given a map reference, with the usual convention of quoting eastings (A, B ··· etc.) before northings (1, 2 ··· etc.). For example the figure +4·0 in table 5.2 column D line (1) is referred to as (D.1). This system is also used to describe a particular arithmetical operation. For example, if we take the figure

Theory of Errors and Survey Adjustment

Table 5.2. Gauss–Doolittle Method of Solution of Normal Equations

		A	B	C	D	E	F		
9			+1·0	−1·0	+2·0			9	
10			0·0	−2·0	+1·0			10	
			x	y	z	Abs.	Sum		
1	x	+4·0	+4·0	−2·0	+4·0	−14·0	−8·0	1	
2			−1·0	+0·5	−1·0	+3·5	+2·0	2	(33a)
3	y	−2·0	+5·0	+4·0	−2·0	+8·0	+10·0	3	
4				−1·0	+0·5	−2·0	−2·5	4	(33b)
5	z	+4·0	−4·0	+14·0	+9·0	−18·0	−9·0	5	
6					−1·0	+2·0	+1·0	6	(33c)
7	Abs.	−14·0	+15·0	−36·0				7	
8	Sum	−8·0	+14·0	−22·0				8	
		A	B	C	D	E	F		

+4·0 in table 5.3 (B.1), find its square root, and place the result in (A.1), we describe the whole process as

$$(B.1)^{\frac{1}{2}} \to (A.1)$$

This system of exposition has been adopted because it has been found in the past that the various steps involved in a solution cannot be described in words satisfactorily.

5.7.2. THE GAUSS–DOOLITTLE METHOD OF SOLUTION

The normals (31b) and their summation check are written in the abridged form, i.e.

```
   x     y      z     Abs.    Sum
  +4    −2     +4    −14      −8
        +5     −4    +15     +14
               +14   −36     −22
```

(i) These are placed in a vertical position in columns A, B, and C according to the scheme

```
  +4
  −2    +5
  +4    −4    +14
 −14   +15   −36
  −8   +14   −22
```

i.e. putting +4·0 in (A.1), −2·0 → (A.3), etc.

(ii) Rewriting the first normal along line (1), i.e.
(A.1) → (B.1);   (A.3) → (C.1);   (A.5) → (D.1);   (A.7) → (E.1) and (A.8) → (F.1).

(iii) The forward solution is carried out within the bold lines. Dividing line (1) by its leading term (+4·0) and changing the signs throughout, writing the results in line (2), i.e.
$$-(B.1)/(B.1) \to (B.2); \quad -(C.1)/(B.1) \to (C.2); \text{ etc.}$$
Line (2) is equation (33a) and its check.

In practice, it is easier to multiply throughout by the reciprocal of the leading term which is often written in brackets.

(iv) Check: (B.2) + (C.2) + (D.2) + (E.2) = (F.2)

(v) Considering the second normal and lines (1) and (2) of the forward solution.

(B.3) + (C.1)(C.2) → (C.3),   i.e. $+5·0 + (-2·0)(+0·5) = +4·0$
(B.5) + (C.1)(D.2) → (D.3),   i.e. $-4·0 + (-2·0)(-1·0) = -2·0$
(B.7) + (C.1)(E.2) → (E.3),   i.e. $+15·0 + (-2·0)(+3·5) = +8·0$
(B.8) + (C.1)(F.2) → (F.3),   i.e. $+14·0 + (-2·0)(+2·0) = +10·0$

(vi) Now dividing line (3) inside the bold lines by its leading term +4·0 and changing the signs of each term, placing the results in line (4), i.e.

$$\frac{-(C.3)}{(C.3)} \to (C.4); \quad \frac{-(D.3)}{(C.3)} \to (D.4); \quad \frac{-(E.3)}{(C.3)} \to (E.4)$$

and $$\frac{-(F.3)}{(C.3)} \to (F.4).$$

Line (4) is equation (33b) and its check.

(vii) Check: (C.4) + (D.4) + (E.4) = (F.4)

(viii) Considering the third normal and lines (1), (2), (3), and (4) of the solution inside the bold lines.

(C.5) + (D.1)(D.2) + (D.3)(D.4) → (D.5),   i.e. $+14.0$
$\qquad\qquad + (+4·0)(-1·0) + (-2·0)(+0·5) = +9·0$
then   (C.7) + (D.1)(E.2) + (D.3)(E.4) → (E.5),   i.e. $-36·0$
$\qquad\qquad + (+4·0)(+3·5) + (-2·0)(-2·0) = -18·0$
and   (C.8) + (D.1)(F.2) + (D.3)(F.4) → (F.5),   i.e. $-22·0$
$\qquad\qquad + (+4·0)(+2·0) + (-2·0)(-2·5) = -9·0$

(ix) Dividing line (5) of the solution within the bold lines by its leading term +9·0 changing the signs throughout and writing the results in line (6)

$$\frac{-(D.5)}{(D.5)} \to (D.6); \quad \frac{-(E.5)}{(D.5)} \to (E.6); \quad \frac{-(F.5)}{(D.5)} \to (F.6).$$

Line (6) is equation (33c) and its check.

(x) Check: (D.6) + (E.6) = (F.6)

This completes the forward solution, to give equations (33). Since only these equations (33) are now to be used for the back solution it is convenient if they are written in a different colour from the rest of the forward solution. The products in stages (v) and (viii) are accumulated on the calculating machine without writing down intermediate steps, and only the final result is rounded off and written down.

The computor will find that a straight edge is useful in guiding the eye to the correct line or column. It is necessary to work to one more figure than is required in the final result. The summation checks in the normals should invariably agree exactly, though small discrepancies of up to 3 in the last place may occur in the solution. As a guide to whether such a disagreement is likely, a figure which has been rounded off from 5 should have an indication $+$ or $-$ alongside it. For example if the figure 3·455 is rounded to 3·46 we would write 3·46⁻. Then if several figures in the summation have been similarly rounded off we can foresee a discrepancy from this source. Normally any discrepancy will be checked to ensure that a mistake has not been made.

5.7.2.1. *The Back Solution*

This consists of substituting the value z in equation (33*b*) to get y, and then both z and y in (33*a*) to give x. The term in (E.6) is the required value of z. The term in (F.6) is $z - 1$, since (D.6) is always -1. As a check on the computation of x and y in the back solution, we also substitute $(z - 1)$ to get $(y - 1)$ from (33*b*), and then $(x - 1)$ from (33*a*), substituting for both $(z - 1)$ and $(y - 1)$. The results of the back solution and its check are placed in lines (9) and (10). These lines are placed at the top of the table to bring the values of the unknowns as close as possible to the equations (33) but Rainsford places them at the bottom of the forward solution†. The back solution is carried out as follows:

(i) (E.6) $= z = +2·0 \to$ (D.9)
(ii) (F.6) $= z - 1 = +1·0 \to$ (D.10)
(iii) Check that (D.9) $- 1 =$ (D.10), i.e. $z - 1 = z - 1$
(iv) (E.4) $+$ (D.4)(D.9) $= y \to$ (C.9), i.e. $-2·0 + (+0·5)(+2·0) = -1·0$
(v) (F.4) $+$ (D.4)(D.10) $= y - 1 \to$ (C.10), i.e. $-2·5 + (+0·5)(+1·0) = -2·0$
(vi) Check that (C.9) $- 1 =$ (C.10)
(vii) (E.2) $+$ (D.2)(D.9) $+$ (C.2)(C.9) $= x \to$ (B.9)
(viii) (F.2) $+$ (D.2)(D.10) $+$ (C.2)(C.10) $= x - 1 \to$ (B.10)
(ix) Check that (B.9) $- 1 =$ (B.10)

This completes the solution by the Gauss–Doolittle method.

5.7.3. THE CHOLESKY METHOD OF SOLUTION OF NORMAL EQUATIONS

The triangular matrix (34) is obtained by the Cholesky Method of solution. Tables of square roots may be found useful, though it is just as quick to evaluate

† H. F. Rainsford, *Survey Adjustments and Least Squares* (London: Constable, 1957).

a square root on the calculating machine by one of the methods described in Chapter 2. The figures in the example have been carefully chosen so that those numbers whose square roots are required are perfect squares, enabling the reader to follow the process more easily.

The abridged normals are written inside the bold lines of table 5.3 according to the system

x	y	z	Abs.	Sum	
+4	−2	+4	−14	−8	Line (1)
	+5	−4	+15	+14	Line (2)
		+14	−36	−22	Line (3)

The equations (34) will be written in a vertical direction reading down columns (A), (B), and (C) outside the dark lines. The procedure for the forward solution is as follows:

(i) $(B.1) \div (B.1)^{\frac{1}{2}} \to (A.1)$; $(C.1) \div (B.1)^{\frac{1}{2}} \to (A.2)$; $(D.1) \div (B.1)^{\frac{1}{2}} \to (A.3)$; $(E.1) \div (B.1)^{\frac{1}{2}} \to (A.4)$; $(F.1) \div (B.1)^{\frac{1}{2}} \to (A.5)$, i.e. $(+4.0) \div (+4.0)^{\frac{1}{2}} = +2.0$; $(-2.0) \div (4.0)^{\frac{1}{2}} = -1.0$, etc. Column (A) is equation (34a) reading vertically, and its summation check.

(ii) Check that $(A.1) + (A.2) + (A.3) + (A.4) = (A.5)$.

(iii) $\{(C.2) - (A.2)^2\}^{\frac{1}{2}} \to (B.2)$ i.e. $\{5.0 - (-1)^2\}^{\frac{1}{2}} = +2.0$

(iv) $\{(D.2) - (A.2)(A.3)\} \div (B.2) \to (B.3)$ i.e. $\{-4.0 - (-1.0)(+2.0)\} \div +2.0 = -1.0$. (N.B. There is no square root in this process.)
$\{(E.2) - (A.2)(A.4)\} \div (B.2) \to (B.4)$ i.e. $\{+15.0 - (-1.0)(-7.0)\} \div +2.0 = +4.0$.
$\{(F.2) - (A.2)(A.5)\} \div (B.2) \to (B.5)$ i.e. $\{+14.0 - (-1.0)(-4.0)\} \div +2.0 = +5.0$.

Column (B) outside the bold lines is equation (33b) and its check written vertically.

(v) Check that $(B.2) + (B.3) + (B.4) = (B.5)$.

(vi) $\{(D.3) - (A.3)^2 - (B.3)^2\}^{\frac{1}{2}} = (C.3)$ i.e. $\{+14.0 - (2.0)^2 - (-1.0)^2\}^{\frac{1}{2}} = +3.0$.

(vii) $\{(E.3) - (A.3)(A.4) - (B.3)(B.4)\} \div (C.3) \to (C.4)$ i.e. $\{-36.0 - (+2.0)(-7.0) - (-1.0)(+4.0)\} \div +3.0 = -6.0$.
Again there is no square root involved.
$\{(F.3) - (A.3)(A.5) - (B.3)(B.5)\} \div (C.3) \to (C.5)$ i.e. $\{-22.0 - (+2.0)(-4.0) - (-1.0)(+5.0)\} \div +3.0 = -3.0$
Column (C) is equation (34c) and its check written vertically.

(viii) Check that $(C.3) + (C.4) = (C.5)$.

This concludes the forward solution.

5.7.3.1. *The Back Solution*

The values of the unknowns x, y, and z will be written in column (G), and in column (H) will be the check values of $(x - 1)$, $(y - 1)$, and $(z - 1)$. The procedure is then as follows:

(i) $\quad z = -(C.4) \div (C.3) \to (G.3)$, i.e. $\quad z = -(-6.0) \div 3.0$
$\quad\quad\quad\quad\quad\quad\quad\quad\quad\quad\quad\quad\quad\quad\quad\quad\quad\quad\quad = +2.0$

Table 5.3. Cholesky Method of Solution of Normal Equations

		A	B	C	D	E	F	G	H	
		x	y	z	abs.	sum				
1	x	+2·0	+4·0	−2·0	+4·0	−14·0	−8·0	+1·0	0·0	1
2	y	−1·0	+2·0	+5·0	−4·0	+15·0	+14·0	−1·0	−2·0	2
3	z	+2·0	−1·0	+3·0	+14·0	−36·0	−22·0	+2·0	+1·0	3
4	Abs.	−7·0	+4·0	−6·0						4
5	Sum	−4·0	+5·0	−3·0						5
Equations		(34a)	(34b)	(34c)						
		A	B	C	D	E	F	G	H	

(ii) $z - 1 = -(C.5) \div (C.3) \rightarrow (H.3)$,   i.e. $z - 1 = -(-3 \cdot 0) \div 3 \cdot 0$
$$= +1 \cdot 0$$
    Check that $(G.3) - 1 = (H.3)$.
(iii) $y = \{(B.4) + (G.3)(B.3)\} \div (B.2) \rightarrow (G.2)$
     i.e. $y = -\{(+4 \cdot 0) + (+2 \cdot 0)(-1 \cdot 0)\} \div +2 \cdot 0 = -1 \cdot 0$
(iv) $y - 1 = -\{(B.5) + (H.3)(B.3)\} \div (B.2) \rightarrow (H.2)$
     i.e. $y - 1 = -\{(+5 \cdot 0) + (+1 \cdot 0)(-1 \cdot 0)\} \div +2 \cdot 0 = -2 \cdot 0$
     Check that $(G.2) - 1 = (H.2)$.
(v) $x = -\{(A.4) + (G.3)(A.3) + (G.2)(A.2)\} \div (A.1) \rightarrow (G.1)$
     i.e. $x = -\{(-7 \cdot 0) + (+2 \cdot 0)(+2 \cdot 0) + (-1 \cdot 0)(-1 \cdot 0)\} \div +2 \cdot 0$
$$= +1 \cdot 0$$
(vi) $x - 1 = -\{(A.5) + (H.3)(A.3) + (H.2)(A.2)\} \div (A.1) \rightarrow (H.1)$
     i.e. $x - 1 = -\{(-4 \cdot 0) + (+1 \cdot 0)(+2 \cdot 0) + (-2 \cdot 0)(-1 \cdot 0)\} \div +2 \cdot 0$
$$= 0 \cdot 0$$
     Check that $(G.1) - 1 = (H.1)$.

This concludes the Cholesky solution.

Although this method is more difficult than the Gauss–Doolittle since a greater variety of operations is involved, it produces a better result because the square rooting process retains the number of significant figures to a consistent accuracy throughout the solution. This is especially important in solutions that tend to instability.

5.8. THE ADJUSTMENT OF OBSERVATIONS OF DIFFERENT WEIGHT

So far the arguments have been confined to observations of equal weight. When observations are of different weight they are reduced to equivalent observations of equal weight by multiplying each residual by the square root of its weight. From §5.4. the most probable values of weighted observations are obtained when the sum of the weighted squares of residuals is a minimum. If the residuals are $v_1 \cdots v_n$ of respective weights $p_1 \cdots p_n$ the m.p.v.'s are obtained when $[pv^2]$ is a minimum, i.e. when $p_1v_1^2 + p_2v_2^2 + \cdots + p_nv_n^2$ is a minimum, i.e. when $v_1p_1^{\frac{1}{2}}v_1p_1^{\frac{1}{2}} + \cdots + v_np_n^{\frac{1}{2}}v_np_n^{\frac{1}{2}}$ is a minimum; and putting $v_1p_1^{\frac{1}{2}} = V_1, \cdots v_np_n^{\frac{1}{2}} = V_n$, when $[V^2]$ is a minimum.

In the least squares adjustment by observation equations, multiply each observation equation by the square root of its weight and proceed as for unweighted observations thereafter.

5.8.1. EXAMPLE OF ADJUSTMENT OF WEIGHTED OBSERVATIONS

If in example §5.6.1 the lengths in miles of the lines of levelling are: AB = 7; AC = 10; AD = 6; BC = 9; BD = 6; CD = 4, and we assume that the errors in the levelling are purely random, the respective weights are inversely proportional to the lengths of lines, for $p \propto 1/(p \cdot e)^2$. If the random error per mile is σ the error accumulated after n miles is $\sigma n^{\frac{1}{2}}$, $\therefore p \propto 1/\sigma^2 n$ $\therefore p \propto 1/n$. In this example, the respective weights are therefore 1/7; 1/10; 1/6; 1/9; 1/6; and 1/4, i.e.

0·142 86;   0·100 00;   0·166 67;   0·111 11;   0·166 67;   and   0·250 00

and the respective square roots of the weights are

0·378 0;   0·316 2;   0·408 2;   0·333 3;   0·408 2;   and   0·500 0

Since weights are purely relative we may multiply them all by any convenient number we choose without altering their relative importance. It is usually most convenient to reduce the weights to about unity. In this case they can be used as they stand. Table 5.4 shows the observation equations (20) (§5.6.1) in tabular form, and table 5.5 gives these equations reduced to equal weight after multiplication by the square roots of the weights. Table 5.6 shows the normals and

Table 5.4. Observation Equations With Weights

Eqn.	$p^{\frac{1}{2}}$	dx	dy	dz	Abs.	v	v rounded
20a	0·3780	+1			0	+0·0476	+0·05
20b	0·3162		+1		0	−0·0389	−0·04
20c	0·4082			+1	0	−0·0173	−0·02
20d	0·3333	−1	+1		+0·08	−0·0065	−0·01
20e	0·4082	−1		+1	+0·11	+0·0451	+0·04
20f	0·5000		−1	+1	−0·04	−0·0184	−0·02
Corrections		+0·0476	−0·0389	−0·0173			
Rounded off		+0·05	−0·04	−0·02			

Table 5.5. Observation Equations Reduced to Same Weight

dx	dy	dz	Abs.	Sum
+0·3780			0	+0·3780
	+0·3162		0	+0·3162
		+0·4082	0	+0·4082
−0·3333	+0·3333		+0·0267	+0·0267
−0·4082		+0·4082	+0·0449	+0·0449
	−0·5000	+0·5000	−0·0200	−0·0200
−0·3635	+0·1459	+1·3164	+0·0516	+1·1540 SUM

their solution. The m.p.v.'s are given by

$$\begin{array}{rl} & \text{m.p.v.} \quad\quad \text{Observed values} \quad v\text{'s} \\ x = P_x + \mathrm{d}x = +12{\cdot}00 + 0{\cdot}05 = +12{\cdot}05 & +12{\cdot}00 \quad +0{\cdot}05 \\ y = P_y + \mathrm{d}y = +69{\cdot}36 - 0{\cdot}04 = +69{\cdot}32 & +69{\cdot}36 \quad -0{\cdot}04 \\ z = P_z + \mathrm{d}z = -1{\cdot}55 - 0{\cdot}02 = -1{\cdot}57 & -1{\cdot}55 \quad -0{\cdot}02 \\ y - x = +69{\cdot}32 - 12{\cdot}05 = +57{\cdot}27 & +57{\cdot}28 \quad -0{\cdot}01 \\ z - x = -1{\cdot}57 - 12{\cdot}05 = -13{\cdot}62 & -13{\cdot}66 \quad +0{\cdot}04 \\ z - y = -1{\cdot}57 - 69{\cdot}32 = -70{\cdot}89 & -70{\cdot}87 \quad -0{\cdot}02 \end{array}$$

As a final check on the work, the residuals are formed in two ways and compared for exact equality. Firstly they are formed as above from the final accepted m.p.v's and the original observed values. Secondly they are derived

Table 5.6. Gauss–Doolittle Solution

Correction	+0·0467	−0·0389	−0·0173			
Check	−0·9522	−1·0387	−1·0167			
	dx	dy	dz	Abs.		Sum
+0·4206	+0·4206 (2·3776)	−0·1111 +0·2642	−0·1666 +0·3961	−0·0272 +0·0647		+0·1157 −0·2751
−0·1111	+0·4611	+0·4317 (2·3164)	−0·2940 +0·6810	+0·0117 −0·0271		+0·1495 −0·3463
−0·1666	−0·2500	+0·5833	+0·3171 (3·1536)	+0·0055 −0·0173		+0·3224 −1·0167
−0·0272	+0·0189	+0·0083				
+0·1157	+0·1189	+0·1749				

N.B. The numbers in brackets are the reciprocals of the numbers above them, i.e. $2{\cdot}3776 = 1/0{\cdot}4206$.

from the table of observation equations table 5.4. The comparison of these two sets of residuals ensures that the observation equations that have been solved were correctly formed in the first place.

To form the residuals from the table 5.4, the values of dx, dy, and dz are substituted in the observation equations, and the amount by which the left hand of these equations fails to equal zero is the residual. For example the residual in equation (20f) is $-0.018\ 4 = (-0.038\ 9)(-1) + (-0.017\ 3)(+1) - 0.04 = -0.02$ when rounded off.

Finally the residuals are adjusted through the various lines of levelling by proportion according to distance, e.g. a point one third of the way along line AB will be adjusted by $+\frac{1}{3}(0.05)$, i.e. by $+0.017$. This same example is worked out by the condition equations method in §5.9.2.5 below.

5.8.2. USE OF WEIGHTS

In practice it is usually found that unless the various weights differ by more than a factor of about seven it is sufficiently accurate to treat the observations as of equal weight. For example the angles of a triangulation network are usually considered to be of equal weight.

5.9. CONDITION EQUATIONS

In §5.5.1 above, a simple example was considered in which the most probable values of angles x, y, and z were obtained by observation equations, and condition equations. In the latter method we concentrated on the *geometrical condition* existing between the m.p.v's, i.e.

$$x + y = z \quad \text{or} \quad x + y - z = 0 \tag{3}$$

If the respective observed values are M_x, M_y, and M_z, and v_x, v_y, and v_z are the residuals, by definition of residuals

$$x - M_x = v_x; \quad y - M_y = v_y; \quad z - M_z = v_z$$

It was found by making $[v^2]$ a minimum that $v_x = -E/3$; $v_y = -E/3$; and $v_z = +E/3$ where $E = M_x + M_y - M_z$. That is, E is the amount by which the observed values fail to satisfy the condition equation.

5.9.1.1. *Method of Correlatives*

Since the method of solution given in §5.5.3 is inconvenient when a large number of equations has to be solved, the following device introduced by Lagrange is adopted.

Consider the condition equation (3) $x + y - z = 0$. This reduces to

$$v_x + v_y - v_z + E = 0 \tag{10}$$

Let C be *any* number, as yet unknown, called a *correlative* or *Lagrangian multiplier*, then from (10)

$$C(v_x + v_y - v_z + E) = 0 \tag{35}$$

Now, the most probable values are obtained when $[v^2]$ is a minimum, which will be the same as making $\frac{1}{2}[v^2]$ is a minimum, i.e. when

$$\frac{1}{2}[v^2] = \frac{1}{2}(v_x^2 + v_y^2 + v_z^2) - C(v_x + v_y - v_z + E) \text{ is a minimum} \quad (36)$$

Since the observed values M_x, M_y, and M_z are independent of each other, the residuals v_x, v_y, and v_z are also independent, hence (36) will be a minimum when the first differentials with respect to v_x, v_y, and v_z are all zero, i.e. when

$$v_x - C = 0$$
$$v_y - C = 0 \quad (37)$$
$$v_z + C = 0$$

Whence $\quad v_x = +C, v_y = +C, \text{ and } v_z = -C \quad (38)$

Thus equations (38) give the values of the unknowns in terms of the correlative C, which itself is as yet unknown. To find C, substitute for the v's in the *original condition equation* (10), whence $+C + C - (-C) + E = 0$, i.e. $C = -E/3$. Now substituting for C in equations (38) we obtain the v's.

$$v_x = -E/3, \quad v_y = -E/3, \quad \text{and } v_z = +E/3$$

equations (37) are called the *correlative normal equations*.

5.9.2. DIRECT FORMATION OF CORRELATIVE NORMAL EQUATIONS (WITH WEIGHTS)

It is unnecessary to form the minimum function such as (36) and therefrom derive the normal equations by partial differentiation. The normals may be derived directly from the condition equations by a simple process which is similar to the case of observation equations. To the beginner this similarity is confusing, hence it is advised that one method should be mastered before attempting the next.

Suppose that two conditions given by theory, e.g. the geometry of a figure, have to be satisfied by three unknowns x, y, and z whose observed values are M_x, M_y, and M_z. Then the residuals are

$$v_x = x - M_x; \quad v_y = y - M_y; \quad v_z = z - M_z$$

Suppose the condition equations to be satisfied are

$$ax + by + cz + k = 0 \quad (39)$$
$$dx + ey + fz + l = 0$$

where a, b, c, d, e, f, k, and l are coefficients given by theory. Substituting for x, y, and z in terms of the observed values and residuals gives

$$(aM_x + bM_y + cM_z + k) + av_x + bv_y + cv_z = 0$$
$$(dM_x + eM_y + fM_z + l) + dv_x + ev_y + fv_z = 0 \quad (40)$$

Let $(aM_x + bM_y + cM_z + k) = K$, and $(dM_x + eM_y + fM_z + l) = L$. Both K and L can be calculated since they are the amounts by which the observed values fail to satisfy the condition equations (39). Thus equations (40)

reduce to
$$av_x + bv_y + cv_z + K = 0$$
$$dv_x + ev_y + fv_z + L = 0 \tag{41}$$

Or in tabular form

v_x	v_y	v_z	Abs.
a	b	c	K
d	e	f	L

5.9.2.1. Weights

Suppose the observed values have respective weights p_x, p_y, and p_z. The most probable values are obtained when $\frac{1}{2}[pv^2]$ is a minimum. Also, let there be two correlatives C_1 and C_2, one for each condition equation (41). Then the m.p.v's are obtained when

$$\tfrac{1}{2}[pv^2] = \tfrac{1}{2}(p_x v_x^2 + p_y v_y^2 + p_z v_z^2) - C_1(av_x + bv_y + cv_z + K)$$
$$- C_2(dv_x + ev_y + fv_z + L) \tag{42}$$

is a minimum. Differentiating (42) with respect to v_x, v_y, and v_z in turn and equating the results to zero

$$p_x v_x - C_1 a - C_2 d = 0$$
$$p_y v_y - C_1 b - C_2 e = 0 \tag{43}$$
$$p_z v_z - C_1 c - C_2 f = 0$$

whence
$$v_x = (1/p_x)(C_1 a + C_2 d)$$
$$v_y = (1/py)(C_1 b + C_2 e) \tag{44}$$
$$v_z = (1/p_z)(C_1 c + C_2 f)$$

and putting $u_x = 1/p_x$; $u_y = 1/p_y$; $u_z = 1/p_z$ equations (44) become

$$v_x = u_x(C_1 a + C_2 d)$$
$$v_y = u_y(C_1 b + C_2 e) \tag{45}$$
$$v_z = u_z(C_1 c + C_2 f)$$

Hence if the values of C_1 and C_2 are found, we obtain the v's from equations (45). Substituting for the v's from (45) in the condition equations (41) gives

$$u_x a(C_1 a + C_2 d) + u_y b(C_1 b + C_2 e) + u_z c(C_1 c + C_2 f) + K = 0$$
$$u_x d(C_1 a + C_2 d) + u_y e(C_1 b + C_2 e) + u_z f(C_1 c + C_2 f) + L = 0$$

Collecting coefficients of the correlatives C_1 and C_2 gives

$$C_1(u_x aa + u_y bb + u_z cc) + C_2(u_x ad + u_y be + u_z cf) + K = 0 \tag{46a}$$
$$C_1(u_x ad + u_y be + u_z cf) + C_2(u_x dd + u_y ee + u_z ff) + L = 0 \tag{46b}$$

Equations (46) are the *correlative normal equations*. It will be seen that the coefficient of C_1 in the second equation equals the coefficient of C_2 in the first equation, i.e. the correlative normals are also symmetrical about the diagonal terms as for the observation equation method. Thus terms to the left of the

Theory of Errors and Survey Adjustment 245

diagonal terms may be omitted. To obtain equation (46b) we read down the second column then along to the absolute term.

The correlative normals are solved for the correlatives in the manner described above in §5.7. Finally the residuals are obtained from equations (45). It will be evident from a study of the condition and correlative normal equations tabulated below how the latter are formed from the former.

5.9.2.2. *Condition Equation (41) and Reciprocals of Weights u*

	v_x u_x	v_y u_y	v_z u_z	Abs.
C_1	a	b	c	K
C_2	d	e	f	L

5.9.2.3 *Correlative Normals in C_1 and C_2*

C_1	C_2	Abs.
$(u_x aa + u_y bb + u_z cc)$	$(u_x ad + u_y be + u_z cf)$	K
	$(u_x dd + u_y ee + u_z ff)$	L

It will be seen that one works along the lines here as compared with down the columns for the observation equations method. The most significant difference is that the absolute terms of the correlative normals are the *same as the absolute terms of the condition equations*.

5.9.2.4. *Summation Checks*

In an identical manner to observation equations, an extra line is added to each set of equations to check the work. This line is the sum of the terms up the columns, this time omitting the absolute term since it is the same in both types of equation. The detailed operation of this check will be illustrated in the example in §5.9.2.5 following.

5.9.2.5. *General Expressions for the Condition and Correlative Normal Equations*

In a similar manner to that described above, the formation pattern of normal equations may be extended to any number of equations. If there are m condition equations in n unknowns, $m < n$, there will be m correlatives and therefore m correlative normals. Consider the m condition equations given below in table 5.7, where the a's b's, c's $\cdots$ m's are coefficients given by theory, and the K's are the absolute terms given by the observed values applied to the condition equations and the u's are the reciprocals of the weights of the observed values, etc., then the correlative normals are as given in table 5.8

$s_1 = a_1 + b_1 + c_1 + \cdots + m_1$ (N.B. Do not include the u's in the sum.)

$[uaa] = u_1 a_1 a_1 + u_2 a_2 a_2 + u_3 a_3 a_3 + \cdots + u_n a_n a_n,$ etc.

$[uas] = u_1 a_1 s_1 + u_2 a_2 s_2 + \cdots + u_n a_n s_n,$ etc.

Table 5.7

| | v_1 | v_2 | v_3 | v_4 | v_5 | ... | v_n | Abs. | Sums omitting K's |
	u_1	u_2	u_3	u_4	u_5	...	u_n		
C_1	a_1	a_2	a_3	a_4	a_5	...	a_n	K_1	S_1
C_2	b_1	b_2	b_3	b_4	b_5	...	b_n	K_2	S_2
C_3	c_1	c_2	c_3	c_4	c_5	...	c_n	K_3	S_3
.	.	.	.	.	.		.	.	.
.	.	.	.	.	.		.	.	.
.	.	.	.	.	.		.	.	.
C_m	m_1	m_2	m_3	m_4	m_5	...	m_n	K_m	S_m
Sum	s_1	s_2	s_3	s_4	s_5	...	s_n	check total sums	

Table 5.8

C_1	C_2	C_3	...	C_m	Abs.	Sum
[uaa]	[uab]	[uac]	...	[uam]	K_1	[uas] + K_1
	[ubb]	[ubc]	...	[ubm]	K_2	[ubs] + K_2
		[ucc]	...	[ucm]	K_3	[ucs] + K_3
			...	.	.	.
				.	.	.
				[umm]	K_m	[ums] + K_m

Finally the residuals are derived from equations of the form of (45), i.e.

$$\begin{aligned}
v_1 &= u_1(a_1C_1 + b_1C_2 + c_1C_3 + \cdots + m_1C_m) \\
v_2 &= u_2(a_2C_1 + b_2C_2 + c_2C_3 + \cdots + m_2C_m) \\
v_3 &= u_3(a_3C_1 + b_3C_2 + c_3C_3 + \cdots + m_3C_m) \\
&\quad \cdot \quad \cdot \quad \text{etc.} \quad \cdot \\
v_n &= u_n(a_nC_1 + b_nC_2 + c_nC_3 + \cdots + m_nC_m)
\end{aligned} \qquad (47)$$

Theory of Errors and Survey Adjustment 247

Note how these equations (47) can easily be formed from the table of the condition equations if the values of the correlatives are placed alongside their respective condition equations. An extra line should be left at the foot of the table of condition equations in which to write the values of the residuals, after the normals have been solved for the correlatives.

Example: We shall now work the example in §5.8.1 by the method of condition equations. Since the weights are inversely proportional to the lengths of the respective lines, the inverses of the weights are directly proportional to the lengths of the lines. Since the concept of weight is purely relative we may reduce all the lengths by the same factor. Therefore we choose $u_1 = 0.7$, $u_2 = 1.0$, $u_3 = 0.6$, $u_4 = 0.9$, $u_5 = 0.6$, and $u_6 = 0.4$, i.e. the respective lengths of the lines AB, AC, AD, BC, BD, and CD multiplied by 0.1. In figure 5.7 the lines have been arrowed in the sense that their height differences have been evaluated, e.g. $H_B - H_A = +12.00$. The condition equations are formed by the closures in height round the various unique circuits, i.e. no circuit is counted twice. To ensure that this is so, the diagram is built up line by line. When a complete circuit is completed a condition equation is formed for that circuit. Thus drawing the lines (1), (3), and (5) gives the condition that

$$(H_B - H_A) + (H_D - H_B) - (H_A - H_D) = 0 \tag{48a}$$

The negative sign is introduced before the third term because we are considering the height difference $H_A - H_D$, whereas it was the height difference $H_D - H_A$ that was tabulated to be -1.55. To ensure that the correct signs are applied, the lines have been arrowed in the sense in which the height differences have been evaluated. In most cases a wrong sign will be obvious from the circuit misclosure. If we denote the *most probable values* of the height differences for lines $1, 2 \cdots 6$, by $\Delta H_1, \Delta H_2, \Delta H_3, \Delta H_4, \Delta H_5, \Delta H_6$, the three condition equations to be satisfied may be written

$$\begin{aligned} \Delta H_1 + \Delta H_5 - \Delta H_3 &= 0 \\ -\Delta H_2 + \Delta H_3 - \Delta H_6 &= 0 \\ +\Delta H_4 - \Delta H_5 + \Delta H_6 &= 0 \end{aligned} \tag{48}$$

Substituting the respective observed values for these height differences in (48)

$$\begin{aligned} +v_1 + v_5 - v_3 - 0.11 &= 0 \\ -v_2 + v_3 - v_6 - 0.04 &= 0 \\ +v_4 - v_5 + v_6 + 0.07 &= 0 \end{aligned} \tag{49}$$

It is possible to choose other circuits and form condition equations for them; but only three are unique. As an exercise the reader may choose three other circuits to adjust, and satisfy himself that an identical answer to that given below is obtained. Table 5.9 gives the condition equations (49) in their tabular form, together with the u's. The figures in the first column are the values of the correlatives C_1, C_2, and C_3, found after solution of the correlative normals in table 5.10. The values of the residuals in the two lines (7) and (8) are found from the condition equations when substituting the values of the correlatives

248 Practical Field Surveying and Computations

in equations of the type of (47). Since the final values of the m.p.v's are to be quoted to two decimal places, the residuals have to rounded off, but at the same time the condition equations must still be satisfied exactly. For this reason, the residual v_5 has to be rounded off to 0·04 instead of 0·05. Sometimes this final rounding off is difficult, and in a few cases impossible. The best way to tackle the process is to round off the obvious values such as 0·048, leaving the doubtful ones to the last. The final most probable values of the height differences are +12·05, +69·32, −1·57, +57·27, −13·62, −70·89 respectively.

Table 5.9. Condition Equations

		v_1	v_2	v_3	v_4	v_5	v_6	
1	Residuals							
2	U's D × 10⁻¹	0·7	1·0	0·6	0·9	0·6	0·4	Abs. Sum
3	c_1 + 0·0679	+1		−1		+1		−0·11 +1
4	c_2 + 0·0390		−1	+1			−1	−0·04 −1
5	c_3 − 0·0072				+1	−1	+1	+0·07 +1
6	Sum	+1	−1	0	+1	0	0	Total sums equal
7	Residuals	+0·0475	−0·0390	−0·0173	−0·0064	+0·0451	−0·0185	
8	Rounded v's	+0·05	−0·04	−0·02	−0·01	+0·04	−0·02	
9	Observed ΔH	+12·00	+69·36	−1·55	+57·28	−13·66	−70·87	
10	m.p.v.	+12·05	+69·32	−1·57	+57·27	−13·62	−70·89	

Table 5.10. Gauss–Doolittle Solution

		0·0679	0·0390	−0·0072		
		−0·9321	−0·9610	−1·0072		
		c_1	c_2	c_3	Abs.	Sum
c_1	+1·9000	+1·9000	−0·6000	−0·6000	−0·1100	+0·5900
		(0·52632)	+0·3158	+0·3158	+0·0579	−0·3105
c_2	−0·6000	+2·000	+1·8105	−0·5895	−0·0747	+1·1463
			(0·55233)	+0·3256	+0·0413	−0·6331
c_3	−0·6000	−0·4000	+1·9000	+1·5186	+0·0109	+1·5295
				(0·65850)	−0·0072	−1·0072
Abs.	−0·1100	−0·0400	+0·0700			
Sum	+0·5900	+0·9600	+0·9700			

N.B.—
(a) The normals are written to the left of and below the bold lines.
$$+1·900\ 0 = (+1)(+1)(0·7) + (−1)(−1)(0·6) + (+1)(+1)(0·6)$$
$$-0·600\ 0 = (−1)(+1)(0·6)$$
$$-0·600\ 0 = (+1)(−1)(0·6)$$

Abs. term same as condition equation, i.e. 0·110 0
$$+0·590\ 00 = (+1)(+1)(0·7) + (−1)(0)(0·6) + (1)(0)(0·6) − 0·110\ 0.$$

(b) The numbers in brackets are the reciprocals of the numbers above them, i.e. (0·526 32) = 1/1·900 0, etc.

(c) The residuals are obtained from Table 5.9 as follows:
$$+0·047\ 5 = 0·7\{(0·067\ 9)(+1)\}$$
$$-0·039\ 0 = 1·0\{(0·039\ 0)(−1)\}$$
$$-0·017\ 3 = 0·6\{(0·067\ 9)(−1) + (0·039\ 0)(+1)\},\ \text{etc.}$$

These values are the same as those already found by the observation equations method.

It must be stressed that although the condition equations are satisfied, the final results may be incorrect, if some mistake has been made in the formation of the condition equations. Therefore the condition equations should be checked independently by a second person before commencing the solution. All processes after the formation of the condition equations are self-checking. The final check on the work is that the various height differences taken round all circuits agree exactly, e.g.

$$\Delta H_1 + \Delta H_4 - \Delta H_2 = 0, \quad \text{i.e.} \quad +12\cdot05 + 57\cdot27 - 69\cdot32 = 0, \quad \text{etc.}$$

5.10. OBSERVATION AND CONDITION EQUATIONS METHODS

There is insufficient space in a book of this range to consider the relative merits of both methods of adjustment in any great detail. However, the following major points should be noted.

(i) In a solution by observation equations there are as many normal equations as there are independent unknowns.

(ii) In a solution by condition equations there are as many normal equations as there are conditions.

Since most of the labour of solution by hand methods is in the solution of the normals, it is important that the number of normals should be as small as possible. The nature of each adjustment problem will decide which is the most convenient method. For instance, if only one new point is to fixed from many fixed points, there are only two independent unknowns—the easting and northing of the point—and therefore only two normals. In such a network there are many conditions relating the fixed points to each other, and therefore a correspondingly large number of normals. Hence we would choose the observation equations method by variation of coordinates (see chapter 6).

(iii) Condition equations are often more difficult to derive than observation equations (see chapter 13).

(iv) Condition equations are more self-checking at the formation stage, because any large misclosure is suspect.

In the course of this book, the reader will find many examples of survey adjustment by these two methods.

5.10.1. SURVEY ADJUSTMENT BY ELECTRONIC COMPUTER

Most large surveying organizations today have access to an electronic computer which can be programmed to carry out all the various stages of adjustment described above. Only the data from the field books needs to be processed by hand to give the mean observed values for adjustment. Various methods of solving normals in addition to the Gauss–Doolittle and Cholesky are in current use or stages of development, and considerable advances in technique are very probable in the next few years.

6 Angular Measurement and Computation

6.1. INTRODUCTION

In this chapter consideration is given to the following aspects of surveying:

(i) The *observation* of horizontal angles.
(ii) The *adjustment* of these angles by both semi-rigorous and least squares methods.
(iii) The *computation* of the coordinates of a survey which consists mainly of angular observations.

Most theory and examples are illustrated by reference to one survey, the data of which is listed in tables 6.1 and 6.2, the diagram of which is figure 6.1. The coordinates of table 6.2 were computed by plane trigonometry from the adjusted angles, without reference to a projection. Points E and F were the starting points whose coordinates were taken as given.

Triangulation consists of three main stages:
Observation or field work, adjustment of angles or coordinates, and the computation of final coordinates, either from adjusted angles or by variation of provisional coordinates. These various stages cannot be isolated from each other. For example, the observer is well advised to consider how his observations are to be computed whilst he is in the process of observation. In this way, some possible omission can be avoided.

When the three angles of a triangle are observed, the triangle is said to be *fully observed*. Computation of positions is carried out by the methods of *intersection* and *resection* explained in §1.1.2.5, or complex combinations of both. Linear measures may also be introduced, particularly with the advent of electromagnetic distance measurement (see Chapter 13).

6.2. PRELIMINARIES

Before actual field work can begin, preliminary planning is required, which will vary according to the complexity and type of survey to be carried out. After a careful reconnaissance, a scheme of observation should be chosen, and an accurate diagram drawn up at a convenient scale, usually as an overlay to an existing map. This will be useful in many ways. It will assist the observer to identify the various stations in the field, it will allow of a unique system of numbering the points, it will help in the compilation of the station descriptions, and may be the basis of the field calculation of spherical excess. Such a diagram should be capable of illustrating all stages of the work as it proceeds, so that an

Table 6.1. Data for Chapter 6 (See Figure 6.1)

No.	Observed angles	Least squares correction	Adjusted angles
1	32° 14' 18".8	−0".6	18".2
2	23 35 17.6	+1.5	19.1
3	17 05 36.2	−0.5	35.7
4	52 04 57.1	−1.5	55.6
5	79 14 33.5	−1.2	32.3
6	55 34 32.3	−1.8	30.5
7	56 00 48.8	−1.5	47.3
8	57 42 28.2	−2.7	25.5
9	40 59 38.9	+4.1	43.0
10	59 20 44.4	+0.4	44.8
11	48 01 23.9	+1.4	25.3
12	27 35 52.1	−4.2	47.9
13	45 02 04.0	−2.0	02.0
14	72 09 20.7	−2.0	18.7
15	53 18 32.5	+1.6	34.1
16	94 27 06.5	+1.2	07.7
17	35 12 51.4	0.0	51.4
18	31 38 05.6	+1.3	06.9
19	66 16 48.9	−1.7	47.2
20	45 10 57.2	0.0	57.2
21	87 14 09.4	+0.2	09.6

Table 6.2. List of Final Coordinates computed on the plane from the Adjusted Angles of Table 6.1 Accepting Values for E and F as Datum. Coordinates are in Metres.

	Easting	Northing		Easting	Northing
A	351 240.22	138 628.80	E	345 780.67	150 394.05
B	356 788.67	144 328.27	F	347 490.50	145 480.79
C	356 442.71	148 778.96	G	351 629.10	144 899.07
D	350 044.25	150 752.70			

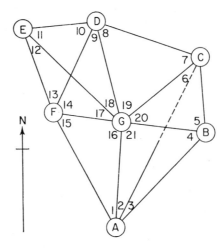

Figure 6.1

instantaneous picture of the current state of the survey is always available (see figure 1.14).

The choice of beacons, theodolites, number of rounds of angles, and methods of fixation, will all depend on the ultimate precision desired present or future. Section 6.12.3 below gives methods which may be employed to assist the surveyor to decide how precise the field work has to be. When a triangulation system has to accord with generally accepted international standards, the following requirements will act as a useful guide:

6.2.1. PRIMARY TRIANGULATION

No triangle misclosure should exceed 3", the average of all must be below 1", and no side equation correction should be greater than $\frac{1}{2}$" per angle.

6.2.2. SECONDARY TRIANGULATION

No triangle misclosure should exceed 5", the average must be below 3", and side equation corrections should be below 1". The words 'primary' and 'secondary' connote the degrees of precision cited above. If a triangulation does not conform to these standards, the terms *major* and *minor* are preferred to primary and secondary since they do not connote any special degree of precision.

Other preliminary activities will be concerned with administration, recruitment of labour, obtaining permission to go on the land, checking of transport and survey equipment, all of which are described in more detail in Chapter 1.

6.3. BEACONS OR SIGNALS

To enable the observer to sight accurately, some form of signal or beacon has to be erected over the points of a survey. Such points may vary from temporary pegs to permanent survey pillars. As the signals must not introduce a significant error into the work, they must be designed to suit the particular

survey in question. The major factors affecting the choice of a beacon are:
  (i) The precision of the survey.
  (ii) The lengths of the sides.
  (iii) The type of country being surveyed.
  (iv) The amount of available funds.
  (v) Availability of labour.
  (vi) Expected weather conditions.
  (vii) Possible future use of the beacons for other surveys.

Whatever its design, a good beacon should possess the following features:
  (i) It must be clearly identifiable by the observer.
  (ii) The part which is to be observed must be symmetrical to the observer.
  (iii) The bisected part must subtend an angle at the observer, which is consistent with the precision of the survey.

The following two approximate relationships are useful in determining the size of a particular beacon:
A foot subtends a second of arc at forty miles.
An inch subtends a minute of arc at 300 feet.
Beacons are of two main types, *opaque* and *illuminated*.

6.3.1. OPAQUE BEACONS

Opaque beacons are more commonly used than illuminated ones, because in most cases they give equally acceptable results, are generally more convenient to operate, and are less expensive as a rule. In primary triangulation, opaque beacons are not normally used. A selection of some successful beacons is given in figure 6.2. These range from the ordinary chaining arrow, which is not employed as often as it might be, to large quadripod beacons built from rough timbers. The points marked A are the actual points of bisection, while those marked B are the identifying features. The materials used in the construction of such beacons vary a great deal; machined aluminium rods are convenient for portable types which will be a permanent part of the surveyor's equipment; $1\frac{1}{2}$ inch diameter galvanised pipe is a useful material for the permanently situated beacon. Turnbuckles or rough tourniquets serve to tighten the stays, and a large bricklayer's level is a great time saver when erecting a beacon.

In all work over long sights, the back ray should always be cleared, if this will silhouette the beacons against the skyline. As a further aid to future identification from afar, the surveyor should anticipate the appearance of the beacon and its location on the hill as seen from the various directions of the survey, and make sketches where possible.

Whatever type of beacon is used, all its dimensions must be recorded in the relevant field book.

Beacons will be painted in colours which best suit their surroundings. In a new area, time is well spent experimenting with this aspect. Generally speaking, two contrasting colours are best; red and yellow being preferred to black and white. Modern self-luminous paints can be useful on solid materials, but do not adhere easily to fabrics.

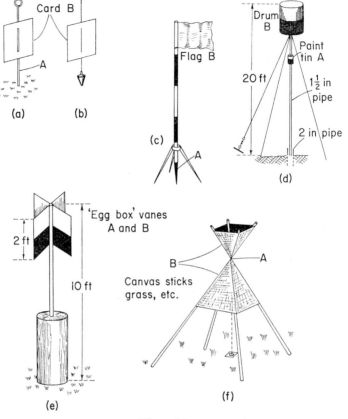

Figure 6.2

6.3.2. PHASE ERROR ON BEACONS

Side illumination of an opaque beacon can give rise to errors of bisection. For example, a square sectioned pillar or quadripod vane that is illuminated by a side light may present an asymmetrical aspect to the observer. Beacons of circular section may also give rise to an error of bisection if they are partially illuminated from the side. There are two theoretical cases:

(i) When the circular surface of the beacon is reflective,
(ii) When this surface is dull and non-reflective.

These cases are illustrated in figures 6.3 and 6.4 respectively.

With the reflective surface, the observer sees a bright line which he bisects. The angular error from the centre of the beacon is given by $+r \cos \frac{1}{2}\alpha / D \sin 1''$ where α is the angle between the survey line and the sun as shown in figure 6.3.

With the dull surface, the observer sees that sector of the pillar which is shaded in figure 6.4, bisection of which introduces an error $+ r \cos^2 \frac{1}{2}\alpha / D \sin 1''$ to the observed direction (figure 6.4).

The angle α can be observed to the sun at the time of observation and the relevant correction duly applied.

Angular Measurement and Computation

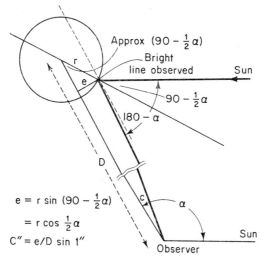

Figure 6.3

In practice, however, the linear amount of the phase effect can often be determined directly by examination of the side illumination of the beacon at the observer's position. The linear amounts $r \cos \frac{1}{2}\alpha$ and $r \cos^2 \frac{1}{2}\alpha$ are measured directly and the consequent angular corrections computed.

In practice it is often difficult to determine which case pertains at a particular moment; and in the case of a wet beacon, drying rapidly in the sun, the first can give way to the latter during the observations.

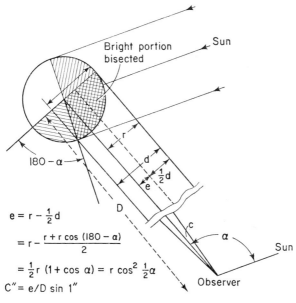

Figure 6.4

In the case of square sectioned beacons, a correction can be applied if the orientation of the faces is known. Such information should be listed in the beacon description if necessary.

6.3.3. ILLUMINATED BEACONS

These are of two types *helios*, and *lamps*; both of which may be employed by day. They have certain advantages over opaque beacons in that they are usually unmistakable, they are portable, they can be used over great distances up to 80 miles or more, and the observation point is clear and may be varied in size according to the length of line observed. The disadvantages are that operators are required at each station of the survey, and that helios are dependent on the sun, and lights are dependent on batteries or fuel. In precise work, the ideal size of light is important, and each surveyor has his personal preference. Three seconds of arc subtended at the observer is a typical light size. To ensure a consistent apparent light size throughout the survey, masks are fitted to the lamps and helios. These masks may be cut in cardboard according to the length of each line and inserted before observation. This procedure is normally only worthwhile on primary and secondary work. Variations in weather conditions are countered by altering the light intensity.

6.3.4. HELIOS

In its simplest form, the *helio* consists of a plane mirror about 5 inches in diameter, which is centred over the survey mark and used to reflect the light of the sun towards the observer. A means of moving the mirror in both horizontal and vertical planes is supplied, to allow for the movement of the sun. Sometimes a second, or *duplex* mirror is required to direct the light in the correct direction, as shown in figure 6.6. The size of light seen by the observer depends on the aperture of the helio; but the degree of alignment required for him to see it at all, is the sun's diameter or about 32' of arc; more if the sky round the sun is very bright. Figures 6.5 and 6.6 show typical pencils of light directed towards a survey point. To assist the pointing, a cross-sight is mounted either on a small arm attached to the helio and its tripod, or to a wooden stand erected at some distance up to 20 feet from it. This latter, or *semaphore board* figure 6.6(*a*), is preferable since it permits more precise alignment, and is easier

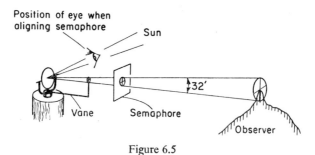

Figure 6.5

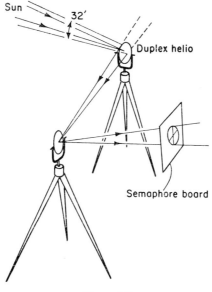

Figure 6.6

to operate. The helio and cross-sight are correctly positioned when the former is plumbed over the survey point and the latter is on line between the helio and the observer. To align the helio and board the best procedure is as follows:

(i) Ensure that the sun's rays can be directed towards the observer's hill. This may require a duplex mirror. If so, care must be taken to ensure that the mirror seen by the observer is plumbed over the survey mark.
(ii) Look into the mirror or mirrors from the approximate direction of the sun, and rotate the mirror centred over the mark until the distant observer's position appears to bisect the central hole of the helio.
(iii) Move the cross-sight until it lies on this line between the helio centre and the image of the observer's position still seen in the mirror. The helio and its sight are then correctly aligned. The alternative method of looking through the helio centre is impossible with many helios.

Figure 6.6(a)

Once aligned, the light keeper keeps the image of the sun centrally on the cross-sight and therefore pointing in the correct direction. The small black dot created by the hole drilled in the helio mirror makes this task easier.

On a cloudy but bright day, it is possible to use a helio effectively over moderate distances of about 6 miles. Light keepers should be made aware of this, since they often think that brilliant sun is essential.

Alignment of helios by day is much simpler than the alignment of lamps by night. Hence the latter are more easily aligned in conjunction with the former. The semaphore board is erected by day, and used at night when shining the lamp. For a lamp and helio of different heights a small adaptor is sometimes required to convert the lower to the same height as the higher.

The obvious drawback of the helio is its dependence on the sun. It also requires a conscientious operator, and is subject to the general disadvantages of organization problems. Its singular advantage is that it can be used for vertical angle observations at noon, i.e. at the optimum time of day; nor is it likely to be mistaken for some other light or survey signal.

6.3.5. LAMPS

Lamps for survey purposes are either of the pressure liquid fuel type such as the 'Tilley' or 'Bi-aladdin', or the electric battery-operated type, similar to a motor car headlamp.

The pressure type provides good signals over distances of about 6 miles maximum and is not subject to problems of alignment. They are reliable if looked after properly and may be used for general purposes when not required as beacons. A simple tripod or pillar adaptor is easily constructed for them, or a plane table on its tripod makes a convenient stand.

Electric lamps specially designed for survey work are obtainable from all the recognized manufacturers of survey equipment. They normally operate on 6 volts, powered by inert dry cells which are activated by filling with water. Brighter lights are easily obtained by introducing bulbs of greater power and suitably altering the input from batteries. For efficient operation, the beam of light should be parallel. This is achieved by altering the position of the mirror; analysing the beam on a misty night or by creating a dusty atmosphere. A semaphore board is normally employed by the light keeper to align the beam. If there is any danger of the wrong light being observed in built up areas, a suitable identification signal should be evolved.

6.3.6. FIELD PROCEDURE WITH OPAQUES, HELIOS, AND LAMPS

Observing to opaque beacons possesses the singular advantage that once they are in position, little by way of organization is required, and the observational sequence is more flexible. If care is taken with the choice of opaque beacon, if they are securely fixed in position, and if no trouble arises from vandals or thieves because of attractive materials used in construction, opaque beacons are preferred to luminous on all but precise surveys. Opaque beacons may move off centre in a high wind or if disturbed by animals. In such cases an eccentric correction described in §6.12.1 may be applied. However, it is not always

Angular Measurement and Computation

certain when the eccentricity occurred, and therefore uncertain whether to apply the correction or not.

Observation to helios and lamps requires much forethought before the observer goes into the field, and is much less flexible once observing has begun. Dependable, well-trained light keepers, capable of effecting minor repairs to lamps, have to be equipped and supplied with food and water at regular intervals. A light keeping party of three is ideal even in most difficult country. The number of parties will depend on the network to be observed, and the available routes throughout the area. Only two light parties are required for traversing, and at least six for a triangulation chain of braced quadrilaterals, if work is to make rapid progress and full use must be taken of what good weather conditions arise.

The complete observing programme should be worked out beforehand and explained to the light keepers. Each light keeper should be given instructions concerning the order in which to shine his lamps and helios, and the size of mask to be used in each case. A timetable of operations is not advisable, but the sequence in which the work will be carried out, should be decided.

Some system of signals is required to tell the light keepers when to change from one stage in the plan to another. The Morse code is much too complicated and unnecessary for survey work. The following simple system has been employed to good effect in various surveys, for lines up to 60 miles or more. For a signal to be of any use at all, it must be reliable, and both parties to a message must be aware that the signal has been correctly sent and received. The observer uses a lamp which is much brighter than those used by the light keepers, because the latter have to see the light with the naked eye. The table below shows the sequence of a typical signal:

Stage	Observer	Light keeper
1	Out	On
2	On	On
3	On	Out
4	Out	Out
5	Out	On

The observer wishes to tell the light keeper to change to the next stage in the programme, e.g. he may want him to change from one direction to the next. At stage 1 the light keeper is shining his lamp. Stage 2: the observer then shines his lamp to the keeper. Stage 3: on seeing the observer's light, the keeper puts his light out. Stage 4: the observer, on seeing this reaction to his signal, puts his light out. Stage 5; finally the keeper puts his light back on again in response to the observers acknowledgement. Thus, both the observer and the light keeper are aware that the other has sent and received a signal. The five stages constitute one unit of a signal, which will have a particular meaning. A double unit will have a different meaning, and so on.

To get the best results from light parties over a long period of time, the individuals concerned must have complete faith in the reliability of the observer, not so much from a technical as from an administrative point of view. One failure to bring water to a light party on the prearranged day, can cause a complete breakdown of the teamwork, so essential to success. Again, a helio operator will soon become inefficient if he feels his helio is not being observed. Hence the observer should make every effort to observe when the helios and lights are 'up', and when finished, tell the helio operator to cease; otherwise his own labourers will soon inform the light keepers that their work is quite often in vain, and a quick drop in efficiency results.

6.4. SURVEY TOWERS

In flat country, or in other special circumstances, such as to give vision over buildings, survey towers are employed to advantage. These consist of an outer tower which holds the observer and his equipment and an inner tower which holds the theodolite; and being entirely independent of the outer, is unaffected by the movement caused by the observer. This independent duality is the basic feature of all survey towers used for angular work. Specially designed portable steel towers capable of erection to 120 feet have been used throughout the world (e.g. the Bilby, and Kenya). These towers are 'portable', in the sense that they can be dismantled and transported in sections by a 5 ton truck, occasionally being manhandled over short distances of up to a mile. A tower of say 75 ft can be erected by six trained labourers in one day, and dismantled in a slightly shorter period. The feet of the towers are normally concreted in position, though on hard soils baulks of heavy timber, such as railway sleepers, may take the load. Great care is required in levelling the feet of the tower, because the separation between the tops of both towers is only of the order of 6 inches. The centering of the inner can be varied by about 9 inches hence the feet of the towers and their verticality must accord with this degree of tolerance. It is easier to construct a ground mark underneath an erected tower than to erect a tower over an existing mark. Erection of the first section, of the Bilby, is best done on its side and later heaved into position. A small auxiliary ladder is very useful at the first stage of erection. Hawser guys are useful on the outer tower, but are a doubtful blessing on the inner, because of the vibration they set up. The nuts of a tower may work slightly loose in a matter of hours, with disastrous results on the stability of the inner tower. The observer is well advised to carry a spare set of spanners, and to tighten up the inner tower before observing. Survey equipment is lifted to the top by a block and tackle.

6.4.1. CENTERING OF TOWERS

The plate which carries the theodolite has a maximum movement of about 9 inches, hence the rough centering at erection must be within these limits. Final centering may be carried out by observations from two theodolites (situated on the ground) nearly at right angles to the base of the tower and about 400 feet from it. This method involves two observers, or a great delay if only one is present. A convenient method is to use a theodolite plumbed over the survey mark as a collimator by pointing it to the zenith and viewing the upper plate of the tower through the diagonal eyepiece. The theodolite has to be carefully levelled. If there is a horizontal collimation error it will be impossible to see the zenith. In this event the theodolite telescope is moved in azimuth through 360° while the cross hairs will appear to trace out a circle into which the tower is centred. Some theodolites such as the Geodetic Tavistock cannot point to the zenith, and are therefore unsuitable for the method described. It is advisable to use some form of protection, such as a net, to protect the observer and theodolite from anything accidentally dropped from above.

6.4.2. BEARING MARKS

It is usual practice to establish at least two permanent ground marks at every tower station, one centred beneath it and the other at say 500 yards distance, both of which are intervisible on the ground. This second mark will be observed as an integral part of the survey, so that a future surveyor will be able to obtain orientation as well as position from the station without erecting a tower.

6.4.3. SITING OF TOWERS

The detailed siting of towers needs care if the most efficient use is to be made of them. Curvature and refraction, as described in chapter 8, have to be allowed for. In practice, a detailed profile of the apparently 'flat' surface has to be drawn, to allow for local topography which can easily upset a theoretical determination based on the assumption of a spherical Earth. An altimetric height traverse with distances obtained from a calibrated vehicle mileometer is a useful method to use. A tall 60 ft ladder is also of great assistance in siting towers.

6.4.4. OTHER TOWERS

Towers may often be constructed of timber cut on the spot. The inner 'tower' may often be a convenient, tall tree around which the outer structure is erected. A length of 2" diameter galvanized pipe has been used successfully as an inner 'tower'. In all cases both 'towers' are independent of each other.

6.5. THE OBSERVATION OF HORIZONTAL ANGLES

6.5.1. INTRODUCTION

The process of setting a theodolite in a level position means that the observed horizontal angles will be recorded on a plane, that of the horizontal circle of the instrument, which is at right angles to the direction of gravity at this point. Hence the angles that are recorded at one station are all contained in one plane, even although the stations observed may differ in altitude. Since this is so, the angles observed round a point should sum to 360°, whereas angles observed in inclined planes would not. Angles are sometimes observed in inclined planes, e.g. by sextant, but these will not be considered in this book.

Since the surface of the Earth is not plane, the geometrical relationships between the three observed angles of a triangle whose apices are at different heights is theoretically complex. However, it is usually sufficient in all but geodetic survey, to consider the Earth as spherical and sometimes as flat over the area covered by the survey. For a consideration of the small effects due to differences in the direction of gravity, heights above sea level, and curves on the surface of a reference spheroid, the reader is referred to Clark's *Plane and Geodetic Surveying*† or Bomford's *Geodesy*.‡

6.5.2. GENERAL OBSERVATIONAL TECHNIQUE

The observation of horizontal angles is now considered. Although the verticals will normally be read at the same visit to the station, the horizontal and vertical circles should not be read at the same pointing of the telescope if the best results are to be obtained. In most instances, they will not be observed at the same time of day, since horizontals are best observed when the air is cool, and verticals when refraction is least, i.e. about noon.

It is assumed that the theodolite is in reasonable adjustment, and that the observer is familiar with its operation and reading system, etc (see chapter 3). Since no instrument can be in perfect adjustment, the technique of observation is designed to eliminate the effects of inevitable small residual errors. It also guards against various errors of observation due to causes external to the theodolite itself. The various factors which observational procedure is intended to remove or render negligible are:

6.5.2.1. *Gross Error*

This is more common than one would suppose. Whatever the standard of the survey, at least two different parts of the graduated circle should be used in recording a direction, i.e. at least *two zeros* should be used, and the included angles checked in the field.

6.5.2.2. *Backlash*

The observer should always make a few preturns of the theodolite in the direction he intends to observe, to eliminate the mechanical looseness between the sighting and the recording devices of the instrument. If an observed beacon is overshot, the theodolite must be swung through a full circle and the bisection repeated in the correct direction. Opinions differ as to whether the last movement of the tangent screw is best made against the spring, even if it does mean contravening the process just described. The ultimate answer will depend on the particular instrument used.

6.5.2.3. *Change of Face*

The usual procedure is to swing clockwise with the vertical circle to the left, i.e. on *face left*, and anticlockwise on *face right*. The reasons for *changing face* are to eliminate the effects of trunnion axis tilt and collimation errors introduced in observing to stations which are at different heights. Errors due to vertical axis tilt are not eliminated by change of face.

† D. Clark, *Plane and Geodetic Surveying* (London: Fourth Edition Constable, 1946).
‡ G. Bomford, *Geodesy* (Oxford: Clarendon Press, 1952).

6.5.2.4. *Change of Zero*

In precise work, the periodic errors of graduation of the horizontal circle (and micrometers) may be significant. Various parts of the circles are therefore used to record the angles, the mean of which will be almost free of errors from this source. With double reading theodolites, two readings of the circle are taken simultaneously at 180° from one another hence the zeros, or settings into which the circle is divided, are spread over 180°. The micrometers, on the other hand, have to be divided through their entire range. In primary triangulation, it is usual to observe 16 different zeros. As an example of the choice of zeros consider the Geodetic Tavistock, whose micrometer range is 10' of arc, and which is of the double reading type. The 16 separate zeros will be in order of observing:

1	00° 01' 05"	9	11° 16' 05"
2	90 08 55	10	101 23 55
3	45 02 10	11	56 17 10
4	135 07 50	12	146 22 50
5	22 33 20	13	33 48 20
6	112 36 40	14	123 51 40
7	67 34 30	15	78 49 30
8	157 35 30	16	168 50 30

The first two zeros are placed at 90° to each other, the next two divide these sectors equally, and so on. If only two zeros are observed, the first two will be used; if only three, the first three, and so on.

There is no best method of employing different zeros. If only eight sets of angles on two faces are to be observed, the first eight zeros may be used, with each one used twice, once on face left and once on face right. This is the normal method because of its convenience both at the observing and the abstracting stages. However, it would be better to record eight faces on the first eight zeros and eight faces on the second eight, but less convenient.

In precise work several bisections of the beacon, and several readings with the micrometers will be taken on one zero to achieve consistence of precision.

6.5.2.5. *Bisection of Beacon*

The beacon is bisected with the cross hairs, the readings booked, and the bisection checked, as it may change as a result of a strong cross wind. If the check shows a bisection error, the readings should be cancelled and repeated.

6.5.2.6. *Lateral Refraction*

In precise work, lateral refraction of a ray passing close to an inclined land surface can be a serious source of error, though there is a tendency on the part of some observers to use it as an excuse for bad results, when in fact such is not the case. However, to eliminate its effects, the angles should be observed in as wide a variety of atmospheric conditions as possible, thus attempting to change a systematic error into a random one. This implies that observations be made at all times of the day and night, and on as many different days as is consistent with economy.

6.5.2.7. *Differential Heating of the Tripod*

Differential heating of the tripod often causes it to twist even when shaded by a survey umbrella, introducing considerable systematic error. In primary work, a concrete pillar will always be used because of this. To reduce the twisting effect to a minimum, not more than five directions should be observed in one round of angles, and these should be carried out at a consistently high speed on both faces, in the hope that the means will be as nearly free of error as possible.

6.5.2.8. *Centering*

Over short sides, the effects of malcentering the theodolite can be appreciable, the relationships in §6.3 being a useful guide to their size.

For a more detailed treatment see chapter 7.

6.5.3. METHODS OF OBSERVING ANGLES

The several directions radiating from a survey station may be observed in several ways. For example, consider the station A figure 6.1 from which there are three directions to survey points at B, G, and F. For the best results each angle should be observed independently, i.e. in this case the angles FAG, GAB, and BAF would be observed separately. This involves making fresh pointings each time, and not accepting the pointing to one station say to G for both angles FAG and GAB. In a similar way the angles FAB and GAF could be observed independently. This procedure gives values of each angle which are truly independent and which therefore may be treated by least squares to achieve adjustment of discrepancies. However, it is wasteful in time. If readings are taken round the various points without resetting each time the accuracy is a little less but time is saved. This method is the *method of rounds*, a further illustration of which is given in the specimen field books of tables 6.3 and 6.4. One direction is chosen as the reference one, the station observed being the *reference object* or the R.O. for short. In table 6.3 the station F is R.O. Face left readings are made successively round the stations from the R.O. initially turning the theodolite clockwise, i.e. in the same sense as the circle is graduated. Opinion differs as to whether a face left reading should be made again on the R.O. in the middle of a round. In the example the R.O. was not reobserved a second time. A change of face is made on station B and readings taken round the stations moving the theodolite anticlockwise, though of course the readings will still increase clockwise.

A more complicated method of observing is Schreiber's in which all combinations are observed between directions at a station, and in addition each direction of the complete survey is observed the same number of times so that a consistent standard of precision is obtained throughout the entire survey. This method

Table 6.3. Booking Horizontal Angles

Station	A, Glastonbury			Field Book	1003/2
Inst:	Microptic No. 2 120 889			Date	28·iv·61
Height of inst:	1ft above pillar			Time	12·45 B.S.T.
	3ft 11ins above Bench mark				
Obsr: B.B.W	Bkr: R.L.S.				
Weather	Overcast with sun patches				

Station	Face	Observed direction	Reduced direction	Mean direction	Remarks
Panborough F	L	00 17 24	00 00 00	00 00 00·0	
Ben Knowle G	L	32 31 43	32 14 19	32 14 18·5	
Pen Hill C	L	56 07 02	55 49 38	55 49 37·5	
Dulcote B	L	73 12 39	72 55 15	72 55 14·0	All observations to opaque beacons
B	R	253 12 44	72 55 13		
C	R	236 07 08	55 49 37		
G	R	212 31 49	32 14 18		
F	R	180 17 31			
F	L	90 07 14	00 00 00	00 00 00	All to opaque beacons
G	L	122 21 37	32 14 23	32 14 22·5	
C	L	145 56 52	55 49 38	55 49 39·0	
B	L	163 02 24	72 55 10	72 55 10·0	
B	R	343 02 10	72 55 10		
C	R	325 56 40	55 49 40		
G	R	302 21 22	32 14 22		
F	R	270 07 00			

Table 6.4. Alternative Layout to 6.3 Using Less Space

Station	Zero Face left	one Face right	Zero Face left	two Face right	Remarks
F Reduced direction	00 17 24 00 00 00	180 17 31 00 00 00	90 07 14 00 00 00	270 07 00 00 00 00	All to opaque beacons
G	32 31 43 32 14 19	212 31 49 32 14 18	122 21 37 32 14 23	302 21 22 32 14 22	
C	56 07 02 55 49 38	236 07 08 55 49 37	145 56 52 55 49 38	325 56 40 55 49 40	
B	73 12 39 72 55 15	253 12 44 72 55 13	163 02 24 72 55 10	343 02 10 72 55 10	

Abstraction by faces

gives rise to a complicated programme prior to the field work which is less adaptable in bad weather, than the less precise methods.†

Generally speaking any of these methods will give acceptable results if careful observations are made, each of which is as genuinely independent as it can be. It is wrong to insist on rigid standards of agreement between successive measures of the same angle if by so doing the observations cease to be independent. Very often a high standard deviation accompanies high accuracy as may be seen from the side equations in an adjustment.

In the method of rounds, the R.O. should be chosen on account of its apparent size at the observer, the accuracy with which it has been centred, and the likelihood of it being the most often visible. Sometimes a false R.O. is used, i.e. a station is especially erected to act as the reference direction but which does not form part of the survey for any other purpose. This procedure is useful in traverses in which only two stations can be occupied at any one time and each station is visited only once; each traverse angle being observed in two halves via an intermediate false R.O.

6.5.4. BOOKING AND ABSTRACTING

If possible a booker should be employed on precise work to enable the observer to concentrate solely on the observations and to lessen the chances of unconsciously giving bias to them. At night, work proceeds faster with a booker. The reader should refer to the general discussion of booking field work given in §1.2.8. Tables 6.3 and 6.4 give examples of booking layouts commonly used, though a special form of booking may often be devised to suit individual circumstances. The various directions should be reduced to the R.O. whilst the observer is still at the station, so that any reading which is suspect may be repeated.

An *abstract* of a field book is a list of all information that will be used by the computor. Table 6.5 shows an abstract of the information of the field books table 6.3 or 6.4, together with other observed values of the angles which are on other pages of the field book. All angles observed should be listed, even those which may be discarded for any reason. If more than one R.O. is used the abstract will show the directions reduced to these separately, the combination being left to the computor who may apply a station adjustment (see §6.6.2).

Since much time and effort goes into the adjustment of the final mean values of angles appearing in an abstract, it is essential that they should be meticulously checked if possible by two persons independently.

6.5.5. NUMBER OF MEASURES OF EACH ANGLE

The number of measures of each angle is calculated from the accuracy required for the final positions of the survey stations. A completely rigorous assessment of the errors of a survey is beyond the scope of this book, and the reader should refer to Rainsford‡ for more information. However, a good

† J. E. S. Bradford, S. Rhodesian Military Surveyors in World War 1939–45 *The Empire Survey Review*, No 67, 1948.
‡ H. F. Rainsford, Survey Adjustments and Least Squares (London: Constable, 1957).

Table 6.5

		Abstract of Horizontal Angles at Stn. A			
		Glastonbury Tor			
Field books 1003/2, 3 and 4			Dates 28, 29 April 1961		
Observer B.B.W.					
Stations	Panborough F	Ben Knowle G	Pen Hill C	Dulcote B	
Zero 1	00 00 00	32°14′ 18·5	55°49′ 37″·5	72°55′ 14″·0	
2		22·5	39·0	10·0	
3	R.O.	20·0	39·0	12·5	
4		19·5	42·0	05·0	
5		23·0	37·0	10·0	
6		22·5	35·5	11·5	
7		18·0	39·5	08·5	
8		17·0	36·0	13·0	
Sum		161·0	305·5	85·0	
Mean		32°14′ 20″·1	55°49′ 38″·2	72°55′ 10″·6	
Range		6″·0	6″·5	8″·5	
		Abstracted by B.B.W.			
		Checked by R.L.S.			

practical guide can be obtained from formulae for the propagation of errors through simplified survey networks, which are given in terms of the standard deviation of observed angles (see §6.12.3). If it is calculated from these formulae that a standard error of a mean observed angle should be E, the number of measures n of each angle is given by $E = \pm e/n^{\frac{1}{2}}$ where e is the standard error of a single angle observed (a) with the particular theodolite in question, (b) under the conditions in question, and (c) by particular observers. Thus e can be derived from sets of observations carried out for the purpose, and to avoid wasted effort these will be at the first station of the survey. A typical value of e for a one second theodolite is $\pm 3''$. Hence if E has to be $1''$, at least nine measures of the angle are required. Attention should be paid to the consideration of *possible error*, i.e. three times the standard error, where important specifications are involved, and attention should be drawn to clients in contracts, etc. that occasionally this limit will be exceeded (see §5.3.10).

6.6. THE ADJUSTMENT OF ANGLES

In all surveys it is found that the final mean values of the observed angles are inconsistent within themselves, and with external data, such as old surveys, to which they have to be related. An adjustment is therefore required to bring about *consistency*. Consistency of data to be used in computation is desirable because it permits the computor to check his work conclusively, for he knows that any discrepancies he might find are due solely to computation and should therefore be rectified. It will be noted that the adjustment process does not *correct* the observations though it may *improve* them. The adjusted angles are still in error.

There are two main methods of adjusting surveys:

(i) By first adjusting the observed angles, etc. and thereafter computing final coordinates.
(ii) By computing provisional coordinates with unadjusted observed angles, and finally adjusting the coordinates.

In both of these methods the adjustment may be achieved by either *semi-rigorous* methods, or by *rigorous* methods (*least squares*). Method (i) involves *condition equations*, and method (ii) *observation equations* if least squares are employed. Summarizing in a tree classification.

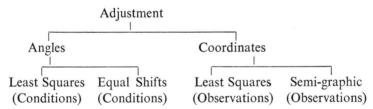

A further distinction may be made according to whether *angles* or *directions* are considered as the variables. Concentration in this book is mainly on angles; a full treatment of both being too long for inclusion here. The preliminary stages of adjustment by semi-rigorous and by least squares methods are identical; the difference being in the manner in which the final adjustments are obtained. Generally speaking, the semi-rigorous methods are incapable of giving a simultaneous adjustment of all observations of a network and one has recourse to a step by step procedure. On the other hand the least squares method permits the complete adjustment to be affected simultaneously, and therefore permits the interaction of all factors. Because of their basic similarity at the early stages, the explanation is not divided between the rigorous and the semi-rigorous. The reader who is unfamiliar with least squares should refer to chapter 5, or disregard its application to the problems described in the following sections, paying attention only to the semi-rigorous methods.

6.6.1. CONDITIONS IN A TRIANGULATION

No matter which method of adjustment is employed, the final angles of a triangulation network must obey certain geometrical conditions:

(i) *Station condition*: (Centre point) The adjusted angles round a point must be unique and their sum must be 360°.
(ii) *Figural condition*: The final angles must satisfy the geometry of the figure. These *figural conditions* are of two types
    (i) *angle conditions*.
    (ii) *side conditions*.
(iii) *Scale condition*: The final survey will have to conform to a particular size or scale.
(iv) *Position condition*: The final coordinates must agree with the positions of any fixed points included in the survey.
(v) *Bearing condition*: Directions of the survey will have to agree with bearings of lines obtained astronomically, suitably corrected for projection distortion if any, or with the bearings of fixed lines of former surveys to which the work is tied.

It will be noted that some of these conditions may be implied in others, e.g. one fixed point and a fixed length and bearing radiating from it implies two fixed points. Hence in the formation of the conditions to be satisfied, care has to be taken to ensure that the same condition has not in effect been applied twice, i.e. that a *redundant equation* has not been included.

Space in this book only permits detailed working of the simple examples which may be encountered in relatively small surveys of an isolated nature. A very full account of the adjustment of large National Surveys is given in Rainsford† to which the reader is referred.

6.6.2. STATION ADJUSTMENT

The example of §5.5.2 is of a station adjustment. As a further example, the angles observed at station A of figure 6.1 are adjusted. Table 6.6 shows the values of angles (1), (2), (3), (1 + 2), (2 + 3), and (1 + 2 + 3) all observed independently on the number of zeros indicated. The weight p of an observation is inversely proportional to the square of its standard deviation e, i.e. $p \propto 1/e^2$. If the angle is observed n times its mean value has a standard deviation E given by $E = e/n^{\frac{1}{2}}$. Hence the weight of the mean angle P is given by

$$P \propto 1/E^2 \propto n$$

Thus the weight of each angle is proportional to the number of times it was

Table 6.6
Station adjustment; by observation equations

Angle	Observed value	Zeros		Angle	Observed value	Zeros	
1	32 14 18·8	4	---(i)	1 + 2	55 49 38·0	2	--(iv)
2	23 35 17·6	4	---(ii)	2 + 3	40 40 51·6	2	--(v)
3	17 05 36·2	4	---(iii)	1+2+3	72 55 14·3	2	--(vi)

† H. F. Rainsford, Survey Adjustments and Least Squares (London: Constable, 1957).

Angular Measurement and Computation

observed, assuming that each individual zero was of the same reliability throughout the measurement. To reduce the figures of the working, the observed values of angles (1), (2), and (3) are chosen as approximate values to which corrections are applied. Thus the approximate values for (1 + 2), (2 + 3), and (1 + 2 + 3) consistent with these approximate values are respectively 55° 49′ 36″·4, 40° 40′ 53″·8, and 72° 55′ 12″·6, giving absolute terms −1″·6, +2″·2, and −1″·7 respectively. The tabulated *observation equations* are given in table 6.7(a), the normals in 6.7(b), and the solution by slide rule showing all the working is given in 6.7(c).

Tables 6.8 show the working for the same example by the *condition equations* method. The angles (1 + 2), (2 + 3), and (1 + 2 + 3) are denoted by (4), (5), and (6) respectively. The condition equations are

$$+v_1 + v_2 - v_4 - 1\cdot 6 = 0$$
$$+v_2 + v_3 - v_5 + 2\cdot 2 = 0$$
$$+v_1 + v_2 + v_3 - v_6 - 1\cdot 7 = 0$$

The formation and solution of the normals are given in tables 6.8(b) and (c) giving the corrections to the approximate values in 6.8(a). The final adjusted angles obtained by each method are 32° 14′ 19″·7, 23° 35′ 17″·6, and 17° 05′ 35″·8.

6.6.3. ADJUSTMENT OF THE THREE ANGLES OF A SINGLE TRIANGLE

The three observed angles of a triangle will not as a rule sum to the correct theoretical amount of 180° for a small triangle, or 180° plus the spherical excess

Table 6.7

(a) Observation equations

	p	1	2	3	k	s
i	4	+1			0	+1·0
ii	4		+1		0	+1·0
iii	4			+1	0	+1·0
iv	2	+1	+1		−1·6	+0·4
v	2		+1	+1	+2·2	+4·2
vi	2	+1	+1	+1	−1·7	+1·3
Σ		+3	+4	+3	+1·1	+8·9

(b) Normal equations

1	2	3	k	s
+8	+4	+2	−6·6	+7·4
	+10	+4	−2·2	+15·8
		+8	+1·0	+15·0

Practical Field Surveying and Computations

Table 6.7. (Continued)

(c) Solution of normals by slide rule (Gauss–Doolittle)

1	2	3	k	s
+8	+4	+2	−6·6	−7·4
−1	−0·5	−0·25	+0·825	−0·925
	+10	+4	−2·2	+15·8
	−2·0	−1·0	+3·300	−3·700
	+8·0	+3·0	+1·1	+12·1
	−1·0	−0·375	−0·1375	−1·5125
		+8·0	+1·0	+15·0
		−0·5	+1·650	−1·850
		−1·125	−0·4125	−4·5375
		+6·375	+2·2375	+8·6125
		−1·000	−0·35	−1·35

Back solution

1	2	3
+0·92	−0·01	−0·35
−0·08	−1·01	−1·35

Angle	Observed value	Correction	Adjusted value
1	32 14 18·8	+0·92	32 14 19·72
2	23 35 17·6	−0·01	23 35 17·59
3	17 05 36·2	−0·35	17 05 35·85

ϵ in a large one. If each angle is of different weight it is theoretically incorrect to assign one third of the misclosure to each angle though in practice this is done on the assumption that all three angles are of equal weight. As a further exemplification of the theory of least squares as applied to condition equations, and to illustrate other aspects, this simple problem is now considered from first principles. In triangle ABC, M_A, M_B, and M_C are the observed values of the angles A, B, and C respectively, p_A, p_B, and p_C are their respective weights, and v_A, v_B, and v_C are the residuals, where $v_A = A - M_A$, etc. The most probable values of the angles must sum to 180°, neglecting spherical excess, i.e. $A + B + C = 180°$ is the condition equation. If

$$M_A + M_B + M_C = 180 + k, \quad v_A + v_B + v_C + k = 0 \qquad (1)$$

Table 6.8. Station Adjustment by Condition Equations

(a) The condition equations are:

	1	2	3	4	5	6	k	s	c
wt w	4	4	4	2	2	2			
1/w	1	1	1	2	2	2			
	+1	+1		−1			−1·6	−0·6	+0·35
		+1	+1		−1		+2·2	+3·2	−0·92
	+1	+1	+1			−1	−1·7	+0·3	+0·56
Σ	+2	+3	+2	−1	−1	−1	−1·1	+2·9	
v	+0·91	−0·01	−0·36	−0·35	+0·92	−0·56			

One check is $[pv^2] = -[ck]$ i.e. $+3·52 = +3·54$

(b) Normal equations

c_1	c_2	c_3	k	s
+4	+1	+2	−1·6	+5·4
	+4	+2	+2·2	+9·2
		+5	−1·7	+7·3

(c) Solution of normals

c_1	c_2	c_3	k	s
+4	+1	+2	−1·6	+5·4
−1	−0·25	−0·5	+0·4	−1·35
	+4	+2	+2·2	+9·2
	−0·25	−0·5	+0·4	−1·35
	+3·75	+1·5	+2·6	+7·85
	−1·00	−0·4	−0·7	−2·1
		+5·0	−1·7	+7·3
		−1·0	+0·8	−2·7
		−0·6	−1·05	−3·1
		+3·4	−1·9	+1·5
		−1·0	+0·56	−0·44

Back solution

c_1	c_2	c_3
+0·35	−0·92	+0·56
−0·65	1·92	−0·44

To obtain a solution, the principle of least squares is invoked, and the residuals have to be such that $[pv^2]$ is a minimum. For convenience in the differentiation later, it is the same thing if we make $[\tfrac{1}{2}pv^2]$ a minimum, i.e.

$$\tfrac{1}{2}pv^2 = \tfrac{1}{2}p_A v_A^2 + \tfrac{1}{2}p_B v_B^2 + \tfrac{1}{2}p_C v_C^2$$

is to be a minimum. The solution now follows the method of correlatives. Let equation (1) be multiplied by an unknown number C_1, then since $C_1(1)$ is zero, the minimum function may be written

$$[\tfrac{1}{2}pv^2] - C_1(v_A + v_B + v_C + k)$$

This will be a minimum when the partial differentials with respect to each unknown are separately zero, i.e. when

Differentiating with respect to A, $\quad p_A v_A - C_1 = 0$
$B, \quad p_B v_B - C_1 = 0$
$C, \quad p_C v_C - C_1 = 0$

whence $\quad v_A = C_1/p_A; \quad v_B = C_1/p_B; \quad v_C = C_1/p_C \quad$ (2)

or putting $\quad u_A = 1/p_A; \quad u_B = 1/p_B; \quad u_C = 1/p_C$

when $\quad v_A = C_1 u_A; \quad v_B = C_1 u_B; \quad v_C = C_1 u_C$

Substituting these values for the residuals in (1) gives

$$C_1 = -\frac{1}{u_A + u_B + u_C} k$$

whence from equations (2)

$$v_A = -\frac{u_A k}{u_A + u_B + u_C}; \quad v_B = -\frac{u_B k}{u_A + u_B + u_C};$$

$$v_C = -\frac{u_C k}{u_A + u_B + u_C} \quad (3)$$

If $p_A = p_B = p_C$, expression (3) reduces to the common rule that the adjustment to each angle of a triangle is minus one third of the misclosure k.

Again
$$(pv^2) = p_A v_A^2 + p_B v_B^2 + p_C v_C^2$$
$$= u_A C_1^2 + u_B C_1^2 + u_C C_1^2$$
$$= C_1^2(u_A + u_B + u_C) = -C_1 k$$
$$[pv^2] = -C_1 k \quad (4)$$

which is a useful expression to check the computation of the sum of the weighted squares of residuals.

Example: The mean observed angles of triangle ABC are given below, together with the number of times each angle was observed. Adjust the angles of the triangle, allowing for the weights of the respective angles.

$$\begin{array}{lllll} M_A & 32° \ 14' \ 18''\!8 & \text{Observed} & 2 \text{ times} \\ M_B & 94 \ 27 \ 06{\cdot}5 & \text{Observed} & 4 \text{ times} \\ M_C & 53 \ 18 \ 32{\cdot}5 & \text{Observed} & 6 \text{ times} \\ \hline \text{Sum} & 179 \ 59 \ 57{\cdot}8 & \therefore \ k = -2{\cdot}2. \end{array}$$

The weights are in proportion to $1:2:3 \therefore u_A + u_B + u_C = 11/6$.

$$\begin{array}{rl} \text{The adjustments are:} & v_A = 6/11 \times 2{\cdot}2 = 1{\cdot}2 \\ & v_B = 6/11 \times \tfrac{1}{2} \times 2{\cdot}2 = 0{\cdot}6 \\ & v_C = 6/11 \times \tfrac{1}{3} \times 2{\cdot}2 = 0{\cdot}4 \\ & \text{Sum } +2{\cdot}2 \end{array}$$

$$\begin{array}{lll} \text{And the finally adjusted angles are } A & 32° \ 14' \ 20''\!0 \\ B & 94 \ 27 \ 07{\cdot}1 \\ C & 53 \ 18 \ 32{\cdot}9 \\ \hline \text{Sum} & 180 \ 00 \ 00{\cdot}0 \end{array}$$

Check: $[pv^2] = 2{\cdot}64$ and $-C_1 k = 2{\cdot}64$.

6.6.4. THE ADJUSTMENT OF A BRACED QUADRILATERAL

In the adjustment of a braced quadrilateral many of the basic concepts involved in the adjustment of large networks are illustrated. As an example, the quadrilateral DEFG of figure 6.1 will be adjusted. It is assumed that spherical excess is negligible. Using triangle DEF as an example, each triangle of the quadrilateral gives rise to an *angle condition* of the form

$$10 + 11 + 12 + 13 = 180$$

in which the numbers 10, 11, etc. denote the most probable values of the observed angle of the same number.

In addition to an angle condition for each triangle, the angles of the braced quadrilateral must satisfy the following: $10 + 11 = 14 + 17$, and $12 + 13 = 9 + 18$; and the sum of the internal angles should be $2n - 4$ right angles where n is the number of sides, i.e. in this case the sum should be 360°. All the possible angle conditions of the quadrilateral are illustrated in table 6.9. However, since some of these equations can be derived from others, all seven angle conditions are not unique. Provided all the angles are represented, only three of the equations need be solved. That this is so can be seen by the interlocking of columns of table 6.9. Thus when three *necessary conditions* are satisfied, all seven will be satisfied. In semi-rigorous adjustment, conditions (1), (2), and (3) of table 6.9 are normally used, and in least squares, conditions for any three triangles are used, though this choice will not affect the result unless an unstable adjustment is being attempted.

It might be thought that there are no other conditions to be satisfied in the quadrilateral. That this is not so can be seen from the figure 6.7. It is possible

Table 6.9. Angle Conditions in a Braced Quadrilateral

Angle conditions in a braced quadrilateral

Equation number	Angle of quadrilateral DEFG								
	9	10	11	12	13	14	17	18	Sum
1	+1	+1	+1	+1	+1	+1	+1	+1	360
2		+1	+1			−1	−1		0
3	+1			−1	−1			+1	0
4		+1	+1	+1	+1				180
5				+1	+1	+1	+1		180
6	+1					+1	+1	+1	180
7	+1	+1	+1					+1	180

to satisfy the three triangle equations without the figure being a closed quadrilateral, and instead of one unique apex there are three at each corner.

Although the figure exaggerates the size of the positional errors of the apexes, it illustrates the facts of survey observations, since the beacons and theodolite will not be centred exactly over the mark during the observations. Thus a further condition has to be satisfied.

If the sides of the figure are computed (i) from DE in triangle EDG to give DG, then (ii) from triangle DGF to give GF, (iii) from triangle GFE to give FE and finally (iv) from triangle FED to give DE again, the two values of DE should be the same. This may be written

$$\frac{ED}{DG} \cdot \frac{DG}{FG} \cdot \frac{FG}{FE} \cdot \frac{FE}{ED} = 1 \tag{1}$$

Since we are concerned not with lengths but with angles, this *side equation* is transformed into angular terms by application of the sine rule in the successive

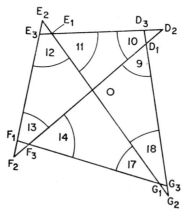

Figure 6.7

Angular Measurement and Computation 277

triangles EDG, DGF, GFE, and FED giving

$$\frac{\sin 18 \sin 14 \sin 12 \sin 10}{\sin 11 \sin 9 \sin 17 \sin 13} = 1 \qquad (2)$$

For convenience in computing, equation (2) is reduced to a linear form by taking logs, to give

$\log \sin 18 + \log \sin 14 + \log \sin 12 + \log \sin 10$
$\qquad -(\log \sin 11 + \log \sin 9 + \log \sin 17 + \log \sin 13) = 0 \quad (3)$

This equation (3) may be written on sight, because successive angles round the figure alternate with positive and negative log sines. For an alternative treatment of the side equation see §6.6.7.

This same equation (2) could have been derived by consideration of the ratios

$$\frac{OD}{OE} \times \frac{OE}{OF} \times \frac{OF}{OG} \times \frac{OG}{OD} = 1$$

and applying the sine rule successively in triangles ODE, OEF, OFG and OGD. Since the angles at point O are not involved, this side equation is said to be derived by 'taking the POLE at O'. By taking poles at D, E, F, or G four other forms of the side equation may be derived, e.g. that obtained by taking the pole at F is

$$\frac{EF}{DF} \times \frac{DF}{FG} \times \frac{FG}{EF} = 1$$

which reduces to

$\log \sin 10 + \log \sin (17 + 18) + \log \sin 12$
$\qquad - [\log \sin (11 + 12) + \log \sin 9 + \log \sin 17] = 0 \quad (4)$

Since the angles at the pole F are not involved, this side equation is useful in two ways;

(i) A suspected bad observation at F may be located
(ii) If there had been no observations at F a side equation may still be obtained.

Only one side equation has to be satisfied in the adjustment. That chosen should involve small angles, because their log sines are more sensitive than those of large angles.

The adjustment of the quadrilateral is now carried out by altering the observed angles by as little as possible, so the four conditions are satisfied. It is usual, in all but the most refined work, to assume that the angles are of equal weight. The quadrilateral DEFG will now be adjusted by the semi-rigorous method, and by least squares.

6.6.5. ADJUSTMENT OF THE QUADRILATERAL BY A SEMI-RIGOROUS METHOD

This method is often called the 'equal shifts method', because an attempt is made to alter each observed angle by an equal amount. The adjustment is undertaken in three stages (Table 6.10).

Table 6.10. Semi-rigorous Adjustment of a Quadrilateral

1 No.	2 Angle	3 1st adj.	4 C^d angle	5 Pairs	6 2^d adj.	7 C^d angle	8 Logsine even angles	9 Δ 10"	10 Logsine odd angles	11 Δ 10"	12 3^d adj.	13 Final angles	14 Final logsine evens	15 Final logsine odds
10	59 20 44·4	−0·2	44·2		+1·0	45·2	9·934 6302	125			−1·1	59 20 44·1	9·934 6288	
11	48 01 23·9	−0·1	23·8	68·0	+0·9	24·7			9·871 2340	189	+1·1	48 01 25·8		9·871 2361
12	27 35 52·1	−0·2	51·9		−2·8	49·1	9·665 8147	403			−1·1	27 35 48·0	9·665 8102	
13	45 02 04·0	−0·1	03·9	55·8	−2·9	01·0			9·849 7396	210	+1·1	45 02 02·1		9·849 7419
14	72 09 20·7	−0·1	20·6		−1·0	19·6	9·978 5874	68			−1·1	72 09 18·5	9·978 5867	
17	35 12 51·4	−0·1	51·3	71·9	−1·0	50·3			9·760 8984	299	+1·1	35 12 51·4		9·760 9017
18	31 38 05·6	−0·1	05·5		+2·9	08·4	9·719 7587	342			−1·1	31 38 07·3	9·719 7550	
9	40 59 38·9	−0·1	38·8	44·3	+2·9	41·7			9·816 8986	242	+1·1	40 59 42·8		9·816 9013
Sum	360 00 01·0	−1·0	00·0		0		9·298 7910	938	9·298 7706	940	−1·1	00·0	9·298 7807	9·298 7810

Diff. 9·298 7706 | 940
204 | 1878

$$C''\text{ per angle} = \frac{204}{1878}\text{ units of }10''$$

i.e. 1"·09 or 1"·1 per angle

Angular Measurement and Computation

Stage one: (Angle equation—Sum of interior angles)

The observed angles are listed in the second column of the table, listing the pairs of opposites together for convenience at stage two. One eighth of the misclosure on to 360° is applied with the correct sign to each angle. Rounding off may be required to give complete agreement. This gives the angles of column 4.

Stage two: (Angle equations—Opposite pairs)

The pairs of angles are added, i.e. 10 + 11, which should equal 14 + 17, and 12 + 13 should equal 9 + 18. To save space, only the seconds have been written, but the degrees and minutes should be mentally checked in case a compensating gross error has passed undetected at stage one. The difference of $2''.9$ has to be assigned to pairs 10 + 11 and 14 + 17, $1''.9$ to the former and $2''.0$ to the latter, assigning the extra $0''.1$ to angle 10 to redress the arbitrary $0''.1$ of stage one. The four triangles should now sum to 180° each. This bears out that the satisfaction of three angle equations automatically satisfies the other four.

Stage three: (Side equation)

The log sines of the angles in column 7, are listed alternately in columns 8 and 10, together with the tabulated differences of their log sines for any convenient change in the angle, in columns 9 and 12. In this case the tabulated differences are for a change of 10" per angle. The sums of the log sine 'evens' differs from the sums of the log sine 'odds' by 204 in the seventh place of logs. This discrepancy has to be removed by altering each angle by the same, numerical amount: the even angles negatively and the odd angles positively in this case. By so doing, the previous adjustments of stages one and two are preserved, and the side equation is made to close very nearly. If all the even angles are changed by 10", the sum of their log sines will change by the sum of the tabular differences for 10", i.e. by 938 in the seventh place of logs in the example. Similarily, the change in the sum of the log sines of the odd angles will change by 940 in the seventh place. A change of 10" negatively for all the even angles together with a change in 10" positively for all the odd angles will alter the difference between the sums of their log sines by 938 + 940 = 1 878. But a discrepancy of only 204 has to be removed, hence each angle need be adjusted by 204/1 878 units of 10", i.e. by $1''.1$ each. This adjustment gives the final values of the angles.

As a final check on the computation, the log sines of the final angles are obtained afresh from tables and their sums checked for near equality.

If an angle is greater than 90° its log sine, though still positive, *decreases* as the angle *increases*; hence this has to be taken into account in the adjustment by applying the tabular difference of the log sine negatively, e.g. if the angle (10) had been 120° 39' $14''.8$ its log sine is still 9·934 630 2 but its tabular difference would be −125, and the sum at the foot of column (9) would then be 688. Otherwise the adjustment is as that given in the example, the actual sign of the angle correction being unchanged i.e. still negative in this case.

Table 6.11. Least Squares Adjustment of Quadrilateral DEFG

(a) Condition equations

C ↓	9	10	11	12	13	14	17	18	K	Sum
−2·936				+1	+1	+1	+1		+8·20	+12·20
+1·696	+1					+1	+1	+1	−3·40	+0·60
+0·987	+1	+1	+1					+1	−7·20	−3·20
−0·783	−1·210	+0·625	−0·945	+2·015	−1·050	+0·340	−1·495	+1·710	+10·70	+10·69
Sum⃗	+0·790	+1·625	+0·055	+3·015	−0·050	+2·340	+0·505	+3·710	+8·300	+20·29
v⃗	+3·63	+0·50	+1·72	−4·51	−2·11	−1·51	−0·07	+1·35		+20·29 ✓

Check $[v^2] = 45.3$ $-[ck] = 45.3$

(b) Formation of normal equations

C_1	C_2	C_3	C_4	K	Sum	
+4·000	+2·000		−0·190	+8·200	+14·010	✓
	+4·000	+2·000	−0·655	−3·400	+3·945	✓
		+4·000	−0·180	−7·200	−1·020	✓
			+13·185	+10·700	+23·220	✓

(c) Solution of normal equations [abridged Gauss–Doolittle]

C_1	C_2	C_3	C_4	K	Sum	
+4·000	+2·000		−0·190	+8·200	+14·010	✓
−1	−0·500		+0·048	−2·050	−3·502	✓
	+3·000	+2·000	−0·559	−7·500	−3·059	✓
	−1	−0·667	+0·186	+2·500	+1·020	✓ (1)
		+2·666	+0·552	−2·200	+1·020	(2) ✓
(0·37509)		−1	−0·207	+0·825	−0·383	(1) ✓
			+12·958	+10·147	+23·103	(2) ✓
	(0·077172)		−1	−0·783	−1·783	✓
−2·936	+1·696	+0·987	−0·783	Back		
−3·937	+0·698	−0·014	−1·783	solution		
(1) ✓	(2) ✓	(1) ✓	✓			

6.6.6. ADJUSTMENT OF THE QUADRILATERAL BY LEAST SQUARES: (CONDITION EQUATIONS)

The semi-rigorous method of adjustment has two main drawbacks. The adjustment process is arbitrary, and it has to be carried out in successive stages. Consequently, slightly different answers are obtained according to the whims of the particular computor; and in large interlocking figures of triangulation, all the various adjustments cannot be dealt with simultaneously and many conditions have to be virtually ignored for convenience.

To enable a unique solution of large interlocking figures to be obtained, the method of least squares is employed. The quadrilateral DEFG will now be adjusted by this method, using condition equations. The conditions that have to be satisfied are exactly as described above in §6.6.4. However for convenience in layout, the three necessary angle conditions that are chosen are any three triangle closures; in this case triangles EFG, FGD, and GDE giving equations (5), (6) and (7) of table 6.9. The side equation is that used in §6.6.5 but whose numerical values will be slightly different because it is formed from the observed and not partly adjusted angles.

6.6.6.1. Formation of Equations: (1) Angle Equations

The angle equations represent the changes that have to be made to the angles of a triangle to make it close on 180°, e.g. in triangle EFG $E + F + G$ should equal 180°. Now, if the actual resulting values from observations are $(E - v_E)$, $(F - v_F)$ and $(G - v_G)$ and k is the triangle misclosure, the condition equation may be written

$$(E - v_E) + (F - v_F) + (G - v_G) = 180 + k$$

then
$$v_E + v_F + v_G = -k$$

In the example

$$v_{12} + v_{(13+14)} + v_{17} + 8 \cdot 2 = 0$$

In separate angles
$$v_{12} + v_{13} + v_{14} + v_{17} + 8 \cdot 2 = 0$$

For simplicity in writing, the v's are omitted and the equation is written

$$12 + 13 + 14 + 17 + 8 \cdot 2 = 0$$

OR in tabular form

(12)	(13)	(14)	(17)	Abs = k
+1	+1	+1	+1	+8·2

The actual triangle misclosure appears in the column k, the absolute term with its sign *unchanged*. Similarly the angle equations are written for the other two triangles FGD and GDE.

(2) *Side equation*

The form of the side equation is identical to that used in the method of equal shifts. In this case the observed angles are used directly. Table 6.13a gives the

Table 6.12. Adjusted Angles of Quadrilateral DEFG

No.	Observed angles	Adjustment Least squares	Adjustment Equal shifts	Final angles Least squares	Final angles Equal shifts
9	40 59 38·9	+3·63	+3·9	42·53	42·8
10	59 20 44·4	+0·50	−0·3	44·90	44·1
11	48 01 23·9	+1·72	+1·9	25·62	25·8
12	27 35 52·1	−4·51	−4·1	47·59	48·0
13	45 02 04·0	−2·11	−1·9	01·89	02·1
14	72 09 20·7	−1·51	−2·2	19·19	18·5
17	35 12 51·4	−0·07	0·0	51·33	51·4
18	31 38 05·6	+1·35	+1·7	06·95	07·3

Table 6.13. Formation of Side Equations

(a) By logarithms

Angle	Logsine	Δ for 1″	Angle	Logsine	Δ for 1″
10	9·934 629 2	12·5	11	9·871 232 5	18·9
12	9·665 826 8	40·3	13	9·849 745 9	21·0
14	9·978 588 2	6·8	17	9·760 901 7	29·9
18	9·719 749 1	34·2	9	9·816 891 8	24·2
Sum	9·298 793 3			9·298 771 9	941

$$\frac{9\cdot298\ 771\ 9}{214}$$ Dividing throughout by 20 gives the equation:

$0\cdot625 v_{10} + 2\cdot015 v_{12} + 0\cdot340 v_{14} + 1\cdot710 v_{18} - (0\cdot945 v_{11} + 1\cdot050 v_{13} + 1\cdot495 v_{17} + 1\cdot210 v_9)$
$+ 10\cdot7 = 0$

(b) By naturals

Angle	Sine	Cot.	Angle	Sine	Cot.
10	0·860 258 92	0·593	11	0·743 416 94	0·900
12	0·463 262 09	1·913	13	0·707 531 74	0·999
14	0·951 893 02	0·322	17	0·576 635 92	1·417
18	0·524 504 45	1·623	9	0·655 981 82	1·151

$$\frac{\sin 10 \sin 12 \sin 14 \sin 18}{\sin 11 \sin 13 \sin 17 \sin 9} = \frac{1\cdot000\ 0491\ 1 = 1 + k}{k \times 206\ 265 = 10\cdot1276}$$

Converting the cotangents by $1\cdot05275 \times 10^{-7} = 1/(2 \times 0\cdot474\ 943 \times 10^7)$ gives
$0\cdot625 v_{10} + 2\cdot016 v_{12} + 0\cdot339 v_{14} + 1\cdot711 v_8 - (0\cdot949 v_{11} + 1\cdot053 v_{13} + 1\cdot494 v_{17} + 1\cdot213 v_9)$
$+ 10\cdot7 = 0$

Angular Measurement and Computation

detailed formation of the side equation of the example. The log sines and tabular differences are listed as before, only in this case the tabular difference is for one second of arc. Since the whole condition equation is unaltered if it is multiplied throughout by a constant, and the solution of the equations is simpler if all coefficients are of similar size, all terms of this side equation are multiplied by 0·05 giving the side equation:

$$+0\cdot625v_{10} + 2\cdot015v_{12} + 0\cdot340v_{14} + 1\cdot710v_{18}$$
$$- (+0\cdot945v_{11} + 1\cdot050v_{13} + 1\cdot495v_{17} + 1\cdot210v_9) + 10\cdot7 = 0$$

The solution of the four condition equations is given in tables 6.11. The final values of the adjustments to each angle are rounded off to two decimals from three, assigning the small residual errors of 0·01 to angles (18) and (11) so that the condition equations are exactly satisfied.

6.6.7. CHECK ON THE FORMATION OF THE SIDE EQUATION

Once the condition equations have been tabulated, their solution is self-checking at all stages. If the equations have been formed wrongly this will not become apparent until the coordinates are computed, and a great deal of work will have been wasted. Where possible, a second person should form the conditions independently and tabulate them. The two sets are then compared for identity. If a second person is not available, the side equations, in which lies most chance of mistakes, can also be computed by natural trigonometrical functions, and the two final equations compared.

Consider the side equation:

$$\frac{\sin 10 \sin 12 \sin 14 \sin 18}{\sin 11 \sin 18 \sin 17 \sin 19} = 1 \qquad (1)$$

The left-hand side of this equation will not normally equal unity using the observed angles. Let its value be $1 + k$. The object of the adjustment is to alter the angles so that this misclosure k is removed. Differentiating (1) logarithmically

$$\cot 10 v_{10} + \cot 12 v_{12} + \cot 14 v_{14} + \cot 18 v_{18}$$
$$- (\cot 11 v_{11} + \cot 13 v_{13} + \cot 17 v_{17} + \cot 19 v_{19}) = 0 \qquad (2)$$

In this case the residuals v are in radians. This differential expression (2) has to be equal to $-k$ to affect the adjustment. Converting the residuals to seconds of arc the final side equation becomes:

$$\cot 10 v_{10}'' + \cot 12 v_{12}'' + \cot 14 v_{14}'' + \cot 18 v_{18}''$$
$$- (\cot 11 v_{11}'' + \cot 13 v_{13}'' + \cot 17 v_{17}'' + \cot 19 v_{19}'') + 206\,265 k = 0 \qquad (3)$$

Table 6.13(b) gives the numerical version of this equation for the example. The two versions of the side equation differ only by a constant, in this case

1·052 75 × 10⁻⁷, as indicated at the foot of the table. This constant is derived as follows

$$\partial \log \sin x = \cot x \, dx'' \sin 1''$$

Also
$$\partial \log \sin x = \Delta \log_e \sin x \, dx''$$
$$= \Delta \log_{10} \sin x \, dx'' \log_e 10$$
$$\therefore \cot x = 206\,265 \times 2·302\,58 \times \Delta \log \sin x$$
$$= 0·474\,943 \times 10^6 \times \Delta \log \sin x$$

and in the example
$$\cot x = 0·474\,943 \times 10^6 \times 20 \,(\Delta \log \sin x/20)$$

i.e.
$$(\Delta \log \sin x/20) = 1·052\,75 \times 10^{-7} \cot x$$

6.7. NUMBER OF CONDITION EQUATIONS IN A FREE NETWORK OF TRIANGULATION

In a complex free network, i.e. one without fixed elements, it is useful to have a formula that will enable the surveyor to determine the number of conditions that are to be satisfied. It will be noted that it is only the *number* of equations and not the equations themselves, that are indicated by the formulae.

6.7.1. ANGLE CONDITIONS

The following symbols will be employed.

The number of lines in the network $= l$
The number of lines observed one way $= l_1$
The total number of stations $= S$
The total number of unoccupied stations $= S_u$
The number of centre points $= C_p$

The number of angle conditions N_A in the network is then given by

$$N_A = (l - l_1) - (S - S_u) + C_p + 1 \qquad (1)$$

which is proved as follows. Let the number of angle equations be given by

$$N_A = Al + Bl_1 + CS + DS_u + E$$

The constants A, B, C, D, and E will be evaluated by substituting for simple figures.

(i) Intersection;

$l = 3; \quad l_1 = 2; \quad S = 3; \quad S_u = 1; \quad N_A = 0$
$$\therefore 0 = 3A + 2B + 3C + D + E \qquad (2)$$

(ii) Fully observed triangle:

$l = 3; \quad l_1 = 0; \quad S = 3; \quad S_u = 0; \quad N_A = 1$
$$\therefore 1 = 3A + 0 + 3C + 0 + E \qquad (3)$$

(iii) Two fully observed triangles:

$l = 5$; $l_1 = 0$; $S = 4$; $S_u = 0$; $N_A = 2$

$$\therefore 2 = 5A + 0 + 4C + D + E \qquad (4)$$

(iv) Braced quadrilateral

$l = 6$; $l_1 = 0$; $S = 4$; $S_u = 0$; $N_A = 3$

$$\therefore 3 = 6A + 0 + 4C + 0 + E \qquad (5)$$

(v) Quadrilateral with one diagonal observed one way.

$l = 6$; $l_1 = 1$; $S = 4$; $S_u = 0$; $N_A = 2$

$$\therefore 2 = 6A + B + 4C + E \qquad (6)$$

Solution of equations (2) to (6) gives $A = +1$, $B = -1$, $C = -1$, $D = +1$, and $E = +1$. The number of centre point conditions is the same as the number of centre points C_p whence we have equation (1).

Example: In figure 6.1, $l = 14$, $l_1 = 1$, $S = 7$, $S_u = 0$, $C_p = 1$ whence $N_A = 13 - 7 + 1 + 1 = 8$.

6.7.2. SIDE CONDITIONS

In the braced quadrilateral there is one side condition to be satisfied. In any closed figure, other than the triangle, a side equation will have to be satisfied, subject to a minimum number of four stations and six lines. Using the same terms as above, N_S, the number of side equations in a network, is given by

$$N_S = l - 2S + 3 \qquad (7)$$

Assume that the relationship is

$$N_S = \alpha l + \beta S + \gamma$$

Considering the figures below in turn we obtain the equations
(1) Braced quadrilateral

$l = 6$; $S = 4$; $N_S = 1$

$$\therefore 1 = 6\alpha + 4\beta + \gamma \qquad (8)$$

(2) Polygon and quadrilateral

$l = 9$; $S = 5$; $N_S = 2$

$$\therefore 2 = 9\alpha + 5\beta + \gamma \qquad (9)$$

(3) Two quadrilaterals and polygon.

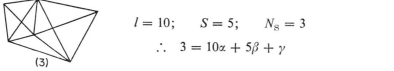

$$l = 10; \quad S = 5; \quad N_S = 3$$
$$\therefore \quad 3 = 10\alpha + 5\beta + \gamma \tag{10}$$

Which on solution gives the formula

$$N_S = l - 2S + 3$$

Example: From figure 6.1

$$l = 14; \quad S = 7$$
$$\therefore \quad N_S = 14 - 14 + 3 = 3$$

Therefore there are 3 side equations in the system.

6.7.3. TOTAL NUMBER OF EQUATIONS

The total number of equations in the system N_T is given by adding equations (1) and (7) together giving

$$N_T = 2l - 3S - l_1 + S_u + C_p + 4 \tag{11}$$

6.7.4. ADDITIONAL CONDITIONS

The above formulae take no account of other possible conditions that may have to be satisfied when the network is not free i.e. conditions of scale, position, and azimuth. This aspect is considered briefly in §6.11 and §13.10.

6.8. FORMATION OF THE CONDITION EQUATIONS IN A NETWORK

Although the number of equations may be assessed by formulae, it is not possible to ensure that the chosen equations are sufficient, other than by meticulous care and a rational procedure in carrying out the work. It may be that, the correct number of equations is chosen and formed but that one merely repeats another, i.e. a redundant equation is used, while a necessary one must have been omitted. To guard against possible mistakes of this nature, it is advisable to have the equations formed separately by a second computer. The two formations will not be checked until the stage of the final tabular form used for solution, because although the equations may be correct, it is easy to tabulate them erroneously.

A simple but effective procedure is to construct a copy of the triangulation figure, line by line. Each time a new condition is required it should be formed and tabulated before adding another line to the diagram.

6.8.1. EXAMPLE OF ADJUSTMENT OF A FREE NETWORK

The observations of figure 6.1 are adjusted by the condition equations method in tables 6.14–6.17, and the final adjusted angles are listed in table 6.1.

6.8.2. ADJUSTMENT BY OBSERVATION EQUATIONS

The adjustment by observation equations and variation of coordinates is given in §6.11 after the various methods of computing these coordinates have been described.

Table 6.14. Adjustment of Network by Condition Equations—Schedule of Equations

Adjustment of network by condition equations – schedule of equations. All of equal weight

No	C's	1	2	3	4	5	6	7	8	9	10	11	12	13	14	15	16	17	18	15	20	21	K	S	Check conditions Rounded
1	−0·457		+1	+1	+1																	+1	+0·30	+4·30	−0·299 −0·3
2	−1·060					+1	+1														+1		+3·00	+6·00	−3·000 −3·0
3	−2·103							+1	+1											+1			+5·90	+8·90	−5·901 −5·9
4	+2·129									+1	+1	+1							+1				−7·20	−3·20	+7·206 +7·2
5	−1·426												+1	+1	+1			+1					+8·20	+12·20	−8·201 −8·2
6	−1·339										+1	+1	+1										+4·40	+8·40	−4·398 −4·4
7	+0·885	+1														+1	+1						−2·20	+0·80	+2·200 +2·2
8	+0·363																+1	+1	+1	+1	+1	+1	−1·00	+5·00	+0·997 +1·0
9	−0·336		−2·42	+3·42	+0·92	+1·12																	+8·20	+8·28	−8·195 −8·2
10	−0·688																	−1·50	+1·71		−2·00	−0·96	+10·70	+10·69	−10·74 −10·7
11	−0·919	+1·67	−1·22	−1·22	+0·82	−0·20	+0·72	−0·71	+0·66	−1·22					+0·34	−0·78							+12·20	+11·06	+12·15 12·2
Sum		+2·67	−2·64	+3·20	+2·74	+1·92	+1·72	+0·29	+1·66	1·44	+2·62	+1·06	+4·02	+0·95	+1·68	+0·22	+2·00	+0·50	+3·71	+2·00		+1·04	+42·50	+72·43 +72·43	✓
v's		0·650 1·477	0·485	1·520	1·253	1·722	1·451	2·710	4·090	0·363	1·437	4·155	2·043	1·972	1·602	1·248	0·031	1·316	1·740	−0·025	0·229				
Rounded v's		−0·6 +1·5	−0·5	−1·5	−1·2	−1·8	−1·5	−2·7	+4·1	+0·4	+1·4	−4·2	−2·0	−2·0	+1·6	+1·2	0	+1·3	−1·7	0	+0·2			

Check $[v^2] = -CK$ $-[CK] = +72\cdot28$
$[v^2] = +72\cdot32$

Table 6.15

Formation of normal equations

	1	2	3	4	5	6	7	8	9	10	11	K	Sum	
1	+4	0	0	0	0	0	0	+1	+0·96	0	−1·62	+0·30	+4·64	✓ 1
2		+3	0	0	0	0	0	+1	−0·88	0	+0·52	+3·00	+6·64	✓ 2
3			+3	0	0	0	0	+1	0	0	−0·05	+5·90	+9·85	✓ 3
4				+4	0	+2	0	+1	0	+0·18	−1·22	−7·20	−1·24	✓ 4
5					+4	+2	0	+1	0	−0·19	+0·34	+8·20	+15·35	✓ 5
6						+4	0	0	0	+0·65	0	+4·40	+13·05	✓ 6
7							+3	+1	0	0	+0·89	−2·20	+2·69	✓ 7
8								+6	−2·96	+0·21	0	−1·00	+8·25	✓ 8
9									+24·58	0	−0·69	+8·20	+29·27̄	✓ 9
10										+13·23	+1·60	+10·70	+26·37(2)	✓ 10
11											+10·15	+12·20	+22·11	✓ 11

Table 6.16

	1	2	3	4	5	6	7	8	9	10	11	k	Sum	
	+4·00	0	0	0	0	0	0	+1	+0·96	0	−1·62	+0·30	+4·64	
	−1								−0·250	−0·240	+0·405	−0·075	−1·160	✓
	(0·3333)	+3	0	0	0	0	0	+1	−0·88	0	+0·52	+3·00	+6·64	
		−1							−0·333	+0·293	−0·173	−1·00	−2·213	✓
		(0·3333)	+3	0	0	0	0	+1	0	0	−0·05	+5·90	+9·85	
			−1						−0·333		+0·017	−1·966	−3·283 (1)	✓
			(0·25)	+4	0	+2	0	+1	0	+0·18	−1·22	−7·20	−1·24	
				−1		−0·500			−0·250	−0·045	+0·305	+1·800	+0·310	✓
Forward solution				(0·25)	+4	+2	0	+1	0	−0·19	+0·34	+8·20	+15·35	
					−1	−0·500			−0·250	+0·048	−0·085	−2·050	−3·838 (1)	✓
					(0·5)	+4	0	0	0	+0·66	+0·44	+3·90	+5·99 (1)	
						−1			+0·500	−0·330	−0·220	−1·950	−3·000	✓
						(0·3333)	+3	+1	0	0	+0·89	−2·20	+2·69	
[Normals could be written vertically							−1		−0·333		−0·297	+0·733	−0·897	✓
in this space. See chapter five]							(0·26667)	+3·75	−2·91	+0·54	+0·39	−1·61	+0·17 (1)	✓
								−1	+0·776	−0·144	−0·104	+0·429	−0·045 (2)	✓
								(0·04581)	+21·63	+0·42	+0·15	+7·76	+30·17 (1)	✓
									−1	−0·019	−0·007	−0·355	−1·382 (1)	✓
									(0·07764)	+12·88	+1·47	+10·21	+24·57 (1)	✓
										−1	−0·114	−0·793	−1·908 (1)	✓
Back solution										(0·11862)	+8·43	+7·75	+16·17 (1)	✓
											−1	−0·919	−1·918 (1)	✓
	−0·457	−1·060	−2·103	+2·129	−1·426	−1·339	+0·885	+0·363	−0·336	−0·688	−0·919			
	−1·456	−2·060	−3·103	+1·131	−2·426	−2·341	−0·115	−0·639	−1·336	−1·689	−1·918			
	✓ (1)	✓	✓	✓ (2)	✓	✓ (2)	✓	✓ (2)	✓	✓ (1)	✓ (1)			

[The numbers in brackets are the reciprocals of the numbers to their immediate right, e.g. 0·11862 = 1/8·43]

6.9. COMPUTATION OF INTERSECTIONS (ANALYTICAL METHODS)

In this section, three commonly used methods of computing intersected points are described. The formulae used to compute a triangle in which only two angles have been observed, are identical to those employed for the fully observed triangle, with the exception that in the former case the third angle is deduced, and in the latter the three angles are first adjusted for closure on 180°. The minimum number of rays required to fix a point by intersection is two. In most surveys, a third ray will be observed to act as a check on gross error in both observation and computation. If the angles are not adjusted, it is often sufficient to adopt arithmetic means of the coordinates obtained by the various fixations or alternatively, to compute provisional coordinates from unadjusted

Table 6.17. Formation of Side Equations of Network

For quadrilateral DEFG see tables (13)

(a) Quadrilateral ABCG by logs

	Angle	Log sin	for 1"	No.	Angle	Log sin	Δ for 1"
(3)	17 05 36·2	9·468 243 9	+68·5	(4+5)	131 19 30·6	9·875 625 0	−18·5
(20+21)	132 25 06·6	9·868 196 1	−19·2	2	23 35 17·6	9·602 234 4	+48·3
(5)	79 14 33·5	9·992 300 0	+4·0	20	45 10 57·2	9·850 864 3	+20·9
Sum		9·328 740 0				9·328 723 7	

$$\frac{723\ 7}{+16\ 3}$$ Divide all by 20 and regroup under separate angles:

$-2·42\ v_2 + 3·42\ v_3 + 0·92\ v_4 + 1·12\ v_5 - 2·00\ v_{20} - 0·96\ v_{21} + 8·2 = 0$

(b) Quadrilateral ABCG by natural tables:

No.	Angle	Sin	Cot	No.	Angle	Sin	Cot
(3)	17 05 36·2	0·293 930 04	+3·25	(4+5)	131 19 30·6	0·750 974 16	−0·879
(20+21)	132 25 06·6	0·738 237 58	−0·914	2	23 35 17·6	0·400 160 65	+2·290
(5)	79 14 33·5	0·982 426 42	+0·190	20	45 10 57·2	0·709 356 17	+0·994

$1 + k = 1·000\ 037\ 8$ Convert by $\frac{8·20}{7·797} = 1·0523$

$k \times 206265 = +7·797$

$-2·41\ v_2 + 3·42\ v_3 + 0·925\ v_4 + 1·125\ v_5 - 2·008\ v_{20} - 0·962\ v_{21} + 8·2 = 0$

(c) Polygon ABCDF by logarithms:

No.	Angle	Log sin	Δ for 1"	No.	Angle	Log sin	Δ for 1"
1	32 14 18·8	9·727 090 1	33·4	(2+3)	40 40 53·8	9·814 151 0	24·5
4	52 04 57·1	9·897 020 1	16·4	5	79 14 33·5	9·992 300 0	4·0
6	55 34 32·3	9·916 387 2	14·4	7	56 00 48·8	9·918 643 5	14·2
8	57 42 28·2	9·927 028 8	13·3	9	40 59 38·9	9·816 891 8	24·3
14	72 09 20·7	9·978 588 2	6·7	15	53 18 32·5	9·904 103 8	15·7

$$\begin{array}{cc} 9·446\ 1144 & 9·446\ 090\ 1 \\ 0901 & \end{array}$$ Divide by 20 and regroup by separate angles.
$+243$

$+1·67\ v_1 - 1·22\ v_2 - 1·22\ v_3 + 0·82\ v_4 - 0·20\ v_5 + 0·72\ v_6 - 0·71\ v_7 + 0·66\ v_8 - 1·22\ v_9 + 0·34\ v_{14}$
$- 0·78\ v_{15} + 12·2 = 0$

(d) Polygon ABCDF by natural tables:

No.	Angle	Sin	Cot	No.	Angle	Sin	Cot
1	32 14 18·8	0·533 445 58	1·586	(2+3)	40 40 53·8	0·651 855 05	1·163
4	52 04 57·1	0·788 896 73	0·779	5	79 14 33·5	0·982 426 42	0·190
6	55 34 32·3	0·824 873 21	0·685	7	56 00 48·8	0·829 169 85	0·674
8	57 42 28·2	0·845 334 88	0·632	9	40 59 38·9	0·655 981 82	1·151
14	72 09 20·7	0·951 893 02	0·322	15	53 18 32·5	0·801 869 80	0·745

$1 + k = 1·000\ 055\ 6$

$k \times 206\ 265 = 11·468$ Convert by $1·053$ to the equation:

$+1·67\ v_1 - 1·22\ v_2 - 1·22\ v_3 + 0·82\ v_4 - 0·20\ v_5 + 0·72\ v_6 - 0·71\ v_7 + 0·67\ v_8 - 1·21\ v_9$
$+ 0·34\ v_{14} - 0·78\ v_{15} + 12·1 = 0$

290   *Practical Field Surveying and Computations*

angles, which will later be adjusted by the method of variation of coordinates described in §6.11.

The triangle EFG of figure 6 is used to illustrate all three methods. The coordinates of the points E and F are known, together with the angles at E and F, here denoted by α and β. The bearings of the lines EG and FG are A and B respectively. It is required to compute the coordinates of the point G. It will be noted that the sequence of lettering the figure is clockwise. It is imperative that this clockwise convention be followed when using the formulae (5), (6), (7), and (8) below.

6.9.1. METHOD (1) SOLUTION OF THE TRIANGLE

The triangle EFG is solved for the sides EG and FG. The bearing of EF is derived from $\tan^{-1}(\Delta E/\Delta N)$ and hence the bearings of EG and FG are derived. Point G is coordinated by bearing and distance from both E and F. The formulae used are

$$G = 180 - (\alpha + \beta); \qquad B_g\text{EF} = \tan^{-1}\frac{\Delta E}{\Delta N}$$

$$\text{EF} = \Delta E \csc B_g\text{EF}; \qquad \text{EG} = \frac{\sin \beta}{\sin G}\text{EF} \qquad \text{FG} = \frac{\sin \alpha}{\sin G}\text{EF}$$

$$A = B_g\text{EF} + \alpha; \qquad E_G = E_E + \text{EG} \sin A; \qquad E_G = E_F + \text{FG} \sin B$$
$$B = B_g\text{FE} - \beta; \qquad N_G = N_E + \text{EG} \cos A; \qquad N_G = N_F + \text{FG} \cos B$$

This method is probably the best to use when computing by logs, although some surveyors even prefer those of methods (2) and (3), which entail at least one antilogging process in the middle.

6.9.2. METHOD (2) BEARINGS METHOD

In this method the coordinates of the point G are obtained directly without solving the triangle. The formulae are well-suited to machine computation, and their use requires very little writing down.

$$\tan A = \frac{E_G - E_E}{N_G - N_E} \quad \text{and} \quad \tan B = \frac{E_G - E_F}{N_G - N_F}$$

whence
$$E_G = E_E + N_G \tan A - N_E \tan A \qquad (1)$$
and
$$E_G = E_F + N_G \tan B - N_F \tan B \qquad (2)$$
$$N_G = N_E + E_G \cot A - E_E \cot A \qquad (3)$$
$$N_G = N_F + E_G \cot B - E_F \cot B \qquad (4)$$

whence
$$N_G = \frac{N_E \tan A - N_F \tan B - E_E + E_F}{\tan A - \tan B} \qquad (5)$$

and
$$E_G = \frac{E_E \cot A - E_F \cot B - N_E + N_F}{\cot A - \cot B} \qquad (6)$$

There are a number of possible selections of the formulae (1)–(6) which will give the required coordinates. If the point G can be computed by a third ray, not shown in the diagram, the simplest sequence is to derive the value of N_G from (5), and, substituting this value in (1) and (2), obtain two values of E_G. If this triangle alone fixes G, the entirely independent formulae (5) and (6) should be used.

6.9.3. METHOD (3) ANGLES METHOD

This method involves the coordination of G using the observed angles α and β and the coordinates of E and F. In figure 6.8 lines ES and PQ are East–West

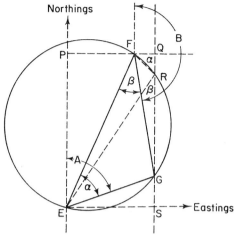

Figure 6.8

grid lines through E and F respectively: SQ is a North–South line through G meeting the circle through E, F, and G, in R.

$$E_G - E_E = ES = RS \tan \beta; \quad E_G - E_F = FQ = QR \tan \alpha$$
$$N_F - N_E = QS = QR + RS = (E_G - E_F) \cot \alpha + (E_G - E_E) \cot \beta$$

whence
$$E_G = \frac{E_E \cot \beta + E_F \cot \alpha - N_E + N_F}{\cot \alpha + \cot \beta} \quad (7)$$

Similarly the formula for the northing of G can be obtained by considering the coordinate axes to be rotated through a right angle, but with the E–W direction necessarily reversed, giving

$$N_G = \frac{N_E \cot \beta + N_F \cot \alpha + E_E - E_F}{\cot \alpha + \cot \beta} \quad (8)$$

This formula (8) may also be proved from first principles by constructing a line East–West through G to meet the circle EFG in a point analogous to R.

It will be seen that formula (7) may be derived from the figure by the application of formula (6) to give the easting of R, $E_R = E_G$, for bearing FR $= 180 - \alpha$ and bearing ER $= \beta$.

6.9.4. USE OF ARBITRARY ORIGIN

Coordinate values that are excessively large for computation may be converted to an arbitrary local origin to reduce the number of figures in the work, with a final return to the correct origin. A different origin will often be chosen for each triangle.

292   *Practical Field Surveying and Computations*

If E_0 is chosen to be $\frac{1}{2}(E_E + E_G)$ the number of figures is reduced to a minimum, often permitting a slide rule solution with ample accuracy.

If the origin is taken at one of the points, one term disappears in the expressions (5)–(8) above, a particularly convenient device when computing by logarithms.

6.9.5. PRACTICAL EXAMPLES

Method (3) will be employed, as it is usually more convenient than method (2) when triangulation is computed; the latter being more convenient in traverse work, and fixation from points which are not intervisible. For example if a fixation of G (Figure 6.1) is from F and C, the formulae of method (2) would be most convenient.

Table 6.18 gives the computation of the coordinates of G, using the adjusted angles.

Table 6.18. Computation of the coordinates of G from E and F using point D to check
(See figure 6.1, and tables 6.1 and 6.2)
No projection system has been used.

Point	Easting m	Northing m	Adjusted angles	Cotangents
F	2 490·50	2 480·79	45° 02′ 02″·0	+0·998 818
E	780·67	7 394·05	75 37 13·2	+0·256 378
D	5 044·25	7·752·70	59 20 44·8	+1·255 196
† +	345	+143	180 00 00·0	
	350 044·25	150 752·70		
F	2 490·50	2 480·79	117 11 20·7	−0·513 689
E	780·67	7 394·05	27 35 47·9	+1·913 10
G	6 629·10	1 899·07	35 12 51·4	+1·399 411
+	345	+143	180 00 00·0	
	351 629·10	144 899·07		
E	780·67	7 394·05	48 01 25·3	+0·899·655
D	5 044·25	7 752·70	100 20 27·8	−0·182 471
G	6 629·10	1 899·07	31 38 06·9	+0·717 184
+	345	+143	180 00 00·0	
	351 629·10	144 899·07		

† A false origin of 345 000 East and 143 000 North was used to reduce the figures of the working

Angular Measurement and Computation

6.9.6. CHECK ON USE OF COTANGENTS

When a point is fixed by only one triangle its coordinates should be checked in some way. A useful check on the cotangents in formulae (6), (7), and (8) of §6.9.2, and the Tienstra formulae of §6.10 is

$$\cot A \cot B + \cot B \cot C + \cot C \cot A = 1 \tag{9}$$

where $A + B + C = 180°$. For $A + B = 180 - C$

$$\cot (A + B) = -\cot C$$

$$\frac{\cot A \cot B - 1}{\cot A + \cot B} + \cot C = 0$$

whence formula (9). The same also applies when $A + B + C = 360°$.

6.10. ANALYTICAL METHODS OF RESECTION COMPUTATION

In this section, three of the many analytical solutions to the resection problem are described in detail. In resection, a unique solution is possible from three rays provided that all four points do not lie on a circle, in which case the problem is insoluble. If more than three points of known coordinates are observed, an additional adjustment is required. Normal practice is to observe four stations, compute the new point from those three which give the best cuts, and the fourth is used solely to check the final result. A multiple ray analytical solution is possible by the method of variation of coordinates, as outlined in §6.11.14 below.

In all three methods described here, A, B, and C are fixed points of known coordinates between which the angles have been observed at the new point P. These angles α, β, and γ are first adjusted to sum to 360°, by distributing one third of the misclosure to each angle.

The three methods are:

(i) *Tienstra*.
(ii) *Pothonot–Snellius*, or $(\phi - 45)$.
(iii) *Collin's point*, or *Bessel's*.

The Tienstra method is the quickest and simplest to apply in practice, whether computing by machine or by logarithms. However, methods (ii) and (iii) are well established, feature in examination syllabuses, and illustrate certain general principles.

6.10.1 (a) THE TIENSTRA METHOD. (or Barycentric METHOD)

Referring to figure 6.9 it will be seen that triangle ABC is lettered in a clockwise manner, and that α is the clockwise angle between the directions PB and PC, β the clockwise angle between the directions PC and PA, and γ the clockwise angle between the directions PA and PB. This sequence of lettering is essential if confusion is to be avoided when P lies outside the triangle. (An anticlockwise convention throughout would of course be permissible.)

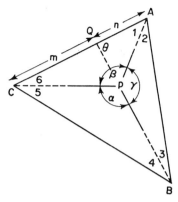

Figure 6.9

The coordinates of P are then given by

$$E_P = \frac{K_1 E_A + K_2 E_B + K_3 E_C}{K_1 + K_2 + K_3} \;;\qquad N_P = \frac{K_1 N_A + K_2 N_B + K_3 N_C}{K_1 + K_2 + K_3}$$

where

$1/K_1 = \cot A - \cot \alpha;\qquad 1/K_2 = \cot B - \cot \beta;\qquad 1/K_3 = \cot C - \cot \gamma$

Check on machine work is $(E_P - E_A)K_1 + (E_P - E_B)K_2 + (E_P - E_C)K_3 = 0$ and similarly for northings.

6.10.1.1. *Proof: Two Theorems in Trigonometry*

The particular proof of the Tienstra resection formula given here, depends on two theorems in trigonometry which are proved before the actual Tienstra formula itself. In triangle ABC (figure 6.10), CD divides angle C into x and y, and AB into m and n, and meets AB at an angle θ on the opposite side of CD from x and A. Then

$$(m + n) \cot \theta = n \cot A - m \cot B \qquad (1)$$

and

$$(m + n) \cot \theta = m \cot x - n \cot y \qquad (2)$$

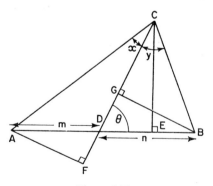

Figure 6.10

These are proved as follows: (1) Since CE is perpendicular to AB, $\cot A = AE/CE$; $\cot B = EB/CE$; and $\cot \theta = DE/CE$.

$$\therefore \quad \frac{m}{n} = \frac{AD}{DB} = \frac{AE - DE}{DE + EB} = \frac{CE(\cot A - \cot \theta)}{CE(\cot B + \cot \theta)}$$

$$\therefore \quad (m + n) \cot \theta = n \cot A - m \cot B \qquad (1)$$

(2) AF and BG are perpendicular to CF,

$$CF/DF = \cot x/\cot \theta \quad \text{and} \quad CG/DG = \cot y/\cot \theta$$

$$\therefore \quad \frac{CF - DF}{DF} = \frac{\cot x - \cot \theta}{\cot \theta}; \quad \text{and} \quad \frac{CG + DG}{DG} = \frac{\cot y + \cot \theta}{\cot \theta}$$

Whence equating values of CD

$$(m + n) \cot \theta = m \cot x - n \cot y \qquad (2)$$

6.10.1.2. *Tienstra Formula*

In figure 6.9, the cotangent formulae for intersection (see §6.9.3) give the following relationships when applied successively to triangles ABP, BCP, and CAP.

$$E_P (\cot 2 + \cot 3) = E_A \cot 3 + E_B \cot 2 - N_A + N_B$$
$$E_P (\cot 4 + \cot 5) = E_B \cot 5 + E_C \cot 4 - N_B + N_C$$
$$E_P (\cot 1 + \cot 6) = E_C \cot 1 + E_A \cot 6 - N_C + N_A$$

Adding these expressions gives:

$$E_P (\cot 1 + \cot 2 + \cot 3 + \cot 4 + \cot 5 + \cot 6)$$
$$E_A (\cot 3 + \cot 6) + E_B (\cot 2 + \cot 5) + E_C (\cot 4 + \cot 1)$$

which may be written

$$E_P(L_1 + L_2 + L_3) = L_1 E_A + L_2 E_B + L_3 E_C$$

where $L_1 = \cot 3 + \cot 6$, etc.

$$\therefore \quad E_P = \frac{L_1 E_A + L_2 E_B + L_3 E_C}{L_1 + L_2 + L_3} \qquad (3)$$

Similarly it can be shown that

$$N_P = \frac{L_1 N_A + L_2 N_B + L_3 N_C}{L_1 + L_2 + L_3} \qquad (4)$$

Since BP meets AC in Q, and divides AC into m and n say, remembering that $CPQ = 180 - \alpha$, and $APQ = 180 - \gamma$,

$$(m + n) \cot \theta = m \cot 4 - n \cot 3 \quad \text{(triangle ABC, theorem 2)}$$
$$(m + n) \cot \theta = n \cot 6 - m \cot 1 \quad \text{(triangle APC, theorem 1)}$$
$$\therefore \quad m(\cot 4 + \cot 1) = n(\cot 3 + \cot 6) \qquad (5)$$

Table 6.19. Tienstra Resection of G from A, B, and C. (See figure 6.1)

Stage (i) Computation of angles A, B, and C. This may not be required since these angles will be listed in survey records

Computation of bearings:

Point	E	N	ΔE	ΔN	Tan/Cot	Bearing	Line
A	351 240·22	138 628·80	+5 548·45	+5 699·47	+0·973 503^T	44° 13' 50"·8	AB
B	356 788·67	144 328·27	− 345·96	+4 450·69	−12·864 75^C	355 33 18·8	BC
C	356 442·71	148 778·96	+5 202·49	+10 150·16	+0·512 552^T	27 08 15·0	AC

Whence A = 17° 05' 35"·8   Check [ΔE] = 0
        B = 131 19 28·0   [ΔN] = 0
        C = 31 34 56·2
        Sum = 180 00 00·0

Stage (ii) Tienstra resection proper.
           Clockwise convention A, C, B

Number	Angle	Cotangent	k	E	N
A	17° 05' 35"·8	+3·251 91		351 240·22	138 628·80
α (20)	45 10 57·5	+0·993 645		−351 000·00	−144 328·27
	1/k_1	+2·258 265	k_1 = +0·442 818	+240·22	−5 699·47
C	31 34 56·2	+1·626 604		356 442·71	148 778·96
β (21)	87 14 09·8	+0·048 277		−351 000·00	−144 328·27
	1/k_2	+1·578 327	k_2 = +0·633 582	+5 442·71	+4 450·69
B	131 19 28·0	−0·879 278		356 788·67	144 328·27
γ (16+17 +18+19)	227 34 52·7	+0·913 724		−351 000·00	144 328·27
	1/k_3	−1·793 002	k_3 = −0·557724	+5 788·67	zero
A+C+B	180 00 00·0		[k] = +0·518 676	+629·10	+570·78
α+β+γ	360 00 00·0			351 000·00	144 328·27
			Coordinates of G	351 629·10	144 899·05

Checks: Cot A cot B + cot B cot C + cot C cot A = 1
        Cot α cot β + cot β cot γ + cot γ cot α = 1
        $(E_G − E_A)k_1 + (E_G − E_C)k_2 + (E_G − E_B)k_3 = 0$
        $(N_G − N_A)k_1 + (N_G − N_C)k_2 + (N_G − N_B)k_3 = 0$

Check bearing to D or use semigraphic method. Angle (19) computed from these coordinates = 66° 16' 37", which checks with observed value for gross error in G

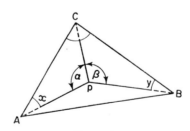

Figure 6.11

Also; $(m + n) \cot \theta = - m \cot \alpha + n \cot \gamma$ (triangle APC, theorem 2)
and $(m + n) \cot \theta = n \cot C - m \cot A$ (triangle ABC, theorem 1)
$$\therefore m(\cot A - \cot \alpha) = n(\cot C - \cot \gamma) \quad (6)$$

From (5) and (6) $\quad \dfrac{\cot 1 + \cot 4}{\cot 3 + \cot 6} = \dfrac{n}{m} = \dfrac{\cot A - \cot \alpha}{\cot C - \cot \gamma}$

$$\therefore \frac{L_3}{L_1} = \frac{K_3}{K_1}$$

for $1/K_1 = \cot A - \cot \alpha$, etc.
In a like manner it can be shown that

$$\frac{L_3}{L_2} = \frac{K_3}{K_2}$$

$$\therefore \frac{L_1}{K_1} = \frac{L_2}{K_2} = \frac{L_3}{K_3} = W$$

say, and $L_1 = K_1 W$, $L_2 = K_2 W$, and $L_3 = K_3 W$
and
$$L_1 + L_2 + L_3 = W(K_1 + K_2 + K_3)$$
$$\therefore L_1/(L_1 + L_2 + L_3) = K_1/(K_1 + K_2 + K_3)$$
$$L_2/(L_1 + L_2 + L_3) = K_2/(K_1 + K_2 + K_3)$$
$$L_3/(L_1 + L_2 + L_3) = K_3/(K_1 + K_2 + K_3)$$

Substitution for these expressions in equations (3) and (4) gives the Tienstra formulae. It will be noted that it is simple matter to reduce the number of figures in the computation by introducing an arbitrary origin. Alternative forms of the coefficients are

$$K_1 = \frac{\sin A \sin \alpha}{\sin (\alpha - A)}; \quad K_2 = \frac{\sin B \sin \beta}{\sin (\beta - B)}; \quad K_3 = \frac{\sin C \sin \gamma}{\sin (\gamma - C)}$$

which are useful for logarithmic computation and for checking.

When the resected point P approaches the danger circle, the K coefficients tend to infinity, and the solution becomes unstable. The method breaks down if the points A, B, and C are collinear. Table 6.19 gives the example of the resection of G from A, B, and C of figure 6.1, using the data of tables 6.1 and 6.2.

6.10.2. THE POTHONOT–SNELLIUS METHOD. (OR ϕ-45 METHOD)

Consider figure 6.11 which shows the point to be resected P lying within the control triangle ABC, and observed angles α and β. If either of the angles x or y can be found, the orientation of a direction from P can be determined, and the coordinates of P found by intersection from any two of A, B, and C. Solution for x and y is as follows:

$$x + y + \alpha + \beta + C = 360°$$
$$\therefore y = [360 - (\alpha + \beta + C)] - x = S - x$$

where S is known. Now

$$\frac{AC \sin x}{\sin \alpha} = PC = \frac{BC \sin y}{\sin \beta}$$

$$\therefore \quad \frac{\sin y}{\sin x} = \frac{AC \sin \beta}{BC \sin \alpha} = K \qquad (1)$$

which can be evaluated. At this stage in the computation, three possible alternative methods of solution arise, two of which are well-suited to machine computation, and the third to logarithmic work.

Method (1)

The value of K in (1) is computed. Then

$$K = \frac{\sin y}{\sin x} = \frac{\sin (S - x)}{\sin x} = \frac{\sin S \cos x - \cos S \sin x}{\sin x}$$

whence
$$\cot x = (K + \cos S)/\sin S$$

Whence the value of x is found, and all the angles of the figure can be deduced. Thereafter P is coordinated by intersection.

Method (2)

The value of K is computed and equated to the tangent of an angle ϕ, i.e.

$$\sin y / \sin x = \tan \phi$$

whence the angle ϕ is derived.
Then

$$\tan (\phi - 45) = \frac{\tan \phi - 1}{1 + \tan \phi} \qquad (2)$$

$$= \left(\frac{\sin y}{\sin x} - 1\right) \div \left(1 + \frac{\sin y}{\sin x}\right) = \frac{\sin y - \sin x}{\sin y + \sin x}$$

$$= \frac{2 \cos \tfrac{1}{2}(y + x) \sin \tfrac{1}{2}(y - x)}{2 \sin \tfrac{1}{2}(y + x) \cos \tfrac{1}{2}(y - x)} = \cot \tfrac{1}{2}(y + x) \tan \tfrac{1}{2}(y - x)$$

$$\therefore \quad \tan \tfrac{1}{2}(y - x) = \tan \tfrac{1}{2}(y + x) \tan (\phi - 45)$$

and since both $(\phi - 45)$ and $\tfrac{1}{2}(y + x)$ are known, $\tfrac{1}{2}(y - x)$ can be evaluated.
Then $\quad \tfrac{1}{2}(y + x) + \tfrac{1}{2}(y - x) = y \quad$ and $\quad \tfrac{1}{2}(y + x) - \tfrac{1}{2}(y - x) = x$

P is coordinated by intersection. This method is well-suited to logarithms since no antilogging is required at an interim stage in the computation.

Method (3)

This is a machine variation of the above which saves one looking up the value of the angle ϕ. Since $\tan \phi$ is known it may be introduced into the computation at stage (2) of method (2), thus a value of $\tan (\phi - 45)$ is obtained directly. Otherwise, the method is as in (2).

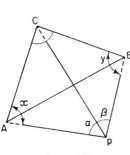
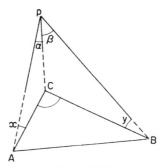

Figure 6.12   Figure 6.13

When the point P is outside the triangle formed by the control points, figures such as 6.12 and 6.13 apply. In these cases the comparable relationships between the angles are respectively.

$$x + y + \alpha + \beta + C = 360 \quad \text{and} \quad x + y + \alpha + \beta - C = 0$$

6.10.3. THE COLLINS' POINT OR BESSEL'S METHOD

In figure 6.14 the circle through the new point P and fixed points B and C is drawn. AP produced meets the circle in the point I. Then since the CBI = $180 - \beta$ and BCI = $180 - \alpha$, the point I can be coordinated from B and C by the intersection formulae, hence

$$E_I = \frac{-E_B \cot \alpha - E_C \cot \beta - N_B + N_C}{-\cot \alpha - \cot \beta};$$

$$N_I = \frac{-N_B \cot \alpha - N_C \cot \beta + E_B - E}{-\cot \alpha - \cot \beta}$$

The bearing of PA is then given by $\tan^{-1}\left(\dfrac{E_A - E_I}{N_A - N_I}\right)$

i.e. by $\tan^{-1}\left(\dfrac{E_B \cot \alpha - E_A \cot \alpha + E_C \cot \beta - E_A \cot \beta + N_B - N_C}{N_B \cot \alpha - N_A \cot \alpha + N_C \cot \beta - N_A \cot \beta - E_B + E_C}\right)$

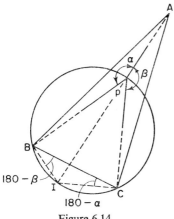
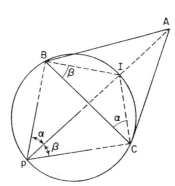

Figure 6.14   Figure 6.15

The bearings of PB and PC are therefore obtained by application of the observed angles α and β, and the point P is coordinated by intersection from B and A, and C and A.

The outside fixation is shown in figure 6.15. In addition to the insolubility if the point P lies on the danger circle, it is likewise troublesome if P lies on the line BC. In the latter case, an acceptable solution is obtained if the circle through P, A, and B or through P, C, and A is drawn. If the line AP is tangential to the circle BPC, the points I and P coincide, though this does not affect the solution. The Collins' Point method is useful on the plane table (see chapter 1).

6.11. VARIATION OF COORDINATES

Very often a new point will be fixed by more than the bare minimum of observations. For example, if the point G, figure 6.1, is intersected from the six points A, B, C, D, E, and F, its coordinates may be computed from various pairs of rays, and generally the values obtained will all differ from one another, if the observed angles have not been adjusted. In some instances, it may be sufficient to compute only two values of the coordinates of G from two pairs of rays which give good intersections at G, and accept the final mean. However if the very best results are desired, every good observation should be used. To achieve this 'best result', two approaches are possible:

(i) The angles will be adjusted for consistency prior to the computation of coordinates, which by virtue of the adjustment will be unique, irrespective of which rays are chosen for the computation.

(ii) An adjustment will be made to provisional coordinates of G, to give final values which best fit all the observations.

If the rigorous, least squares method of adjustment is adopted in either case, an identical answer will result; with semi-rigorous methods of adjustment answers vary within small limits.

6.11.1. PRINCIPLE OF THE METHOD OF VARIATION OF COORDINATES

An approximate position of the new point is obtained from an analytical solution involving the minimum number of observations, e.g. from the two 'best' rays of an intersection, or the three 'best' rays of a resection. These *provisional* coordinates are adjusted to give final values. Suppose the provisional coordinates of G are E_G', N_G', and dE and dN are the corrections to be added to them to give the final coordinates $E_G N_G$, then

$$E_G = E_G' + dE \quad \text{and} \quad N_G = N_G' + dN \tag{1}$$

Although it is possible to obtain E_G' and N_G' from a diagram instead of a precise computation, it is advisable to have dE and dN as small as possible to avoid successive approximation. There are two methods of deriving the corrections dE and dN

(i) *The semi-graphic method.*
(ii) *The analytical method.*

The latter often involves the application of least squares.

Angular Measurement and Computation 301

6.11.2. SEMI-GRAPHIC METHODS

In semi-graphic methods, the observed rays are plotted on a diagram showing the vicinity of the new point. The origin of the diagram is the provisional position of the new point; and its scale should be just sufficient to reflect the precision of the observations. For example if the computation involves angles to the nearest second of arc and rays of 10 km at most, the diagram should be capable of showing a positional change of about 0·05 m. A scale of 1 mm ≡ 0·01 m would be adequate. If the scale is too large the final selection of the new point is difficult, and if too small, the required precision cannot be shown; also plotting errors affect the result significantly. The diagrams are conveniently drawn on good quality graph paper.

6.11.3. INTERSECTION

To illustrate the method, the coordinates of G (figure 6.1) will be found by intersection from A, B, C, D, E, and F as follows:

(i) A preliminary adjustment of the directions observed to G is carried out. For example, since the sum of the angles (1) + (2) + (3) must equal angle FAB any small misclosure is distributed equally. In practice, this stage is often considered to be unnecessary; however for completeness the 'observed bearings' of table 6.20 have been adjusted in this way.

Table 6.20

	1	2	3	4	5	6	7	8	9
1	Ray	$\Delta E'$	$\Delta N'$	Tan Cot	α_C	α_O	da''	Distance km	Shift dm
2	AG	+388·78	+6 270·20	T +0·06200	03° 32' 52·8	56·9	+4·1	6·3	+0·125
3	BG	−5 159·67	+570·73	C −0·11061	276 18 43·5	46·6	+3·1	5·2	+0·078
4	CG	−4 813·71	−3 879·96	T +1·24066	231 07 49·4	49·6	+0·2	6·2	+0·006
5	DG	+1 584·75	−5 853·70	T −0·270726	164 51 06·0	05·9	−0·1	6·1	−0·003
6	EG	+5 848·33	−5 495·05	T −1·06429	133 12 58·0	52·3	−5·7	8·0	−0·221
7	FG	+4 138·50	−581·79	C −0·14058	98 00 08·0	06·6	−1·4	4·2	−0·028

(ii) A provisional position of G, say G', is found by computation of two rays giving a good intersection at G, e.g. in this case the rays CG and DG give the provisional coordinates of G

$$E_G' = 351\ 629\cdot 0; \quad N_G' = 144\ 899\cdot 0$$

(iii) From these provisional coordinates, bearings α_C are computed for all the rays, i.e. for AG', BG' ··· FG' (see table 6.20 column 5).

(iv) From the observed angles at the fixed points, the 'observed' bearings are also known. The differences between the respective 'observed' and 'computed' bearings are found, and listed in the (O − C) column 7.

(v) The approximate lengths of the rays are scaled from a working diagram, or may be computed from the approximate coordinates.

(vi) Consider figure 6.16 for the ray DG.

α_C = computed bearing; α_0 = observed bearing

$d\alpha = \alpha_0 - \alpha_C$; $\Delta E' = E_G' - E_D$; $\Delta N' = N_G' - N_G$

The 'shift' is the amount by which G' changes its position as a result of a swing in the bearing DG' from its approximate to observed position DG, i.e. through the angle dα. Provided G' is close to G, the rays DG' and DG may be considered parallel. In practice, the angle dα should not

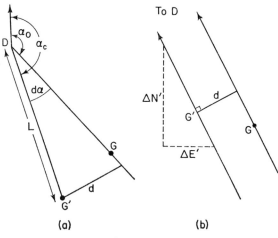

Figure 6.16

exceed one minute of arc which is unplottable. Hence it is possible to draw a large scale figure 6.16(b) of the vicinity of G' and G, showing G' and the directions DG' and DG parallel to each other and separated by a distance $d = L\, d\alpha''\sin 1''$. The side of DG' on which DG lies is obvious from the (O—C) term. The direction DG' is plotted using $\Delta E'$ and $\Delta N'$ at any convenient scale, and checked by protractor; DG is plotted parallel to DC' at a distance d at a large scale, e.g. 1 mm = 0·01 m. Since G' lies on DG' we may say that DG' is a *position line* for G'; hence DG is a position line for G.

(vii) In the same manner, position lines AG, BG $\cdots$ FG are drawn on the figure 6.17 using the data of table 6.20. If there is no error in the observed rays, all these position lines for G meet at one point, G. In practice an error figure is produced which indicates the precision of the work. There are a number of ingenious graphical solutions of the error figure, such as Bertot's, but in our opinion these are of doubtful practical value. If a precise solution is required, the least squares method should be employed (see §6.11.14). The final coordinates of G are chosen after a careful inspection of the error of figure 6.17, a process which becomes more efficient with experience. Normally the final answer is never in doubt and accords with the precision of the

Angular Measurement and Computation

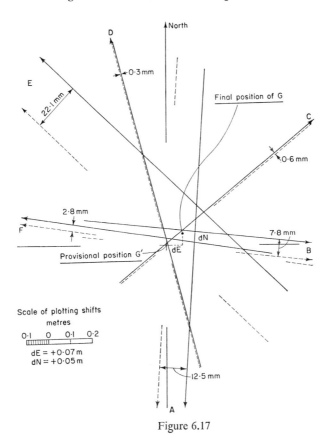

Figure 6.17

observations. If a gross error in observation is present, this is at once evident from the graph. If there are only three rays, a unique answer is possible on the assumption of an equal error in each observed ray (see Chapter 1—plane tabling).

6.11.4. RESECTION

The solution of a resection may be carried out in the same manner as for intersection to the stage at which the error figure is solved. In the resection problem this error figure is largely due to a faulty orientation of all the rays, arising from the error in the provisional coordinates. The final position of the point will therefore be decided in accordance with Lehmann's rules (Chapter 1—plane tabling).

The following alternative method has the advantage of eliminating this systematic effect from the error figure and is useful in solving the fully observed triangle. The theory also enables one to estimate the strength of a resection.

6.11.5. THEORETICAL BASIS

Consider figure 6.18. G′ is the provisional position of the point G, A, and B are fixed points, the angle γ_0 is observed at G between A, and B, γ_C is the

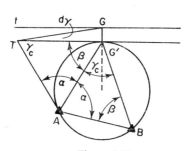
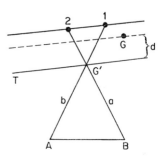

Figure 6.18   Figure 6.19

computed angle BG'A, $d\gamma = \gamma_0 - \gamma_C$, TG' is the tangent to the circle ABG' at G', AT is constructed so that angle TAG' = α, triangles ABG' and AG'T are similar, and Gt is parallel to G'T at a distance d from it. As G' moves round the circle ABG' the angle γ_C remains unchanged. Therefore the circle ABG' is the position circle for G'. *For small distances close to G' the tangent G'T will be indistinguishable from the arc of the position circle, hence it is a position line for G'.* If γ_C is changed to the observed angle γ_0 Gt is the position line for G, parallel to G'T, at a distance d from it. Since $d\gamma$ is small, we may consider Gt coincident with GT in the vicinity of G. On a diagram plot the tangent G'T and GT, the position line for G, parallel to it, a distance d away at any convenient scale. The shift d is given by $d = (ab/c \sin 1'') d\gamma''$ for

for
$$d = TG' \, d\gamma = \frac{\sin \alpha}{\sin \gamma_c} G'A \, d\gamma = \frac{ab}{c} d\gamma$$
$$= \frac{ab}{c} d\gamma'' \sin 1'' = \left(\frac{ab}{c} \sin 1''\right) d\gamma''$$

The direction of the tangent G'T may be plotted by protractor; or mark off two points (1) and (2) of figure 6.19 on b and a respectively and if possible on the same side of G'T as G, such that G'(1):G'(2) as $a:b$. Then the line through (1) and (2) gives the direction of G'T, and therefore of Gt. Position lines for G are drawn for all observed angles to fixed points of which there must be a minimum of three. Their intersection gives the position of G. It will be seen that there is one position line for each observed angle.

6.11.6. PRACTICAL EXAMPLE

Refer to table 6.21 giving computation details: The point G is to be resected from observations to A, B, and C using the observed angles 20 and 21 of table 6.1 and coordinates from table 6.2. The procedure is as follows:

(i) Plot points A, B, and C on a large scale diagram (figure 6.20(a)). Plot angles 20 and 21 on a piece of tracing paper and fit the rays over the control points A, B, and C giving the first approximate position of G, i.e. G', whose coordinates scaled from the diagram are E_G' 351 639, N_G' 144 908.

Angular Measurement and Computation

Table 6.21. Semi-Graphic Resection (See figure 6.20(a))

1	1	2	3	4	5	6	7	8
2		First approximation			Second approximation			
3	Coords of G	351 639 ; 144 908			351 629 ; 144 899			
4	Ray	$\Delta E'$	$\Delta N'$	Tan/Cot	Bearing	$\Delta E''$	$\Delta N''$	Tan/Cot	Bearing
5	G'A	−398·78	−6279·20	+0·063 51	03°38′00″	−388·78	−6270·20	+0·062 00	03°32′52″·8
6	G'B	+5149·67	−579·73	−0·112 58	276°25 25	+5159·67	−570·73	−0·110 61	276 18 43·5
7	G'C	+4803·71	+3870·96	+1·24096	231°08 15	+4813·71	+3879·96	+1·240 66	231 07 49·4
				Angle 20		Angle 21		Angle 20 + 21	
8	Approx. side lengths		6·2	5·2	5·2	6·3	6·2	6·3
9	in km a b c		4·5		7·9		11·4	
10	$K = \frac{ab}{c} \sin 1''$		0·035		0·020		0·017	
11	First shift m		−13·06		+1·90		−4·73	
12	dα seconds		−373		+95		−278	
13	dα		− 6′ 13″		+ 01′ 35″		− 04′ 38″	
14	Computed angle		45 17 10		87 12 35		132 29 45	
15	Observed angle		45 10 57·5		87 14 09·8		132 25 07·3	
16	Computed angle		45 10 54·1		87 14 09·3		132 25 03·4	
17	dα seconds		+3·4		+0·5		+3·9	
18	Second shift m		+0·119		0·010		+0·066	
19	Corrections from graphs							
20	First coords G' E;N		351 639			144 908		
21	First corrections		−10			−9		
22	Second coords G″		351 629			144 899		
23	Second corrections		+0·10			+0·06		
24	Final coords		351 629·10			144 899·06		

(ii) Using these approximate coordinates, the bearings G'A, G'B, and G'C are computed (columns 1–4). Subtracted in pairs, these bearings give the computed values of the angles 20 and 21 (line 14).

(iii) From the observed values of these angles in line 15 the dα's are obtained (line 12).

(iv) The approximate lengths of the rays are scaled from figure 6.1 (lines 8 and 9).

(v) The respective shifts (ab/c) dα″ sin 1″ are calculated (line 11).

(vi) The tangents are plotted on figure 6.20(a) with the assistance of the directions G'A, G'B, and G'C already plotted. For example, to plot the tangent for the angle 2–0 proceed as follows: Points 1 and 2 are plotted along G'C and G'B respectively of length 5·2 cm and 6·2 cm. Hence triangles G'1 2 and G'BC are similar. (Check that line 1–2 is the 4·5 cm equivalent to BC.) The broken line 1–2 gives the direction of the

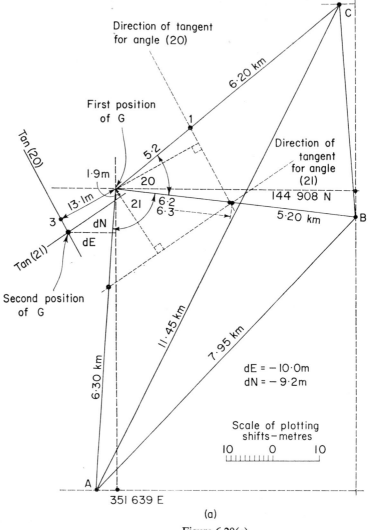

Figure 6.20(a)

required tangent. In this case it was not possible to plot this direction on the same side of G' as the final position of G.

(vii) The shift was plotted at a scale of 0·1 inch to 1 metre, i.e. G'3 is perpendicular to 1–2 and 1·31 inches long. The tangent for angle 2–0 is drawn through 3 parallel to line 1–2.

(viii) In a similar manner, the tangent for angle 2–1 is plotted. The intersection of the tangents gives the second position of G, i.e. G''. The corrections to the coordinates of G' are

$dE = -1\cdot 0$ inches $\equiv -10$ metres; $dN = -0\cdot 9$ inches $\equiv -9$ metres

whence the coordinates of G'' are 351 629, 144 899.

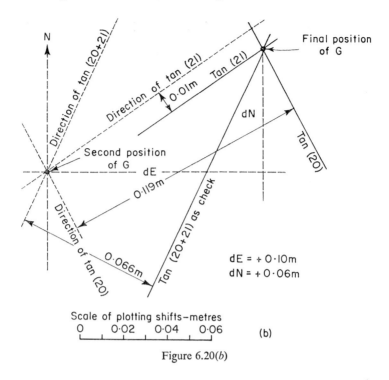

Figure 6.20(b)

(ix) To obtain more accurate coordinates of G, the complete process is repeated including the tangent for (20 + 21) as a check. See columns 5, 6, 7, and 8 and lines 15, 16, and 17 of table 6.21, and figure 6.20(b) in which the scale of plotting the shifts was 1 cm ≡ 0·01 m, i.e. true to size. Final coordinates of G are

351 629·10;   144 899·06

With practice it is convenient to plot all diagrams on one sheet of paper, transferring the directions of the observed rays by parallel ruler to a clear part of the paper.

This particular example required two approximations to obtain the final result because the whole computation was by the semigraphic method. However, since the method is normally used only for multiple ray solutions, the first position will normally be derived by an analytical method. This is particularly simple if, as is often the case, one inward ray has been observed from a fixed point to the new station. With more than three rays, an error figure results, from which the final position is chosen by inspection.

6.11.7. THE FULLY OBSERVED TRIANGLE

If all three angles of a triangle are used to fix a new point, the semigraphic computation will combine the intersection and resection methods described above, i.e. there will be a line for each inward observed direction and a tangent for each observed angle at the new point. If linear measures have also been made, a position line representing them may also be shown on the error diagram (see Chapter 13). The final position of the point will be estimated from the error figure. Generally speaking, if such a complicated adjustment is desired it will be simpler to employ analytical methods and least squares.

6.11.8. POSITIONAL ERRORS OF POINTS

The semi-graphic method can be used to give a quick estimate of the positional error of a fixation by either intersection or resection. The shifts $\pm d_1$ and $\pm d_2$ corresponding to errors $\pm d\alpha_1$ and $\pm d\alpha_2$ in the observed angles give the area of error in the vicinity of the computed point, as shown in figure 6.21. The maximum likely positional error x is of most interest where safety margins have not

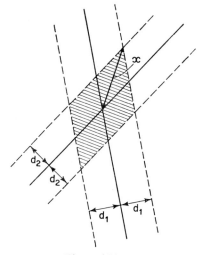

Figure 6.21

to be exceeded, for example in setting out work. It is at once obvious from this approach that the smallest positional error from given angular errors depends on

(i) The length of the sides producing the shifts d.
(ii) The angle at which the position lines cut each other; the best solution being when they meet at right angles.

6.11.9. ANALYTICAL METHOD

As an alternative to the semi-graphic solution, the adjustment of position by the method of variation of coordinates may also be carried out entirely analytically. As in the above sections, the method is capable of handling both intersection and resection, or combinations of them, and the addition of linear measures may also be treated conveniently (see §13.9.6).

6.11.10. BASIC THEORY

Consider the equation connecting the coordinates of points A and G with the bearing of AG, α. i.e.

$$E_G - E_A = (N_G - N_A) \tan \alpha$$

or $$\log \tan \alpha = \log (E_G - E_A) - \log (N_G - N_A) \quad (1)$$

Differentiating gives:
$$d\alpha/\sin \alpha \cos \alpha = \frac{dE_G - dE_A}{E_G - E_A} - \frac{dN_G - dN_A}{N_G - N_A}$$

If the length of AG $= S$, and putting $E_G - E_A = \Delta E$ and $N_G - N_A = \Delta N$, after a little reduction

$$d\alpha = (dE_G - dE_A)\frac{\Delta N}{S^2} - (dN_G - dN_A)\frac{\Delta E}{S^2}$$

or for in seconds of arc

$$d\alpha'' = (dE_G - dE_A)\frac{\Delta N}{S^2 \sin 1''} - (dN_G - dN_A)\frac{\Delta E}{S^2 \sin 1''}$$

$$= (dE_G - dE_A)P - (dN_G - dN_A)Q \quad (2)$$

Equation (2) is the basic differential equation for the direction AG. The coefficients $P = \Delta N/S^2 \sin 1''$ and $Q = \Delta E/S^2 \sin 1''$ are the *direction coefficients* of the equation, evaluated from the best available values of ΔE, ΔN, and S, which may initially be scaled from a diagram. An alternative method of evaluating P and Q is given below in §6.11.13.

If a point has a fixed position the variations to its coordinates are zero, i.e. $dE = dN = 0$.

6.11.11. INTERSECTION

Consider the point G to be fixed by intersection from fixed points A and B, figure 6.1. Since $dE_A = dN_A = dE_B = dN_B = 0$, the two direction equations of the form of (2) representing the directions AG and BG are

$$d\alpha_1 = P_1 dE_G - Q_1 dN_G \quad \text{and} \quad d\alpha_2 = P_2 dE_G - Q_2 dN_G \quad (3)$$

where P_1 and Q_1, P_2 and Q_2 are the direction coefficients for AG and BG respectively. If the provisional coordinates of G are E_G', N_G', values α_C may be computed for the bearings AG and BG, whence the values of $d\alpha$ for each bearing are obtained from $d\alpha = \alpha_O - \alpha_C$. Substituting these values of $d\alpha$ in equations (3) leaves two equations in two unknowns dE_G and dN_G. Solution for these corrections enables the final values of the coordinates of G to be calculated from $E_G = E_G' + dE$ and $N_G = N_G' + dN$.

Example: Using the data of table 6.1 and table 6.2, i.e. for angles ABG and GAB, and the coordinates of A and B, and assuming the position of G to be 351 629 E, 144 899 N, compute the coordinates of G.

The bearing of AB $= \tan^{-1} 1(5\ 548 \cdot 45/5\ 699 \cdot 47) = 44° 13' 51''$. The computation of the other quantities is tabulated as follows.

Ray	$\Delta E'$	$\Delta N'$	$S^2 \times 10^{-6}$	P	Q	Tan Cot	bearing α_c	α_c	α_o	$\alpha_o - \alpha_c$
AG	+388·78	+6270·20	39·47	+32·77	+2·03	+0·062 004 T	03°32'52·8"	57"		+4"
BG	−5159·67	+570·73	26·95	+4·77	−39·50	−0·110 614 C	276 18 43·5	48		+4".5

Whence equations (3) are

$$+32\cdot77\,dE - 2\cdot03\,dN = 4\cdot0$$
$$+4\cdot37\,dE + 39\cdot50\,dN = 4\cdot5$$

which on solution give $dE = +0\cdot13$, $dN = +0\cdot09$, and final coordinates of G 351 629·13, 144·899·09. The only difficulty in the method is to decide the sign of dα. This is however a simple matter when one remembers the observed bearing is that which gives the correct position, hence

$$\text{Observed} = \left. \begin{array}{l} \text{Computed or} \\ \text{Provisional} \end{array} \right\} + \text{Correction}$$

$$\alpha_0 = \alpha_P + d\alpha \quad \text{or} \quad \alpha_C + d\alpha$$

therefore $\quad \alpha(O - P) = d\alpha \quad \text{or} \quad \alpha(C - O) = -d\alpha.$

If dα is placed on the left-hand side of the equation (2) the final form is

$$(dE_G - dE_A)P - (dN_G - dN_A)Q + \alpha(C - O) = 0$$

6.11.12. RESECTION

In the intersection problem, the quantity dα'' was obtained directly from

$$\text{Observed} - \text{computed bearing} = d\alpha''$$

In the resection problem, since orientation of the directions at the observation point is unknown, the observed bearings cannot be obtained directly. A process of successive approximation is also required for the observed bearings. An approximate value of one bearing is assumed, to which the others are related via the observed angles. All the bearings will therefore have the same error of orientation, usually denoted by z

Then $\quad d\alpha'' = $ (Observed $-$ Computed) bearing

i.e. $\quad d\alpha'' = $ Provisional observed bearing $+ z'' - $ Computed bearing

Since z is unknown we write

$$d\alpha'' - z'' = \text{(Provisional Observed—Computed) bearing}$$

the right-hand side of which is known. The equation (2) must therefore be altered by the subtraction of z from both sides to give

$$d\alpha'' - z'' = P\,dE - Q\,dN - z'' \qquad (4)$$

If only three directions are observed, the solution of the values of the three unknowns dE, dN, and z, is unique.

Example: The point G is to be resected from the fixed points A, B, and C from the data given below. The assumed position of G is 351 629 E 144 899 N and the computed bearing of GA is chosen to be the 'provisional observed bearing' of GA. The provisional observed bearings of GB and GC are derived from this bearing GA via the observed angles AGB and BGC, whose values are respectively, 87° 14′ 09″·7 and 45° 10′ 57″·5. The coordinates of the control points are

$$A \quad 351\ 240{\cdot}22 \text{ E}; \quad 138\ 628{\cdot}80 \text{ N}$$
$$B \quad 356\ 788{\cdot}67 \quad\quad\quad 144\ 328{\cdot}27$$
$$C \quad 356\ 442{\cdot}71 \quad\quad\quad 148\ 778{\cdot}96$$

Ray	ΔE	ΔN	$s^2 \times 10^{-6}$	Tan Cot Bearing	P	Q	Bearing C	Bearing O	O−C
AG	+388·78	+6 270·20	39·47	+0·062004 T	+32·77	+2·03	03 32 52·8	52·8	0
BG	−5 159·67	+570·73	26·95	−0·110614 C	+4·37	−39·50	276 18 43·5	43·1	−0·4
CG	−4 813·71	−3 879·96	38·22	+1·240660 T	−20·94	−25·98	231 07 49·4	45·6	−3·8

The directions AG have been considered instead of the directions GA, etc., in order to use equation (2) for both intersection and resection. The equations for the directions are:

Direction $\overrightarrow{GA}$ $\quad +32{\cdot}77 \ dE - 2{\cdot}03 \ dN - z = 0$ (a)
Direction GB $\quad +4{\cdot}37 \ dE + 39{\cdot}50 \ dN - z = -0{\cdot}4$ (b)
Direction GC $\quad -20{\cdot}94 \ dE + 25{\cdot}98 \ dN - z = -3{\cdot}8$ (c)
(a)−(b) $\quad +28{\cdot}40 \ dE - 41{\cdot}53 \ dN = +0{\cdot}4$ (d)
(b)−(c) $\quad +25{\cdot}31 \ dE + 13{\cdot}52 \ dN = +3{\cdot}4$ (e)

Equations (d) and (e) could have been obtained directly by considering the variations of the angles AGB and BGC; however there is less chance of error if the directions are considered separately, and the equations for the angles formed thereafter. Solution of equations (d) and (e) gives $dE = +0{\cdot}10$, $dN = +0{\cdot}06$. The final coordinates of G are then 351 629·10 E; 144 899·06 N.

6.11.13. CHECK ON THE FORMATION OF THE DIRECTION COEFFICIENTS

The direction coefficients P and Q can be checked by logarithms. Consider the equation relating coordinates and bearing $\tan \alpha = (E_G - E_A)/(N_G - N_A)$. Taking logs we have

$$\log \tan \alpha = \log (E_G - E_A) - \log (N_G - N_A)$$

Then for small changes in bearing and coordinates

$$d(\log \tan \alpha) = d[\log (E_G - E_A)] - d[\log (N_G - N_A)]$$
$$\therefore \ \Delta_\alpha \ d\alpha'' = \Delta_{\Delta E}(dE_G - dE_A) - \Delta_{\Delta N}(dN_G - dN_A)$$

where Δ_α is the tabulated difference in the log tan α in any convenient unit of α, usually 1" of arc; and $\Delta_{\Delta E}$, $\Delta_{\Delta N}$ are the tabulated differences in log $(E_G - E_A)$ and log $(N_G - N_A)$ respectively for differences of one unit, usually one metre or one foot. Then

$$d\alpha'' = (dE_G - dE_A) \Delta_{\Delta E}/\Delta_\alpha - (dN_G - dN_A) \Delta_{\Delta N}/\Delta_\alpha$$

which is of the same form as equation of

$$\therefore \ d\alpha'' = P(dE_G - dE_A) - Q(dN_G - dN_A)$$

whence $\quad P = \Delta_{\Delta E}/\Delta_\alpha \ \text{ and } \ Q = \Delta_{\Delta N}/\Delta_\alpha$

The example in the next section shows the detailed working of this alternative method of deriving P and Q. It is essential that these coefficients P and Q

312   *Practical Field Surveying and Computations*

are checked carefully before solving the equations, especially when a large network is involved.

6.11.14. SOLUTION OF MULTIPLE RAY PROBLEM

The solution of a multiple ray intersection will now be carried out. The point G of figure 6.1 is to be fixed from the directions observed at A, B, C, D, E, and F whose coordinates are to be held fixed.

A preliminary computation of the angles of the figure formed by the fixed points is required, so that the observed angles may sum to the correct amount in sympathy with the fixed coordinates. For example the observed angles (1) + (2) + (3) must sum to the value of FAB derived from the fixed coordinates of F, A, and B. The small misclosures are assigned equally to each observed angle, i.e. in this case to (1) and to (2 + 3). Hence the 'observed' bearings AG, etc. are determined. These are listed in column (9) of table 6.22. The computed bearings obtained from the provisional coordinates of G and the fixed coordinates, are listed in column (8).

The direction coefficients P and Q for each direction are formed by the first method in table 6.22 and, by the second, in table 6.23. An observation equation

Table 6.22

Intersection of G by Variations of Coordinates

Equations are of form:

$$\frac{\Delta N}{S^2 \sin 1''} dE - \frac{\Delta E}{S^2 \sin 1''} dN + (C-O)'' = 0$$

Provisional coordinates of G   351 629·00; 144 899·00

1	2	3	4	5	6	7	8	9	10
	ΔE	ΔN	$S^2 \times 10^{-6}$	$\tan \int B_q$ $\cot$ C	$\frac{\Delta E \cos\text{ec } 1''}{S^2}$	$\frac{\Delta N \cos\text{ec } 1''}{S^2}$	Bearing C	Bearing O	C–O
AG	+388·78	+6 270·20	39·468	Tan +0·062 004	+32·77	+2·032	03°32'52"8	03°32'56"9	−4"1
BG	−5 159·67	+570·73	26·947	Cot −0·110 614	+4·369	−39·50	276 18 43·5	276 18 46·6	−3·1
CG	−4 813·71	−3 879·96	38·224	Tan +1·240 660	−20·94	−25·98	231 07 49·4	231 07 49·6	−0·2
DG	+1 584·75	−5 853·70	36·777	Tan −0·270 726	−32·84	+8·890	164 51 06·0	164 51 05·9	+0·1
EG	+5 848·33	−5 495·05	64·399	Tan −1·064 291	−17·603	+18·735	133 12 58·0	133 12 52·3	+5·7
FG	+4 138·50	−581·79	17·466	Cot −0·140 580	−6·872	+48·88	98 00 08·0	98 00 06·6	+1·4

Normals				Observation Equations	$dE \times 10$	$dN \times 10$	K	Sum
$dE \times 10$	$dN \times 10$	K	S					
+29·669	+5·196	−25·694	+9·171	I	+3·277	−0·203	−4·100	−1·026
	+50·588	−29·546	+26·238	II	+0·437	+3·950	−3·100	+1·287
(0·03370)	Solution			III	−2·094	+2·598	−0·200	+0·304
+29·669	+5·196	−25·694	+9·171	IV	−3·284	−0·889	+0·100	−4·073
−1	−0·1751	+0·8659	−0·3091	V	−1·760	−1·874	+5·700	+2·066
(0·020130)	+49·678	−25·047	+24·632	VI	−0·687	−4·888	+1·400	−4·175
	−1	+0·5042	−0·4958		−4·111	−1·306	−0·200	−5·802
+0·7776	+0·5042			$dE = +0·08$				−5·802
−0·2223	−0·4958			$dN = +0·05$				
				Adjusted value of G		351 629·08; 144 899·05		

Angular Measurement and Computation

Table 6.23

Check on the formation of the direction coefficients of table 6.22 by logarithms.
Figures are in the seventh place of the logarithms.

$$P = \Delta_{\log \Delta E}/\Delta_{\log \tan \alpha}; \quad Q = \Delta_{\log \Delta N}/\Delta_{\log \tan \alpha}$$

Ray	$\Delta_{\log \tan \alpha}$	$\Delta_{\log \Delta E}$	$\Delta_{\log \Delta N}$	P	Q
AG	+340	+11 155	+693	+32·8	+2·04
BG	−192·7	−842	+7603	+4·37	−39·4
CG	+43·1	−902	−1 119	−20·93	−25·96
DG	−83·4	+2 739	−741	−32·84	+8·885
EG	−42·2	+743	−790	−17·61	+18·72
FG	−152·8	+1 049	−7458	−6·86	+48·81

is then obtained for each direction, as shown in table 6.22 in a tabulated form, e.g. equation (6) means

$$-6·87 \, dE + 48·88 \, dN + 1·4 = 0$$

Since there are six equations in only two unknowns, dE and dN, the principle of least squares assists solution. Hence the two normal equations are formed, as explained in Chapter 5, solution of which gives finally that the most probable values of the corrections are

$$dE = +0·08 \text{ m}; \quad dN = +0·05 \text{ m}$$

It will be noticed that the coefficients P and Q have all been divided by 10 to bring them to the same magnitude as the absolute terms (C − O). Hence the final values of 0·7776 and 0·5042 have also to be divided by 10 to bring them to the correct units. The final coordinates of G are 351 629·08, 144 899·05.

6.11.15. ADJUSTMENT OF NETWORKS

The adjustment of the coordinates of a network such as that in figure 6.1 consists of the combination of the intersection and resection methods described above, the latter also involving least squares. Observed lengths may also be treated as described in Chapter 13. Fixed points, fixed lengths, and fixed bearings are also handled simply and conveniently by this method, which contrasts with the difficulties they cause in the condition equations method.

6.11.15.1. *Fixed Points*

This aspect has already been considered above. If the point is fixed $dE = dN = 0$, and the various equations involved are correspondingly simplified.

6.11.15.2. *Fixed Lengths*

A fixed length introduces two factors:

(i) The chosen provisional coordinates must be consistent with this fixed length.

(ii) One of the variations is dependent on the other three and is therefore omitted ultimately in the equations.

Consider the length of line AB as fixed equal to L say. The provisional co-ordinates of A, (E_A', N_A') and of B, (E_B', N_B') must be chosen such that

$$(E_B' - E_A')^2 + (N_B' - N_A')^2 = L^2 = \text{a constant}$$

Differentiating gives

$$2(E_B' - E_A')(dE_B - dE_A) + 2(N_B' - N_A')(dN_B - dN_A) = 0$$

$$\therefore \quad dE_B = dE_A - \frac{\Delta N'}{\Delta E'}(dN_B - dN_A) \qquad (5)$$

This expression for dE_B is substituted wherever it occurs in the observation equations prior to adjustment, and its final value is obtained from (5) when the adjusted values of the three variables dE_A, dN_A, and dN_B are known.

6.11.15.3. *Fixed Bearings*

In a similar way to a fixed length, a fixed bearing imposes two new factors:

(i) the provisional coordinates must accord with the fixed bearing.
(ii) one of the variables is omitted from the equations.

Using the same example as in §6.11.15.2, the provisional coordinates must be such that

$$\therefore \quad E_B' - E_A' = (N_A' - N_A') \tan \alpha$$

where $\tan \alpha$ is constant, say K.

Differentiating $\quad dE_B - dE_A = K(dN_B - dN_A)$

$$\therefore \quad dE_B = dE_A + K(dN_B - dN_A) \qquad (6)$$

The similarity between (6) and (5) should be noted. This expression for dE_B is substituted in a similar way as in §6.11.15.2.

Example: Space does not permit giving a worked example of this complete application of the variation of coordinates to a complicated network. As an exercise, the student may adjust the network of this chapter; assuming co-ordinates within 0·5 metre of the final values of table 6.2; holding D and E fixed.

6.11.16. COMPLEX PROBLEMS

The variation of coordinates method permits the surveyor to solve complex survey figures, without having to solve the problem analytically. For example the coordinates of points A, B, C, and D can be obtained readily from the

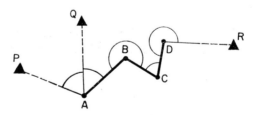

Figure 6.22

fixed points P, Q, and R of figure 6.22, in which ABCD is a traverse, and the observed angles are as marked. Provisional coordinates are obtained from a diagram; and equations are formed and solved for the corrections to them.

6.12. THE INACCESSIBLE BASE, OR TWO POINT PROBLEM

In figures 6.23(a) and (b), A and B are fixed points visible from C whose position is required. If A and B are inaccessible, for example they may be church spires fixed by intersection, direct resection of C is impossible. The problem may be solved by introducing a fourth point D visible from C and

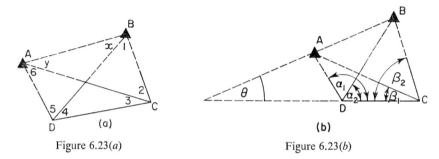

Figure 6.23(a)　　　　　　　　　　Figure 6.23(b)

from which A, B, and C are visible. The angles 2, 3, 4, and 5 are observed, and 1 and 6 deduced. The coordinates of C may then be derived in a number of ways:

(i) By variation of coordinates.
(ii) By assuming a length for the base DC, scaled from a diagram; computing the coordinates of A and B by intersection formulae from D and C on an assumed origin at D, and coordinates of C (DC, O), and obtaining the final coordinates of C on the original system by transformation formulae (see Chapter 2).
(iii) By special formula. There are two common formulae for the two point problem.

In figure 6.23(a)　$\dfrac{\sin x}{\sin y} = \dfrac{\sin 1}{\sin 2} \times \dfrac{\sin 3}{\sin 4} \times \dfrac{\sin 5}{\sin 6} = K$ or $\tan \phi$

Whence solution for x and y is carried out as for the $\phi - 45$ resection method of §6.10.2.

Alternatively, in figure 6.23(b) the angle θ is given by

$$\cot \theta = \frac{\cot \alpha_2 \cot \beta_1 - \cot \alpha_1 \cot \beta_2}{\cot \alpha_1 - \cot \alpha_2 + \cot \beta_1 - \cot \beta_2}$$

which can be proved as follows. Assume the coordinates of D to be $(0, 0)$ and of C to be $(E_C, 0)$. The coordinates of A and B on this system are by the cotangent formulae:

$$E_A = \frac{E_C \cot \alpha_1}{\cot \alpha_1 + \cot \beta_1} ; \quad N_A = \frac{E_C}{\cot \alpha_1 + \cot \beta_1}$$

and
$$E_B = \frac{E_C \cot \alpha_2}{\cot \alpha_2 + \cot \beta_2} \; ; \quad N_B = \frac{E_C}{\cot \alpha_2 + \cot \beta_2}$$

Now $\cot \theta = (E_B - E_A)/(N_B - N_A)$ whence on substitution for E_A, N_A, E_B, and N_B the required formula is obtained. Once θ is known, the angles at A and B may be deduced and C coordinated by intersection from A and B. Data for an example may be selected from tables 6.1 and 6.2.

6.12.1. REDUCTION TO CENTRE: OR SATELLITE STATIONS

It often happens that the theodolite and the beacon cannot be centred over the same mark; for example, if a flagpole is used as a beacon. In figure 6.24 theodolite observations are made at A to stations B and D, and to the centre C

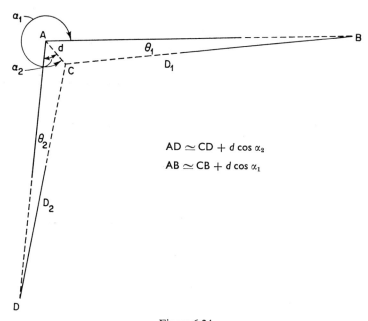

$AD \simeq CD + d \cos \alpha_2$
$AB \simeq CB + d \cos \alpha_1$

Figure 6.24

at which the beacon is placed, and to which the observations from B and D are made. Prior to computation of coordinates, either the angles at A are reduced to what they would have been at the centre C, or the inward directions to C are reduced to the satellite station A. In both cases the angular corrections are θ_1 and θ_2. Since $d(=AC)$ is normally very small compared with the sides $CB(=D_1)$ and $CD(=D_2)$, it can be assumed with sufficient accuracy that $AB = CB$, and that $CD = AD$ for the purpose of evaluating θ_1 and θ_2. In triangle ABC

$$\sin \theta_1 = d \sin \alpha_1 / D_1$$

i.e.
$$\theta_1 - \frac{\theta_1^3}{3!} + \cdots \text{etc.} = d \sin \alpha_1 / D_1 \quad (\theta_1 \text{ in radians})$$

$$\therefore \theta_1'' = d \sin \alpha_1 \operatorname{cosec} 1''/D_1 + (\theta_1'' \sin 1'')^3 \operatorname{cosec} 1''/3!$$
$$\simeq K \sin \alpha_1 / D_1$$

where $K = d \operatorname{cosec} 1''$ and θ_1 is small. Similarly $\theta_2'' = K \sin \alpha_2 / D_2$.

Angular Measurement and Computation

The sign of the correction θ takes the sign of sin α provided the directions at the station A are referred to the centre C as the R.O., as in this case. If the directions observed at B and D are corrected to the satellite station A, θ_1 and θ_2 are of opposite sign to those of the reduction to centre C of observations at A.

Example: Reduce the following directions observed to stations A, B, and C at station X to what they would have been at station Y.

Point	Observed direction	α	θ	Final directions
A	00° 00′ 00″	310° 54′ 15″	−01′ 59″	359° 58′ 01″
B	182 21 13	133 15 28	+37″	182 21 50
C	247 48 50	198 43 05	−31″	247 48 19
Y centre	49 05 45	00 00 00		

$d = 8\cdot 2$ feet $\log d = 0\cdot 913\ 81$
$\log \operatorname{cosec} 1″ = 5\cdot 314\ 43$ $\therefore \log k = 6\cdot 228\ 24$

Station	A	B	C	
Distance D	10 759 ft	32 985	17 631	$\therefore \theta_A = -119″$
Colog D	5·968 22	5·481 69	5·753 72	
Log K	6·228 24	6·228 24	6·228 24	$\theta_B = +37″$
Log sin α	N 9·878 41	9·862 30	N 9·506 39	
Log θ	N 2·074 87	1·572 23	N 1·488 35	$\theta = -31″$

The approximations used in the satellite formula impose restrictions on its use as follows:

(i) The second term in the expansion of sin θ must not be significant. For example if the correction θ is not to be in error by more than 0″5, if D is 1 mile, the length of d cannot exceed 120 feet at the worst position. At worst d is perpendicular to D, therefore:

$$\theta^3 = 6 \times 0\cdot 5 / 206\ 265 \quad \therefore \quad \theta = \tfrac{1}{41} \text{ radian}$$
$$\therefore \quad d = D\theta = 5\ 280/41 = 130$$

The ratio $d/D \simeq \tfrac{1}{40}$ is often taken as the limit of applying the approximate formula.

(ii) Accuracy of d. The distance d has to be measured with sufficient accuracy to enable the error in the correction θ to fall below a certain limit. The worst case is again when d is perpendicular to D. If the error in the correction is E_θ and the error in d is E_d, $E_d = D \times E_\theta$. Substituting the values of (i) for E_θ and D gives

$$E_d = (5\ 280 \times 0\cdot 5 \times 12)/206\ 265 = 0\cdot 15 \text{ in}$$

(iii) Accuracy of α. An error E_α in α has greatest effect when A, C, and B are nearly collinear, i.e. when

$$E_\alpha = DE_\theta/d$$

Again substituting the same values

$$E_\alpha = 5\,280 \times 0.5/130 = 20''$$

At an intermediate angle α, $E_\alpha = DE_\theta \sec \alpha/d$. When α is a multiple of 90° its accuracy need not be high.

(iv) Accuracy in the distance D.

$$\theta = k \sin \alpha/D$$

Differentiating logarithmically, $\log \theta = \log k + \log \sin \alpha - \log D$

$$\therefore \quad d\theta/\theta = -dD/D$$

i.e. $\quad d\theta = -(\theta/D)\,dD$

If $d\theta = \frac{1}{2}''$, $D = 5\,280$ ft, $d = 120$ ft, the maximum value of θ occurs when $\sin \alpha = 1$, whence

$$-dD = D\,d\theta/\theta = D^2\,d\theta''\sin 1''/k \sin \alpha = 68 \text{ ft}$$

The error in the distance D has to be less than about 70 ft. In the formula the distance XA was assumed to be equal to YA for practical purposes. If these two sides are not sufficiently equal to make the assumption that they are, the one can be reduced to the other by the application of a correction $c = d \cos \alpha$, the sign of which is evident from the figure in question.

The effect of probable errors in d, D, and α is best considered by the method of logarithmic differences, since a monomial function is involved.

6.12.2. PROBABLE ERROR IN SATELLITE CORRECTION; METHOD OF LOGARITHMIC DIFFERENCES

The probable error in a quantity which is a monomial function of the variables is best derived as follows.

The satellite correction θ is given by

$$\theta = d \csc 1'' \sin \alpha/D$$

$$\therefore \quad \log \theta = \log d + \log \csc 1'' + \log \sin \alpha - \log D$$

Differentiating

$$\partial(\log \theta) = \partial(\log d) + \partial(\log \sin \alpha) - \partial(\log D)$$

Now $\partial(\log \theta) = \Delta_\theta\,\partial\theta$, where Δ_θ is the tabular difference in $\log \theta$ for any unit of θ, and $\partial\theta$ is in these same units. One uses this fact when interpolating in the log tables for intermediate values. Denoting the respective probable errors of θ, d, $\log \sin \alpha$, and D by E_θ, E_d, E_α, and E_D, and the tabular difference of $\log \sin \alpha$ by Δ_α

$$\Delta_\theta^2 E_\theta^2 = (\Delta_d E_d)^2 + (\Delta_\alpha E_\alpha^2) + (\Delta_D E_D)^2$$

Which is a convenient formula for the computation of E_θ, and which proceeds at the same time as the computation of θ itself, and with very little extra working.

Example: Consider the satellite correction to A of the above example in §6.12.1, and let the respective probable errors be

$$E_d = \pm 0{\cdot}1 \text{ ft}, \ E_\alpha = \pm 2', \text{ and } \ E_D = \pm 30 \text{ ft}$$

The complete computation of the correction and its p.e. is tabulated as follows:

	Log	Δ	E	ΔE	Δ²E²
d	0·9138	53	1	53	2 809
Cosec 1"	5·3144				Sum = 2 957
Sin α	N 9·8784	1	2	2	4
Sum	6·1066				
D	4·0319	4	3	12	144
θ	2·0747	36	E_θ''	$36E_\theta$	$1\,296 E_\theta^2$
θ	−118·8		2 957/1 296 = 2·3

The tabular log difference for d is 53 in the fourth place for a change in d of 0·1 ft. The p.e. of d is one such unit. The tabular difference for θ cannot be evaluated until θ itself is known. The final error is given by $E_\theta = (2\cdot3)^{\frac{1}{2}} = 1''{\cdot}5$.

6.12.3. PROPAGATION OF ERRORS IN TRIANGULATION

An accurate assessment of the precision of a triangulation network can be made only after the corrections of a least squares adjustment of the observations have been obtained. However, the formulae given here are much used to give approximate estimates of the precision of work already done, but more so, of work being planned. In this way, the required precision of the field work can be assessed with reasonable accuracy. The usual assumption is that the triangulation consists of a chain of single triangles each affecting the other, so that the error after n triangles is $n^{\frac{1}{2}}$ times the error propagated through each individual triangle. This is true of scale, though bearing error depends on different considerations (see §7.12). Consider the triangle GEF, whose sides are g, e, and f, and let σ_D, σ_E, and σ_F; and σ_d, σ_e, and σ_f be the respective standard errors in the angles G, E, and F, and the sides g, e, and f. It is assumed that the side $FG = e$ is the base of the triangulation. The side g calculated from the triangle is given by

$$g = \sin G e / \sin E$$

$$\therefore \quad \log g = \log \sin G + \log e - \log \sin E \qquad (1)$$

Differentiating (1)

$$\Delta_g \, dg = \Delta_G \, dG + \Delta_e \, de - \Delta_E \, dE \qquad (2)$$

where Δ_g denotes the tabular change in $\log g$ for one unit of g, and Δ_G denotes the tabular change in $\log \sin G$ for one second of G. $\partial(\log \sin G)$ may also be written

$$\cot G \, dG = \cot G \, dG'' \sin 1''$$

hence
$$\Delta_G \, dG'' = \cot G \, dG'' \sin 1'' \tag{3}$$

The standard error σ_g in g depends on the standard error σ_e in the line e, and on the standard errors of the particular angles measured. There are three possible cases for the latter, to each of which the former is common. The effect of σ_e is simply given by

$$(\Delta_g \sigma_g)^2 = (\Delta_e \sigma_e)^2$$
$$\therefore \sigma_g = \Delta_e \sigma_e / \Delta_g$$

or since
$$\frac{d}{dg}(\log g) = \frac{1}{g} = \Delta_g, \quad \sigma_g = g\sigma_e/e. \tag{4}$$

The triangle may be solved if any of the following three combinations of angles is observed, assuming $\sigma_e = 0$

(i) G and E, or F and E which is identical in theory and is not considered.
(ii) G and F, the intersection.
(iii) All three angles G, E, and F which are adjusted to 180° prior to computation.

In case (i) the angles G and E are independent, hence the standard error σ_g due solely to errors in the observed angles, is given by:

$$(\Delta_g \sigma_g)^2 = (\Delta_G \sigma_G)^2 + (\Delta_E \sigma_E)^2$$

or
$$\left(\frac{\sigma_g}{g}\right)^2 = (\Delta_G \sigma_G)^2 + (\Delta_E \sigma_E)^2$$

If $\sigma_G = \sigma_E = \sigma$ which is usual,

$$\left(\frac{\sigma_g}{g}\right)^2 = \sigma^2(\Delta_G^2 + \Delta_E^2)$$

or remembering (3)
$$\sigma_g^2 = g^2(\sigma'' \sin 1'')^2(\cot^2 G + \cot^2 E) \tag{5}$$

In case (ii) the angle E is derived from both the observed angles G and F, i.e. from $E = 180° - (G + F)$

$$\therefore \quad dE = -dG - dF$$

Substituting in (2)
$$\Delta_g \, dg = dG(\Delta_G + \Delta_E) + dF \, \Delta_E$$

which gives d_g in terms of independent errors dG and dF hence

$$(\Delta_g \sigma_g)^2 = \sigma_G^2(\Delta_G + \Delta_E)^2 + \sigma_F^2 \Delta_E^2 = \sigma^2\{(\Delta_G + \Delta_E)^2 + \Delta_E^2\}$$

if $\sigma_G = \sigma_F = \sigma$ or from (2)

$$\sigma_g^2 = g^2(\sigma'' \sin 1'')^2\{(\cot G + \cot E)^2 + \cot^2 E\} \tag{6}$$

Angular Measurement and Computation

In case (iii) none of the three angles used in computation is entirely independent of the other observed angles. Let the observed values of angles E, F, and G be M_E, M_F, and M_G, and let

$$M_E + M_F + M_G = 180 + k$$

Assuming equal weights for the observed angles, one third of the misclosure k will be assigned to each angle in the course of the adjustment, hence

$$E = M_E - \tfrac{1}{3}k, \quad F = M_F - \tfrac{1}{3}k, \quad G = M_G - \tfrac{1}{3}k$$

therefore

$$dE = dM_E - \tfrac{1}{3}dk; \quad dF = dM_F - \tfrac{1}{3}dk; \quad dG = dM_G - \tfrac{1}{3}dk$$

and substituting in (2)

$$\Delta_g\, dg = \Delta_G(dM_G - \tfrac{1}{3}dk) - \Delta_E(dM_E - \tfrac{1}{3}dk)$$

which reduces to

$$\Delta_g\, dg = dM_G(\tfrac{2}{3}\Delta_G + \tfrac{1}{3}\Delta_E) - dM_E(\tfrac{1}{3}\Delta_G + \tfrac{2}{3}\Delta_E) - dM_F(\tfrac{1}{3}\Delta_G - \tfrac{1}{3}\Delta_E)$$

Now since dg is in terms of independent errors

$$(\Delta_g \sigma_g)^2 = \sigma_G^2(\tfrac{2}{3}\Delta_G + \tfrac{1}{3}\Delta_E)^2 + \sigma_E^2(\tfrac{1}{3}\Delta_G + \tfrac{2}{3}\Delta_E)^2 + \sigma_F^2(\tfrac{1}{3}\Delta_G - \tfrac{1}{3}\Delta_E)^2$$

and if $\sigma_G = \sigma_E = \sigma_F = \sigma$ which is usual

$$(\Delta_g \sigma_g)^2 = \tfrac{2}{3}(\Delta_G^2 + \Delta_G\Delta_E + \Delta_E^2)\sigma$$

or

$$\sigma_g^2 = g^2(\sigma'' \sin 1'')\tfrac{2}{3}(\cot^2 G + \cot G \cot E + \cot^2 E) \tag{7}$$

The error in the final side of a chain of n triangles which have been fully observed and adjusted is given by

$$\sigma_g^2 = g^2(\sigma'' \sin 1'')\tfrac{2}{3}\{[\cot^2 G] + [\cot G \cot E] + [\cot^2 E]\}$$

where $[\cot^2 G]$ denotes the summation of 'G-type' angles. It must be remembered that all the above formulae take into account neither the effects of interlocking figures, nor the side equations and must therefore give approximate estimates only.

6.12.4. ERRORS IN ANGLES

If the triangular misclosures of n triangles of a network are $e_1, e_2, e_3, \ldots e_n$, the standard error of a triangle misclosure is given by $\sigma_T^2 = [e^2]/n$

Hence the standard error of one angle of a triangle is

$$\sigma_A = \sigma_T/3^{\frac{1}{2}} = ([e^2]/3n)^{\frac{1}{2}} \tag{8}$$

which is Ferrero's formula. Ferrero's formula does not consider the effect of adjusting each triangle to 180°.

Using the same notation as above, the adjusted angle E is given by

$$E = M_E - \tfrac{1}{3}(M_E + M_F + M_G)$$

and $dE = dM_E - \frac{1}{3}dM_E - \frac{1}{3}dM_F - \frac{1}{3}dM_G = \frac{2}{3}dM_E - \frac{1}{3}dM_F - \frac{1}{3}dM_G$

Therefore its standard error S_E is given by

$$S_E{}^2 = \tfrac{4}{9}\sigma_E{}^2 + \tfrac{1}{9}\sigma_F{}^2 + \tfrac{1}{9}\sigma_G{}^2$$

and if $\sigma_E = \sigma_F = \sigma_G = \sigma$

$$S_E = \sigma(\tfrac{2}{3})^{\frac{1}{2}}$$

whence on substituting from (8)

$$S_E = (2[e^2]/9n)^{\frac{1}{2}} \tag{9}$$

The formulae (8) and (9) can be used in estimating the effect of the bearing error accumulated through a chain of triangles, by considering a traverse through the chain and substituting in formulae (4) of §7.12.2.

6.13. THE COMPUTATION OF TRIANGULATION ON THE SPHEROID AND ON A SURVEY PROJECTION

An example of the computation of a triangle on the surface of a reference spheroid using Legendre's theorem and the Mid-latitude formulae is given in §§4.1.10 and 4.1.13; whilst the same triangle is computed on the Universal Transverse Mercator projection in §4.2.7.

7 Traverses

7.1. INTRODUCTION

A survey traverse consists of an orderly sequence of determinations of the lengths and directions of lines between points on the earth made to determine the positions of the points. According to the requirements of a particular survey the character of a traverse varies considerably. Traverses are identified in a variety of ways; according to the methods employed, as, tellurometer traverse; quality of results, as first-order traverse; purpose served, as cadstral traverse; according to form, as, closed traverse. In this chapter it is intended to consider only traverses where angles are measured by theodolite and distances either with precise electronic equipment, tapes or with subtense equipment.

7.1.1. LINEAR MEASUREMENT WITH A STEEL TAPE

The steel tape or band consists of a suitably graduated uniformly narrow strip of steel. Tapes are of various lengths from 50 feet to 500 feet or 10 metres to 200 metres; the width varies from $\frac{1}{16}$ inch to $\frac{3}{4}$ inch, or 2 mm to 1 cm; and thickness varies from 0·01 inch to 0·03 inch, or 0·2 mm to 0·6 mm. The heavier tapes are usually wider and the nominal length is between the outer edges of the handles; with the lighter tape the zero and end graduations are on the tape. Intermediate graduations are usually marked by small brass studs, various symbols being used. Sometimes the first 10 feet or 3 metres of the tape is more closely graduated than the rest of the tape. Tapes are usually wound on special steel crosses or metal reels and care should always be taken in unwinding and winding the tape on to its reel or cross.

7.1.2. METHOD OF USE

Two methods are commonly employed, termed here *ground taping* and *catenary taping*. The latter is described in §7.5.2 and is commonly employed when an accurate determination of the length of a survey line is required.

Ground taping involves only two men, one at each end of the tape. The rear man or 'follower' aligns the forward man or 'leader' on a ranging rod placed at the far end of the line; he then checks that the tape is straight and holds the rear zero graduation at the fine mark of the starting point. After checking that the tape is correctly positioned he shouts 'mark' and the 'leader' marks the forward end of the tape. Arrows are the most convenient method of marking this point, an arrow being a 12 inch long piece of wrought iron or steel wire pointed at one end and shaped into a circle about 1 in in diameter at the other end. A small piece of coloured rag tied to the ring makes the arrow easily visible. The next tape length (bay) is now measured in a similar manner, the 'follower' holding the zero mark against the arrow.

When the follower cannot see the forward ranging rod the leader should place himself on the line defined by a ranging rod placed at the starting point and the position of the previously placed arrow.

As the end of the tape is moved forward past the arrow the follower collects it and the number of arrows he holds will indicate the number of whole tape lengths measured. If the line is long the leader should start with 11 arrows 10 of these will have passed to the follower after 1 000 feet and are transferred forward after the 11th arrow has been placed. Each transfer will represent 1 000 feet measured.

Booking is illustrated by figures 7.10 and 7.11, the latter applying to catenary measurement.

Accuracy may be improved if the tape is used at the tension of standardization, the temperature at the time of measurement is recorded, and if the slope of the ground is measured. Tension is applied with a normal spring balance, temperature is measured with a thermometer preferably enclosed in a metal tube to provide some protection against damage, and the angle of slope of the ground is measured at each change of slope. In catenary taping the thermometer should be tied to a forked stick stuck into the ground half way along the tape length. The angle of slope is measured either by theodolite or if the accuracy sought is low with a hand clinometer, alternatively the slope correction may be deduced from a profile of the line obtained by spirit levelling.

7.1.3. RELATIVE ACCURACY

Ground taping over rough ground without corrections for tension and temperature but with a slope correction will give an accuracy of about 1/1 000. On specially prepared ground and if corrections are applied, see below, an accuracy of 1/10 000 may be reached. If catenary taping is employed as described in §7.5.2 an accuracy of 1/30 000 can be obtained without much difficulty. Catenary taping is slower than normal ground taping.

The most accurate method of measurement with a tape is with special base line equipment; For a description of such methods the reader is referred to Clarke's *Plane and Geodetic Surveying for Engineers* (London: Constable, 1964) where methods capable of reaching accuracies comparable with electromagnetic distance measuring equipment (see Chapter 13) are described. The methods are excessively slow and tedious.

7.1.4. ERRORS IN LINEAR MEASUREMENT

Details of the various sources of error in linear measurement are given in §7.2.3.

7.1.5. CORRECTIONS TO MEASURED DISTANCES

Any length measured with a steel tape may require correction for the following:

(i) Standard.
(ii) Temperature.
(iii) Tension.

(iv) Slope.
(v) Alignment.
(vi) Height above sea level.
(vii) Sag.
(viii) Scale of projection.

These corrections may be combined or treated separately; and it is essential that they are dealt with systematically.

7.1.5.1. *Correction for Standard and Temperature*

Two methods of standardizing a tape are in general use, in one the true length of the tape is given at a certain specified temperature, in the other the temperature is given at which the tape is its reputed length. For instance a tape may be said to be:

$$99 \cdot 992 \text{ feet at } 62°F \quad \text{or} \quad 100 \cdot 000 \text{ feet at } 74°F$$

In the former case the standard correction of $-0 \cdot 008$ is applied to each full tape length and a small correction in simple proportion for each part of a tape length. A temperature correction is also necessary although this correction will be small if the temperature of standardization is close to the normal field temperature.

In the latter case only a temperature correction is necessary, this correction arises as a result of the expansion of steel. The introduction of invar steel has reduced the significance of temperature corrections and any possible errors in estimation of temperature. The coefficients of expansion for 1 degF are about

$$0 \cdot 000\ 006\ 25 \text{ for ordinary steel}$$
$$0 \cdot 000\ 006\ 61 \text{ for Chesterman steel tapes}$$
$$\left. \begin{array}{l} 0 \cdot 000\ 000\ 3 \\ \text{to } 0 \cdot 000\ 000\ 4 \end{array} \right\} \text{for invar steel}$$

When a steel tape is used at a temperature in excess of its standard temperature a correction is *added* to the measured length, conversely when the temperature is below standard a correction is *subtracted*. For instance if when using the tape mentioned above to measure a line recorded as 300·000 feet the field temperature was 78°F then the true length of the line is given by:

Standard correction:
$$3 \times -0 \cdot 008 = -0 \cdot 024$$

Temperature correction:
$$0 \cdot 000\ 006\ 61 \times (78 - 62) \times 300 = +0 \cdot 031\ 728$$

True length:
$$300 \cdot 000 - 0 \cdot 024 + 0 \cdot 032 = 300 \cdot 008$$

or

Temperature correction:
$$0 \cdot 000\ 006\ 61 \times (78 - 74) \times 300 = +0 \cdot 007\ 932$$

True length:
$$300 \cdot 000 + 0 \cdot 008 = 300 \cdot 008$$

7.1.5.2. Correction for Tension

When a tape is standardized a tension is applied to the tape to maintain it straight along the standard base. If a different tension is used in the field a correction must be applied. When the tension is in excess of the standard tension a correction is *added* to the measured length, conversely when the tension is below standard a correction is *subtracted*.

By Hooke's law, if the elastic limit of the material is not exceeded, the ratio stress/strain, is equal to a constant, which is Young's modulus (E) of elasticity. An average value of $28 \cdot 5 \times 10^6$ lb/in² is usually taken for E.

By definition

$$E = \frac{T/A}{b/B}$$

where T is the tension in lb, B the unstretched length, b the stretch and A the cross sectional area in square inches (in²).

Hence for a change of tension from T_s to T_f

$$b = \frac{B}{AE}(T_f - T_s)$$

As AE is large compared to T_f the recorded length may be taken for the unstretched length without sensible error then

$$b = \frac{B_f}{AE}(T_f - T_s) \tag{1}$$

where b is the stretch, B_f the recorded length, T_f the field tension, T_s the standard tension, and A and E are as defined above.

7.1.5.3. Correction for Slope

All linear measurements are reduced to the horizontal, in practice they are rarely made on a perfectly horizontal plane and using normal trigonometrical relationships a line of recorded length B on a line sloping at an angle of θ is reduced to $B \cos \theta$. The correction is given by $B(1 - \cos \theta)$

or
$$2B \sin^2 \tfrac{1}{2}\theta \tag{2}$$

The expression $(1 - \cos \theta)$ is termed versine and tables of this function are readily available.

The slope correction is always *subtracted* from the recorded length

If instead of measuring the angle of slope the difference in height between two ends of the tape is determined the formula for correction takes the form

$$\frac{h^2}{2B} + \frac{h^4}{8B^3} \tag{3}$$

This expression is derived as follows:

$$B^2 = L^2 + h^2$$

where B is the recorded length, L the horizontal length and h the difference in height. Then $L - B$ is the required correction.

$$L/B = \left(1 - \frac{h^2}{B^2}\right)^{\frac{1}{2}}$$

$$= 1 - \frac{1}{2}\frac{h^2}{B^2} - \frac{1}{8}\frac{h^4}{B^4} - \cdots$$

$$\therefore \quad L - B = -h^2/2B - h^4/8B^3$$

as H is small compared to B.

7.1.5.4. *Correction for Alignment*

In this the angle between the line and the measured bay is measured and the correct length along the line obtained by a traverse computation treating the direction of the line as origin. The total ΔN will thereby be the correct length. See §7.6 for details of traverse computation.

7.1.5.5. *Correction for Height Above Sea Level*

It is customary for all precise work to reduce linear measurements to the spheroid. This effectively amounts to reducing all measurements to sea level.

If B is the recorded length at a height H above sea level, L the required length at sea level, and R is the radius of the spheroid at and in the direction of the measured line, then

$$B/(R + H) = L/R$$

$$B - L = B - \frac{BR}{R + H} = \frac{BH}{R + H} = \text{required correction} \qquad (4)$$

The correction for height above sea level is always *subtracted* from the recorded length.

7.1.5.6. *Correction for Sag*

When a tape is suspended between two points it takes up a catenary and the recorded length will be too great, this increase in length amounts to approximately:

$$\frac{w^2 B^3}{24 T^2} \qquad (5)$$

where w is the weight of one unit length of the tape, B the length supported in the same units, and T the tension in the same weight units as w, the correction is then given in the same units of length as B.

For example if w is in lb/ft of the tape, B the supported length in feet, T is the tension in lb, then the correction is in feet.

The correction for sag is always *subtracted* from the recorded length.

The correction may be derived as follows:

Let ACB (figure 7.1) represent a uniform tape supported at A and B two points at the same level, C being the lowest point of the tape.

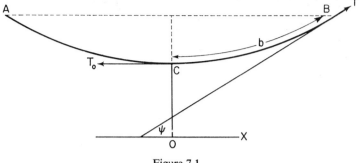

Figure 7.1

Let T_0 and T be the tensions at C and B respectively.

Resolving horizontally $\quad T \cos \psi = T_0$

Resolving vertically $\quad T \sin \psi = wb$

If we put $T_0 = wc$, where c is some constant

$$\tan \psi = b/c \quad \text{or} \quad b = c \tan \psi$$

hence $\quad db/d\psi = c \sec^2 \psi.$

Considering the curve ACB

$$dy/dx = \tan \psi \quad \text{and} \quad dy/db = \sin \psi$$
$$\therefore \quad dy/d\psi = \sin \psi \, c \sec^2 \psi = c \sec \psi \tan \psi$$

Integrating $\quad y = c \sec \psi + \text{constant}$

If $c = y$ when $\psi = 0$ the constant vanishes and

$$y^2 = c^2 \sec^2 \psi = c^2(1 + \tan^2 \psi)$$
$$\therefore \quad y^2 = c^2 + b^2 \tag{6}$$

Also $\quad T \cos \psi = T_0 = wc \quad \text{and} \quad T \sin \psi = wb$

Integrating $\quad T^2 = w^2(c^2 + b^2) = w^2 y^2$

and $\quad T = wy \tag{7}$

Also as $\quad dy/dx = \tan \psi = b/c = (y^2 - c^2)^{\frac{1}{2}}/c$

$$dy/(y^2 - c^2)^{\frac{1}{2}} = dx/c$$
$$\therefore \quad \cosh^{-1}(y/c) = (x/c) + \text{constant } A$$

Since $y = c$ when $x = 0$, $A = 0$

$$y = c \cosh(x/c) \tag{8}$$

From the above

$$db/dx = \sec \psi = (1 + \tan^2 \psi)^{\frac{1}{2}} = (c^2 + b^2)^{\frac{1}{2}}/c$$
$$db/(c^2 + b^2)^{\frac{1}{2}} = dx/c$$
$$\therefore \quad \sinh^{-1}(b/c) = (x/c) + \text{constant } B$$

Since $x = 0$ when $b = 0$, $B = 0$

$$b = c \sinh(x/c) \qquad (9)$$

Equations (6), (7), (8), and (9) are the intrinsic equations of the catenary.

From (9) $\qquad (x/c) = \sinh^{-1}(b/c)$

$$= b/c - b^3/6c^3 + 3b^5/40c^5$$

hence $\qquad x = b - b^3/6c^2 + 3b^5/40c^4 \qquad (10)$

From (6) and (7)

$$c^2 = T^2/w^2 - b^2 = (T^2/w^2)(1 - w^2 b^2/T^2)$$

Substituting this value for c^2 in (10)

$$x = b - (b^3 w^2/6T^2)(1 - w^2 b^2/T^2)^{-1} + 3b^5 w^4/40T^4(1 - w^2 b^2/T^2)^{-2}$$
$$\therefore \quad x = b - (b^3 w^2/6T^2)(1 + w^2 b^2/T^2 + \cdots) + 3b^5 w^4/40T^4(1 + 2w^2 b^2/T^2 + \cdots$$
$$= b - b^3 w^2/6T^2 - 20b^5 w^4/120T^4 + 9b^5 w^4/120T^4 - \cdots$$
$$= b - b^3 w^2/6T^2 - 11b^5 w^4/120T^4 \quad \text{to the first three terms.}$$

If the total chord length is X and the total tape length B

then $\qquad \tfrac{1}{2}X = \tfrac{1}{2}B - B^3 w^2/48T^2 - 11B^5 w^4/3\,840\,T^4$

or $\qquad X - B = -B^3 w^2/24T^2 - 11B^5 w^4/1\,920\,T^4$

The last term in this expression is negligible in practice except for the most precise work.

By a similar treatment it can be shown that the sag of the tape is $B^2 w/8T + B^4 w^3/128T^3$.

Determination of the sag factor

This may be established by weighing the tape without the handles and computing the values for different lengths using formula (5). Alternatively the following field method may be used.

Consider a 300 foot tape:

(i) On level ground suspend the tape between two stakes about 300 feet apart in three catenaries each of 100 feet. Let the recorded length be L_1.
(ii) Again measure the distance between the stakes but this time in two catenaries of 150 feet. Let the recorded length be L_2.
(iii) Again measure the distance between the stakes but this time in two catenaries, 1 of 100 feet, and 1 of 200 feet. Let the recorded length be L_3.

If X is the chord length between the stakes

$$X = L_1 - 3(w^2 100^3/24T^2) \quad \text{without sensible error}$$
$$X = L_2 - 2(w^2 150^3/24T^2)$$
and $\qquad X = L_3 - w^2 100^3/24T^2 - w^2 200^3/24T^2.$

Let the required factor $w^2 100^3/24T^2 = K$.

Then $X = L_1 - 3K = L_2 - 6\cdot 75K = L_3 - K - 8K$ from which K can be found and a check obtained.

7.1.5.7. *Correction for Scale of Projection*

Details of this correction are given in §4.2.7.1

7.1.6. COMPUTATION OF CORRECTIONS

The method of reduction of linear measurement is illustrated in the examples of booking given in figures 7.10 and 7.11.

7.2. SOURCES OF ERROR

7.2.1. INTRODUCTION

Errors arise in each line (leg) of a theodolite transverse from the measurement and adjustment of the angles, and from the measurement and adjustment of the distances. From the true position angular errors produce a displacement at right angles to the direction of the line and linear errors produce a displacement along the direction of the line.

7.2.2. SOURCES OF ERROR IN THE MEASUREMENT OF TRAVERSE ANGLES

7.2.2.1. *Defective Centring of the Theodolite.*

See figure 7.2, where an angle BA'C is measured instead of angle BAC.

If A' falls within the sector bAc the error is the sum of the angles subtended by A'A at B and C and is positive in sign.

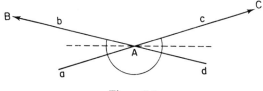

Figure 7.2

If A' falls within the sector aAd the error is again the sum of the subtended angles but is negative in sign.

If A' falls within the sectors bAa and cAd the error is the difference of the subtended angles, being positive or negative according to which side of the arc through BAC it falls.

For any position on the arc BAC, angle BAC = angle BA'C and the observation is error free.

The subtended angles are inversely proportional to the lengths AB and BC, hence the importance of accurate centring when lines are short.

7.2.2.2. *Defective Centring of the Signal*

This may be treated in the same manner as §7.2.2.1 above, the error being proportional to the length of the line and being the perpendicular displacement of the signal from the correct line of sight.

7.2.2.3. Defective Levelling

This error was discussed in greater detail in §3.10.1 under the effects of maladjustments in the theodolite. Where I_v is the component of the displacement of the main axis of rotation at right angles to the line of sight, the error introduced in the horizontal circle reading is $I_v \tan H$ where H is the vertical angle. It is important to note that this error is not cancelled out with change of face and that the greater the difference in elevation between two pointings the greater will be the error introduced into the horizontal angle.

7.2.2.4. Defective Operation of the Theodolite

This may be due to:
  (i) The theodolite not secured to the tripod.
  (ii) The shifting head being unclamped.
  (iii) The lower plate being unclamped.
  (iv) The use of the wrong tangent screw.

If any of these circumstances values of different faces will not agree.

7.2.2.5. Residual Parallax

An error will arise if the adjustment to eliminate parallax in the telescope is not completed. With different positioning of the eye, different objects can be bisected with the cross hairs. The magnitude of the error is inversely proportional to the length of the line. Whilst it is possible for this error to be compensating this should not be relied upon and the parallax adjustment must be completed before observing.

7.2.2.6. Residual Errors of Adjustment

These are removed by adoption of a suitable observing procedure (see §3.11). In brief, by changing face, swinging in different directions and use of different zeros.

7.2.2.7. Observational Errors

These arise from:
  (i) Inaccurate bisection of the signal.
  (ii) Non-verticality of the signal.
  (iii) Displacement of the ground mark.

The magnitude of these errors is inversely proportional to the length of line.

  (iv) Errors of circle reading.
  (v) Errors of booking.

These should become apparent during the reduction of the observations which should be done in the field.

7.2.2.8. Errors Due to Natural Causes

These arise from:
  (i) Shimmer.
  (ii) Refraction.
  (iii) Wind.
  (iv) Unequal heating of the instrument and tripod.

7.2.3. SOURCES OF ERROR IN THE MEASUREMENT OF DISTANCE WITH A STEEL TAPE

7.2.3.1. Errors of Deformation

If the line AB is (figure 7.3) measured along AD, DB the error introduced is

$$(b_1 - x_1) + (b_2 - x_2) = \frac{d^2}{2b_1} + \frac{d^2}{2b_2} + \cdots$$

if $b_1 = b_2 = \frac{1}{2}B$ $\qquad (b_1 - x_1) + (b_2 - x_2) \simeq 2d^2/B$

This error will always make the recorded measurement too great being constant in sign but variable in magnitude.

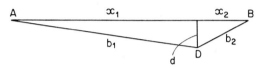

Figure 7.3

7.2.3.2. Errors in Measurement of Slope

The correction for slope is given by $b(1 - \cos S)$, where S is the slope angle. Let dS_1 be the error in S arising from a constant error in the instrument measuring the slope angle, then

$$\text{Error due to } dS_1 = b \sin S \, dS_1 \sin 1''$$

Let dS_2 be the error in S arising from random errors in measurement of the slope angle, then

$$\text{Error due to } dS_2 = b \sin S \, dS_2 \sin 1''$$

Errors arising from dS_1 will be constant in sign but varying in magnitude with changes in b and S. For different instruments dS_1 may change in sign and magnitude.

Errors arising from dS_2 will be variable in sign and magnitude.

7.2.3.3. Errors in Measurement of Temperature

The correction for temperature is given by bct where c is the coefficient of expansion of the material of the tape, and t is the difference of temperature from that of standardisation of the tape.

Let dt_1 be the error in t arising from a constant error in the thermometer, then

$$\text{Error due to } dt_1 = bc \, dt_1$$

Let dt_2 be the error in t arising from random errors in reading the thermometer, then

$$\text{Error due to } dt_2 = bc \, dt_2$$

Errors arising from dt_1 will be constant in sign but varying in magnitude according to changes in b. For different thermometers dt_1 may change in sign and magnitude.

Errors arising from dt_2 will be variable in sign and magnitude.

7.2.3.4. *Errors in Measurement of Tension*

The correction for tension is given by bF/AE, where F is the difference of tension from that of standardization, A the cross sectional area of the tape, and E Youngs modulus of elasticity. Let dF_1 be the error in F arising from a constant error in the equipment applying the tension, then

$$\text{Error due to } dF_1 = b\, dF_1/AE$$

Let dF_2 be the error in F of a random nature, then

$$\text{Error due to } dF_2 = b\, dF_2/AE$$

Errors arising from dF_1 will be constant in sign but varying in magnitude according to changes in b, A, and E. For different equipment dF_1 may change in sign and magnitude.

Errors arising from dF_2 will be variable in sign and magnitude.

If measurement is made in catenary the correction is given by $w^2b^3/24F^2$ and a second error of the form $(w^2b^3/12F^3)\, dF$ will arise for both dF_1 and dF_2. Here w stands for the weight per unit length of the tape.

7.2.3.5. *Errors Due to Faulty Standardization*

These errors will be constant and of the same sign for one tape until re-standardization.

Let db be the error per unit of measurement, then in a line of length B the error due to db is $B\, db$.

7.2.3.6. *Errors Due to Faulty End Reading and Setting*

These errors are usually variable in sign and magnitude, magnitude depending upon the methods of reading used. The possibility of an error constant in sign and magnitude does exist unless care is taken during reading and setting.

7.2.3.7. *Errors in booking*

These are variable in sign and magnitude. Check taping of the line in different units is essential if accurate results are required. The danger of omitting a complete tape length should be carefully guarded against.

7.3. EQUIPMENT FOR TRAVERSING

Consideration of the sources of error in traverses has affected the design of equipment for traversing. To reduce errors arising from defective centring of theodolite and signal the three tripod equipment was devised. Here the tribrach of the theodolite can be separated from the main body of the instrument (see figure 7.4). In place of the theodolite a signal can be placed in the tribrach, accuracy of register being automatic. Similarly the signal and an optical plummet are interchangable. The fine mark at the centre of the signal is at the same height above the tribrach as the transit axis of the theodolite. If taping is to be made in catenary a measuring head will fit into the tribrach in the same manner as the theodolite or accessories. In exchanging the theodolite and

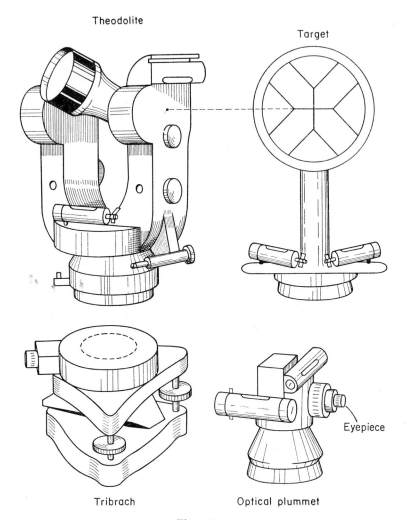

Figure 7.4

accessories there is no necessity to do anything which might disturb the tripod. The procedure being to slacken a single clamp, lift out the theodolite, replace it with the accessory and tighten a single clamp.

A suggested list of three tripod equipment is as follows:

A theodolite with detachable tribrach
A tripod for the theodolite
Four extra tripods
Four extra tribrachs
Four signals

Additional equipment is required if precise taping in catenary is to be carried out. In using this equipment care must be taken to ensure that the signals do not become twisted.

It will often happen that three tripod equipment is not available. In this case good results can be obtained if angular observations are taken to plumb bob strings threaded through strips of paper (see figure 7.5). A danger is in letting the string become slack so that the ground mark is supporting the bob, heavy bobs are prone to this fault. Wind may disturb the bob and make observations difficult, stability can be achieved by packing a few stones around the bob after centring.

The ground mark may be designed to receive a signal; a length of 1½ inch gauge steel pipe set in a concrete block makes an excellent mark into which a ranging rod can be fitted.

If angular observations are made to ranging rods the rod must be plumbed and centred over the ground mark; special supports can be purchased, which

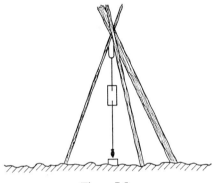

Figure 7.5

are useful when traversing along roads, on rough ground a support made of three forked sticks is easier to use. Where possible an arrow threaded through a piece of paper and held on the ground mark makes a good target. For very short lines a matchstick held in position over the ground mark with plasticine can be used, a rod nearby may help in identification.

The equipment required for taping depends upon the precision sought. If taping is to be in catenary and the precision is to be high the following additions to the three tripod equipment listed above are suggested:

Three extra tripods
Three extra tribrachs
Two straining trestles
A field weight
Three measuring heads
Four thermometers
Two tape grips
A 20 metre field tape graduated throughout and finely graduated at one end
A 100 metre field tape finely graduated at both ends
A 100 foot field tape graduated throughout and finely graduated at one end
A 300 foot field tape finely graduated at both ends

Six 8 foot ranging rods, two being fitted with sliding brackets for intermediate support of the tape

A supply of woven chord for applying tension.

It will often happen that the first five items listed above are not available. In this case the taping will be carried over marks about 3 feet above ground level. These marks may consist of either stout wooden posts driven into the ground and strutted along the line of taping (figure 7.6) or plane table tripods with a short wooden mark tapped lightly into the recess for the locking screw. In either case tension is applied by spring balance, usually the theodolite is plumbed over one of the posts and measurement is made to the theodolite

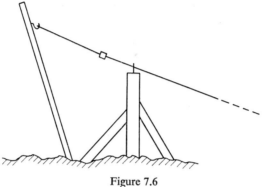

Figure 7.6

transit axis. The tape need not be finely graduated at the ends but should be coarsely graduated throughout; parts of a coarse graduation, say 50 cm, are measured with a scale held against the transit axis.

If taping is made along the ground, taping arrows are required to mark the end of each tape length, tension is applied with spring balance or by hand (approximately 10 lb) according to the precision sought. Thermometers are not required unless the surface over which the taping is taking place is sufficiently smooth to enable a high precision to be obtained in the measurement. The tape should be standardized at a temperature around that being used in the field.

7.4. FIELD WORK IN THEODOLITE TRAVERSING

7.4.1. ACCURACY OF THE TRAVERSE

The final accuracy of a traverse depends upon the methods and equipment used and upon the care and honesty of the surveyor. Having settled the accuracy required, the surveyor should choose methods such as to make the standards of linear and angular measurement about the same. For instance, if ground taping is proposed, tension to be applied by hand, then temperature corrections are meaningless, equally angles observed to 1" are unnecessary. In angular work a useful relationship to remember is that: 1 minute of arc subtends 1 inch at 300 feet.

Some caution is necessary however as an error in a traverse angle is propogated throughout the whole traverse. For further treatment of this point the reader is referred to the §7.12 on propogation of error in traverses.

7.4.2. RECONNAISSANCE

The following precepts are adopted for all traverses:

(i) It is of the utmost importance to keep lines as long as possible. If this is not done no amount of care in angular work will give accurate results.
(ii) If a traverse is to be run for a specific purpose say for detail survey, the stations of the traverse should be placed in positions favourable for the subsequent work, even if this may mean having lines shorter than possible.
(iii) Lines should be suitable for linear measurement.
(iv) Stations should be located on firm level ground.

During the reconnaissance a clear diagram is drawn approximately to scale. Stations are given a unique reference number which is maintained throughout the survey. As soon as the position of a station is decided a written description and a diagram is prepared. Descriptions should be in good English, and understandable without reference to the diagram (see figure 7.7).

The nature of the ground mark depends upon the purpose of the survey. In soft ground a 9 inch long wooden peg, $1\frac{1}{2}$ inch square will remain in position for several months. In roads a 3 inch pipe nail driven flush is equally suitable and may remain in position until resurfacing takes place. In tropical countries wooden pegs may be destroyed by termites in a few days and instead a hole should be drilled and filled with concrete into which a nail is placed.

Where permanent marks are required these should be of concrete reinforced with a steel pipe or bar, three types are illustrated in figure 7.8. The station reference number can be imprinted in the wet concrete.

7.4.3. OBSERVATIONS AND MEASUREMENTS

For a small traverse it is preferable to observe all the angles before taping begins. On a long traverse separate parties measure the angles and distances, or if this is not practicable one party will measure both angles and distances, as outlined below. When distances are measured with electronic equipment this is the usual practice and the procedure is discussed later in §7.11 on precise traversing.

The reader is referred to the relevant sections of Chapters 6 and 13 on how the angular and linear measurements are made. Good organization of the field work is essential and the surveyor should aim to be working continuously at skilled work. The progress of work in the field can best be shown by an example.

Example: Figure 7.9 represents a typical portion of a traverse, taping being in catenary to wooden posts from the transit axis of the theodolite. The party consists of the surveyor S, two chainmen C1 and C2 and a minimum of four labourers.

338   *Practical Field Surveying and Computations*

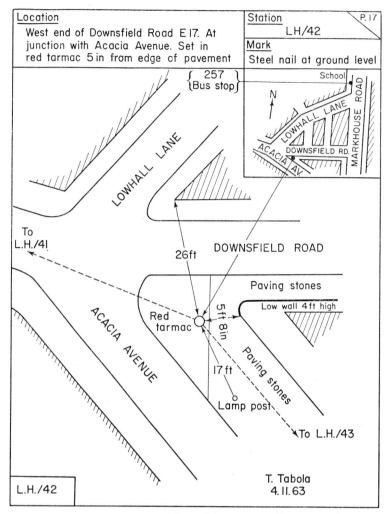

Figure 7.7

The angle at 8 has just been observed by S line 7 to 8, previously had been measured by S and C1, a tripod has been set up at 9 by C2.
What follows is

  (i) S aligns posts at 8/1, 8/2, and 8/3, a tripod is left at 8/2.
  (ii) A labourer at 6 removes the signal and tripod and brings them forward to 8.
  (iii) C2 takes a tripod, optical plummet and signal forward to 10.
  (iv) On completion of (i) S measures the slope to 8/1 and with C1 measures bay 8 to 8/1.
  (v) S replaces the theodolite at 8 with a signal and sets the theodolite over the peg at 8/2.
  (vi) S measures the slope to 8/1 and with C1 measures bay 8/2 to 8/1

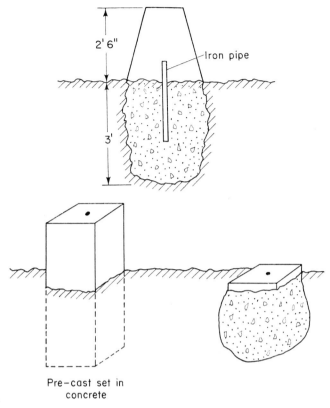

Figure 7.8

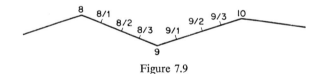

Figure 7.9

(vii) Similarly bay 8/2 to 8/3 is measured.
(viii) S moves to 9 and replaces the signal with the theodolite.
(ix) S measures the slope to 8/3 and with C1 measures bay 9 to 8/3.
(x) Only now is the tripod at 8/2 moved forward to the approximate position of 9/2.
(xi) S observes the angle at 9.

The procedure is now repeated for line 9 to 10

7.4.4. RECORDING AND REDUCTION OF OBSERVATIONS

Figure 7.10 illustrates a page of a typical field book where ground taping is being carried out along a carefully prepared surface. Figure 7.11 illustrates the page of a typical taping field book where taping is made in catenary. It will be noted that in both cases the reduction of the observations and measurements is

| TRAVERSE | TS 64 | SURVEYOR | JRH | BOOKER | KAK | INSTRUMENT | CTS 16117 | TAPE | C 24 | DATE | 24.10.62. | PAGE 8 |

HORIZONTAL ANGLES			STNS 27 and 28		VERTICAL ANGLES			DISTANCES	Lines 27→28		Ground Taping 28→29			
Station	Face	Reading ° ′ ″	Mean ° ′ ″	Reduced to R.O. ° ′ ″	Face	Reading ° ′ ″	Reduced Angle ° ′ ″	Approx. Dist	GRID			Back	Fore	Difference	Temp
At 27.26	L	90 26 48	90 26 45		L	27→ 00 17	c/s (a) 00 17	165m	ⓐ 24·1	30·0	30·0	5·00	20·24	15·24	65
28	L	157 35 42	157 35 36	67 08 51	R	179 45	00 15	㉗	ⓑ 12·4	30·0	30·0	1st measure		165·24m	
28	R	337 35 30				27 c/s(a)	00 16		(metres)			corrections		−0·155	
26	R	270 26 42				c/s(a)→c/s(b)								165·085	
					L	03 16 30	03 16 30		c/s						
					R	176 46 00	03 14 00		(a) 27	24·1	Check				
26	L	05 38 24	05 38 27			c/s(a) c/s(b)	03 15 15		c/s(a) c/s(b) 28	18·3	165·2				
28	L	72 47 24	72 47 22·5	67 08 53·5		c/s(b)→28			c/s(b)			2nd measure		542·13 ft	68
28	R	252 47 21			L	357 26 40	02 33 20	㉘	100	100	(feet) 100	20·00	62·13	42·13	
26	R	185 38 30			R	182 35 00	02 35 00					542·13 500	500 42	152·400 12·802	
						c/s(b)→28	02 34 10		Corrections 1st		2nd		0·13	·040	
		Mean							Slope −ve	0·000	0·000	2nd measure corrections		165·242m −0·152	
26		00 00 00							Slope	0·029	0·029				
28		67 08 52							Slope	0·123	0·123			165·090	
									Temp	−0·003	Nil				
						27→28 mean	165·09		Sum	0·155	0·152				
At 28.27	L							261m							
29	L														

Figure 7.10

Bay	Slope		Measure	Sag	Temp	Length	Corrections	Line
8	L 84°16'20"	05°43' 40"	(ft)					
↓ 8/1	R 95°43'30"	05°43' 30"	299·913	$\frac{3}{100}$	78°	Std and Temp	+ 0·0198	⑧
8/1	L 87° 27'	2° 33'				Slope	−1·4970	
↓ 8/2	R 92° 33'	2° 23'	299·971	$\frac{3}{100}$	79°	Sag	−0·0378	
8/2	L 89° 21'	0° 39'		$\frac{1}{100}$		Std and Temp	+ 0·0218	
↓ 9	R 90°39'	0° 39'	146·172	$\frac{1}{45}$	79°	Slope	−0·0873	⑨
						Sag	−0·0378	
			746·056			Std and Temp	+ 0·0106	
			−1·641			Slope	−0·0192	
			744·415			Sag	−0·0139	
						Sum	−1·6408	
						744·42		

Figure 7.11

carried out in the field book. A check is easily carried out by abstracting the field observations onto sheets printed identical to the field books but overprinted 'check'.

7.5. THE CADASTRAL OR TOWN SURVEY CONTROL TRAVERSE

Here a misclosure vector in feet of $0·007L^{\frac{1}{2}}$ would be aimed at, with maximum permissible value of $0·014L^{\frac{1}{2}}$. (L being the length of the traverse in feet.) Concrete pillars or nails driven into the road surface, mark the stations in both cases full descriptions are prepared.

7.5.1. EQUIPMENT

(i) Theodolite—Mean reading, glass arc, 1", optical micrometer, equipped for star observations.
(ii) Tapes—300 ft $\frac{1}{8}$" steel graduated throughout, and 30 metre $\frac{1}{4}$" steel graduated throughout.
(iii) Eleven taping arrows, tape grips, spring balance, 1 ft steel rule graduated in 1/100ths of a foot.
(iv) Four plumb bobs.
(v) Six ranging rods.
(vi) Two thermometers.
(vii) Quantity of 18", 1" × 1" wooden pegs, mallet, and nails.

It is assumed that all traverse stations have been pillared.

7.5.2. METHOD OF WORK

7.5.2.1. *Taping*

The wooden pegs are driven in the ground every 300 feet or slightly under 300 feet. These pegs are aligned using the theodolite and a fine mark made on the peg on line. The slope angle to the peg is measured, the tape unwound, tension is to be applied to the tape at the theodolite and the zero mark held against the fine mark on the peg. Two ranging rods are placed at the 100 ft and 200 ft marks and aligned so that the lowest point of an attached loop of chord supporting the tape lies on the line between the theodolite and the peg. When the zero mark on the tape is in coincidence with the fine mark on the peg the forward chainman chants a continuous 'on-n-n-'; the surveyor at the instrument end now checks that the required tension is being applied (at both ends of the tape straining is achieved by use of ranging rods), and then marks the tape with a pencil using the centre of the transit axis as a reference. Using the nearest whole foot mark on the tape and the 1 ft steel rule the parts of a foot are determined and the length of the bay recorded. Temperature is recorded, the theodolite moved forward 600 feet and the process repeated.

As a check the line is measured on the ground with the metre tape.

7.5.2.2. *Angular Observations*

These are taken to plumb bob strings unless the lines are very long when a larger signal is permissible. (The average length of line will be about 1 000 feet and no line should be under 400 feet.)

A minimum of two zeros are required and on each zero two pointings one on each face are made. When one line is short an extra zero should be observed after recentring theodolite and signals. The means of each zero should agree to within 10″ otherwise an extra zero should be taken and the overall mean determined.

7.5.2.3. *Azimuth Control*

Exmeridian or hour angle methods are consistent with the standard required. Sun observations are not satisfactory, ideally four pairs of stars (four East and four West) should be taken but if time is short two pairs will suffice.

Azimuths should be observed after about 20 stations or more frequently if lines are short. Observations should be taken to and from any triangulation stations near the line of the traverse.

7.6. TRAVERSE COMPUTATION

Once the reduction of the field observations is completed an abstract can be prepared. This is only necessary if there are many observations. Mistakes may be made in any of the copying stages involved and it is frequently better to work from the original field books which have been compared with the checking sheets.

For a surveyor accustomed to computing triangulation the computation of a traverse requires a change of approach. Triangulation, except for base measurement, is almost self-checking and it is almost impossible for gross errors to occur. With traverses this is very different, the computation of a 30 line traverse involves before a check is obtained 60 entries into natural tables, the determination of 60 products and the addition of 2 columns each of 30 figures.

It is important to notice the difference between a computation which is self-checking and one which is checked by some form of duplicate computation. Agreement in the first case is more significant particularly when the duplicate computation is performed consecutively by the same computor. It is quite pointless to attempt to check a traverse by working through the figures of the original computation. For instance if in meaning the taped distances 233·717, 233·724, 233·726 the original computer had written 223·722 the checker will frequently mark this as correct, his attention being directed to the less significant last two figures. This is particularly liable if the magnitude of the error is unknown. Even with truly duplicate computations there are certain types of error which are liable to be made time after time. For instance if there are 20 or 30 bays of 299· .... followed by one of 279· .... this bay will frequently and repeatedly be written as 299· ...

In duplicate computations the following points should be noted:

(i) Each set of computations should originate from the field records.
(ii) Original and check computations should be by different computors in different coloured inks.
(iii) Comparisons should be allowed only at the end of definite stages.
(iv) Corrections should be made in ink of a third colour.
(v) Wherever possible the check computation should be by a different method to the original computation.

The steps in the computation of a traverse are:

(i) Establishment of bearing for the first and last line of the traverse.
(ii) Computation of bearings of the intermediate lines.
(iii) Adjustment of bearings of the intermediate lines.
(iv) Computation of differences in eastings and northings for each line.
(v) Check computation of (iv).
(vi) Adjustment of the values obtained in (iv).
(vii) Determination of adjusted coordinates.

Considering these stages in terms of the computation of the closed traverse loop A, B, C, D, E, F, G, H, and the traverse D, I, K, H between two fixed lines CD and HG see figure 7.12.

Step i, see Example 3, the bearings of lines C to D and H to G have been computed from the coordinates of these points

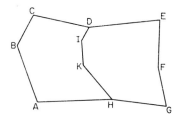

Figure 7.12

Step ii, see Example 3, the bearings of the intermediate lines are determined as follows:

Bearing D to C 272° 04' 27"	(a)
Clockwise angle CDI 279° 53' 01"	(b)
Unadjusted bearing D to I 191° 57' 28"	(a) + (b)
Unadjusted bearing I to D 11° 57' 28"	(a) + (b) − 180°
Clockwise angle DIK 166° 26' 46"	
Unadjusted bearing I to K 178° 24' 14"	

With a little practice the fourth line can be omitted. It should be remembered that if 180 is subtracted from an angle less than 180 the result is the same as if 180 were added to the angle.

Step iii, the bearing of the closing leg will usually differ from the value brought through the traverse. If the difference is e'' and there are n lines the adjustment to the first bearing is e/n'', to the second $2e/n''$ and so on. This is the same as adjusting each angle by e/n''. In Example 3 the misclosure was 15" over four stations, the adjustments to the bearings are rounded off to the nearest second.

Example 1. Loop Traverse

Line	Length	At	Back station	Fore station	° ' " Clockwise angle
AB	1 180·79	A	H	B	241 48 11
BC	1 379·34	B	A	C	237 09 54
CD	2 988·49	C	B	D	243 24 54
DE	3 185·24	D	C	E	173 25 54
EF	1 489·82	E	D	F	288 09 47
FG	1 085·29	F	E	G	137 39 53
GH	3 267·45	G	F	H	299 07 41
HA	3 162·23	H	G	A	179 13 28

If the traverse closes on itself as in Example 1 the internal angles will sum to $n \cdot 180 - 2 \cdot 180$ and the external angles will sum to $n \cdot 360 - (n \cdot 180 - 2 \cdot 180)$. Consequently the angles can be adjusted before the bearings are derived. The exact closure of the bearings after the adjustment of the angles will prove the completeness of the adjustment.

It is usual to spread the error equally between the observed angles but it should be borne in mind that an angle measured between two stations one or both of which are relatively close is more open to doubt than one which is observed between two distant points.

Step iv, the computation of a traverse is the reverse process to that of finding bearings and distances from coordinates.

Let difference in eastings be ΔE, difference in northings ΔN, the length of line s, and the bearing of line a. Then $\Delta E = s \sin a$ and $\Delta N = s \cos a$.

The machine computation is illustrated by Examples 1 and 3.

Step v, the check computation of a traverse conveniently performed by use of auxiliary bearings. The method is equally suitable for both logarithmic or machine computation.

Example 1. Closed Traverse Loop; Adjustment by Bowditch

At	Angle	Bearing	Length (ft)	Sine bearing / Cosine bearing	Difference eastings +	Difference eastings −	Difference northings +	Difference northings −	Eastings (ft)	Northings (ft)		Data
A	241° 48' 11" +2"	331° 29' 48" (Data)	1 180.79	−0.477 210 / +0.878 789		563.48 +0.07	1037.66 +0.08		131 006.41	206 474.89	A	
B	237° 09' 54" +3"	28° 39' 45"	1 379.34	+0.479 650 / +0.877 460	661.60 +0.08		1 210.32 +0.09		130 443.00	207 512.63	B	
C	243° 24' 54" +2"	92° 04' 41"	2 988.49	+0.999 343 / −0.036 261	2 986.53 +0.18			108.37 +0.20	131 104.68	208 723.04	C	
D	173° 25' 54" +2"	85° 30' 37"	3 185.24	+0.996 932 / +0.078 280	3 175.47 +0.19		249.34 +0.22		134 091.39	208 614.87	D	
E	288° 09' 47" +2"	193° 40' 26"	1 489.82	−0.236 395 / −0.971 657		352.19 +0.09		1 447.59 +0.10	137 267.05	208 864.43	E	
F	137° 39' 53" +3"	151° 20' 22"	1 065.29	+0.479 619 / −0.877 477	520.53 +0.07			952.32 +0.08	136 914.95	207 416.94	F	
G	299° 07' 41" +2"	270° 28' 05"	3 267.45	−0.999 966 / +0.008 169		3 267.34 +0.20	26.69 +0.22		137 435.55	206 464.70	G	
H	179° 13' 28" +2"	269° 41' 35"	3 162.23	−0.999 986 / −0.005 357		3 162.19 +0.19		16.94 +0.22	134 168.41	206 491.61	H	
Sum	1799° 59' 42"	A 241° 48' 13"	17 738.65		7 344.13	7 345.20	2 524.01	2 525.22	131 006.41	206 474.89	A	
Required	1800° 00' 00"	A → B 331°29'48"		Misclosure	+1.07	1.07	+1.21	1.21				
Misclose	18"			Adjustment								
Adjustment	+2".25			Bowditch factor per 1000 ft	0.0503		0.0682					

Total misclosure $(1^2.07 + 1^2.21)^{1/2} = 1.62$ Proportional misclosure $1/10,950$

Computed by J.R.H.
4.12.63.

Required angle sum
$8 \times 360°$
$-6 \times 180°$
$= 10 \times 180°$
$= 1,800°$

Diagram — External angles

Example 2. Closed Traverse Loop; Adjustment by Transit

At	Difference in eastings (ft) +	Difference in eastings (ft) −	Difference in northings (ft) +	Difference in northings (ft) −	Eastings (ft)	Northings (ft)	To	
					131 006·41	206 474·89	A	Data
A		563·48 +0·04	1 037·66 +0·25		130 442·97	207 512·80	B	
B	661·60 +0·05		1 210·32 +0·29		131 104·62	208 723·41	C	
C	2 986·53 +0·22			108·37 +0·03	134 091·37	208 615·07	D	
D	3 175·47 +0·23		249·34 +0·06		137 267·07	208 864·47	E	
E		352·19 +0·02		1 447·59 +0·34	136 914·90	207 417·22	F	
F	520·53 +0·04			952·32 +0·23	137 435·47	206 465·13	G	
G		3 267·34 +0·24	26·69 +0·01		134 168·37	206 491·83	H	
H		3 162·19 +0·23		16·94 0·00	131 006·41	206 474·89	A	

Sum 7 344·13 7 345·26 2 524·01 2 525·22
Misclosure   1·07   1·21
Adjustment   +1·07   +1·21
Transit factor +0·0728   +0·2396   Computed by J.R.H.
per 1,000 ft   5.12.63

Total misclosure 1·62   Proportional misclose $\frac{1}{10,950}$

Example 3. Traverse between Fixed Lines

Line	Length	At	Back station	Fore station	Clockwise angle ° ′ ″
D I	1,202·04	D I	C D	I K	279 53 01 166 26 46
I K	448·33	K	I	H	149 30 37
K H	589·74	H	K	G	122 33 13

Bearing A→B	331° 29′ 48″	
Coordinates of A	Eastings (ft)	Northings (ft)
	131 006·41	206 474·89

Example 3. Traverse between Fixed Lines; Adjustment by Bowditch

At	Angle	Bearing	Length (ft)	Sine bearing / Cosine bearing	Difference eastings +	Difference eastings −	Difference northings +	Difference northings −	Eastings (ft)	Northings (ft)	To
C		92° 04' 27" (Data)							134 091·39	208 614·87 (Data)	D
D	279° 53' 01"	191° 57' 28" +04" 191° 57' 32"	1 202·04	−0·207 210 −0·978 297		249·07 +0·20		1175·95 +0·28	133 842·52	207 439·20	I
I	166° 26' 46"	178° 24' 14" +08" 178° 24' 22"	448·33	+0·027 815 −0·999 613	12·47 +0·08			448·16 +0·11	133 855·07	206 991·15	K
K	149° 30' 37"	147° 54' 51" +11" 147° 55' 02"	589·74	+0·531 144 −0·847 281	313·24 +0·10			499·68 +0·14	134 168·41 (check) 134 168·41 (data)	206 491·61 (check) 206 491·61 (data)	H
H	122° 33' 13"	90° 28' 04" +15" 90° 28' 19" (Data)									G
Sum	718° 23' 37" 358° 23' 37" 92° 04' 27"	Misclosure 15" Adjustment +03"·75	2 240·11		325·71 +76·64−77·02 = −0·38 +0·38 0·1696	249·07 −0·38		2123·79 −2123·79+2123·26 = −0·53 +0·53 0·2366	+77·02	−2123·26	
Sum	90° 28' 04"			Bowditch factor per 1,000 ft							

Total misclose $(0^2·38 + 0^2·53)^{\frac{1}{2}} = 0·65$ Proportional misclose $\frac{1}{3500}$

$\ell_1(\sin\alpha_1, -\cos\alpha_1) + \ell_2(\sin\alpha_2, -\cos\alpha_2) + \ldots + \ell_n(\sin\alpha_n, -\cos\alpha_n) = +2,200·417$

Total difference eastings = −Total difference northings = +2,200·43

Computed by J.R.H
15·12·63

347

348   *Practical Field Surveying and Computations*

Let $S = s2^{-\frac{1}{2}} \sin(a + 45)$

and $C = s2^{-\frac{1}{2}} \cos(a + 45)$

Expanding $S = s2^{-\frac{1}{2}}(\sin a \cos 45 + \cos a \sin 45)$

hence $S = s/2(\sin a + \cos a)$

Similarly $C = s/2(\cos a - \sin a)$

$$S + C = s \cos a = \Delta N$$

and $S - C = s \sin s = \Delta E$

Example 4 illustrates the computation, the method being to add 45° to the starting bearing and then derive new bearings each 45° greater than used in the original computation. Each length is multiplied by 0·707 106 8 or if logarithmic methods are used 9·849 485 0 is added to log s.

It is not necessary to work out both S and C as

$$\Delta N + \Delta E = 2 \quad \text{or} \quad \Delta N - \Delta E = 2C.$$

Example 4. Check Computation of Closed Traverse Loop. Use of Auxiliary Bearings.

At	Angle	Auxiliary bearing	Length (ft) Length/√2	Sin aux B Cosine aux B	S +	S −	C +	C −	S−C +	S−C −	S+C +	S+C −	To
A	Data 241° 48' 11"	331° 29' 48" +45° 00' 00" 16° 29' 48"	1180·79 834·94	+0·283 960 +0·958 837	237·09		800·57		563·48		1037·66		B
B	237° 09' 54"	73° 39' 42" +3" 73° 39' 45"	1379·34 975·34	+0·959 622 +0·281 295	935·96		274·36		661·60		1210·32		C
C	243° 24' 54"	137° 04' 36" +5" 137° 04' 41"	2988·49 2113·18	+0·681 001 −0·732 282	1439·08			1547·45	2986·53			108·37	D
D	173° 15' 54"	130° 30' 30" +7" 130° 30' 37"	3185·24 2252·30 (2252·305)	+0·760 289 −0·649 585	1712·40 (1712·403)			1463·06 (1463·064)	3175·46 (3175·467)		249·34 (249·339)		E
E	268° 09' 47"	238° 40' 17" +9" 238° 40' 26"	1489·82 1053·46	−0·854 222 −0·519 908		899·89		547·70		352·19		1447·59	F
F	137° 39' 53"	196° 20' 10" +12" 196° 20' 22"	1085·29 767·42 (767·416)	−0·281 327 −0·959 612		215·90 (215·895)		736·43 (736·422)	520·53 (520·527)			952·33 (952·317)	G
G	299° 07' 41"	315° 27' 51" +14" 315° 28' 05"	3267·45 2310·44 (2310·436)	−0·701 307 +0·712 860	1620·33 (1620·325)		1647·02 (1647·017)		3267·35 (3267·342)		26·69 (26·692)		H
H	179° 13' 28"	314° 41' 19" +16" 314° 41' 35"	3162·23 2236·03 (2236·034)	−0·710 885 +0·703 308	1589·56 (1589·563)		1572·62 (1572·621)		3162·18 (3162·184)		16·94		A
	A Bearing A→B Misclose Adjustment	241° 48' 11" 16° 29' 30" +18" + 2"·25	Sum l sin (α +45) Sum l cos (α +45) −(1·07 +1·21) 1/√2 (+1·07 −1·21) 1/√2	−1·616 −0·086 −1·612 −0·099	Quick check						Computed by J.R.H. Date 4 12 63			

For a whole traverse

$$\sum \Delta N + \sum \Delta E = 2/2^{\frac{1}{2}} \sum s \sin(a + 45)$$

or
$$2^{-\frac{1}{2}}(\sum \Delta N + \sum \Delta E) = \sum s \sin(a + 45)$$

and
$$2^{-\frac{1}{2}}(\sum \Delta N - \sum \Delta E) = \sum s \cos(a + 45)$$

The determination of $\sum s \sin(a + 45)$ can easily be performed on a calculating machine with only s and $\sin(a + 45)$ being tabulated. This is illustrated at the foot of column 4 of Example 4.

Step vi, the adjustment of the misclosures in the differences of eastings and northings must be made using a method which will ideally be simple, disturb the observed quantities by a minimum amount, and be correctly related to the probable displacements caused by errors in the angular and linear measurements.

Unfortunately no such method exists, if the closing error is small theoretical objections to a method should be discounted in favour of simplicity, if the closing error is large no method of adjustment will improve the quality of the field work.

Following the treatment of Bowditch let $s_1, s_2, s_3, s_4, \ldots$ denote the measured lengths and $a_1, a_2, a_3, a_4, \ldots$ denote the bearings of the lines of a closed traverse, and $x_1, x_2, x_3, x_4 \ldots$; $y_1, y_2, y_3, y_4, \ldots$ be respectively their most probable corrections.

Since the corrected ΔE and ΔN must balance two condition equations are formed

$$[s \cos a] + [x \cos a] - [sy \sin a] = 0$$
$$[s \sin a] + [x \sin a] + [sy \cos a] = 0$$

with a least square condition that

$$[px^2] + [qy^2] = \text{a minimum}$$

where p and q are the weights of x and y respectively.

By the usual processes of combination and differentiation correlate equations are of the form

$$+C_1 \cos a_1 + C_2 \sin a_1 = p_1 x_1$$
$$-C_1 s_1 \sin a_1 + C_2 s_1 \cos a_1 = q_1 y_1$$

If we assume that $\quad p_1 = 1/s_1, \quad p_2 = 1/s_2, \ldots$

and $\quad q_1 = s_1, \quad q_2 = s_2, \ldots$

the normal equations formed from the correlate equations reduce to

$$C_1[s] = -[s \cos a]$$
$$C_2[s] = -[s \sin a]$$

The correction in eastings reduces to

$$e_1 = x_1 \sin a_1 + s_1 y_1 \cos a_1$$

By substitution in the correlate equations

$$x_1 = -\frac{s_1[s\cos a]\cos a_1}{[s]} - \frac{s_1[s\sin a]\sin a_1}{[s]}$$

and

$$y_1 = +\frac{[s\cos a]\sin a_1}{[s]} - \frac{[s\sin a]\cos a_1}{[s]}$$

Substituting these expressions we find

$$e_1 = -\frac{s_1[s\sin a]}{[s]}$$

Similarly the correction to northing reduces to

$$n_1 = x_1 \cos a_1 - s_1 y_1 \sin a_1$$

from which by the same process we find that

$$n_1 = -\frac{s_1[s\cos a]}{[s]}$$

As $[s\sin a]$ and $[s\cos a]$ are the misclosures in a closed traverse the corrections appear as the Bowditch rule given below for the adjustment of a traverse.

Bowditch's method. The correction to be applied to the differences in eastings and northings is as follows:

Correction to difference in eastings equals misclosure in eastings multiplied by length of corresponding side divided by total length of traverse.

Correction to difference in northings equals misclosure in northings multiplied by length of corresponding side divided by total length of traverse.

Bowditch's rule depends for it's validity upon the truth or otherwise of the relationships:

$$p_1 = 1/s_1 \quad \text{and} \quad q_1 = s_1$$

This corresponds to the assumption that errors of measurement of the lengths are directly proportional to the square root of the length of the line and that errors in the determination of bearing are inversely proportional to the square root of the length of the line.

Expressed in terms of figure 7.13

$$ds \propto s^{\frac{1}{2}} \quad \text{and} \quad da \propto 1/s^{\frac{1}{2}}$$

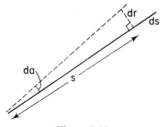

Figure 7.13

but $\qquad da = dr/s$

hence $\qquad dr \propto s^{\frac{1}{2}}$ and $dr \propto ds$

Whilst most surveyors are prepared to accept the assumption that in taping errors of measurement of a random nature are proportional to $s^{\frac{1}{2}}$ it is most unlikely that any would accept that errors of bearing are proportional to $1/s^{\frac{1}{2}}$. With the measurement of lines by electronic equipment it is doubtful if the error of measurement is ever proportional to $s^{\frac{1}{2}}$.

Transit method For this method there is no theoretical basis comparable to the Bowditch method, but this should not be regarded as a serious objection as the basis of the Bowditch method is suspect. The transit method does not disturb the adjusted bearings to the same extent as the Bowditch method which may be regarded as disturbing the unadjusted bearings by an unjustifiably large amount. The correction to be applied to the differences in eastings and northings is as follows:

Correction to difference in eastings equals misclosure in eastings multiplied by eastings of that side divided by the total eastings of the traverse.

Correction to difference in northings equals misclosure in northings multiplied by northings of that side divided by the total northings of the traverse.

Comparison of Bowditch and transit methods Example 1 illustrates the adjustment of a closed loop by the Bowditch method, the same differences in eastings and northings have been adjusted in example 2 by the transit method.

The Bowditch method produces a marked change in the bearings of East–West or North–South lines, this does not happen in the transit method. The Bowditch method will always produce the same linear movement of a point regardless of the bearing of a line, this is not the case in the transit method. For simplicity the Bowditch method has a marginal advantage and is usually preferred.

Step vii, once the differences in eastings and northings have been adjusted the coordinates are obtained. The correctness of the adjustment is demonstrated by the traverse closing exactly onto the fixed point or in a closed loop onto the starting point (see examples 1, 2, and 3).

7.7. PERMISSIBLE CLOSURES AND EXPRESSIONS OF ACCURACY

In expressing the accuracy of a traverse two quantities are considered, the angular misclosure and the linear misclosure. The former should be expressed in the following manner: 18″ over 8 stations, adjustment 2″.25 per station.

Permissible closures should be in terms of the square root of the number of stations of the form $An^{\frac{1}{2}}$. A brief explanation of this point is pertinent. With reasonable care and reasonably long lines systematic errors should be small.

Let E be the standard error of the measured angle, and r the standard error of the nth bearing, then following the usual treatment

$$r = En^{\frac{1}{2}}$$

A is usually taken as twice E on closed surrounds and three times for traverses between fixed lines or astronomical azimuths. For normal traverses A varies from 3″ to 30″ according to the precision of the work.

Linear misclosures are commonly expressed as a ratio of total misclosure or vector, to total length of traverse. In Example 1 1·62 ft in 17 738 ft or more commonly as 1 in 10 950. This is quite satisfactory if the total length is quoted, but 1 in 15 000 representing a 2 foot misclosure in a closed traverse of 6 miles indicates hardly the same quality work as 1 in 15 000 representing 0·18 feet misclosure in a surround of 2 700 feet.

A means of comparing traverses of different lengths is to determine a factor Q where $Q = t/L^{\frac{1}{2}}$; t being total linear misclosure and L being the length of the traverse. In the case above

$$Q = 0\cdot011\ 2 \text{ for the 6 mile traverse}$$

$$Q = 0\cdot003\ 5 \text{ for the 2 700 feet traverse}$$

Permissible values for t may be given by $QL^{\frac{1}{2}}$ where Q varies in feet from 0·005 for precise work to 0·030 for low quality work. If L and consequently t are in metres the value for the constant Q will vary from 0·003 for precise work to 0·017 for low quality work.

The Ordnance Survey minor control traverses which as a rule do not exceed 3 000 metres in length are required to close within limits expressed by:

$$M = 0\cdot124 + 0\cdot092\ 2L$$

where M is the misclosure in metres, 0·124 the relative terminal error in metres, and 0·092 2L the systematic error in metres (L being the length of the traverse in kilometres).

7.8. MISCELLANEOUS PROBLEMS IN TRAVERSING

7.8.1. COMPUTATION OF A TRAVERSE BETWEEN TWO FIXED POINTS WHEN NO STARTING OR CLOSING BEARING CAN BE DERIVED

The traverse is computed on an assumed bearing, usually 0°, for the first leg. On the basis of this traverse the bearing between the two fixed points is computed; the true bearing between the two fixed points is also computed. The difference between these two bearings indicates the amount by which the traverse has to be swung. This is applied to all the bearings in the original traverse computation and the traverse recomputed. This misclosure will now be small and can be adjusted in the usual way.

A diagram will ensure that the bearing correction is applied in the correct direction. The difference between the distances computed from coordinates of the two fixed stations and the distance computed using the terminal points of the traverse computed on the assumed bearing in a measure of the misclosure of the traverse. If this quantity is treated as a change of scale and the difference in bearings as a change in the direction of the axes a coordinate transformation may be performed instead of recomputation of the traverse.

Traverses

If A and B are the true coordinates of the terminals and A′ and B′ are the temporary coordinates of the terminals using an assumed bearing for the first line. Then let $AB/A'B' = m$ (change of scale) and bearing A′ to B′ − bearing A to B = C (swing) then if $P = m \cos C$ and $Q = m \sin C$

and
$$E_p = E_A + (e_p - e_a)P - (n_p - n_a)Q$$
$$N_p = N_A + (n_p - n_a)P + (e_p - e_a)Q$$
where $e_a, e_b, e_c \cdots e_p$ and $n_a, n_b, n_c \cdots n_p$

are temporary coordinates for station A, B, C, ... P. See formulae (3) and (4) §2.10.5.

7.8.2. LOCATION OF GROSS ERRORS IN TRAVERSING

7.8.2.1. Linear Errors.

The traverse is computed in the usual way. On finding an excessively large misclosure with only a small misclosure in bearings an error in taping is suspected. The bearing of the misclosure is computed paying attention to signs, a small diagram will often help. The gross linear error may lie on a line whose bearings is nearly the same as or about 180° different from the bearing of the misclosure. Lines that fall in this category are retaped. This method works well if there is only one error and all of the rest of the work is of a fairly high quality. Gross linear errors are frequently 1, 10, or 100 units in size, or represent a reversal of part of a tape length, for instance in a line of 364·8feet measured with a hundred foot tape the recorded value may be 300 feet plus 35·2 feet due to a misreading of the end portion. Equally frequent this distance may be recorded as 44·8 particularly when a chain is used.

7.8.2.2. Angular Errors

These will appear as a bad misclosure in angles or bearings. The traverse is computed from each end or in each direction and, unadjusted coordinates are obtained. The station where the two sets of coordinates differ by a normal misclosure it is reasonable to suspect will be the point in error. If the error is large a plot by scale and protractor is all that is needed, the traverse is plotted from each end and where the two plots differ by a normal misclosure is the location of the error. Both of these methods will break down if there is more than one error.

7.8.3. OMITTED MEASUREMENTS

It is bad practice in traversing to omit any quantity that it is physically possible to measure. There are six possible problems capable of solution, these are when the following quantities have not been measured or determined.

(i) Length of one line.
(ii) Bearing of one line.
(iii) Length and bearing of one line.
(iv) Lengths of two lines.
(v) Bearings of two lines.
(vi) Length of one line and the bearing of another.

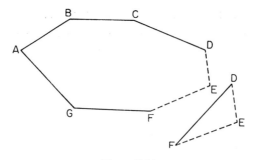

Figure 7.14

Considering cases (i), (ii), (iii), and (iv). Two equations can be established:

$$s_1 \cos a_1 + s_2 \cos a_2 + \cdots + s_0 \cos a_0 + s_n \cos a_n = [\Delta N]$$

$$s_1 \sin a_1 + s_2 \sin a_2 + \cdots + s_0 \sin a_0 + s_n \sin a_n = [\Delta E]$$

in which for (i) s_0 only is unknown; for (ii) $\cos a_0$ and $\sin a_0$ are unknown; for (iii) s_0, $\cos a_0$, and $\sin a_0$ are unknown; and for (iv) s_0 and s_0' are unknown. In cases (i) and (ii) a check exists there being two equations with effectively one unknown. In cases (iii) and (iv) no check exists, all the error in the traverse is thrown into the omitted measurements. For problem (iv) no solution can be obtained if the missing lines are parallel. Considering case (v). In figure 7.14 the bearings of lines DE and EF are unknown. If the unadjusted coordinates of D and F are obtained the bearing and distance DF can be computed. From the expression of the form $\tan \tfrac{1}{2} A = [(s - b)(s - c)/s(s - a)]^{\tfrac{1}{2}}$ the various angles of the triangle DEF are found and consequently the bearings DE and EF.

Considering case (vi) illustrated by Example 5 and figure 7.15. When the two incomplete sides do not adjoin each other a triangle requiring solution can still be formed by shifting the intervening line or lines parallel to themselves along one of the unknowns. In the example line 34 has been shifted along 23 to 24'. Triangle 4'54 is now solved, in which the angle at 5 and the sides 4'5 and 4'4 are known.

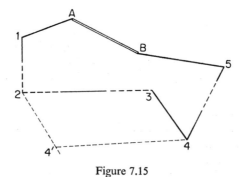

Figure 7.15

Example 5. Omitted Measurements

Data from Traverse Computation (not shown)

Line	Bearing	Length (ft)	Difference eastings (ft) +	Difference eastings (ft) −	Difference northings (ft) +	Difference northings (ft) −	Remarks
A-1				799·2	305·0		
1-2			129·0		1183·7		
2-3		1125·7					Angles at 2 and 3 not measured
3-4				662·5		872·5	
4-5	55° 46′ 15″						Distance 4-5 not measured
5-B				1630·4	332·3		

Data

Stn	Coordinates Eastings (ft)	Coordinates Northings (ft)
A	549 671·8	796 104·3
B	550 311·7	795 230·7

Omitted Measurements Computation of Coordinates

At	Difference eastings +	Difference eastings −	Difference northings +	Difference northings −	Provisional eastings (ft)	Provisional northings (ft)	To	Final eastings (ft)	Final northings (ft)	Stn
					549 671·8 (Data)	796 104·3 (Data)	A	549 671·8 (Data)	796 104·3 (Data)	A
A		799·2	305·0		548 872·6	795 799·3	1	548 872·6	795 799·3	1
1	129·0		1183·7		549 001·6	794 615·6	2	549 001·6	794 615·6	2
2		662·5		872·5	549 664·1	793 743·1	4′	550 075·9	794 952·0	3
								550 738·4	794 079·5	4
					551 942·1	794 898·4	5	551 942·1	794 898·4	5
5	1630·4		332·3		550 311·7 (Data)	795 230·7 (Data)	B	550 311·7 (Data)	795 230·7 (Data)	B

Bearing and distance
From 5 (A) to 4′ (B)
E_A 551 942·1 N_A 794 898·4
E_B 549 664·1 N_B 793 743·1
− 2 278·0   − 1 155·3
Cot a +0·507 155
Sin a   0·891 859
Bearing 243° 06′ 28″
Distance 2554·2

Solution triangle 544′
Bearing 5 to 4 235° 46′ 15″
Bearing 5 to 4′ 243° 06′ 28″
Angle at 5   7° 20′ 13″
Distance 4′5   2 554·2
Sine 5   +0·127 704
+Distance 4′4 (23) 1125·7
÷ Sine 4   0·289 759
Angle 4   163° 09′ 23″

Angle 4   163° 09′ 23″
Angle 5   7° 20′ 13″
∴ Angle 4′   9° 30′ 24″
Check   180° 00′ 00″
Distance 4′5   2554·2
Sine 4′   0·165 162
÷ Sine 4   0·289 759
Distance   1455·9

Bearing 2 to 3
Bearing 4′ to 5   63° 06′ 28″
Angle 4′   9° 30′ 24″
Bearing 4′ to 4   72° 36′ 52″
Bearing 2 to 3   72° 36′ 52″

Computed by J.R.H.
12.1.64

Omitted Measurements Traverse Computation

Line	Length (ft)	Bearing	Sine Cosine	Difference eastings (ft) +	Difference eastings (ft) −	Difference northings (ft) +	Difference northings (ft) −
2-3	1125·7	72° 36′ 52″	+0·954 316 +0·298 800	1074·3			336·4
4-5	1455·9	55° 46′ 15″	+0·826 795 +0·562 504	1203·7		818·9	

7.8.4. COMPUTATION OF CORRECTIONS IN EASTINGS AND NORTHINGS FOR SMALL CHANGES IN BEARING AND LENGTH

In a traverse $\Delta E = s \sin a$ and $\Delta N = s \cos a$. Let $y = \Delta E$ and $x = \Delta N$, then by differentiation

$$dy = ds \sin a + s \cos a \, da$$

and
$$dx = ds \cos a - s \sin a \, da$$

dy being the correction to ΔE and dx the correction to ΔN. The expressions are only required in this form if both s and a are being corrected. Usually a changes as result of a new azimuth determination, then

$$dy = s \cos a \, da = \Delta N \, da'' \sin 1''$$

$$dx = -s \sin a \, da = -\Delta E \, da'' \sin 1''$$

Careful attention must be paid to the signs of ΔN, ΔE, and da. The sign of da is given by 'new bearing' minus 'old bearing'.

356   *Practical Field Surveying and Computations*

Example 6. Corrections to ΔE and ΔN for a Change in Bearing

Line	Original +	ΔE −	Correction +	−	Correct ΔE +	−	Original ΔN +	−	Correction +	−	Correct ΔN +	−
AB		799·21	+0·12			799·09	305·00			−0·32		305·32
BC	128·98		+0·48		129·46			1183·81	+0·05			1183·76
CD	1074·31			−0·14	1074·17		336·12		+0·43		336·55	
DE	662·47		+0·35		662·82			872·52	+0·27			872·25
EF	1203·81			−0·33	1203·48		818·97		+0·49		819·46	
FG		1630·40		−0·13		1630·53	332·30			−0·66	331·64	

Computed by J.R.H.   Original bearing F to G   281° 31' 15"
16-11-63   Correct bearing F to G   281° 29' 52"
                                    da   −1' 23" (83")

It will be seen that $da'' \sin 1''$ is a constant throughout the computation which is easily performed on a slide rule. Example 6 illustrates a computation for change of bearing.

A reverse problem may arise when it is required to find the effect of an adjustment of a traverse on the original bearings and distances.

Example 7. Correction to Bearings. Data Abstracted from Traverse Example 1

Line	dy (E)	Cos a	dx (N)	Sin a	s	da"	
CD	+0·18	−0·0363	+0·20	+0·9993	2988	−14"·2	Computed by J.R.H.
GH	+0·20	+0·0082	+0·22	−0·9999	3267	+13"·9	17-11-63

We find that
$$ds = dy \sin a + dx \cos a$$
or
$$ds = dy \, \Delta E/s + dx \, \Delta N/s$$

which can easily be performed on a slide rule or calculating machine. Similarly $da'' = (dy \cos a - dx \sin a) \operatorname{cosec} 1''/s$. Example 7 is given to illustrate this computation when performed on a calculating machine.

Example 8. Adjustment to Bearings in a Deviation

At	RO	To	Observed angle	Bearing	o|r	Bearing	Remarks
D	C	E	268° 44' 27"	D.E. as 00° 00' 00"		D.E. as 106° 06' 26"	from main
D	C	1	271° 06' 51"	02° 22' 24"(04)28"		108° 28' 50"(04)54"	adjustment
1	D	2	134° 17' 16"	316° 39' 40"(09)49"		62° 46' 06"(09) 15"	E.D.I=2° 22' 24"
2	1	3	258° 31' 26"	35° 11' 06"(13)19"		141° 17' 32"(13) 45"	
3	2	4	100° 01' 04"	315° 12' 10"(17) 27"		61° 18' 36"(17) 53"	
4	3	5	104° 16' 17"	239° 28' 27"(22) 49"		345° 34' 53"(22) 15" (35')	
5	4	6	347° 27' 10"	46° 55' 37"(26) 03"		153° 02' 03"(26) 29"	
6	5	7	110° 48' 47"	337°44' 24"(30) 54"		83° 50' 50"(30) 20" (51')	
7	6	E	226° 51' 58"	24° 36' 22"(35) 57"		130° 42' 48"(35) 23"	E.7D
E	D	7	(335° 22' 59") 24° 37' 01"	179° 59' 21"(39) 00"		286° 05' 47"(39) 26"	335° 22' 59"
		F	184° 06' 38"				

39" ÷ 9   39" ÷ 9
4"·3 per station   4"·3 per station
( ) adjustment to each bearing

7.8.5. DEVIATIONS

The errors arising from short lines in traversing can frequently be minimized by bypassing the short lines and so retaining the bearing in the main traverse (see figure 7.16). The main traverse ABCDEFG enters bad ground at D and taping follows the line 1234567, the line DE is however intervisible and is observed. The short lines would vitiate the adjustment of the bearings in the main traverse, so the angles CDE and DEF are used in closing the traverse for

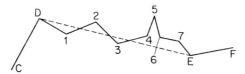

Figure 7.16

bearing and the adjusted bearing DE obtained. If we are not interested in the positions of the stations 1 to 7 the distance DE can be obtained by assuming the bearing DE to be due North, then the sum of ΔN for the traverse D to E via $1 \cdots 7$ is the required length. If the positions of the stations 1 to 7 are required the value of the adjusted bearing DE is held fixed and bearings for D1, 12, 23, etc. are obtained and adjusted. The traverse is now computed through C, D, 1, 2, ... , 7, E, F, ... and the linear misclosure adjusted in the usual way. The permissible angular misclosure in a deviation is usually twice that of the main traverse from which it deviates.

7.9. ADJUSTMENT OF A TRAVERSE NETWORK

7.9.1. ADJUSTMENT BY LEAST SQUARES

The practical details of the adjustment of a traverse network by the method of least squares using condition equations are given below. Figure 7.17 illustrates

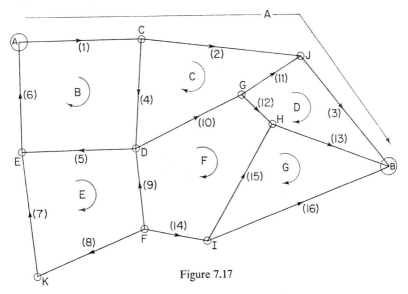

Figure 7.17

358 Practical Field Surveying and Computations

Example 9. Normal Equations

E	1	2	3	4	5	6	7	K_e	Sum	K_n	Sum
1	+1·76	−0·52	0	−0·34	0	0	+0·36	−0·06	+1·20	+0·10	+1·36
2		+2·95	−0·59	0	−0·70	0	+1·14	+0·10	+2·38	−0·11	+2·17
3			+2·28	0	−0·41	−0·64	+0·64	−0·09	+1·19	+0·22	+1·50
4				+2·38	−0·72	0	0	+0·12	+1·44	−0·03	+1·29
5					+3·28	−1·21	0	+0·19	+0·43	+0·60	+0·84
6						+3·41	0	−0·05	+1·51	−0·13	+1·43
7							+2·14	−0·32	+3·96	+0·23	+4·51
								−0·11	+12·11	+0·88	+13·10

Example 9. Forward Solution

1	2	3	4	5	6	7	K_e	Sum	K_n	Sum	Check e	Check n	Remarks
+1·76	−0·52	0	−0·34	0	0	+0·36	−0·06	+1·20	+0·10	+1·36			1st normal
(0·56818)													
−1	+0·2954	0	+0·1932	0	0	−0·2045	+0·0341	−0·6818	−0·0568	−0·7727	0	0	
	+2·95	−0·59	0	−0·70	0	+1·14	+0·10	+2·38	−0·11	+2·17			2nd normal
	+2·7964	−0·59	−0·1005	−0·70	0	+1·2463	+0·0823	+2·7345	−0·0805	+2·5718	0	1	
	(0·35760)												
	−1	+0·2110	+0·0359	+0·2503	0	−0·4457	−0·0294	−0·9779	+0·0288	−0·9197	0	0	
		+2·28	0	−0·41	−0·64	+0·64	−0·09	+1·19	+0·22	+1·50			3rd normal
		+2·1555	−0·0212	−0·5577	−0·64	+0·9030	−0·0727	+1·7670	+0·2030	+2·0426	1	0	
		(0·46393)											
		−1	+0·0098	+0·2587	+0·2969	−0·4189	+0·0337	−0·8198	−0·0942	−0·9476	0	1	
			+2·38	−0·72	0	0	+0·12	+1·44	−0·03	+1·29			4th normal
			+2·3105	−0·7506	−0·0063	+0·1232	+0·1106	+1·7875	−0·0116	+1·6652	1	0	
			(0·4328!)										
			−1	+0·3249	+0·0027	−0·0533	−0·0479	−0·7736	+0·0050	−0·7207	0	0	
				+3·28	−1·21	0	+0·19	+0·43	+0·60	+0·84			5th normal
				+2·7166	−1·3776	+0·5856	+0·2277	+2·1524	+0·6286	+2·5532	1	0	
				(0·36811)									
				−1	+0·5071	−0·2156	−0·0838	−0·7923	−0·2314	−0·9399	0	0	
					+3·41	0	−0·05	+1·51	−0·13	+1·43			6th normal
					+2·5214	+0·5654	+0·0442	+3·1310	+0·2490	+3·3358	0	0	
					(0·39661)								
					−1	−0·2242	−0·0175	−1·2418	−0·0988	−1·3230	1	0	
						+2·14	−0·32	+3·96	+0·23	+4·51			7th normal
						+0·8731	−0·3788	+0·4941	−0·0304	+0·8427	2	0	
						(1·14534)							
						−1	+0·4338	−0·5659	+0·0348	−0·9652	3	0	

Example 9. Back Solution

	C_1	C_2	C_3	C_4	C_5	C_6	C_7
Solution E	−0·1832	−0·3386	−0·2445	−0·1478	−0·2355	−0·1148	+0·4338
Check E	−1·1833	−1·3387	−1·2446	−1·1479	−1·2357	−1·1149	−0·5659
Solution N	−0·1140	−0·1092	−0·2171	−0·0923	−0·2930	−0·1066	+0·0348
Check N	−1·1140	−1·1091	−1·2170	−1·0923	−1·2930	−1·1066	−0·9652
Solution E−(Check +1)	+0·0001	+0·0001	+0·0001	+0·0001	+0·0002	+0·0001	−0·0003
Solution N−(Check +1)	0·0000	−0·0001	−0·0001	0·0000	0·0000	0·0000	0·0000

the network, there are two fixed points A and B and the individual sections numbered (1) to (16) are between the ten junction points and K an abrupt change in direction. The differences in eastings and northings tabulated in Example 9 have been obtained after adjustment of the bearings between azimuths or weighted mean bearings at the junction points (see §7.9.2).

The assignment of weights to the various sections presents many theoretical difficulties but, since a least squares adjustment of a traverse network is only undertaken when the work is of a high quality, the simplest solution taking the weight as proportional to the inverse of the length is in most cases as satisfactory as any other. If the reader is interested in the theoretical problem of weights in traversing reference may be made to §7.12 on propagation of error in traverses, to Clarke's *Plane and Geodetic Surveying for Engineers* (London: Constable

Example 9. Traverse Network Adjustment

Line	No	Difference eastings (ft)	Difference northings (ft)	Length (ft)	Weight (p)
AC	1	+2 146.81	+ 7.16	3 643	2.74
CJ	2	+7 179.26	− 104.24	11 396	0.88
JB	3	+2 089.32	−2 172.17	6 423	1.56
CD	4	− 346.27	−3 226.12	5 236	1.91
DE	5	−1 779.46	− 42.68	3 404	2.94
EA	6	− 21.14	+3 261.74	5 381	1.86
KE	7	+ 49.24	+4 926.86	8 575	1.17
FK	8	−2 355.76	− 837.01	4 601	2.17
FD	9	− 527.18	+4 132.56	7 156	1.40
DG	10	+4 098.62	+2 000.88	6 983	1.43
GJ	11	+3 426.81	+1 121.11	5 942	1.68
GH	12	+2 176.21	− 326.41	4 103	2.44
HB	13	+3 340.01	− 724.87	6 414	1.56
FI	14	+1 474.98	− 120.68	2 438	4.10
IH	15	+4 272.48	+5 927.11	12 061	0.83
IB	16	+7 612.54	+5 202.37	15 600	0.64

	Fixed points	
	Eastings (ft)	Northings (ft)
A	121 176.49	364 398.72
B	132 592.20	362 129.24

Table of Condition Equations and Final Corrections

C_e	C_n	Line	1	2	3	4	5	6	7	8	9	10	11	12	13	14	15	16	K_e	K_n	Σ/p	Check
		$1/p$	0.36	1.14	0.64	0.52	0.34	0.54	0.86	0.46	0.72	0.70	0.59	0.41	0.64	0.24	1.21	1.56				
−0.1832		E1	+1			+1	+1	+1											−0.06	+1.75	+0.06
	−0.1140																			+0.10		−0.10
−0.3386		E2		+1		−1					−1	−1							+0.10	−0.67	−0.10
	−0.1092																			−0.11		+0.11
−0.2445		E3			+1								+1	−1	−1				−0.09	+0.18	+0.09
	−0.2171																			+0.22		−0.22
−0.1478		E4				−1		+1	+1	−1									+0.12	+0.26	−0.12
	−0.0923																			−0.03		+0.03
−0.2355		E5							+1	+1			+1			−1	−1		+0.19	+0.38	−0.19
	−0.2930																			+0.60		−0.60
−0.1148		E6													+1	+1	−1		−0.05	+0.29	+0.05
	−0.1066																			−0.13		+0.13
+0.4338		E7	+1	+1	+1														−0.32	+2.14	+0.32
	+0.0348																			+0.23		−0.23
		Σ/p	+0.72	+2.28	+1.28	0	0	+0.54	+0.86	+0.46	0	0	0	0	0	−0.24	0	−1.56			+4.34	✓
Correction to ΔE			+0.090	+0.108	+0.121	+0.081	−0.012	−0.099	−0.127	−0.068	−0.063	+0.072	+0.056	+0.004	+0.083	+0.056	+0.146	+0.179				
Correction to ΔN			−0.028	−0.005	−0.117	−0.002	−0.007	−0.052	−0.079	−0.042	−0.144	−0.129	−0.064	−0.031	+0.071	+0.070	+0.226	+0.166				

1944) Vol. II, pp. 307–9 and *Records of the Ghana Survey Department*, Accra: Survey Dept. 1931 Vol. III, pp. 40–59.

Since there are six loops in Example 9 there are six conditions to be satisfied in the free network. With two fixed points there is an extra condition to be satisfied to ensure that the free network will fit between these points. The best criterion of the quality of the network is given by the misclosure of the free network onto A and B. Once the free network conditions have been satisfied the circuit closures will be zero whatever route is taken. By placing last the one condition equation for closure onto the second fixed point the value of this closure in terms of the adjusted free network is given by the absolute term in the last reduced normal equation of the Gauss–Doolittle solution used in Example 9.

In the example this closure is 0.38 in eastings and 0.03 in northings. If a Cholesky method of solution had been used the absolute term in the last reduced normal equation would be multiplied by the coefficient of the diagonal term to give the values for the closures.

It will be noted that in the solution there are two identical sets of condition equations with different absolute and check terms. The solution can be performed as one calculation with extra columns for the additional absolute and check values.

Once the adjusted values for the differences in eastings and northings of each line have been obtained, final coordinates for the stations A to K can be determined. Within each section the misclosure, represented by the adjustment for that section, is distributed to the individual lines in proportion to their lengths or in some other systematic logical manner. With modern equipment the misclosure will be small and a Bowditch adjustment is as suitable as any other. In the adjustment of a traverse network no attempt is made to retain the bearings of the lines as established during the first bearing adjustment which preceded the least squares adjustment.

There are other methods of treating a traverse network but the inherent simplicity of the method illustrated by Example 9 makes it preferable. An approximate method using successive weighted means may easily be devised but the order of selection of the weighted points will give variations in the answer from one computor to another.

7.9.2. WEIGHTED MEAN BEARINGS AT JUNCTION POINTS IN TRAVERSE NETWORKS

It is not always practicable to observe an azimuth at a junction point of a traverse network. One line is selected at each junction point and azimuths carried to it from the nearest astronomical stations or fixed lines. In a small local network these may be previously fixed weighted mean lines.

The layout of the computation and it's execution is illustrated by Example 10

Example 10. Weighted Mean Bearing

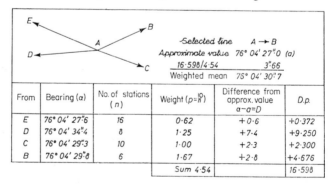

From	Bearing (a)	No. of stations (n)	Weight ($p = \frac{10}{n}$)	Difference from approx. value $a-o=D$	D.p.
E	76° 04′ 27″6	16	0·62	+0·6	+0·372
D	76° 04′ 34″4	8	1·25	+7·4	+9·250
C	76° 04′ 29″3	10	1·00	+2·3	+2·300
B	76° 04′ 29″8	6	1·67	+2·8	+4·676
			Sum 4·54		16·598

Selected line A → B
Approximate value 76° 04′ 27″0 (o)
16·598/4·54   3″66
Weighted mean 76° 04′ 30″7

where weights have been taken as the inverse of the number of angles between the selected line and the fixed line or star.

7.10. SUBSTENSE TRAVERSING

7.10.1. METHOD

In a traverse distance may be measured by subtense methods. This involves measurement by theodolite of the angle subtended by a short taped base or bar

Traverses

of known length positioned approximately or exactly at right angles to the line to be measured. The precision of the measurement is largely dependent upon the precision of the angular measurement and the length of line from the short base or bar to the theodolite. One of the advantages of the subtense method is the ability to vary precision according to the requirements of a particular survey and thereby to increase the speed of measurement. Furthermore the horizontal distance is directly obtained.

7.10.2. POSITIONS FOR THE SUBTENSE BASE OR BAR

7.10.2.1. *Direct Measurement*

Here it is positioned at the end of the line and normal to the line (see figure 7.18).

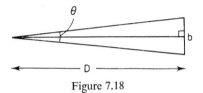

Figure 7.18

If D is the length of the line, b the length of the base or bar, and θ the size of the subtense angle, then, as $b/2D = \tan \frac{1}{2}\theta$,

$$D = b/2 \cot \tfrac{1}{2}\theta. \qquad (1)$$

7.10.2.2. *Bar or Base Mid-positioned*

Here it is positioned in the middle of and normal to the line (see figure 7.19).

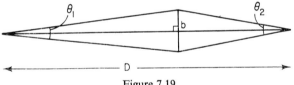

Figure 7.19

With the notation as in (1) above

$$D = b/2(\cot \tfrac{1}{2}\theta_1 + \cot \tfrac{1}{2}\theta_2) \qquad (2)$$

7.10.2.3. *Auxiliary Base*

Here an auxiliary base is formed at the end of the line by a direct measurement. This auxiliary base is at an angle to the direction of the line (see figure 7.20).

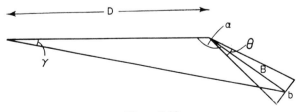

Figure 7.20

If α is the direction of the auxiliary base to the direction of the line, γ is the angle subtended by the auxiliary base, and with the notation as in (i) above

$$B = b/2 \cot \tfrac{1}{2}\theta \quad \text{and} \quad D = B \sin(\alpha + \gamma)/\sin \gamma \qquad (3)$$

7.10.2.4. Auxiliary Base Mid-positioned

Here the base is positioned in the middle of the line and normal to the line (see figure 7.21).

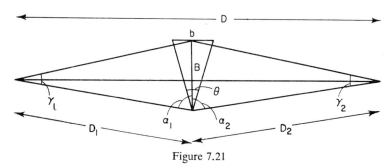

Figure 7.21

With the notation as in §7.10.2.1 and 7.10.2.3

$$B = b/2 \cot \tfrac{1}{2}\theta, \quad D_1 = B \sin(\alpha_1 + \gamma_1)/\sin \gamma_1$$
$$D_2 = B \sin(\alpha_2 + \gamma_2)/\sin \gamma_2$$

and
$$D = D_1 \sin \alpha_1 + D_2 \sin \alpha_2 \qquad (4)$$

7.10.3. THE SUBSTENSE BAR

The subtense bar is most conveniently used as an adjunct to the usual three tripod equipment. The Hilger and Watts subtense bar described here is available with an adjustable centring base which allows the bar to be interchanged with targets or the Microptic Theodolite No. 1 or No. 2.

The 2 metre subtense bar consists of two hinged arms which fold together. The arms are rectangular hollow tubes and within each tube there is an invar rod with a target attached. When the bar is opened out studs at the hinge end of the invar rod are brought into abutment and the bar is then clamped. The bar can then be turned to face any direction and a sighting telescope on one arm enables the bar to be positioned at right angles to the line of sight of the telescope and hence at right angles to the line to be measured. A bubble on one arm enables the bar to be set in a horizontal position.

In addition the Hilger and Watts equipment makes provision for mounting an additional target at the middle of the bar, for taking traverse angles and for auxiliary base methods. All targets can be illuminated for night work.

Before use the equipment should be carefully tested. The length of the bar should be checked with a beam compass and diagonal scale and any error noted. The sighting telescope on the bar should be checked to insure that the bar is positioned at right angles to its line of sight. If a right angle is set out by theodolite and the bar interchanged for the theodolite on the levelling head this check can easily be carried out.

Field Booking for Traversing with Subtense Bar

Instrument *Watts invar*   Bar length ..2m....   Date. 7-7-65....   Remarks........
Observer ......J.R.H......   Bar correction.. *Nil*   Weather.. *Fair* ...   Booker. A.L.A.

Station	Object	1st. Zero					2nd. Zero					Sketch
		°	′	Swing R ″	Swing L ″	Mean ″	°	′	Swing R ″	Swing L ″	Mean ″	
Rock				*Traverse angle*								P.H.
	Pen Hill	207	07	37	35	36	344	20	19	21	20	
	A	82	35	22	18	20	219	48	06	06	06	
		235	27			44	235	27			46	Rock
		Mean traverse angle 235°27′45″										(1)
Rock				*Subtense angle*								
	Bar (1)	231	21	28	31	29·5	203	15	51	48	49·5	
	A	237	07	22	22	22	209	01	42	41	41·5	
		5	45			52·5	5	45			52·0	
	Bar (1)	53	12	39	41	40						
	A	58	58	29	31	30						
		5	45			50						
		Mean subtense angle 5°45′51·5″										
A		*Subtense angle (1) 05°05′11·0″*										
	Bar L	172	15	59	59	59·0	137	58	10	10	10	
	Bar R	177	21	11	12	11·5	143	03	20	21	20·5	(1)
		05	05			12·5	05	05			10·5	
	Bar L	59	03	55	54	54·5						Rock (2)
	Bar R	64	09	06	03	04·5						A
		05	05			10						B
				Traverse angle								
A	Rock	59	20	06	08	07	145	01	15	15	15	
B	B	263	20	21	22	21·5	349	01	29	30	29·5	
		204	00			14·5	204	00			14·5	
		Mean traverse angle 204°00′14·5″										
		Auxiliary base angle										
A	Rock	59	20	06	10	08 }	80	05	58			
	(1)	139	26	04	08	06 }						
		Subtense angle (2) 02 57 14·5										
A	Bar L	300	03	13	15	14·0	08	44	27	30	28·5	
	Bar R	303	00	25	25	25·0	11	41	48	48	48·0	
		02	57			11·0	02	57			19·5	
	Bar L	31	42	53	55	54·0	83	11	39	38	38·5	
	Bar R	34	40	13	12	12·5	86	08	47	48	47·5	
		02	57			18·5	02	57			09	

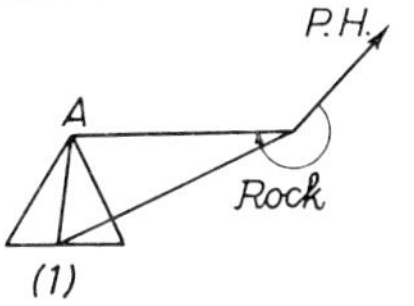

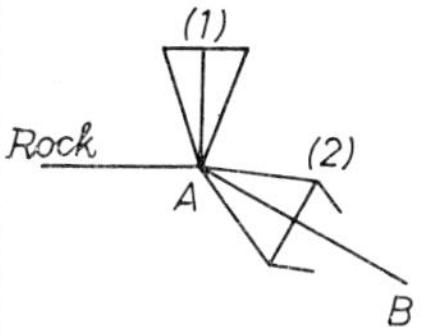

Figure 7.22

7.10.4. FIELD WORK AND REDUCTION OF OBSERVATIONS

A good reconnaissance is essential and a diagram of survey is drawn approximately to scale. For an accuracy of 1 in 10 000 with a 2 metre bar and an allowable error in the subtense angle of ±1" the following limiting lengths for each of the four main positions of the bar apply.

Bar position	Maximum length	Remarks
Direction	40 m	
Mid position	120 m	
Auxiliary base	450 m	B to equal $(bD)^{\frac{1}{2}}$
Mid-auxiliary base	1000 m	B to equal $0.6\,(bD)^{\frac{1}{2}}$

A typical subtense traverse is illustrated in figure 7.23. In this case some of the limiting lengths have been exceeded for purely practical reasons and is reasonable

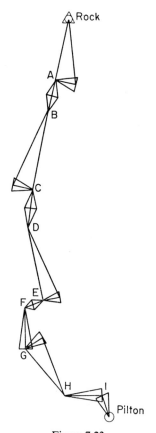

Figure 7.23

Traverses 365

to suppose that the traverse may not achieve 1 in 10 000 in closure unless the number of measurements of the subtense angle is increased. As well the traverse has been closed onto fixed points which may themselves be in error.

As there are many angles to be measured the booking sheets must be carefully prepared and supported by diagrams, a suggested booking sheet is given in figure 7.22. Care is taken during the angular measurements and if the differences between the reduced values are large extra measurements are made.

The reduction and computation of the traverse illustrated is given, in part, in Example 11. All the field data is given and the whole traverse can be computed as an exercise.

Example 11.

Traverse angles

At	Back station	Fore station	Clockwise angle
A	Rock	B	204° 00' 14".5
B	A	C	199° 29' 24".5
C	B	D	192° 39' 29".0
D	C	E	141° 23' 42".5
E	D	F	239° 46' 51".5
F	E	G	97° 32' 18".5
G	F	H	166° 17' 59".0
H	G	J	152° 09' 54".0
J	H	Pilton	252° 20' 36".0

Bar direct abstract and reductions

	Base at A	A to B($\frac{1}{2}$)	B to A($\frac{1}{2}$)	Base at C	
θ	05° 05' 11".0	02° 57' 14".5	03° 16' 37".5	03° 37' 31".7	
$\frac{1}{2}\theta$	02° 32' 35".5	01° 28' 37".2	01° 38' 18".7	01° 48' 45".8	
$\cot \frac{1}{2}\theta$	22.514	38.784	34.958	31.597	
$\frac{1}{2}$	1.000	1.000	1.000	1.000	(metres)
D (B)	22.514 m	38.784 m	34.958 m	31.597 m	

Remaining values are left for the reader to compute

	C to D ($\frac{1}{2}$)	D to C ($\frac{1}{2}$)	Base at E	E to F ($\frac{1}{2}$)	F to E ($\frac{1}{2}$)
θ	01° 30' 26".1	02° 01' 08".3	03° 45' 32".0	02° 22' 20".4	02° 39' 43".5

	F to G	Base at G	H to J	J to Pilton
	01° 46' 18".0	04° 03' 42".0	01° 50' 55".1	02° 34' 32".8

Auxiliary base distances

	Rock to A	B to C	D to E	H to G	
γ	05° 45' 51".5	04° 35' 19".0	03° 31' 20".3	03° 56' 36".4	
α	80° 05' 58"	91° 19' 08"	58° 33' 46"	80° 34' 11"	
$\gamma + \alpha$	85° 51' 49".5	95° 54' 27"	62° 05' 06".3	84° 30' 47".4	
Cosec γ	9.956 59				to nearest 10"
Sin($\gamma + \alpha$)	0.997 40				
B	22.514	31.597			
D	223.580 m				

The values for BC, DE and HG are left for the reader to compute and with the data below the traverse can be computed as an example of a traverse with no starting or closing bearing. (see § 7.8.1.)

Coordinates

	Eastings (m)	Northings (m)	Height above sea level
Rock	360 373.08	142 760.04	420
Pilton	359 324.62	141 439.18	380

Local scale factor
K = 0.99962

7.10.5. USES AND VARIATIONS OF THE SUBTENSE TRAVERSE

The subtense traverse is suitable for:

(i) Open rolling country where taping might prove slower although practicable.
(ii) Congested areas where taping is impracticable due to traffic, or constructional work being in progress.
(iii) Cultivated areas where taping would cause damage to crops with resultant claims for compensation.

According to the ground conditions the standard four positions are varied. Figure 7.24 shows some possible variations.

Case (i) illustrates a river crossing where a longer than usual auxiliary base is formed with the base at the mid-point. Case (ii) illustrates the use of a fully observed triangle to change direction without an additional auxiliary base. Case (iii) is a connection to a triangulated spire with a subtense base. The angles to the spire at U and V have been measured and when the distance UV has been calculated the triangle UV spire can be solved.

7.10.6. ERRORS INVOLVED IN SUBTENSE MEASUREMENT

In the treatment of this aspect of subtense measurement θ, the subtense angle θ, is taken to be small and the invar bar length to be 2 metres.

7.10.6.1. Position (i); Direct Measurement

As $D = b/2 \cot \frac{1}{2}\theta$ then with the approximation mentioned above $D = b/\theta$.

$$\therefore \log D = \log b - \log \theta$$

By differentiation

$$\frac{dD}{D} = \frac{db}{b} - \frac{d\theta}{\theta}$$

Let E_D, E_b, and E_θ be the errors in D, b, and respectively, then

$$\left(\frac{E_D}{D}\right)^2 = \left(\frac{E_b}{b}\right)^2 + \left(\frac{E_\theta}{\theta}\right)^2 \tag{5}$$

Considering limiting errors in b, if $E_0/D = 1/10\,000$ and assuming no error in θ then, if b is 2 metres, $E_b = \pm 0.2$ mm. This is the equivalent of a temperature change of 100 degC. Hence errors arising from neglect of a temperature correction in an invar bar can be regarded as negligible.

If the bar is not exactly at right angles to the base by an amount ϵ. Then $E_b = b(1 - \cos \epsilon) \simeq \frac{1}{2}b\epsilon^2$,

hence $\qquad \epsilon' = (0.000\,2)^{\frac{1}{2}} \operatorname{cosec} 1' \simeq \pm 48'5$.

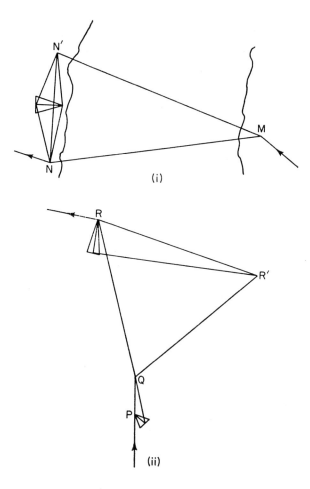

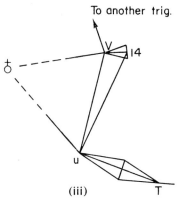

Figure 7.24

Errors arising from non-perpendicularity of the bar are not significant if the sighting device is correct to within 20′, but this error is systematic and any mispositioning of the bar will make the measure of the line too great.

Considering limiting errors in θ, if $E_D/D = 1/10\,000$ and assuming no error in b then, if $D = 40$ metres, $\theta = b/D = 1/20$, and $E_D/D = 20E_\theta$, $E_\theta \simeq \pm 1''$.

The expression
$$E_D = (E_\theta''/b \sin 1'')D^2 \tag{6}$$

which follows from (5) above, should be noted. It is important to realize that the error in a direct measurement by subtense methods increases as the square of the length of the line, or put another way the inaccuracy of angular measurement affects the result with the square of the distance.

7.10.6.2. Position (ii): Bar Mid-positioned

As $D = D_1 + D_2$, then, if errors in b are assumed to be negligible,
$$E_D{}^2 = E_{D_1}{}^2 + E_{D_2}{}^2$$
where
$$E_{D_1}/D_1 = E_{\theta_1}/\theta_1 \quad \text{and} \quad E_{D_2}/D_2 = E_{\theta_2}/\theta_2$$

Then
$$E_D{}^2 = (E_{\theta_1} D_1/\theta_1)^2 + (E_{\theta_2} D_2/\theta_2)^2$$

If the bar is exactly at the mid-point of the line, $\theta_1 = \theta_2$ and $D_1 = D_2 = \tfrac{1}{2}D$.

Let
$$E_{\theta_1} = E_{\theta_2} = E_\theta$$

then
$$E_D = 2\left(\frac{E_\theta{}^2}{b^2} \frac{D^2}{4} \frac{D^2}{4}\right)^{\tfrac{1}{2}} = \left(\frac{E_\theta''}{b 8^{\tfrac{1}{2}}} \sin 1''\right) D^2 \tag{7}$$

This indicates that by splitting the distance into two equal parts the error is reduced by 1/2·8 (compare equations (6) and (7)).

To find a limiting distance if $E_D/D = 1/10\,000$ and $E_\theta'' = \pm 1''$, then from (7)
$$D = \frac{1}{10\,000} \frac{2(8)^{\tfrac{1}{2}}}{1} \operatorname{cosec} 1'' = 113 \text{ metres}$$

D may be up to say 120 metres.

7.10.6.3. Position (iii): Auxiliary Base

As $D = B \sin(\alpha + \gamma)/\sin \gamma$ and $B = b/2 \cot \tfrac{1}{2}\theta$, then with the approximation mentioned above
$$D = B \sin(\alpha + \gamma)/\gamma \quad \text{and} \quad B = b/\theta$$
$$\therefore \quad \log D = \log b - \log \theta + \log \sin(\alpha + \gamma) - \log \gamma$$

By differentiation
$$\frac{dD}{D} = \frac{db}{b} - \frac{d\theta}{\theta} + \frac{d(\alpha + \gamma)}{\tan(\alpha + \gamma)} - \frac{d\gamma}{\gamma}$$

If errors in b are assumed to be negligible then with the usual notation and taking $(\alpha + \gamma) = \alpha$

$$\left(\frac{E_D}{D}\right)^2 = \left(\frac{E_\theta}{\theta}\right)^2 + \left(\frac{E_\gamma}{\gamma}\right)^2 + \left(\frac{E_\alpha}{\tan \alpha}\right)^2 \tag{8}$$

If $\alpha = 90°$ and $E_\theta = E_\gamma$, then

$$\gamma = \frac{B}{D} = \frac{b}{D\theta}$$

and
$$\left(\frac{E_D}{D}\right)^2 = \left(\frac{E_\theta}{\theta}\right)^2 + \left(\frac{E_\theta D\theta}{b}\right)^2 = E_\theta^2 \left(\frac{1}{\theta^2} + \frac{D^2\theta^2}{b^2}\right)$$

For this to be a minimum $E_\theta = 0$,

or
$$-\frac{2}{\theta^3} + \frac{2D^2\theta}{b^2} = 0$$

The latter relationship is achieved if $\theta = \gamma$, as $D/b = 1/\gamma\theta$, then $D = B^2/b$ or

$$B = (bD)^{\frac{1}{2}} \tag{9}$$

This indicates a minimum length of auxiliary base for a particular length of line. However it should be noted that the expression (9) only holds if error in the base angle γ equals the error in the bar angle θ and the base is at right angles to the line.

If $\alpha \neq 90°$ the requirements for E_α will vary according to the value of α. For instance if $E_\alpha = \pm 3'$ and $\alpha = 83°$ and supposing no error in θ or γ,

then
$$\frac{E_D}{D} = \frac{E_\alpha}{\tan \alpha} \simeq \frac{1}{9,300}$$

If $E_\alpha = \pm 30''$ for
$$\frac{E_D}{D} = \frac{1}{10\,000}$$

$$\tan \alpha = 10\,000.30 \sin 1'' \simeq 1\cdot 5$$

$$\alpha = 56°$$

This should be regarded as the lower limit for the angle between the base and the line. This means that one base can be used if the traverse angle is between 110° and 250° provided that the angle of the base is measured to within 30".

To find a limiting distance if $E_D/D = 1/10\,000$ and $E_\theta = \pm 1''$ then when $\alpha = 90°$ and $\theta = \gamma$,

$$\left(\frac{E_D}{D}\right)^2 = 2\left(\frac{E_\theta}{\theta}\right)^2 = \frac{2E_\theta^2 B^2}{b^2}$$

But from (9) above $B = (bD)^{\frac{1}{2}}$

$$\therefore \quad 1/10\,000 = \sin 1'' D^{\frac{1}{2}}$$

$$D = 425 \text{ metres}$$

If α is not 90° a lengthening of the base is desirable.

7.10.6.4. Position (iv) Auxiliary Base Mid-positioned

As $B = b/2 \cot \tfrac{1}{2}\theta$ then with the usual approximations

$$B = b/\theta$$

With the notation of formula (5) *et seq.*

$$\left(\frac{E_B}{B}\right)^2 = \left(\frac{E_b}{b}\right)^2 + \left(\frac{E_\theta}{\theta}\right)^2$$

and if $\alpha_1 + \gamma_1 = \alpha_2 + \gamma_2 = 90°$

$$\frac{E_{D_1}}{D_1} = \frac{E_{\gamma_1}}{\gamma_1} \quad \text{and} \quad \frac{E_{D_2}}{D_2} = \frac{E_{\gamma_2}}{\gamma_2}$$

Combining errors in the usual way and ignoring errors in b

$$E_D^2 = \left(\frac{E_\theta}{\theta}\right)^2 D^2 + \left(\frac{E_{\gamma_1}}{\gamma_1}\right)^2 D_1^2 + \left(\frac{E_{\gamma_2}}{\gamma_2}\right)^2 D_2^2 \tag{10}$$

Considering the relationship of B and D if $E_\theta = E_{\gamma_1} = E_{\gamma_2}$, $D_1 = D_2 = \tfrac{1}{2}D$

Then as $\quad\gamma_1 = \gamma_2 = \gamma = 2B/D \quad \text{and} \quad b/B = \theta$

then $\quad\gamma = 2b/D\theta$

Hence $\quad E_D^2 = E_\theta^2 \left(\dfrac{D^2}{\theta^2} + \dfrac{2D^2}{4\gamma^2}\right)$

or $\quad\left(\dfrac{E_D}{D}\right)^2 = E_\theta^2\left(\dfrac{1}{\theta^2} + \dfrac{1}{2\gamma^2}\right) = E_\theta^2\left(\dfrac{1}{\theta^2} + \dfrac{D^2\theta^2}{8b^2}\right)$

For this to be a minimum $E_\theta = 0$

or $\quad \dfrac{1}{\theta^2} + \dfrac{D^2\theta^2}{8b^2}$ is to be a minimum

i.e. $\quad -\dfrac{2}{\theta^3} + \dfrac{2D^2\theta}{8b^2} = 0$

or $\quad \theta^4 = \dfrac{8b^2}{D^2} = 2\gamma^2\theta^2$

This is achieved if $\theta = (2)^{\frac{1}{2}}\gamma$ or substituting for θ and γ in terms of B, b, and D

$$b/B = 2B(2)^{\frac{1}{2}}/D$$
$$B = 0{\cdot}6(bD)^{\frac{1}{2}} \tag{11}$$

The reader is left to verify the limiting value of D given in §7.10.4.

7.11. PRECISE TRAVERSING

7.11.1. METHOD

The *U.S. Coast and Geodetic Survey, Special Publication* (U.S. Govt. Printing Office: 1928) No. 137 recommends that first-order traverses should be used where the cost of triangulation will exceed twice the cost of first-order traversing. It is considered that traversing lacks the geometric checks obtained in triangulation and however precise the work it will always lag behind the first order or geodetic triangulation. The inherent weakness in a traverse is that the direction of each measured line is determined by a single series of angular observations, further any error in the angle will effect not only that line but all subsequent lines to a greater or lesser extent according to their lengths. A number of points arise in a comparison of traversing and triangulation.

7.11.1.1. *Location of Points*

Triangulation covers a whole belt of country establishing many points on hilltops difficult of access but well placed if marked with permanent beacons for plane tabling and resection for photogrammetric control.

Traversing merely establishes a line of points, frequently following a road, railway, or river of easy access but of little value to plane tabling. As these points are close to the more important detail they are well suited for photogrammetric control.

7.11.1.2. *Reconnaissance*

Triangulation demands a careful reconnaissance and in first order work it is essential that this be completed before observing begins. In forest country, which is a high proportion of the earths surface, the reconnaissance is extremely difficult and involves heavy clearing to test intervisibility.

Traversing, if sensibly planned, calls for less trial clearing as only one line has to go from point to point. Further if the traverse follows a road or railway the location of many points can be made using motor transport. Observations can begin on a traverse soon after the first turning point has been selected. It is unlikely that the surveyor doing the reconnaissance will retrace his steps by more than one or two stations on meeting bad ground.

7.11.1.3. *Supply and Communications*

Triangulation involves a number of small detached parties working over quite large area. These parties must be provisioned and in arid conditions the supply of water can be a major problem.

Traversing does not present the same problems, there will be at the most four separate parties all moving along the same route. These parties may be linked by road or track. If no track is available helicopters can link the parties and the whole operation can be carried out from a central base.

7.11.1.4. *Delays from Weather*

Triangulation is most prone to delays from bad visibility which applies not only to the observing but also to the reconnaissance.

Traversing under modern conditions will require equally good visibility for the angular work as the lines will be as long as sides in triangulation. The linear measurements can be made when visibility is poor and as there are fewer angles to measure than in triangulation there will be fewer delays due to poor visibility. An exception to this are delays arising from lack of clear nights for the astronomical observation.

7.11.1.5. *Errors and Mistakes*

Triangulation has the great merit that the individual closures of the triangles are available to the parties in the field. Other even more rigorous checks can quickly be applied for each figure as the work proceeds.

Traversing suffers from the great defect that mistakes made in the field are not readily appreciated and their precise location is difficult. In the days of linear measurement with tapes this was a very serious objection. Today there are fewer measured quantities and a comparison of various measures of a line with, say, the tellurometer are in themselves a check on the level of precision.

7.11.1.6. *Propogation of Error*

Triangulation has always been held to be very strong in maintaining scale and azimuth. In the triangulation of the U.K., where the complex network was believed to ensure little change of scale, there are indications that a scale error of the order of 1 in 40 000 may have been accumulated in parts of the network.

Traversing is always regarded as being prone to errors of swing (azimuth). This may well be due to insufficient attention being paid to the angular work and to control by azimuth. Scale is well controlled by the frequent linear measurements, provided some form of standardization is carried out. Closures between triangulation stations point as much to errors in the triangulation as to errors in the precise traverse.

7.11.2. CONDITIONS UNDER WHICH PRECISE TRAVERSING IS NECESSARY

(i) In flat country where buildings or trees would demand for triangulation high towers and short triangulation sides.
(ii) In forest country where clearing is required on every hilltop.
(iii) Where climatic conditions prohibit long sides.
(iv) Where the immediate need for control is great and outweighs any possible advantages of triangulation.
(v) In built up areas where triangulations is impracticable and a high degree of positional accuracy is required.

7.11.3. METHOD OF WORK

As a high degree of accuracy is required the basic principles of good traversing outlined earlier become more significant.

7.11.3.1. *Linear Measurement*

Reference should be made to Chapter 13 on the detailed operation of the instruments mentioned below.

In all modern precise traverses electronic distance measuring equipment is used. These instruments provide the slant distance to an accuracy of better than 1 in 100 000. Vertical angles are measured at the two ends of the measured line. Each line is measured twice; with the tellurometer MRA 1 the master and remote instruments exchange position; with the geodimeter two reflectors are used, one at the back station and one at the forward station. It is inconvenient to keep changing the master and remote tellurometers around and an improvement in efficiency is obtained if there are two master sets with a remote set between which is operated by the angle party. With the MRA 2 and 3 tellurometers the line can be measured using each set as a master in turn. It is still important to separate each measure of the line by between 24 and 48 hours. If the reduced transit differ from the mean by more than 1 part in 100 000 a third determination should be made.

A line of 30 kilometres is considered ideal for a tellurometer traverse, although lines as short as 3 kilometres or as long as 60 kilometres can easily be measured. Short lines however waste time and excessively long lines vitiate the angular work. Almost any line can be measured with the tellurometer but it will not reach the required precision unless attention is paid to the dangers of excessive ground swing and to the problems of determining the meteorological conditions for the whole line.

With the Model 4 geodimeter shorter lines are measured, meteorological conditions are less critical and ground swing does not occur. Consequently the reconnaissance is much easier but progress is slower than with the tellurometer.

It is unlikely that the lines of a precise traverse will ever in the future be taped. The salient features of the methods used in the past are:

(i) Taping was done in catenary with invar wires or tapes.
(ii) The procedure was identical to base measurement with wires or tapes except that one end is anchored to the straining trestle.
(iii) Great care was taken in standardization.
(iv) Booking was done in duplicate by two bookers.
(v) Several readings were taken for each bay.

Linear measurements are particularly prone to systematic errors and it is these that the methods used aimed to eliminate. Casual errors, which will in a base line effect the scale of the triangulation, tend in a precise traverse to cancel out. Consequently a relaxation of the procedure for base measurement is permitted in respect of those precautions guarding against casual error.

7.11.3.2. *Angular Observations*

The selection of stations suitable for the tellurometer, i.e. lines producing little ground swing, means that frequently the lines are bad from an angular point of view. High clean lines are unwelcome in a modern tellurometer traverse and errors from lateral refraction may arise from deliberately selected grazing rays.

It is not unreasonable to suppose that the atmospheric conditions will be such as to cause a reversal of the refraction conditions from day to night. The implications of this are that two sets of observations are required. A suggested procedure is two sets of eight zeros, the first by day, the second by night. Should the means of the two sets differ by more than $1''.5$, two more sets will be required.

To organize this will require a late afternoon observation followed by an early evening observation; if these fail to agree a second two sets may be taken before and after dawn.

Vertical angles are required to reduce slant distances to the horizontal and to obtain heights of the stations above datum to reduce the lines to the spheroid. Eight pointings, four on each face, are worth doing although less will suffice for reduction of the linear measurements.

7.11.3.3. *Astronomical Observations*

The transmission of azimuth through a traverse has already been described as weaker than through a triangulation chain. Both theory and practice indicate the need for frequent azimuths.

Considering a straight traverse of n lines equal in length. The standard error $\pm e_L$ of linear measurement is taken as being $\pm s/200\,000$ where s is the length of the line. The standard error of angular measurement e_a is taken as $\pm 1''$ that is $\pm 1/200\,000$.

It will be shown later that the standard error in position arising from angular errors at the end of a straight traverse of n legs is approximately of the form $Se_a(n/12)^{\frac{1}{2}}$, where $ns = S$ the total length of the traverse.

The total standard error in position E_p is given by:

$$E_p^2 = ne_L^2 + n^2 s^2 e_a^2 n/12$$

hence

$$\frac{E_p^2}{n^2 s^2} = \frac{e_L^2}{ns^2} + \frac{e_a^2 n}{12}$$

$$\therefore \frac{E_p}{S} = \frac{1}{200\,000}\left(\frac{1}{n} + \frac{n}{12}\right)^{\frac{1}{2}}$$

This is a minimum when $n = 3$ and 4. Alternatively if $E_p/S = 1/200\,000$ then $1/n + n/12$ is equal to 1 and n is approximately 11 or 1.

This treatment indicates that azimuths are required at least at every eleven stations and preferably more frequently, say, every eight stations.

All astronomical observations depend upon indications of the vertical made with a spirit level. The deviation of the vertical from its theoretical (geodetic) position can be resolved into two components; deviation in latitude and deviation in longitude. The effect of deviation in latitude is cancelled out by effective pairing. The effect of deviation in longitude is given by the Laplace equation.

(Astro-Geodetic) azimuth = (Astro-Geodetic) longitude × sin geodetic latitude. The signs depending upon the convention North latitudes and East longitudes positive. This requires an astronomical value for both azimuth and longitude and theory indicates that this should be about every ten stations. The disadvantage of such a specification is that time spent on observing and computing at the Laplace station may quite easily exceed that spent on the intervening traverse. If the Laplace correction is ignored systematic errors of the order of $\pm 5'' \sin \phi$ must be faced; viz. $\pm(1/40\,000) \sin \phi$. If the standard error in determination of astronomical longitude is of the order of $\pm\frac{1}{2}''$ of arc a random error of $\pm(1/400\,000) \sin \phi$ is present. In most latitudes this can be regarded as negligible and difference of astronomical and geodetic longitude at each azimuth station can be obtained by interpolation through the traverse.

Reference must be made to the methods used by the Division of National Mapping, Australia. For a full explanation the reader is referred to the papers and publications of that department. The practice is to include Sigma Octantis in each round of horizontal angles at a traverse station. The field parties do not make latitude and longitude observations during the course of the traverse but at selected stations a second visit is made to determine longitude. No azimuth is included in the adjustment unless an astronomical longitude has been determined at that station. Such a method, which is practicable in latitudes where the expectation of clear nights is high, will provide ample data for a most rigorous adjustment of bearings throughout the traverse.

7.11.4. COMPUTATION OF PRECISE TRAVERSES

Precise traverses can be computed in terms of geographical coordinates or rectangular coordinates (see Example 12 and figure 7.25) such as the U.T.M. projection. With the almost universal use of electronic distance measuring equipment the average length of precise traverse lines are much greater than in the past and the approximations formerly permissible are no longer valid. These approximations are however permissible if a high accuracy is not sought.

The stages in computation in terms of U.T.M. rectangular coordinates are given below.

7.11.4.1. *Approximate Convergence*

Calculate convergence at the initial azimuth station and compute the traverse on the plane using the observed angles and the spheroidal distances. An approximate value for the convergence at the intervening azimuth stations can be obtained from $c'' = d\lambda'' \sin \phi$ and by using values so obtained the observed angles can be provisionally adjusted. Values for $d\lambda$ and ϕ can be scaled from a map. The approximate coordinates obtained are used to calculate the different corrections mentioned below.

7.11.4.2. *Exact Convergence*

Using the provisional coordinates calculate exact values for the convergence at the azimuth stations. Using a provisional latitude and longitude obtained from the provisional coordinates (see §4.2.3.2 for this computation), calculate a correction to the observed azimuths for deviation of the vertical (Laplace equation).

7.11.4.3. *Scale Factor*

Calculate the scale factor K for each line. For short lines (up to 15 km) this can be taken as the value for the mid-point of the line.

i.e. $$K_{AB} = K_m = K_0 \left\{ 1 + \frac{(E_m')^2}{2\rho\nu} \right\}$$

Example 12. Traverse Computation on the Projection

Data

From	To	Observed direction	Spheroidal distance (m)	Observed azimuth
A	X			341° 19′ 03″.62
A	X	00° 00′ 00″.00	23 789.408	
	P	91° 06′ 31″.03	20 697.082	
P	A	00° 00′ 00″.00		
	Q	183° 27′ 49″.48	28 274.282	
Q	P	00° 00′ 00″.00		
	R	304° 35′ 17″.82	18 123.752	
R	Q	00° 00′ 00″.00		
	S	179° 29′ 16″.73	17 482.561	
S	R	00° 00′ 00″.00		
	T	16° 12′ 36″.77	25 762.065	
S	T			36° 32′ 58″.79

Station	Eastings (m)	Northings (m)
A	253 650.82	5566 384.41

(1) Provisional coordinates

Station	Eastings (m)	Northings (m)	
X	247 082.07	5589 248.95	
A	253 650.82	5566 384.41	Data
P	273 649.98	5571 713.86	
Q	301 360.64	5577 331.14	
R	294 241.98	5560 663.96	
S	287 519.48	5544 525.56	
T	303 670.21	5564 596.38	

Using U.T.M. tables (Clarke 1866)

(2) Convergence $C'' = (XV)q - (XVI)q^3 + F_5$

At A	Eastings 253 651	Northings 55 66 384
$E' \times 10^{-6} = q = 0.246\ 349$	$q^3 = 0.014\ 950$	$\phi' = 50°15′05″.7$
$(XV) = 38\ 823.26$	$(XVI) = 774.7$	$F_5 = 0″.02$
$C'' = 9564″.07\ -$	$11″.58\ +$	$0″.02$
At A $C'' = 9552″.51$	$C = 2° 39′ 12″.51$	

Also at S $C = 2° 16′ 24″.65$

(3) Scale factor line AX $K_{AX} = \frac{1}{6}(K_A + 4K_M + K_X)$

$K = K_0(1+(XVIII)q^2 + 0.00003q^4)$		$E' \times 10^6 = q$	$K_0 = 0.9996$	$(XVIII) = 0.012285$	
Station	q	q^2	q^4	$(XVIII)q^2$	$0.00003q^4$	K
X	0.252 918	0.063 968	0.004 092	0.000 785 (91)	0.000 000 (12)	1.000 38571
Mid	0.249 634	0.062 317	0.003 883	0.000 765 (62)	0.000 000 (12)	1.000 36543
A	0.246 349	0.060 688	0.003 683	0.000 745 (61)	0.000 000 (11)	1.000 34542

Similarly K_{AP} 1.000 2865
K_{PQ} 1.000 1554
K_{QR} 1.000 1022
K_{RS} 1.000 1372
K_{ST} 1.000 1134

K_{AX} 1.000 3654

Example 12. Continued

(4) $(t-T)$ Station A $(t-T)_{AB} = +Q(N_A - N_B)10^{-3}$

Line	$(2E'_A + E'_B)10^{-3}$	Q w- er	$(N_A - N_B)10^{-3}$	$(t-T)$
AX	745·616	−0·634 56	−22·864	+14·51
AP	719·048	−0·61197	−5·329	+3·26

Q having been interpolated from tables against argument of $(2E'_A + E'_B)$
Similarly for the other stations see (5) below for values

(5) Correction to observed directions

From	To	Observed directions	$(t-T)''$	Plane grid direction	Plane grid angle
A	X	00° 00' 00"·00	+14"·51	00° 00' 14"·51	
	P	91° 06' 31"·03	+3"·26	91° 06' 34"·29	91° 06' 19"·78
P	A	00° 00' 00"·00	−3"·17	359° 59' 56"·83	
	Q	183° 27' 49"·48	+3"·11	183° 27' 52"·59	183° 27' 55"·76
Q	P	00° 00' 00"·00	−2"·98	359° 59' 57"·02	
	R	304° 35' 17"·82	−8"·56	304° 35' 09"·26	304° 35' 12"·24
R	Q	00° 00' 00"·00	+8"·66	00° 00' 08"·66	
	S	179° 29' 16"·73	−8"·57	179° 29' 08"·16	179° 28' 59"·50
S	R	00° 00' 00"·00	+8"·66	00° 00' 08"·66	
	T	16° 12' 36"·77	+10·61	16° 12' 47"·38	16° 12' 38"·72

(6) Bearing of azimuth lines

At A 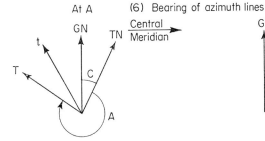 At S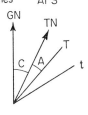

Grid bearing = $C + A + (t-T)$

A	341° 19' 03"·62
C	2° 39' 12"·51
$(t-T)$	14"·51
	343° 58' 30"·64

Grid bearing = $C + A + (t-T)$

A	36° 32' 58"·79
C	2° 16' 24"·65
$(t-T)$	10"·61
	38° 49' 34"·05

where $E'_m = \frac{1}{2}(E'_A + E'_B)$, and K_0 the central scale factor, and E' the grid distance from the central meridian. For long lines (over 15 km)

$$K_{AB} = \tfrac{1}{6}(K_A + 4K_m + K_B) = K_0\left\{1 + \frac{E_A'^2 + E_A'E_B' + E_B'^2}{6\rho\nu}\right\}$$

In the case of the U.T.M. projection $E' = E - 500\,000$ for points east of the central meridian and $E' = 500\,000 - E$ for points west of the central meridian.

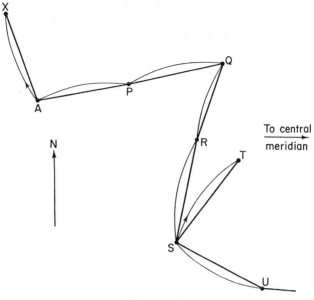

Figure 7.25

7.11.4.4. $t - T$ Correction

Calculate for each line the $t - T$ value (see §4.2.7.3), using for short lines;

$$(t - T)'' = \frac{(N_A - N_B)E_m'}{2\rho\nu \sin 1''}$$

and for long lines

$$(t - T)'' = \frac{(N_A - N_B)(2E_A' + E_B')}{6\rho\nu \sin 1''}$$

7.11.4.5. Grid Angles

Using the $t - T$ values a correction is applied to the measured angles, being the difference of the $t - T$ values for the adjacent lines, thereby giving grid angles. Where an azimuth line forms part of the traverse its grid bearing is only obtained after application of the $t - T$ and convergence corrections.

A diagram of the route of the traverse is used to determine the sign of the corrections and the manner in which the corrections should be combined to correct the angles.

7.11.4.6. Closing Error

With the grid bearing of each azimuth line and the grid angles a closing error is obtained between each pair of azimuths and this is distributed in a systematic way. Usually the error is so small that an equal correction to each angle is justified, alternatively some system of weighting may be used based on the length of the adjacent lines. Most systems of weighting in a well planned and executed traverse result in an equal distribution of the error.

With the adjusted angles final bearings for each line are obtained.

7.11.4.7. *Computation*

Using the bearings obtained in §7.11.4.6 and distances corrected for scale by application of the factors found in §7.11.4 the traverse is computed in the usual way. As many of the calculations are not self checking duplicated computations by different computers are desirable.

7.11.5. COMPUTATION OF LONG SECONDARY AND MINOR TRAVERSES

The treatment is generally as in §7.11.4 but as high accuracy is not sought certain approximations are permissible.

(i) Azimuths are corrected for convergence computed from geographical coordinates scaled from a map. The approximate formula given above $c'' = d\lambda'' \sin \phi$ being used.
(ii) No corrections are applied for deviation of the vertical.
(iii) A mean scale factor for each 20 km portion of the traverse is computed and applied to the measured lengths.
(iv) If the lines have been taped they are reduced to the spheroid by use of a mean correction factor for each line, or for several lines if the lines are short.
(v) Regardless of the length of line the $t - T$ corrections are ignored and the bearings carried forward using the observed angles. The misclosure of bearings at the forward azimuth is distributed equally to the measured angles.

7.12. PROPAGATION OF ERROR IN THEODOLITE TRAVERSES

7.12.1. INTRODUCTION

An investigation of the propagation of error in theodolite traverses is important for three reasons. Firstly, to enable the surveyor to distinguish between discrepancies which are reasonable and those which indicate gross errors or mistakes. Secondly, to determine what degree of accuracy is required in the field work to achieve a certain required accuracy in the final results. Thirdly, to ensure that the adjustment of the discrepancies shall not be entirely arbitrary.

Considering a traverse in which the error in position of one station relative to the previous station on the traverse shall not exceed $s/5\,000$ where s is the length of the line (e.g. 0·1 metre on lines of length 500 metres). The closure of such a traverse when of n lines of equal length should not exceed $(n)^{\frac{1}{2}}$ of this quantity, viz. $n^{\frac{1}{2}}s/5\,000$. As a proportion of the total length of the traverse this is $n^{\frac{1}{2}}s/(5\,000\,ns)$ or $1/(5\,000\,n^{\frac{1}{2}})$.

Such a treatment assumes that the traverse is straight, that there are equal lengths for all lines and that the errors obey a random distribution. In practice such conditions never exist but as an estimate of the permissible closing error of a traverse where the error in position of one station to its adjacent stations should not exceed $s/5\,000$ it is fair and reasonable. It must however be emphasized that a closure which accords with this figure does not prove that this specification has been fulfilled although it does indicate that in all probability it has been fulfilled.

7.12.2. EFFECT OF ERRORS IN THE MEASURED ANGLES

Consider a traverse section PQ which may be regarded as straight and to have n lines each of length s (see figure 7.26).

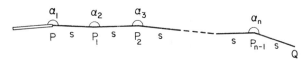

Figure 7.26

If the traverse is unadjusted an error in angle α_1 of magnitude $\delta\alpha_1$ will produce a displacement at Q at right angles to the line of the traverse of magnitude $ns\delta\alpha_1$.

Similarly if α_2 is in error by $\delta\alpha_2$ an additional displacement of magnitude $(n-1)s\delta\alpha_2$ is produced.

The total displacement resulting from an error in each angle of $\delta\alpha_1, \delta\alpha_2 \cdots \delta\alpha_n$ will be

$$ns\,\delta\alpha_1 + (n-1)s\,\delta\alpha_2 + (n-2)s\,\delta\alpha_3 + \cdots + 2s\,\delta\alpha_{(n-1)} + s\,\delta\alpha_n$$

If $\delta\alpha_1 = \delta\alpha_2 = \cdots = \delta\alpha_n = \delta\alpha$

the total displacement is $s\,\delta\alpha\{n + (n-1) + \cdots + 2 + 1\}$

$$s\,\delta\alpha\tfrac{1}{2}n(n+1) \tag{1}$$

If e is the standard error of each measured angle the standard error in the determined position of Q arising from this source alone is

$$\{n^2s^2e^2 + (n-1)^2s^2e^2 + \cdots + 2^2s^2e^2 + 1^2s^2e^2\}^{\frac{1}{2}}$$
$$= \pm se\{n^2 + (n-1)^2 + \cdots + 2^2 + 1^2\}^{\frac{1}{2}}$$
$$= \pm se\{\tfrac{1}{6}n(n+1)(2n+1)\}^{\frac{1}{2}} \tag{2}$$

If the traverse is adjusted as the result of an azimuth observed at Q there will now be $n+1$ angles. The misclosure in bearing E will be $\delta\alpha_1 + \delta\alpha_2 + \cdots + \delta\alpha_n + \delta\alpha_{(n+1)}$ and this will be distributed equally to each measured angle. The correction to each angle is $E/(n+1)$ and will have the effect of swinging the position of Q back to its correct position (from that given by equation (1)) by an amount $s\{E/(n+1)\} \times \tfrac{1}{2}n(n+1)$ that is $\tfrac{1}{2}nsE$.

The standard error in this 'swing back' is thus

$$[\tfrac{1}{4}n^2s^2e^2 + \tfrac{1}{4}n^2s^2e^2 + \cdots + \tfrac{1}{4}n^2s^2e^2]^{\frac{1}{2}}$$

there being $n+1$ terms this reduces to

$$\pm\tfrac{1}{2}nse(n+1)^{\frac{1}{2}} \tag{3}$$

Consequently the standard error in the position of Q after adjustment of the angles will be from (2) and (3) $\pm\{[se\{\tfrac{1}{6}n(n+1)(2n+1)\}^{\frac{1}{2}}]^2 - [\tfrac{1}{2}nse(n+1)^{\frac{1}{2}}]^2\}^{\frac{1}{2}}$

assuming an improvement in the survey. This is

$$\pm se[\tfrac{1}{2}n(n+1)\{\tfrac{1}{3}(2n+1)-\tfrac{1}{2}n\}]^{\frac{1}{2}} = se\{\tfrac{1}{12}n(n+1)(n+2)\}^{\frac{1}{2}} \quad (4)$$

Expressing the total length of the traverse as S then (4) may be written as

$$Se\{(n+1)(n+2)/12n\}^{\frac{1}{2}} \quad (5)$$

For practical purposes this may be simplified as

$$\pm Se\{\tfrac{1}{12}(n+3)\}^{\frac{1}{2}} \quad \text{or} \quad \pm Se(\tfrac{1}{12}n)^{\frac{1}{2}}$$

Expression (2) for the unadjusted traverse may be written as

$$\pm Se\{(n+1)(2n+1)/6n\}^{\frac{1}{2}}$$

which is simplified to $\pm Se\{\tfrac{1}{12}(4n+6)\}^{\frac{1}{2}}$ or $\pm 2Se(\tfrac{1}{12}n)^{\frac{1}{2}}$.

It will be seen that adjustment of the angles of a traverse effectively halves the probable displacement of the terminal point.

The following method of determining the standard error of the angles of a theodolite traverse is suggested.

If M is the total misclosure between azimuths after all corrections have been applied and n is the number of lines in the section including one azimuth line.

Then if there are N sections, i.e. portions of a traverse between azimuths, the standard error of the measured angles may be estimated from

$$\left\{\frac{M^2/(n+1)}{N}\right\}^{\frac{1}{2}} \quad (6)$$

This is illustrated by Example 13.

Example 13. Error of Measured Angles

Section	n	M"	Misclosure" per station	$M^2/{n+1}$
A	15	+4·3	+0·29	1·2
B	17	+14·7	+0·86	12·0
C	14	−3·1	−0·22	0·6
D	16	−11·8	−0·74	8·2
E	15	−8·7	−0·58	4·7
F	13	+1·4	+0·11	0·1
				26·8

$$[M^2/{n+1}] = 26·8$$

Standard error $= \sqrt{26·8/6} = 2\overset{\shortmid\shortmid}{\cdot}2$

7.12.3. EFFECT OF ERRORS IN THE AZIMUTH DETERMINATIONS

Consider a traverse of n lines of length s as in §7.12.2 except that now the angles are considered to be error free.

If there is no closure the displacement arising from an error ϵ in the initial azimuth is $ns\epsilon$.

If the next azimuth was error free the misclosure would be ϵ and the adjustment per angle would be $\epsilon/(n+1)$. The displacement at the end of the section would be reduced by

$$\tfrac{1}{2}ns(n+1)\epsilon/(n+1) \quad \text{to} \quad \tfrac{1}{2}ns\epsilon \quad \text{or} \quad \tfrac{1}{2}S\epsilon. \tag{7}$$

The case when an azimuth is observed between two error free azimuths is worth noting (see figure 7.27).

If A and C are error free azimuths and B is in error by ϵ the displacement of C from its correct position relative to B is $\tfrac{1}{2}\epsilon S_2$ and for A is $\tfrac{1}{2}\epsilon S_1$. Consequently the displacement of C from its correct position relative to A is $\tfrac{1}{2}\epsilon(S_1 + S_2)$.

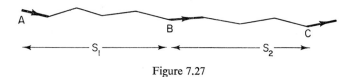

Figure 7.27

7.12.4. EFFECT OF ERRORS IN TAPING

Consider a traverse in which the following standard errors occur:

(i) An error of magnitude r in each tape length of random origin.
(ii) An error of magnitude r' in each tape length constant in sign and magnitude through the whole traverse.
(iii) An error of magnitude r'' in each tape length changing in a random manner after every q bays (say after standardization).

The combined standard error after n bays is

$$R^2 = \{(n)^{\tfrac{1}{2}}r\}^2 + (nr')^2 + \{(nq)^{\tfrac{1}{2}}r''\}^2$$

The significance of these terms may now be considered if $r = \pm 0.01$ ft in a 300 ft tape or 1/30 000, $r' = \pm 0.001$ ft in a 300 ft tape or 1/300 000, $r'' = \pm 0.003$ ft in a 300 ft tape or 1/100 000, $n = 2\,000$, and $q = 400$, then $ns = 600\,000$ ft, $n^{\tfrac{1}{2}}r = \pm 0.45$, $nr' = \pm 2.00$, $(nq)^{\tfrac{1}{2}}r'' = \pm 2.69$, and $R = \pm 3.38$ ft.

It is apparent that the significant portion of the error arises from the systematic effects and the random errors can safely be regarded as cancelling out.

7.12.5. EFFECT ON TRAVERSES OF ERRORS IN ELECTRONIC DISTANCE MEASURING EQUIPMENT

These errors will be propagated according to the number of lines. As a result random errors are more significant than in a taped traverse but the systematic portion of the total error is still the more serious and must be continually guarded against. Particular attention must be paid to index error and uncertainty in the frequency of the transmitted signal.

7.12.6. SHAPE OF TRAVERSE

So far we have considered straight traverses with lines of approximately equal lengths. It is important to compare the probable misclosure of a small surround traverse with a straight traverse of the same length.

Traverses

In the surround taping errors will tend to cancel out whereas in the straight traverse they will produce an error vector along the line of the traverse.

The angular errors in the surround will produce two probable displacements at right angles to each other, at worst each half the probable displacement of the straight traverse.

If the probable displacement due to angular errors in the straight traverse is taken to be $2W$ then the probable displacement in the surround due to angular errors will be, at worst, $(2^{\frac{1}{2}})W$. Therefore due to angular errors alone the displacement in the straight traverse is 1·4 times greater and if the linear errors are also considered it is fair to say that the straight traverse will have a probable displacement of its end point at least twice that of a surround of equal length. In addition a straight traverse must connect two points already fixed which will themselves be in error.

This result is important, there is always a tendency to overestimate the accuracy of traverses that close on themselves and consequently unjustly cast suspicion on internal traverses in a network.

8 *Vertical Control*

8.1. INTRODUCTION

The *vertical* is the direction which a plumb line takes up when it hangs freely under the effect of the Earth's gravitational pull. A *horizontal surface* is a surface at right angles to the vertical. The *height* of a point may be defined as the *linear distance* from the point up or down the vertical through the point, to a reference horizontal surface or *datum*. Later in this chapter a different definition of height is considered (see §8.3.9).

If the horizontal datum surface covers a small portion of the Earth's surface it may be sufficient to consider it as a plane surface throughout its entirety, but when considering the surface of the Earth as a whole, the horizontal surface will be curved and everywhere be at right angles to the verticals, as in figure 8.1

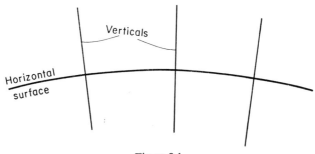

Figure 8.1

The horizontal datum surface usually taken as the reference surface for heights is that which closely approximates to *mean sea level*, taken from observations on tide gauges over a period of years.

Since most vertical angles between ground stations are of the order of 1° or less, computational approximations are usually permissible, though on occasion quite large angles up to 10° may be encountered.

Most heighting processes are concerned with differences in height measured over comparatively short distances carried forward over long distances by a chain of many measurements. For this reason, small *systematic errors* may often accumulate in serious proportions and must therefore be avoided where possible.

A knowledge of the height of a point on the surface on the Earth is required for all or some of the following purposes:

(i) To enable survey measurements to be reduced to the horizontal at sea level.

(ii) To enable the surveyor to calculate whether two points are intervisible on the ground.
(iii) To enable drainage works to be surveyed so that water may flow in the desired directions.
(iv) To enable relief to be depicted on topographical and air maps.
(v) To enable roads and railways to be constructed in such a manner that steep hills are avoided.
(vi) To provide information for scientists concerning the shape and structure of the Earth, and movements in its surface.

8.2. METHODS OF DETERMINING RELATIVE HEIGHTS

The following five ground methods are used to determine the relative heights of survey points:

(i) Hydrostatic levelling.
(ii) Spirit levelling.
(iii) Trigonometrical heighting.
(iv) Barometric heighting.
(v) Tacheometric heighting. (This is considered in Chapter 9 only.)

8.2.1. HYDROSTATIC LEVELLING

This is probably the oldest form of levelling. A tube containing water will readily give two points A and B at the same height due to the balance set up by the water under the force of gravity (see figure 8.2). This method of levelling is

Figure 8.2

much used in Holland where the differences in height are very small and great accuracy is essential. Pipes of up to 10 km long have been used to determine points at the same height. For convenience, flexible pipes are laid out along the bed of a canal by a survey ship. The heights found by this method are 'dynamic' heights (see §8.3.9).

The main problem associated with this method is that of eliminating air bubbles from the pipe when filling it with water, and when joining up sections of the pipe.

8.2.2. SPIRIT LEVELLING

In this method of height determination, a horizontal line is established by the observer with the aid of a spirit bubble, or a plumb line or a freely suspended compensator system, which enables him to sight through a telescope in a horizontal direction. For example, in figure 8.3 the points A, B, and C are

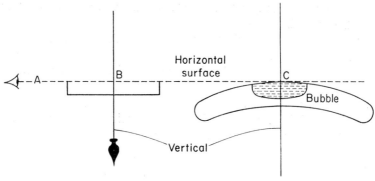

Figure 8.3

seen by the observer to be at the same height. Although spirit bubbles are giving way to compensators, the term *spirit levelling* is retained here.

For ease of operation, spirit levelling is used in conjunction with one or two graduated *rods* or *staves* to determine the difference in height between two points. In figure 8.4 the height of B with respect to that of A is found by taking

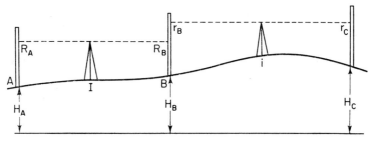

Figure 8.4

readings R_A and R_B on the vertical staves at A and B. For

$$R_A + H_A = R_B + H_B$$

whence
$$H_B = H_A + R_A - R_B \tag{1}$$

Since A is the backward point if we are moving from A to B, the reading on A, i.e. R_A, is called the 'back sight reading' or simple the *backsight*, and that on B, i.e. R_B, is called the 'fore sight reading' or simply the *foresight*. It will be noted that the height difference between A and B is given both in *magnitude* and *sign* by the expression (1). We always consider the height difference in the sense

$$\text{Difference in height} = \text{Backsight} - \text{Foresight}$$

After making a reading R_B on B with the instrument in position I, the staff at B is moved round to face the new instrument position at i, and the staff at A is moved to the forward position C. Thus the backsight from position i is r_B and the foresight is r_C. The foot of the staff at B should not change height in turning the staff round to face i; hence it is necessary that some point whose height is well defined should be chosen for the *change point* B. Such a point would be the head of a rivet on a drain cover, a kerb stone, etc. If a point is heighted but not used as a change point, it is called an *intermediate point*, and the sight to it is an *intermediate sight* (I.S.).

8.2.3. TRIGONOMETRICAL HEIGHTING

In figure 8.5 the difference in height between the points A and B is given by

$$\Delta H = S \tan \theta \tag{2}$$

where S is the horizontal distance AB', and θ is the vertical angle recorded by a theodolite circle against a horizontal datum determined by a bubble or compensator. Since $\tan \theta$ is negative when θ is negative, and positive when θ is positive

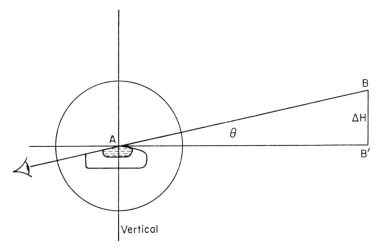

Figure 8.5

for angles less than 90°, equation (2) gives both the magnitude and sign of ΔH. If either the height of A or the height of B is known, the height of the other station may be calculated. In precise work, the vertical angle is observed at both ends of the line, i.e. at both A and B. The length of sights used in this method may be anything up to 50 km, though accurate results are limited to lines of 10 km or less.

8.2.4. BAROMETRIC HEIGHTING

This method of measuring height differences depends on the variation of atmospheric pressure with height above sea level. Since pressure P is related to the acceleration due to gravity g, the density of the air ρ, and the height h by the expression

$$P = f(g, \rho, h) \tag{3}$$

if P is measured at two places, where g and ρ are sensibly constant the pressure–height relationship for two points is

$$h_1/h_2 = k \log_e(P_1/P_2) \qquad (4)$$

Whence if one height is known we may find the other.

This method of heighting is especially convenient where points are not intervisible, such as in forested country, or where rapid, but relatively inaccurate results are required.

8.2.5. ACCURACIES OF THESE METHODS OF HEIGHTING

It is impossible to give figures for the accuracy of a survey method which will apply to all circumstances, since the accuracy of a method depends to some extent on the care with which the measurements are carried out, the time spent on the operation, and various imponderable factors such as the weather conditions pertaining at the time of the measurements. However, as a guide to the student, the following accuracies may be expected if normal technique is adopted.

(i) Hydrostatic and geodetic levelling: a probable error of $0·005\ (M)^{\frac{1}{2}}$ feet, where M is the length of the line in miles.
(ii) Ordinary spirit levelling: a p.e. of $0·02\ (M)^{\frac{1}{2}}$ feet.
(iii) Trigonometrical heighting: a p.e. of $0·20\ (M)^{\frac{1}{2}}$ feet, though accuracies as high as for geodetic spirit levelling can be obtained with special care.
(iv) Barometric heighting: about 5 feet provided operations are limited to an area of about five miles square, and a difference of elevation of not more than 1 000 feet.

Spirit levelling, trigonometrical heighting, and barometric heighting, are considered in the remainder of this chapter. Hydrostatic levelling is not considered, since it is of a highly specialized nature, and is not, as yet, widely used by surveyors.

8.2.6. BENCH MARKS

As described in Chapter 1 (§1.2.6), a *bench mark* is a permanent survey mark whose height is known, and from which levelling and heighting are controlled. Wherever possible a line of levels or series of trig heights should be tied into bench marks; but it is essential that barometer heighting is controlled by bench marks.

8.3. SPIRIT LEVELLING

The principle of levelling with the aid of a staff or staves is explained above in §8.2.2. The same simple technique is common to all types of spirit levelling irrespective of the accuracy required. The following section explains the basic field work involved, the manner in which the observations are booked in the field, and the various simple checks that are applied to the arithmetical operations required to produce the final heights of the stations. In many books, the subject of geodetic levelling is described after that of simple or engineering

Vertical Control

levelling. Since these have so much in common, we shall treat the most accurate work first and point out those operations that are unnecessary if a lower standard of result is sufficient.

8.3.1. TYPES OF LEVEL

The various types of level in common use have been described in Chapter 3. It should not be forgotten that a theodolite may be used as a level on occasion.

8.3.2. TYPES OF STAFF

A large number of different types of staff are available to the surveyor, as will be seen from figure 8.6. Folding or telescopic staves are easier to carry whereas

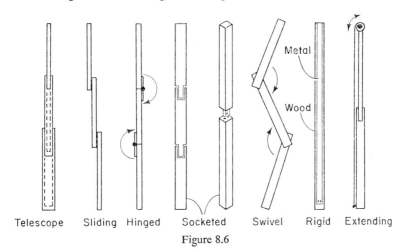

Figure 8.6

rigid staves are more accurate since they maintain their length better over a period of time. Geodetic levelling staves are constructed of wood with the graduations on a strip of invar held securely at the bottom of the staff but which is free to expand elsewhere along its length. The coefficient of expansion of invar is so small that temperature corrections need not be applied as a rule. The figuring of staves also varies from type to type. Some geodetic staves have two different scales side by side so that a gross error in reading the staff may be detected at once and the reading repeated. To ensure that the staff height at a change point is not altered when it is turned round, a special steel foot may be used as shown in figures 8.7. The staff rests on its dome shaped top.

8.3.3. BOOKING OF READINGS

Figure 8.8(a) shows a cross-section along the route taken by a surveyor levelling from point 1 to point 11. The instrument positions are at A, B, C, and D, and the change points are at 5, 9, and 10. Figure 8.8(b) shows the approximate plan positions of the points shown in figure 8.8(a). Table 8.1 shows a typical set of readings for the levelling illustrated in figures 8.8. The algebraic version given on the left of table 8.1 is to assist in the explanation. The instrument is set up at A and carefully levelled. A staff is held at point 1

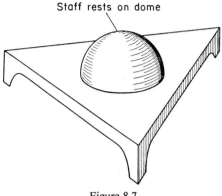

Figure 8.7

and the reading R_1 is taken. The staffman then moves in turn to points 2, 3, 4, and (5) at which readings R_2, R_3, R_4, and R_5 are made. The intermediate sights are booked in the second column of the field page, with the change points in columns 1 and 3 for backsights and foresights respectively. The differences in height between each point are booked in the 'rise' and 'fall' columns where appropriate. The instrument is then changed to position B while the staff is kept at point (5). Readings from position B are denoted by r_5, r_6, r_7, r_8, and r_9 with the staff at points 5, 6, 7, 8, and 9 respectively. The backsight reading is r_5

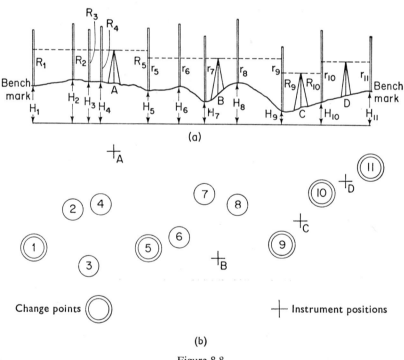

Figure 8.8

Table 8.1. Ordinary Levelling with Intermediate Points

ALGEBRAIC EXAMPLE						NUMERICAL EXAMPLE						
Back sight	Intermediate sight	Fore sight	Rise	Fall	Reduced level	B.S.	I.S.	F.S.	Rise	Fall	R.L.	Notes
R_1					H_1	3·93					119·65	Bench mark
	R_2		R_1-R_2		H_2		2·66		1·27		120·92	
	R_3			R_2-R_3	H_3		4·28			1·62	119·30	
	R_4		R_3-R_4		H_4		4·21		0·07		119·37	
r_5		R_5		R_4-R_5	H_5	3·02		5·21		1·00	118·37	Change point
	r_6		r_5-r_6		H_6		2·90		0·12		118·49	
	r_7			r_6-r_7	H_7		3·64			0·74	117·75	
	r_8		r_7-r_8		H_8		2·15		1·49		119·24	
R_9		r_9		r_8-r_9	H_9	4·57		6·88		4·73	114·51	Change point
r_{10}		R_{10}	R_9-R_{10}		H_{10}	5·26		3·33	1·24		115·75	Change point
	r_{11}		$r_{10}-r_{11}$		H_{11}		2·91	2·35			118·10	Bench mark

Checks:—
$\Delta h = [BS] - [FS]$
$\Delta h = H_{11} - H_1$
$\Delta h = [Rise] - [Fall]$

Checks:—
$\Delta h = 16·78 - 18·33 = -1·55$
$\Delta h = -119·65 + 118·10 = -1·55$ ✓
$\Delta h = 6·54 - 8·09 = -1·55$ ✓

and the foresight reading is r_9. It will be noted that only the change points have two readings taken to them. The reduced level of each point in turn is obtained by adding the rise or subtracting the fall from the height of the point to which the difference of height was related, i.e. the point above it in the field book. At the foot of each field book page an arithmetical check is made. The following four columns are summed separately and the totals entered at the foot of each column:

(i) The backsight readings.
(ii) The foresight readings.
(iii) The rises.
(iv) The falls.

The difference in height between points 1 and 11 is noted, i.e. $H_{11} - H_1 = -1·55$

Table 8.2. Height of Collimation Method

ALGEBRAIC EXAMPLE					ARITHMETIC EXAMPLE			
Back sight	Inter mediate sight	Fore sight	Height of collimation	Reduced level	B.S.	I.S.	F.S.	H of c	R.L.
R_1			$H_1 + R_1$	H_1	3·93			123·58	119·65
	R_2		$H ofc - R_2 = $	H_2		2·66			120·92
	R_3			H_3		4·28			119·30
	R_4			H_4		4·21			119·37
r_5		R_5	$H_5 + r_5$	H_5	3·02		5·21	121·39	118·37
	r_6			H_6		2·90			118·49
	r_7			H_7		3·64			117·75
	r_8			H_8		2·15			119·24
R_9		r_9	$H_9 + R_9$	H_9	4·57		6·88	119·08	114·51
r_{10}		R_{10}	$H_{10} + r_{10}$	H_{10}	5·26		3·33	121·01	115·75
		r_{11}		H_{11}			2·91		118·10

Checks:-
[B.S.]−[F.S.] = ΔH = $H_{11} - H_1$

16·78   18·33   119·65
            16·78   −1·55
            −1·55

in this example. The various checks then are:

$H_{11} - H_1$ = The sum of the backsights − The sum of the foresights
$H_{11} - H_1$ = The sum of the rises − The sum of the falls

This can easily be seen from the algebraic example as follows:

$$H_{11} = H_1 + (R_1 - R_2) + (R_2 - R_3) + \cdots + (R_4 - R_5) + (r_5 - r_6) + $$
$$+ (r_6 - r_7) + \cdots + (r_8 - r_9) + (R_9 - R_{10}) + (r_{10} - r_{11}) \qquad (5)$$

i.e. $\quad H_{11} - H_1 = (R_1 - R_5) + (r_5 - r_9) + (R_9 - R_{10}) + (r_{10} - r_{11})$
$$= (R_1 + r_5 + R_9 + r_{10}) - (R_5 + r_9 + R_{10} + r_{11}) \qquad (6)$$

The right-hand side of equation (5) is the algebraic sum of the differences in height, i.e. it is the sum of the rises minus the sum of the falls, and the right-hand side equation (6) is the sum of the backsights minus the sum of the foresights.

The totals at the foot of the levelling page are then carried forward to the top of the next page paying great care not to make a mistake in this transfer. It must be stressed that these checks in no way check the readings but merely their abstraction into heights. Normally the line of levels will start and close on a bench mark, though a loop may be run which begins and ends on the same point. The adjustment of any misclosure is considered below in §8.5.1.

The method of booking given above is the 'rise and fall' method, which is generally preferred to all other methods. One other worthy of mention is the *height of collimation* method. In this method we book the height of the line of sight and from it subtract the individual readings to obtain each ground height. For example, in figure 8.8(a) the line of sight of the instrument in position A is $H_1 + R_1$. A typical booking by height of collimation of the readings of table 8.1 is given in table 8.2. The arithmetical check on the booking is that

The sum of the backsights − Sum of the foresights = $H_{11} - H_1$

There is no immediate check on the heights of the intermediate points. For this reason, this method is less convenient than the rise and fall method. It is, however, more convenient in the *setting out* of height points on the ground.

8.3.4. ERRORS IN LEVELLING

The following are the various sources of error in levelling.

8.3.4.1. *Gross Reading Error*

In precise work each staff has two scales engraved upon it so that a gross reading error will be detected by the booker at once. Alternatively, the readings to the stadia hairs are used for the same purpose (see the specimen field books—tables 8.4 and 8.5).

8.3.4.2. *Staff Errors*

Datum error

Each staff may have a small datum or zero error. This means that the complete scale on the staff is moved through a small amount with respect to the base of the staff. If a staff becomes badly worn at the base, such an error is introduced. If one staff only is used to height two points, the staff error from this source will not affect the difference in height since it is common to both readings. If two staves are used for speed, as in geodetic levelling, the error in the difference in height of one set up is the difference between the two datum errors of the staves. After every *pair* of set-ups the error is eliminated if the back staff is leapfrogged to the forward position. Figure 8.9 shows how the effect of the combined staff error is zero at every other staff point. Hence if two staves are used, there should be an *even number* of set-ups between each bench mark.

Graduation error

In precise work the errors in the graduations of each staff are determined and applied to the staff readings where applicable. Each staff is placed on a specially designed bench along which is a very accurately divided tape or scale and against whose graduations those of the staff are compared, with the assistance of a travelling microscope. For example in a geodetic staff calibration, sample

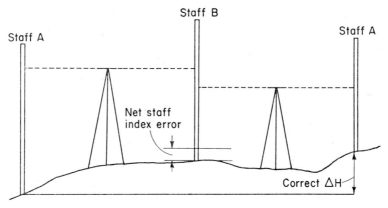

Figure 8.9

readings were as follows:

Staff graduation	Corrected tape readings
3·02 feet	3·020 5 feet
3·04	3·040 5
3·06	3·060 5
etc.	

Hence a staff reading in the field of say 3·039 4 has to be corrected by 0·000 5 to 0·039 9.

Warpage of the staff

If the staff becomes warped a reading becomes too high. For example if the staff in figure 8.10 has a circular warpage about its mid-point, and the deviation

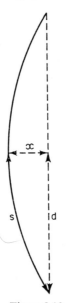

Figure 8.10

from a straight line at its mid-point is x ft, the mid-point reading S is too high by $2x^2/3S$. If $x = 0.1$ ft and $S = 5$ ft this amounts to about 0·001 ft which is serious in geodetic work. The effect is different according to the size of the staff reading R, and in practice is complicated because the staff warpage is not as simple as that assumed above. Most survey organizations insist that the staff should not depart from a straight line anywhere throughout its length by more than $\frac{1}{4}$ inch. Checks are made for warpage once per week.

Non-verticallity of the staff

If the staff in figure 8.11 is held at an angle θ to the vertical the reading is S instead of D. Then

$$D = S \cos \theta = S(1 - \tfrac{1}{2}\theta^2 \cdots) \simeq S - \tfrac{1}{2}S\theta^2$$

therefore the error

$$S - D = \tfrac{1}{2}S\theta^2$$

If $\theta = 2°$ and $S = 5$ ft the error is approximately 0·01 ft. This is a serious source of error in all types of levelling. In ordinary work, the staffman swings

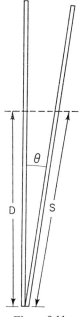

Figure 8.11

his staff backwards and forwards about its base so that it moves through the vertical position at some stage. The observer then obtains a correct minimum reading at this point. In some staves a bubble is fitted to the back of the staff to enable the staffman to hold it vertically. The adjustment of a staff bubble should also be checked every week against a plumb line. A geodetic levelling stave is provided with two rods which enable the staffman to hold it vertically for some time.

Temperature errors

In refined work the graduations are engraved or painted on a strip of invar attached to the bottom of the staff. Thus the scale is free to expand and contract without stresses being set up. Since the coefficient of linear expansion of invar is very small (of the order of $7 \times 10^{-7}/\deg F$) temperature corrections need not be applied as a rule, though in the tropics the correction may be significant.

Staff illumination

It has been found recently that engraved staves may exhibit a small systematic error under certain conditions. If one staff is constantly illuminated by the sun, whilst the other is constantly shaded, a small error of about 0·000 2 ft per shot can arise. The reading on the illuminated staff is too low because the bottom portion of the graduation is not seen. If a line of levelling is run from South to North, the South-facing staves, the foresight staves, are consistently read too low and the difference in height (Back − Fore) is therefore too large if positive and too small if negative. Thus the northern bench marks are given too high a height. The same effect will be obtained when running a line of levels East to West, though the effect will tend to be balanced if the line is levelled in one direction throughout the day. Painted staves do not exhibit this effect to any marked extent.

8.3.4.3. Instrumental Errors

Bubble sensitivity

In a level, the bubble should be sufficiently sensitive to permit staff readings of the required accuracy to be read, whilst at the same time they should not be so sensitive that much time is wasted in levelling the bubble. In a geodetic level the bubble is capable of being levelled with a probable error of $0\rlap{.}''25$, which is equivalent to 0·000 25 ft at about 200 ft.

Temperature effects

Most good instruments are designed so that temperature does not affect their performance to any extent. However, in precise work, the instrument is always shaded from the sun by a survey umbrella to prevent errors arising from differential heating.

Collimation errors

If a level has a collimation error of θ the reading error on a staff will depend on the length of the line of sight. In figure 8.12 the backsight distance B_S is

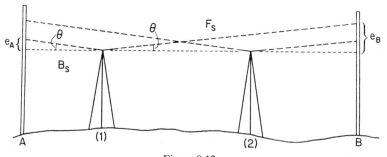

Figure 8.12

Vertical Control 397

less than the foresight distance F_S, with the result that the staff errors in reading are

$$e_A = B_S \theta \quad \text{and} \quad e_B = F_S \theta$$

The error $e_{\Delta H}$ in the height difference $H_B - H_A$ is $e_A - e_B$, i.e.

$$e_{\Delta H} = \theta(B_S - F_S) \tag{7}$$

If the lengths of the back and fore sights are equal there is no error in the height difference. Hence in all types of levelling these sights should be equated where possible, and the collimation error should be made as small as possible by adjusting the level as described in §3.16.

Since it is impossible to eliminate the collimation error completely, its value may be determined and a correction applied to the difference in the total lengths of the backsights and foresights for the whole line. In precise work this difference is kept to ± 2 ft so that any correction will be negligible.

The collimation error of the level may be determined in the following manner. The instrument is set up as in figure 8.12 close to the back staff at A. Readings are taken to both staves together with stadia readings B_S and F_S. The stadia constant is neglected. Let the error in the difference of height $H_B - H_A$ be $e_{\Delta H_1}$. Then

$$e_{\Delta H_1} = (R_B - R_F)_1 - (H_B - H_F)$$

where R_B and R_F are the staff readings at A and B, and $H_B - H_A$ is the true height difference, and the suffix 1 refers to the first instrument position.

$$\therefore \quad H_B - H_A = (R_B - R_F)_1 - e_{\Delta H_1} \tag{8}$$

The instrument is set up again in position 2 close to the forestaff B and the same readings repeated as in position 1. Then:

$$H_B - H_A = (R_B - R_F)_2 - e_{\Delta H_2} \tag{9}$$

Equating the true difference in height from (8) and (9)

$$(R_B - R_F) - e_{\Delta H_2} = (R_B - R_F)_2 - e_{\Delta H_2}$$

i.e. $(R_B - R_F)_1 - (R_B - R_F)_2 = e_{\Delta H_1} - e_{\Delta H_2}$

$$= \theta(S_B - S_F)_1 - \theta(S_B - S_F)_2$$

whence

$$\theta = \frac{(R_B - R_F)_1 - (R_B - R_F)_2}{(S_B - S_F)_1 - (S_B - S_F)_2} \tag{10}$$

This expression (10) is often written

$$\theta = \frac{\text{Sum of near readings} - \text{Sum of far readings}}{\text{Sum of near stadia intercepts} - \text{Sum of far stadia intercepts}}$$

Once θ has been found a correction $-\theta$ may be applied to the observed difference in height $R_B - R_F$ as follows

$$\text{Correction} = -\theta(S_B - S_F)$$

where $S_B - S_F$ is the difference between the total sum of the backsight distances and the total sum of the foresight distances, for the distance between the bench marks, or over which the levelling is carried out.

Example: Table 8.3 shows a set of readings taken to establish the collimation error of a geodetic level, together with a set of readings taken at the mid-point

Table 8.3. Determination of the vertical Collimation of a Geodetic Level
(C.T.S. Staves)

First set-up		Second set-up	
S_B	0·20	S_B	2·80
S_F	2·80	S_F	0·20
S_B-S_F	−2·60	S_B-S_F	+2·60
R_B	4·795	R_B	2·044
R_F	7·664	R_F	4·884
R_B-R_F	−2·869	R_B-R_F	−2·840

$$\theta = \frac{-2\cdot869 + 2\cdot840}{-2\cdot60 - 2\cdot60} = +0\cdot00558$$

Third set-up mid-way between staves

$$S_B = 1\cdot50 \quad R_B = 2\cdot330$$
$$S_F = 1\cdot50 \quad R_F = 5\cdot184$$
$$\therefore \quad R_B - R_F = -2\cdot854$$

of the line as a check. The stadia constant of 100 can be neglected throughout the calculation. From formula (10)

$$\theta = \frac{(4\cdot795 - 7\cdot664) - (2\cdot044 - 4\cdot884)}{(0\cdot20 - 2\cdot60) - (2\cdot80 - 0\cdot20)} = \frac{-0\cdot029}{-5\cdot20}$$

which gives a collimation error of $\theta = +0\cdot005\ 58$. Applying this correction to the set-ups close to the instrument we have the corrected difference in height of

First set-up

$$c'' = -0\cdot005\ 58(-2\cdot60) = +0\cdot014\ 5$$
$$\therefore \quad H_B - H_A = -2\cdot869 + 0\cdot014\ 5 = -2\cdot854$$

Second set-up

and
$$H_B - H_A = -2\cdot840 - 0\cdot014\ 5 = -2\cdot854$$

which agree with that already found with the instrument set up in the normal position at the mid-point of the distance between the two staves.

8.3.4.4. Other Errors

Sinking or rising of the staff

If the staff is set up on soft ground it will gradually sink lower whilst the observer is working. In figure 8.13 the levelling staff was at position F when the reading R_F was taken from position I, but it had sunk to position F' by the time that the reading was taken from position i to give a reading r_B which is in error by an amount e. Consequently the difference in height $r_B - r_F$ is in error by e.

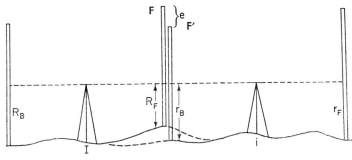

Figure 8.13

There is therefore a strong argument for reading on to staff in quick succession to minimize the time during which the staff may sink. As will be seen below, this conflicts with the effect of the sinkage of the instrument. Both effects can be avoided if all precise work is carried out along a hard surface such as the kerb stones of a road. In hot weather a tarred surface will permit the staff to sink, whilst pegs driven into the ground will rise for a time after being driven, producing the opposite effect.

Sinking of the instrument

In a similar manner to staff sinkage, the instrument may sink into the ground whilst the observations are being made. In figure 8.14 the instrument sinks a

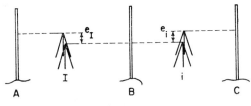

Figure 8.14

distance e_I in position I and e_i in position i. If the same degree of sinking occurs at both positions and the observer takes the same time to carry out his reading, there is a strong argument for reading first to the backstaff at A then to the forestaff at B, during which time the error e_I has occurred, then after changing to i, the first reading should be to the forestaff at C and finally to the backstaff at B, during which the error e_i has occurred. The chances are that these errors e_I and e_i will be nearly equal and therefore the height difference between A and C is correct. This again means that there should be an even number of set-ups between the bench marks. Since it is argued that the instrument is more likely to sink than the staff on account of its greater weight, the procedure is to read alternately to back and fore staves first in precise work, or which is the same thing, always read to the same staff first, the staves being leap-frogged forward.

Effect of the Earth's curvature

In figure 8.15 AB' is a horizontal line of sight at A. Due to the curvature of the surface of the Earth, a staff reading at B will be too great by BB'. This is a very small amount compared with the radius of the Earth as is indicated in

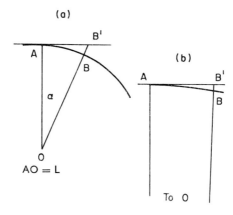

Figure 8.15

figure 8.15(b). To explain the geometry of the figure, refer to figure 8.15(a), but it must be remembered that the true dimensions are those of figure 8.15(b). Let the distance $AB = AB' = S$. In triangle AB'O, $B'O^2 = AO^2 + S^2$

i.e. $(L + BB')^2 = L^2 + S^2$

i.e. $L^2 + 2(BB')L + (BB')^2 = L^2 + S^2$

since $(BB')^2$ is negligible,

$$\therefore \quad BB' \doteqdot S^2/2L \qquad (11)$$

It will be obvious that if the instrument is set up halfway between the two staves both the backsight and foresight readings will be in error by the same amount, and therefore the height difference is free of error. However, the instrument need not be set up each time at the mid-point between the staves to eliminate this error; if the total sum of the backsights equals the total sum of the foresights, the total curvature error over the whole line of levelling will be nil.

Effect of Refraction

B and F are two points in figure 8.16 at the same height, and R_B, S_B, etc. are the quantities described above. We neglect the curvature of the Earth. Since the light ray from the instrument to the staff is refracted as it passes throughout the atmosphere, a recorded staff reading is too low. The errors due

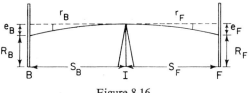

Figure 8.16

to refraction are e_B and e_F on the backsight and foresight respectively. If α is the angle subtended by $BI = S_B$ at the centre of the Earth, whose radius is L (figure 8.15), and the coefficient of refraction is K, the angle of refraction $r_B = K\alpha = KS_B/L$

$$\therefore \quad e_B = r_B S_B = KS_B^2/L \tag{12}$$

The coefficient of refraction K is about 0·07 though its value varies a great deal. The main factor affecting it is the rate of change of temperature with height above the ground; the greater this rate of change, the smaller the refraction. If K is constant for the time that both sights are observed, the effect on the difference in height will be nil if the instrument is positioned mid-way between the staves. But in many cases K is not constant over a given period hence a small systematic error will be introduced into the line of levels. If a line is levelled from early morning to noon, the refraction becomes progressively less; or if a line is levelled up a steady incline the foresight is always closer to the ground and refraction of the foresight is again less. In the first case, always reading to the same staff first will help to eliminate the refraction error since the change in refraction will alternately affect each height difference with an opposite sign, each even number of height differences being relatively free of error. In the latter case, the effect will be systematic and the uphill bench mark will be too low. Wherever possible, the line should be run in the same conditions for both the forward and back levelling so that the mean difference in height will be almost free from error. If the total sum of the backsights equals the total sum of the foresights, but the individual set-ups are not mid-way between the staff positions, the error due to refraction will not be eliminated as a rule because of the change in refraction with time. The average value of refraction for the backsights will not be equal to the average value for the foresights unless these sights are equated as closely as possible in time, i.e. at every set-up.

8.3.5. PRACTICAL LEVELLING

As a result of considering the above sources of errors in levelling, it is apparent that for all types of work most errors will be avoided if the following three rules are adopted:

(i) A line should be run between two points of known height, or twice over the same line, to detect gross errors, or a staff with two scales may also be used for this purpose.
(ii) The staff should be vertical at all times when a reading is taken, either by swinging through the vertical position, or with the aid of a bubble.
(iii) The length of the backsight should equal the length of the foresight at every set-up.

In ordinary work on engineering sites it is not feasible to have backsights and foresights exactly equal, and in intermediate sights no compensation for curvature, refraction, and collimation takes place. Hence it is imperative that the level has very little collimation error, otherwise, tedious corrections will have to be applied. Errors due to curvature and refraction are negligible for most work with sights not exceeding 300 ft. The combined formula for curvature and refraction is obtained from equations (11) and (12), viz. $(1 - 2K)S^2/2L$, which reduces to

$$4S^2/7 \text{ feet, where } S \text{ is in miles and } K = 0.07$$

This error is 0·1 ft for $S = 2\,000$ ft.

Again it is important to avoid soft muddy ground, but for obvious reasons this is not possible on an engineering site.

8.3.6. GEODETIC LEVELLING

In geodetic levelling all the above sources of error have to be taken into account and a field procedure adopted which will eliminate their effect as much as possible. Usually, very long lines of up to 50 km are run, so that small systematic errors can accumulate into quite large proportions. Where possible the work is repeated under varying conditions to change systematic errors into random errors and thus bring about some degree of compensation. Lines of geodetic levelling are always run twice, back and fore along the line, and before the work is accepted a certain tolerance must be reached between the two measures. If the discrepancy between forward and back levelling is d the residuals derived from the mean are $+\frac{1}{2}d$ and $-\frac{1}{2}d$, whence the standard deviation for the line is given by

$$\sigma^2 = \frac{(\tfrac{1}{2}d)^2 + (\tfrac{1}{2}d)^2}{2} = \tfrac{1}{4}d \qquad \therefore \quad \sigma = \pm\tfrac{1}{2}d$$

Or the probable deviation is $\pm\tfrac{1}{3}d$ (P.E. $\simeq \tfrac{2}{3}\sigma$). This estimate of standard deviation is very poor because only two independent measures are available. From the analysis of large numbers of results, it has been found that the probable deviation in geodetic levelling is given by $\pm 0.005(M)^{\frac{1}{2}}$, where M is the distance run in miles. This will be achieved if the discrepancy between two measures does not exceed

$$3 \times 0.005(M)^{\frac{1}{2}} = 0.015(M)^{\frac{1}{2}}$$

If the average length of sight is 125 ft, there will be about say 25 set-ups per mile, and the permissible discrepancy between two measures of the height difference at each set up is $0.015/5 = 0.003$ ft. Normally two measures of the height difference are taken as outlined in §8.3.7, which are repeated until the required tolerance is met.

To equate the lengths of the sights at each set-up to within 1 foot, the staff and instrument positions are set out by a separate party moving ahead of the leveller. This setting out party may require an abney level and staff to ensure that a sight is possible on sloping ground. The booker must always check that, although each set-up is at the mid-point of the staff distance the positions are not consistently biased in one direction so that the total lengths of the backsights and foresights differ by more than the allowable 2 feet. If he finds this is happening he will make slight adjustments to his position at each set-up.

There are no intermediate sights in geodetic levelling. To avoid misreading or booking in the wrong column each staffman should have some distinctive dress. This enables the observer to point first at the same man, and therefore at the same staff at each set-up. Before booking the first reading of a set-up, the booker will enter an arrow pointing up the page if the distinctive man is being observed. Since alternate back and foresights should be read, the arrows will appear alternately in the back and fore columns. If this pattern is broken, the booker draws the attention of the observer to his mistake.

To prevent errors arising from differential heating of the instrument it should be shaded by an umbrella. Since two staves are invariably used in geodetic work, and because the effects of sinking of the instrument and changes in refraction will be partially eliminated, there must always be an even number of set-ups between bench marks.

8.3.7. BOOKING GEODETIC LEVELLING

The type of booking will depend on the type of staff used. In table 8.4 a specimen field book is given for the Watts staff which has two scales painted side by side on the staff, and in table 8.5, the same readings are given for the C.T.S. staff which has one scale only, but in which unlike the Watts, the stadia readings form an important part of the heighting process.

The order in which the readings are taken is given in the 'algebraic' version of the booking, which is the same for both types of stave. The object is to read the centre line on each staff as quickly as possible without delays in reading the stadia. The bubble should be checked immediately prior to and after each centre line reading. Between readings (6) and (7), the bubble is thrown off centre and the instrument relevelled to ensure that two independent values of the height difference are obtained.

The checks on the booking are as follows:

(i) To check against gross error, $(3) - (4) = (8) - (7)$ of the Watts staff; and $(3) - (4) = (1) - (5)$ of the C.T.S. staff, for which reason three figures are read on the stadia.
(ii) To check the difference in level, $(3) - (4) = (8) - (7)$ in both cases to 0·003, and $(3) - (8) = (4) - (7)$ again to 0·003 ft. In the Watts staff the value of $(3) - (8)$ is the scale shift between the two scales of the staves. This may not be exactly the same for each staff, but once determined, these differences should agree with the staff constant to 0·003.
(iii) The individual sight lengths are kept to an equality of 1 foot and the running totals in the 'distance' column to 2 feet.

The field book is checked by summing the totals of the various columns as already described in §8.3.3.

Table 8.4. Watts Staff (Double Scale)

(a) Algebraic version

Notes	Backsight		Foresight		Distance		Diff. level	
	Stadia	Centre	Centre	Stadia	Back	Fore	Rise	Fall
B.M 61/266 ↑	(1)	(3)	(4)	(5)			(3) − (4)	
STAFF A	(2)	(8)	(7)	(6)			(8) − (7)	
	(1) − (2)	(3) − (8)	(4) − (7)	(5) − (6)	(1) − (2)	(5) − (6)	Mean	
	(13)	(12)	(11)	(9) ↑				(12) − (11)
	(14)	(15)	(16)	(10)	(1) − (2)	(5) − (6)		(15) − (16)
	(13) − (14)	(12) − (15)	(11) − (16)	(9) − (10)	(13) $\pm$ (14)	(9) $\pm$ (10)		Mean
↑			etc.					

(b) Numerical example

	Backsight		Foresight		Total distance		Diff. level	
	Stadia	Centre	Centre	Stadia	Back	Fore	Rise	Fall
BM 61/266 ↑	18·53	18·1762	12·4490	12·80			5·7272	
STAFF A	17·82	7·6660	1·9404	12·10			5·7256	
	0·71	10·5102	10·5086	0·70	0·71	0·70	5·7264	
	12·22	11·9218	19·7571	20·06 ↑	0·60	0·60		7·8353
	11·62	1·4106	9·2471	19·46				7·8365
	0·60	10·5112	10·5100	0·60	1·31	1·30		7·8359
↑			ETC.					

8.3.8. LEVELLING ACROSS A WIDE GAP

It sometimes happens that a line of levels has to be carried across a wide water gap of from 300 ft to 5 miles. Under these circumstances it is impossible to read the staff from the instrument end with the required degree of precision. One method of overcoming the problem is to fit special targets to the staff which are moved up and down until the observer signals that the target is bisected by his cross hair. The staffman then reads the staff. A successful target has been a band of brown paper tightly wrapped round the staff. This method is less precise than those now described.

8.3.8.1. *By Gradienter*

The line of sight of a level may be tilted through a small angle by a micrometer tilting screw or *gradienter* which is calibrated in terms of the tangent of the angle through which it is tilted. Readings are in parts per thousand, i.e. a reading r represents an angle whose tangent is $r/1\,000$. A target is placed at

Vertical Control

Table 8.5. C.T.S. Staff (Single Scale)

Notes	Backsight		Foresight		Distance		Diff. level	
	Stadia	Centre	Centre	Stadia	Back	Fore	Rise	Fall
B.M.61/266	↑ (1)	(3)	(4)	(5)			(3) − (4)	
STAFF A	(2)	(8)	(7)	(6)			(8) − (7)	
	(1) − (2)	(3) − (8)	(4) − (7)	(5) − (6)	(1) − (2)	(5) − (6)	Mean	
	(13)	(12)	(11)	(9) ↑				(12) − (11)
	(14)	(15)	(16)	(10)	(13) − (14)	(9) − (10)		(15) − (16)
	(13) − (14)	(12) − (15)	(11) − (16)	(9) − (10)	Total	Total		Mean
↑				etc.				

BM61/266	↑18·528	18·1762	12·4490	12·799			5·7272	
STAFF A	17·822	18·1770	12·4511	12·099			5·7259	
	0·706	0·0008	0·0021	0·700			5·7266	
	12·218	11·9218	19·7571	20·057 ↑				7·8353
	11·623	11·9200	19·7544	19·458				7·8344
	0·595	0·0018	0·0027	0·599				7·8348
↑			ETC.					

some convenient mark on the staff and the *gradienter* reading r taken. A reading on the gradienter r_0 is also taken when the bubble is level. Then for small angles since $\theta \simeq \tan \theta$,

$$\tan \theta = \frac{r - r_0}{1\,000} \quad \text{and} \quad \Delta H = \frac{s(r - r_0)}{1\,000} \tag{13}$$

If possible, a second reading should be taken to a target placed below the horizontal line from the level to the staff so that a negative gradienter reading is obtained. In figure 8.17, R_1 and R_2 are the readings taken to the targets in two successive positions on the staff at B, a distance s from A. If θ_1 and θ_2

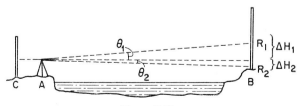

Figure 8.17

are the angles of elevation and depression for readings R_1 and R_2 respectively,

$$\Delta H_1 = \frac{s(r_1 - r_0)}{1\,000} \quad \text{and} \quad \Delta H_2 = \frac{s(r_2 - r_0)}{1\,000}$$

the centre line readings would then be $R_1 - \Delta H_1$ and $R_2 - \Delta H_2$ with a mean value of $\frac{1}{2}\{R_1 + R_2 - (r_1 + r_2 - 2r_0)s/1\,000\}$ which is used in the normal way as a staff reading.

The signs of ΔH_1 and ΔH_2 are taken into consideration in the formula since the reading on a depression is negative. The distance s is obtained by triangulation, subtense traverse or any other survey method. The tellurometer is especially useful since its radio communication is also invaluable. The other reading to a close staff at C is taken in the normal way. It is thus clear that the backsight and foresight distances cannot be equated and therefore the various errors arising from this lack of equality will be considerable. If the level is now moved to a position close to B and the total backsights and foresights thus equated, all errors other than refraction are eliminated by accepting the mean of the two height determinations. Refraction is not eliminated because of the time that elapses in moving the instrument from one bank to the other. To eliminate this effect, two levels are used to take simultaneous reciprocal observations. Then all equipment is interchanged and the process repeated to eliminate the effect of two instruments and observers. As many as twenty complete determinations of the height difference may be required to achieve the desired precision. The effect of refraction is not entirely eliminated since the radius of curvature of the light path at A, may be different from the radius of curvature at B, i.e. the curve is asymmetrical. However the procedure of simultaneous reciprocal levelling will eliminate most of the error.

8.3.8.2. *By Geodetic Theodolite*

As a result of a very comprehensive series of tests involving height determinations over some 45 lines varying in length from 0·25 to 5 miles, the Ordnance Survey of Great Britain has produced results of the same accuracy as geodetic levelling for height transfers by observation of truly simultaneous reciprocal vertical angles using geodetic theodolites at either end of the line (see §8.5.3).

8.3.9. ORTHOMETRIC AND DYNAMIC HEIGHTS

In §8.1 we defined the height of a point to be the *linear distance* along the vertical through the point from a datum surface at mean sea level to the point. However, the concept of 'height' may be considered from a different point of view in terms of the force of gravity acting on a mass situated near the surface of the Earth.

The *gravitational potential* P at a point at a distance D from the centre of the Earth is given by $P = gD$, where g is the acceleration due to gravity. An *equipotential surface* is one on which all points have the same gravitational potential, i.e. the potential P is constant, or gD is constant for any point on an equipotential surface. If g were constant over the surface of the Earth, an equipotential surface would be equidistant from the centre of the Earth. However, since g varies with latitude, being greater at the poles than at the equator, an equipotential surface is closer to the centre of the Earth at the poles than at the equator. Figure 8.18 shows the shape of two equipotential surfaces defined by their potentials P_1 and P_2. If H_0, H_{45}, and H_{90} are the linear separations between these equipotential surfaces at the latitudes of 0°, 45°, and 90° respectively; g_0, g_{45}, and g_{90} their respective gravities; and D_0, D_{45}, and D_{90} the distances at

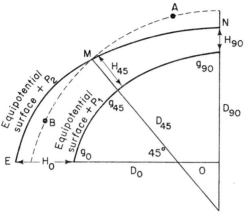

Figure 8.18

these points from surface P_1 to the centre of the earth O, we have P_1 is constant, i.e. $g_0 D_0 = g_{45} D_{45} = g_{90} D_{90}$, and P_2 is constant, i.e.

$$(H_0 + D_0)g_0 = (H_{45} + D_{45})g_{45} = (H_{90} + D_{90})g_{90}$$

whence
$$g_0 H_0 = g_{45} H_{45} = g_{90} H_{90} \tag{14}$$

Since no work is done against gravity if a point moves over an equipotential surface, such a surface is dynamically flat. For example if a pipe line follows P_2 no water would flow along it under the action of gravity. But if the pipe line takes the path of the broken line of figure 8.18 which is equidistant from P_1, water would flow from A to B under the force of gravity due to the difference of potential at those points. Thus, although A and B are at the same linear height above P_1 in a dynamic sense there is a slope from A to B. Points defined by linear distances along the vertical are called *orthometric heights*, and those defined in terms of potential are called *dynamic heights*.

To relate the concept of potential to a linear system of height, a standard latitude is chosen (45° for a world system, 53° for England and Wales) at which both types of height are defined to be equal; e.g. in figure 8.18 the linear separation h between P_1 and P_2 at latitude 45° is used to define the potential surface P_2 with respect to P_1 as datum in linear units. Hence the *dynamic* height of points E, M, and N is h, the *orthometric* height of M above P_1 at the standard latitude 45°. The orthometric height of E is H_0, and of N is H_{90}. The orthometric height at the equator is greater than the dynamic height, and *vice versa* at the poles, the two being equal at 45°.

Now from (14)
$$h = H_{45} = g_0 H_0 / g_{45}$$
and it can be shown that gravity at latitude ϕ, g_ϕ is related to the gravity at the equator g_0 by the expression
$$g_\phi \simeq g_0(1 + 0.005\,3 \sin^2 \phi) \quad (15)$$
whence $g_{45} = g_0(1 \cdot 002\,65)$
$$\therefore \quad h = H_0/1 \cdot 002\,65$$
and for any point at an intermediate latitude ϕ
$$h = \frac{g_\phi H_\phi}{g_{45}} = \frac{H_\phi(1 + 0.005\,3 \sin^2 \phi)}{1 \cdot 002\,65} \quad (16)$$

If the orthometric height H_ϕ is found, its equivalent dynamic height h may be derived from (16).

Example: If the orthometric height of a point on the equator is 1 000 ft its dynamic height is 997·4 ft, i.e. there is a difference of 2·6 ft.

8.3.10. ORTHOMETRIC HEIGHTS FROM LEVELLING

In the actual process of levelling, the bubble sets itself tangential to the equipotential surface through the instrument, whilst the readings on the staves are orthometric operations. In figure 8.19, the divergence between the orthometric and the dynamic surfaces through the instrument gives rise to an error

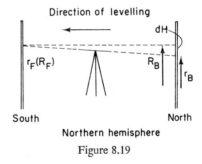

Figure 8.19

dH in the backsight reading r_B such that the correct reading is $R_B = r_B + dH$, when levelling from North to South in the northern hemisphere. Defining the difference in latitude over the line $d\phi = \phi_B + \phi_F$, i.e. in the same sense as ΔH, $d\phi$ is positive in this case, and for a line run from South to North it is negative.

For the equipotential surface, $gH = $ constant. Differentiating
$$H\,dg + g\,dH = 0$$
i.e.
$$dH = -H\,dg/g \quad (17)$$
But
$$g = g_0 (1 + 0.005\,3 \sin^2 \phi)$$
$$dg = g_0 \times 0.005\,3 \sin 2\phi\, d\phi$$

$d\phi$ reckoned positive when increasing from South to North; and in the convention of this section,

$$dg = -g_0 \times 0.005\,3 \sin 2\phi\, d\phi$$

whence substituting in (17), and since $g \simeq g_0$,

$$dH = +H \,.\, 0.005\,3 \sin 2\phi\, d\phi$$

Thus to obtain the orthometric height difference from the dynamic height difference the indicated $r_B - r_F$ is corrected by $+dH$. Since this correction is very small for one set-up, it is normally applied to lines of levels at about every five miles or at convenient section points along the line. The total effect for the line is the summation of all the individual corrections, i.e. the corrections are applied by numerical integration over the whole line.

Example: If the average height of a section of levels 5 miles long run from South to North at latitude 10°N is 1 000 ft, and the indicated height of the northern bench mark was 1 026·938 4 ft, calculate its orthometric and dynamic heights, assuming that the height of the southern bench mark was orthometric. In this case $d\phi$ is negative hence,

$$dH = -(1\,000 \times 0.005\,3 \times 0.342\,0 \times 5)/396\,0$$
$$= -0.002\,3 \text{ ft}$$

Hence the orthometric height of the required bench mark is 1 026·936 1 ft. And from equation (16) the dynamic height is given by

$$h = 1\,026.936\,1(1 + 0.005\,3 \times 0.030\,14)/1.002\,65$$
$$= 1\,024.385\,5 \text{ ft}$$

8.3.11. GEOPOTENTIAL NUMBERS

Since the acceleration due to gravity varies not only with latitude ϕ, but with height above sea level h, and with the composition of the Earth, allowance is now made for these factors; and a new system of heights, called *geopotential numbers*, has been introduced. The system was devised by the International Association of Geodesy, who defined the geopotential number of a point A to be

$$C_A = \int_0^A g\, \Delta h$$

The units to be used are the kilogal/metre, i.e. g is in kilogals and Δh in metres above sea level. It is desirable that observed values of g are used, though theoretical values based on some hypothesis concerning the structure of the Earth may be used for want of observational data.

8.3.12. ADJUSTMENT OF LEVEL POINTS

This aspect of the subject is considered in §8.5 together with the adjustment of trigonometrical heights.

8.4. TRIGONOMETRICAL HEIGHTING

As outlined in §8.2.3 above, a common method of determining height difference ΔH is to observe the vertical angle θ to a survey beacon, and compute the difference in height between the theodolite and the top of the beacon by the relationship $\Delta H = S \tan \theta$, where S is the horizontal distance between the stations; or from $\Delta H = S \sin \theta$, where S is the slant range as measured by a tellurometer, etc.

In figure 8.20, H_A and H_B are the heights of the stations A and B above a horizontal datum line DE which is considered straight meantime, i_A is the height of the trunnion axis of a theodolite at A referred to some reference mark on the

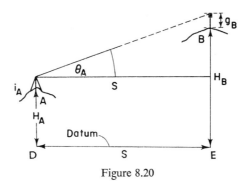

Figure 8.20

station at A, g_B the height of the signal at station B, the vertical angle observed at A is θ_A, and the horizontal distance DE is S. These quantities are related by the expression

$$H_A + i_A + S \tan \theta_A = g_B + H_B \tag{18}$$

Or putting $H_B - H_A = \Delta H$

$$\Delta H = S \tan \theta_A + i_A - g_B \tag{19}$$

Thus, given the quantities on the right-hand side of equation (19), the difference in height ΔH can be evaluated. Note that equations (18) and (19) hold for all values provided the following conventions are adopted.

   (i) Heights *above* a station are considered *positive*; those *below* are negative.
   (ii) The height difference ΔH from A to B is defined as $H_B - H_A$.
   (iii) An angle of *elevation* θ_A is *positive*, and a *depression* is *negative*.

These conventions are commonly used in everyday life and should not cause any difficulty. No diagrams are necessary when computing a height difference, on the contrary, a diagram will often confuse the issue.

Example: If $H_A = 100$ ft, $S = 1\,000$ ft, $\theta_A = +1°\,25'\,57''$, i.e. an elevation, $i_A = 4\cdot00$ ft and $g_B = 10\cdot00$ ft, the height H_B is derived as follows:

$$\tan \theta_A = +0\cdot025\,0$$
$$\therefore \quad \Delta H = +25\cdot00 + 4\cdot00 - 10\cdot00 = +19\cdot00$$
$$\therefore \quad H_B = H_A + \Delta H = 119\cdot00 \text{ ft}$$

It should be noted that although i_A and g_b are normally positive it is possible to have negative values if an instrument or beacon is erected below the reference mark on a pillar.

Example: Consider the same figures above with θ_A, i_A, and g_B all negative, then

$$\Delta H = -25 \cdot 00 - 4 \cdot 00 + 10 \cdot 00 = -19 \cdot 00 \text{ ft}$$
$$\therefore \; H_B = H_A + \Delta H = 100 - 19 \cdot 00 = 81 \text{ ft}$$

It is usual to observe a vertical angle on face left and on fact right so that the effect of the vertical index error of the theodolite is eliminated on taking the mean. Observation should whenever possible be taken from A to B, and from B to A. It is always better to compute the height difference by two separate computations using equation (19), than to use a formula which combines the observations taken at both ends. The reason for this is that any gross error may be detected from any discrepancy. Errors in recording beacon and instrument heights are very numerous in practical field work and great care should be taken to avoid them. The following example serves to indicate the value of two separate computations.

Example: Vertical angles were observed from A to B, and from B to A. If the heights of instrument, beacons, etc are as given below, compute the difference in height between A and B.

$$\theta_A - 00° \, 10' \, 45'', \quad i_A + 4 \cdot 00 \text{ ft}, \quad g_A + 10 \cdot 00 \text{ ft}$$
$$\theta_B + 00° \, 28' \, 00'', \quad i_B + 5 \cdot 00 \text{ ft}, \quad g_B + 9 \cdot 00 \text{ ft}$$

Horizontal distance AB is 2 000 ft.

At station	Observed angle tangent	$S \tan \theta$	i	g	ΔH
θ_A	$-00° \, 10' \, 45''$ $-0 \cdot 003 \, 127$	$-6 \cdot 25$	$+4 \cdot 00$	$+9 \cdot 00$	$-11 \cdot 25$
θ_B	$+00° \, 28' \, 00''$ $+0 \cdot 008 \, 145$	$+16 \cdot 29$	$+5 \cdot 00$	$+10 \cdot 00$	$+11 \cdot 29$

If the angle at B had been misread by 10' to give $+00° \, 38' \, 00''$ the second height difference would be $+17 \cdot 11$ ft, indicative that an error has been made. The magnitude of the error is given approximately from $(17 \cdot 11 - 11 \cdot 25)/2 \, 000 = 2 \cdot 93 \times 10^{-3}$ in radians, i.e. in seconds of arc, $2 \cdot 93 \times 10^{-3} \times 206 \, 265 \simeq 600$.

8.4.1. EARTH'S CURVATURE

The reference datum line DE is not straight as we supposed, but is curved as in figure 8.21. The linear amount of the Earth's curvature $FE = x$ is given by $x = S^2/2R$, where R is the radius of the Earth (see 8.3.4.4). If the datum line DE is considered as an equivalent straight line datum DF, the problem is reduced to that of §8.4.

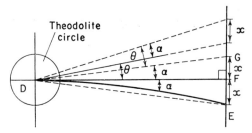

Figure 8.21

If we consider E raised to F, the angle θ has to be considered as increased by the curvature angle α, which is given with ample accuracy by $\alpha'' = 206\,265\,S/2R$

or
$$\alpha'' = PS$$

where P is a constant $206\,265/2R \simeq 5{\cdot}07 \times 10^{-3}\,R$ in feet. Hence the observed angle θ referred to a curved datum DE, can be considered with respect to the straight datum DF if it is increased by α. The angle α will always be positive, though θ may be of either sign. If θ is negative and less than α numerically the sign of the *effective* observed angle $\theta + \alpha$ will be changed.

8.4.2. REFRACTION

In figure 8.22, the light path from D to G is shown as a curved line instead of a straight line. This is so in practice, due to the refractive effect of the

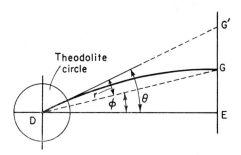

Figure 8.22

atmosphere. The telescope at D points along DG' instead of DG, and therefore records an angle θ which is greater than that used in computation. Hence to obtain the equivalent angle free from refraction, the angle r has to be subtracted from the observed angle θ. From §8.3.4.4 the refraction angle r is given by $r = KS/R$

$$\therefore\ r \text{ seconds} = KS\,206\,265/R = 2KPS$$

Where P is the same constant as in the curvature angle, and the coefficient of refraction K is about 0·07, though it varies according to the density of the air through which the light passes.

8.4.3. CURVATURE AND REFRACTION

To reduce an observed angle θ to its equivalent plane angle ϕ we have:

Corrected plane angle = Observed angle + Curvature − Refraction

i.e.
$$\phi = \theta + PS - 2KPS$$
$$= \theta + PS(1 - 2K) \qquad (20)$$

and putting $K = 0\cdot07$, $\phi = \theta + 0\cdot004\,36S$. Since S is usually from 1–20 miles long, it is more convenient to work in terms of units of 1 000 ft, thus $PS(1 - 2K) = 4\cdot36$ seconds of arc per 1 000 feet of S. This combined curvature and refraction correction is always positive because the curvature correction exceeds the refraction correction. Since the latter varies with the prevalent air conditions, the combined correction will also vary from place to place and with time of day. Since refraction is least at the mid-day period, usually taken from 12·00 hrs to about 15·00 hrs, and since its value during this period varies little from one day to another, vertical angles are best observed during these hours. However, if two observers are situated at each end of the line, truly simultaneous vertical angles may be observed, which enables the mean value to be almost free of refraction error. Such observations may be taken at almost any time of day, but see §8.5.3 below for further details.

In practical computation of trigonometrical heights a few reliable observations taken over several long lines are used to compute the combined curvature and refraction correction in seconds of arc per thousand feet, and thereafter this value is used to compute all heights by single rays.

Example: Table 8.6 shows the necessary extracts from field books taken at the stations D, E, F, and G, together with the relevant horizontal distances in feet. The longest line EG is chosen as that from which to compute the combined curvature and refraction correction. The height difference from E to G is computed without applying correction for curvature and refraction. This is repeated for the direction G to E, i.e.

			S tan	i	g	ΔH
θ_{EG}	+ 00° 07′ 39″	tan +0·002 225	+58·58	+ 0·68	−0·00	+ 59·26
θ_{GE}	− 00° 11′ 33″	tan −0·003 360	−88·46	+ 0·83	−0·00	− 87·63

Since these two values for the difference in height should be numerically equal, the difference is due to twice the effect of curvature and refraction, which is then calculated from

$$C/R = (28\cdot37 \times 206\,265)/2 \times 26\,328\cdot4 = 111''$$

which is the correction to be applied to a single ray observation over this line. This gives the value of the correction, to be $4\rlap{.}''2198$ per 1 000 feet for the area. As a check, these two observations are recomputed using this value for the

414 Practical Field Surveying and Computations

Table 8.6. Extract from Field Books of Vertical Observations

At station D	Peg	$i = +4.96$ ft	$g = 10.08$ ft	Top
To E	$-02°38'00''$	Top pillar		
F	$-02°17'22''$	Top skirt		
G	$-01°39'24''$	Top pillar		

At station E	Pillar	$i = +0.68$ ft	$g = 0$	
To D	$+02°36'57''$	Top beacon		
F	$-00°25'25''$	Top beacon		
G	$+00°07'39''$	Top pillar		

At station F	Peg	$i = +4.80$ ft	10.42 Top beacon / 9.24 Top skirt	
To D	$+02°16'27''$	Top beacon		
E	$+00°23'48''$	Top pillar		
G	$+00°48'46''$	Top pillar		

At station G	Pillar	$i = +0.83$ ft	$g = 0$	
To D	$+01°37'27''$	Top beacon		
E	$-00°11'33''$	Top pillar		
F	$-00°49'25''$	Top beacon		

Computation data for above observations

Line	Distance (ft)	Observed height diff. feet	Adjusted height diff.
DE	14 037.6	-636.295	-636.255
DF	19 211.8	-765.015	-765.226
DG	19 896.3	-562.695	-562.533
EF	17 064.9	-128.840	-128.971
EG	26 328.4	$+73.445$	$+73.722$
FG	13 706.8	$+202.950$	$+202.693$

combined correction as follows:

	θ	S	i	g	ΔH
EG	$+00°07'39''$	26 328.4	$+0.68$	0.00	
C^n	$+00°01'51''$	$\tan\theta$	$S\tan\theta$		
$C^d\theta$	$00°09'30''$	$+0.002\,763$	$+72.74$		$+73.42$
GE	$-00°11'33''$		$+0.83$	0.00	
C^n	$+00°01'51''$				
$C^d\theta$	$00°09'42''$	$-0.002\,822$	-74.30		-73.47

Whence the accepted mean difference in height is 73·445 feet. It will be noticed that this same value is obtained from the mean of the two values 59·26 and 87·63 computed without recourse to the correction at all. To adopt this latter method of computing as dangerous because one does not have a clear indication that the results are free from gross error. The computation of the height differences for the lines DE, DF, DG, EF, and FG is left to the reader. The results are given in table 8.6.

8.4.4. THE COEFFICIENT OF REFRACTION K

If a value of K is desired, it may be obtained from the formula for the combined correction (20) thus: Combined correction from the above examples is $4''219\ 8/1\ 000$ ft

$$\therefore\quad 4\cdot219\ 8S/1\ 000 = PS(1 - 2K)$$
but
$$P = 5\cdot07 \times 10^{-3}$$
$$\therefore\quad 2K = 1 - (4\cdot219\ 8/5\cdot07) = 0\cdot178$$
$$\therefore\quad K = 0\cdot089$$

8.4.5. LINEAR VALUE OF THE CURVATURE AND REFRACTION CORRECTION

It is useful to evaluate the linear effect of the combined curvature and refraction over the distance S as follows: The angular value is given by

$$PS\,(1 - 2K)'' = \frac{S}{2R}(1 - 2K)\ \text{radians}$$

Hence the linear amount subtended by S is $S^2(1 - 2K)/2R$. If S is in miles and the correction is in feet, we have the convenient approximation

$$C^n = \tfrac{4}{7}S^2 \tag{21}$$

This is very useful in working out problems of intervisibility (see §8.5.2). For example, if the distance S is 5 miles, the combined correction is $100/7 = 14$ ft approximately.

8.4.6. PRECISION OF THE MEASUREMENTS

We now consider briefly the precision required in the various measured quantities to achieve a given result in the difference in height. Denote $\tan \theta$ by t, the distance by S, and the height difference by ΔH, i.e.

$$\Delta H = St \tag{22}$$

Since the height of instrument and beacon enter directly into the value obtained for ΔH, they must be measured to at least the same precision as is desired in the final result. In normal work, these quantities are measured to 0·01 ft, by taking readings to a tape held vertically as with a level staff, or by special observations close to the beacon from a short base (see §8.5.6).

Consider small changes dt in t, dS in S, and $d(\Delta H)$ in ΔH, related by

$$d(\Delta H) = S\,dt + t\,dS \tag{23}$$

If $dt = 0$, and $\theta = 1°$, and if $d(\Delta H)$ is to be $\leq 0\cdot 1$ ft, $dS \simeq 57 \times 0\cdot 1 \simeq 6$ ft. Hence if the theodolite or helio is set up within this distance from the trig station a sufficiently accurate result is obtained using the intertrig distance in the computation.

Again, if $dS = 0$, and if $S = 10\,000$ ft

$$d(\Delta H) = S\,dt = S \sec^2 \theta\, d\theta$$

if $\theta = 1°$, $\sec \theta \simeq 1$

$$\simeq S\,d\theta'' \sin 1''$$

whence $d\theta \simeq 2''$.

If a *probable error* of $\pm 0\cdot 14$ ft in ΔH is required, a probable linear error of ± 6 ft in S, and $\pm 2''$ in θ will suffice. If the p.e. of a single measure of a vertical angle is $\pm 5''$, and n is the number of such measures required to give a p.e. of the mean of $\pm 2''$,

$$n = 25/4 \simeq 6$$

To guard against gross errors of reading the vertical angle, one reading should be taken on each of the stadia hairs, thus obtaining an independent check from the mean.

8.5. MISCELLANEOUS ASPECTS OF HEIGHTING

8.5.1. THE ADJUSTMENT OF HEIGHTS

The adjustments of trigonometrical heights and of level networks are carried out in the same way. The difference in height brought through a line of levels is treated in the same manner as a height difference computed for a long line from vertical angle observations. The ends of such a line of levels are called 'junction points'. The method is to obtain adjusted values for these junction points or trig stations, and in the case of a line of levels, to adjust the individual points along the line by proportion if required.

Section 5.6.1 gives the adjustment of a small network of levels by the methods of condition and observation equations. If the height differences quoted had been obtained from trigonometrical heighting, the adjustment would proceed in the same way. As an additional example the heights obtained from the observations of table 8.6 are adjusted in table 8.7, with the lines weighted inversely as their respective lengths. The observed and adjusted height differences are given in table 8.6. The reader should note that since the coefficients of the unknowns (the height differences) are all unity, there is no need to avoid triple multiplications such as *pab* which is *p*.1.1, *p* being the reciprocal of the weight in the condition equations method. In a large network, a diagram showing the observed height differences is drawn. This is then copied line by line, and as each circuit or triangle is closed, the appropriate condition equation is formed. In this way no equation will be omitted, nor will a redundant equation be introduced. The number of condition equations C in a network of L lines and P new points is given by $C = L - P$. At least one point must be fixed as a datum.

8.5.1.1. *Adjustment of Intermediate Points*

In levelling, the bench marks intermediate between the junction points are finally adjusted by a proportional amount of the adjustment for the line. For

Table 8.7. Adjustment of Heights (See Figure 8.23)

(a) Determination of absolute terms in condition equations

```
       −(1)  −636·295       +(1)  −636·295       +(2)  −128·840
       +(2)  −128·840       +(4)   +73·445       +(6)  +202·950
       −(3)  +765·015       −(5)  +562·695       −(4)   −73·445
       Abs.   −0·120        Abs.   −0·155        Abs.   +0·665
```

(b) Condition equations

	1	2	3	4	5	6	Abs.	Sum
$U = 1/p = D$	1·404	1·706	1·921	2·633	1·990	1·371		
C_1 +0·1099	+1	+1	−1				−0·120	+0·880
C_2 −0·0815	+1			+1	−1		−0·155	+0·845
C_3 −0·1868		+1		−1		+1	+0·665	+1·665
Sum	+2	+2	−1	0	−1	+1	+0·390	+3·390
Residual v	+0·0399	−0·1312	−0·2111	+0·2772	+0·1622	−0·2561		
Residual v_2	+0·040	−0·131	−0·211	+0·277	+0·162	−0·257		

(c) Solution of normals [Gauss–Doolittle]

			+0·1099	−0·0815	−0·1868			
			−0·8901	−1·0814	−1·1868			
C_1	+5·031	+5·031	+1·404	+1·706	−0·120		+8·021	
		(0·198768)†	−0·27907	−0·33910	+0·02385		−1·59442	
C_2	+1·404	+6·027	+5·635	−3·109	−0·122		+2·404	
			(0·177462)	+0·5517	+0·02165		−0·4266	
C_3	+1·706	−2·633	+5·710	+3·416	+0·638		+4·054	
				(0·292 740)	−0·1868		−1·1868	
Abs.	−0·120	−0·155	+0·665					
Sum	+8·021	+4·643	+5·448					

† Numbers in brackets are the reciprocals of those above them

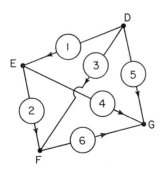

Figure 8.23

example, if D is the datum of the above example, a bench mark 9 000 ft from D along line DE will receive an adjustment of

$$+(0{\cdot}040 \times 9\,000)/14\,038 = +0{\cdot}026 \text{ feet}$$

In ordinary levelling, the adjustment is often apportioned according to the number of set-ups along the line, which amounts to very much the same procedure as above.

8.5.2. INTERVISIBILITY PROBLEMS

Figure 8.24(a) shows two stations A and C whose heights above a horizontal line datum are H_A and H_C respectively. Considering for the moment that the line of sight from A to C is straight we wish to establish the height H_B of an

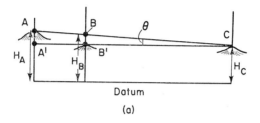

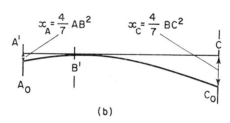

(b)

Figure 8.24

intermediate point B lying on this line. This problem arises when a survey reconnaissance is being carried out, or when a tower has to be erected to a computed height above datum. Equating values of tan θ

$$\frac{H_A - H_C}{A'C} = \frac{H_A - H_B}{A'B'} = \frac{H_B - H_C}{B'C} \qquad (24)$$

or values of sin θ

$$\frac{H_A - H_C}{AC} = \frac{H_A - H_B}{AB} = \frac{H_B - H_C}{BC} \qquad (25)$$

In practice sin θ and tan θ will be very nearly equal.

Example: If $H_A = 100$ ft, $H_C = 50$ ft, AC = 10 000 ft, and BC = 8 000 ft, the height of B is given by

$$H_B - H_C = BC(H_A - H_C)/AC = 40$$

i.e. H_B is 90 ft above datum. If the hill at B lies below this height, the line of sight will pass from A to C, and *vice versa*.

In practice, because of the refraction of the line of sight and the curvature of the datum surface, the simple case of above has to be modified. In figure 8.24(b) it will be seen the net effect of both these factors, curvature and refraction, is to lower the datum at A and C with respect to B by $\frac{4}{7}AB^2$ and $\frac{4}{7}BC^2$ feet, where the distances are in miles (see §8.4.5). Hence if the heights of A and C are decreased by their respective corrections, the problem is reduced to that of §8.5.2 above.

Example: If the heights of A, B, and C are respectively 1369 ft, 896 ft, and 704 ft; and AB = 15·10 miles, BC = 7·60 miles, what height of the tower should be erected at C so that the line of sight to the top from a theodolite at A, 4 feet above ground, will clear the ground at B by 20 ft?

The effective height at B is 916 ft. The curvature and refraction corrections to be applied to H_A and H_C are $x_A = \frac{4}{7}(15\cdot1)^2 = 130$ ft, and $x_C = \frac{4}{7}(7\cdot60)^2 = 33$ ft. The effective height at A is then $136\,9 + 4 - 130 = 124\,3$ ft. The effective height H_C of C is then given by

$$H_B - H_C = 327 \times BC/AB = 164\cdot5$$

$$\therefore H_C \simeq 752 \text{ ft}$$

Since the ground height of H_C is 704 ft above datum and $x_C = 33$ ft, the tower will require to be 81 ft tall.

In practice, a section has to be drawn from map contours and a likely position for a station is chosen. Any doubtful intervisibility problems are solved as above. If no map is available a profile of the proposed route of the survey can be plotted with the aid of altimeter heights and distances obtained from a vehicle mileometer.

8.5.3. GEODETIC LEVELLING TRANSFERS BY THEODOLITE

When transferring a line of levels across a wide gap, usually over water, the gradienter screw of the level may be used in conjunction with targets on the staves. For lines up to about 1·5 miles this is a satisfactory procedure, but for longer lines the effect of the Earth's curvature becomes troublesome since the target has to be separated from the ground station by something other than a staff. The actual terrain can also render the use of the level very difficult. On one transfer over 3·83 miles, the Ordnance Survey obtained an unsatisfactory agreement with the geodetic levelling of 0·415 ft. It was then decided to use truly simultaneous reciprocal vertical angle observations over this line as a test, the result of which was a much better agreement to 0·057 ft. Since 1955 the Ordnance Survey has adopted the trigonometrical method as the standard method of levelling across wide gaps, although a few are still carried out with a level and gradienter. The field procedure is to have each observer taking reciprocal vertical observations at the same time to within about 10 seconds. Contact by radio or carefully synchronized watches is essential. A total of 96 double faced observations is spread over six days, with the observers changed

420 Practical Field Surveying and Computations

from end to end for half of the work. For lines under 3 miles an accuracy of $0.015(M)^{\frac{1}{2}}$ can be obtained where M is the length of line in miles; and for long lines of 10 miles an accuracy of 0·3 ft may be expected. The increase in error over long lines is due to uncertain refraction and deviation of the vertical, in addition to normal observational errors.

8.5.4. THEODOLITE HEIGHT TRAVERSING

Sometimes an approximate height to within a foot is required for a station from which altimeter heights are to be determined. Such stations are usually required at the bottom of a valley and in thick impenetrable forest, or other inaccessible terrain. A rapid accurate height may be obtained by running a traverse from a trigonometrical station often over distances of several miles. No horizontal angles are measured, only verticals are recorded at every other station. Distances may be taken with a tape or by tacheometry. To safeguard against gross errors in reading the vertical angles, the stadia hairs must be used, or failing that, observations should be taken to a second point at some set distance (say a foot) above or below the standard height which is usually a mark on a ranging pole, or a particular staff graduation. The results are reduced immediately to safeguard against gross error, and the traverse should close on another trig point if possible, or at least back on the original point.

8.5.5. HEIGHTING FROM AIR PHOTO DISTANCES

In good open country, such as plateau Africa with kopjes, the surveyor can identify hills from afar. Vertical and horizontal angles are taken to these points, and later the distances are scaled from a slotted template assembly or other photogrammetric plot. Thus the heights of these points may be computed and used to control photogrammetric contouring. Alternatively, and more accurately, the surveyor takes a vertical observation to a trig beacon from some point which he has identified on the photography. Only one trig need be observed

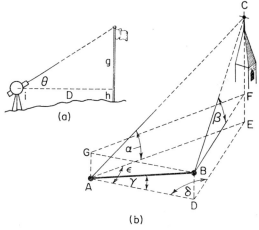

Figure 8.25

and the distance scaled from a photo plot. Needless to say more than one trig would be observed if possible.

8.5.6. OBSERVATIONS FOR BEACON HEIGHTS

When the height of a survey beacon cannot be measured directly by tape, an arrangement such as that shown in figure 8.25(a) in which θ, D, and h are measured, will enable the beacon height to be computed. If the beacon is a tall object like a church spire whose top has been intersected, a base line may have to be set out some distance back from the church, and angles α, β, γ, δ, and ϵ observed, together with the length AB (figure 8.25(b).) Solution for CE gives the height of the spire relative to A, with CE + BD as a check. In both these determinations the various angles should lie between 30° and 60° to give well-conditioned figures and accurate results. In figure 8.25(b), solution is much simpler if A, B, and C lie in the same plane.

8.6. BAROMETRIC HEIGHTING

The surveying aneroid is widely used today to provide height control for aerial mapping in areas that are otherwise inaccessible. If due care is taken heights may be obtained (with modern precision instruments) to an accuracy of about 5 feet. Stable atmospheric conditions are required to give the best results, and for this reason, tropical areas are very suitable for aneroid work. All accurate barometric heighting must be controlled by bench marks supplied by levelling or theodolite. In barometer work, these control bench marks are called *bases*. Mercury barometers are not used because they are bulky and fragile, though a good mercury barometer installed permanently at survey headquarters is useful for calibration. In barometric heighting, the pressure of the atmosphere is recorded and converted to equivalent height via the equation

$$P = f(g, \rho, h) \qquad (26)$$

where P is the atmospheric pressure, g the acceleration due to gravity, ρ the density of the air, and h the height of the place above datum. Although P is a function of three variables, the height h is the major one. The gravity and density contribution to variation in pressure is about 10%.

8.6.1. MEASUREMENT OF PRESSURE

Atmospheric pressure is measured in terms of the height of the column of mercury that will balance it in a barometer, i.e. a height of about 76 cm. The equivalent column of air, whose density at 0°C and pressure 76 cmHg is 0·001 293 g/cm³, will be given by equating pressures $g\rho h$,

$$g\rho_A h_A = g\rho_M h_M \quad \therefore \quad h_A = \frac{13 \cdot 596 \times 76}{0 \cdot 001\ 293}\ \text{cm} \quad \begin{array}{l}(\rho_M = 13.596\ \text{g/cm}^3)\\ (M = \text{mercury},\ A = \text{air})\end{array}$$

Hence $h_A \simeq 26\ 000$ ft. It will be noted that this does not mean that the atmosphere ends at this height, but merely that it would do so if its density remained the same as that close to the ground, whereas in fact it decreases away from the Earth into space. Thus 30 inches of mercury are equivalent to about 26 000 ft or 0·1 inch is equivalent to 90 ft. Occasionally in surveying the unit of measurement of pressure is the bar = 10^6 dynes/cm². To relate inches or centimetres of mercury (inHg or cmHg) to this unit of pressure:

$$P = g\rho h = 980·62 \times 13·596 \times 76 \quad (\text{Average value of } g = 980·62 \text{ cm/sec}^2)$$

Whence 76 cmHg are equivalent to about 101 3 millibars. (One millibar 10^{-3} bar), or 1 000 mb ≡ 75 cmHg (29·53 inHg).

8.6.2. SURVEYING ANEROIDS

An aneroid barometer consists of three essential parts: *The capsule*, the *linkage mechanism*, and the *scale and pointer*. The capsule is a hermetically sealed box from which almost all air has been evacuated, and which is prevented from collapsing by balance springs. The sides of the capsule move in response to changes in the atmospheric pressure outside. These movements are exceedingly small and require to be magnified many times to enable one to read them. Capsules are normally made of beryllium copper and their faces are corrugated to increase the sensitivity. The linkage mechanism is the mechanical or electrical means of detecting the capsule movements. The scale of a surveying aneroid is most conveniently graduated in height units according to some theoretical relationship. Until recently such a height scale was always graduated in an anticlockwise sense. For consistent precision, the scale should be uniformly graduated throughout its range. Recently a digital reading system in millibars has appeared in the Baromec instrument.

Surveying aneroids may be classified in two groups;

(i) The *direct reading* instruments, such as those made by Wallace and Tiernan, and

(ii) The *indirect reading* instruments, such as the Paulin or Baromec.

Figure 8.26 shows a schematic diagram of type (i), and figure 8.27 of type (2). The essential difference lies in the fact that the capsule supplies the mechanical

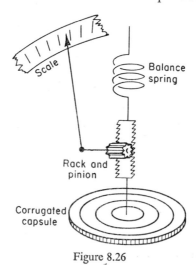

Figure 8.26

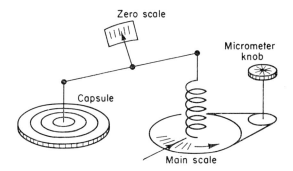

Figure 8.27

force to move the linkage and pointer in type (i), whereas in type (ii), this force is supplied by the surveyor by moving a micrometer knob. In the production of a sensitive aneroid capable of a precision of 1 foot, the following problems have to be solved:

(i) A capsule has to be designed so that the height scale is reasonably uniform throughout its range. The sensitivity has to be 1 foot, and the total range of the scale has to be a practical one such as 4 000 ft. The Wallace and Tiernan (W and T) instruments have a double scale, the pointer rotating twice over the full range of the instrument. A small rough scale indicates on which main scale the reading should be made.
(ii) Friction of pivots, etc. has to be overcome by precise engineering and design. A good aneroid does not require tapping.
(iii) The instrument should be compensated for variations in temperature, i.e. for a given pressure, the reading should remain constant irrespective of the temperature of the aneroid. Such *temperature compensated* instruments usually cover normal temperature ranges.
(iv) The instrument should exhibit very little hysteresis, i.e. the reading at a given height should not depend on whether the last height reading was higher or lower than the current one.
(v) There should be no parallax in reading the scale. This error is usually overcome by silvering the scale behind the pointer. As one looks down on the pointer a direct and mirror image is seen. If these images coincide, the observer's eye is vertically above the scale and the reading will be free of parallax error.

The following are additional desirable features:

(i) Some means of setting the pointer to a fixed value to enable several instruments to be calibrated at approximately similar readings.
(ii) There should be some means of sealing off the instrument case for transit by air.
(iii) There should be a means of calibrating several altimeters over their operational range without the aid of a pressure chamber (see §8.7.1).

(iv) Where required the instrument should allow for readings below sea level. This is normally effected by graduating the instrument from $-1\,000$ ft, and introducing a datum shift of $1\,000$ ft in numbering the scale. For example, a reading at about sea level would be $1\,000$ ft. This avoids having negative readings.

(v) The mechanism should be mounted in shock-proof supports and a desiccant supplied to absorb any moisture.

8.6.3. SELF-RECORDING INSTRUMENTS

A number of self-recording instruments are available on the market. They are of two types—the *barograph* type in which readings are recorded on a moving drum by pens, or the *photographic* type in which a normal altimeter is photographed automatically at regular preset intervals. These instruments are not used widely since they cannot normally be left unguarded at a field station. It is then cheaper and simpler to have someone read a standard instrument.

8.6.4. THEORETICAL ASPECTS OF BAROMETER HEIGHTING

Before proceeding to the practical use of altimeters, the theory of their application will be considered briefly. It has been stated above that the atmospheric pressure P is a function of three variables gravity g, density ρ, and height h, and that height is the most significant. Thus we may write

$$P = f(g, \rho, h) \tag{26}$$

If the function can be found relating these four quantities any subsequent pressure readings may be converted to height via this function. There are two ways of establishing what this function is: *empirically*, i.e. by sampling actual cases, and *theoretically*, i.e. by analysing the various factors which contribute to equation (26). In practice the empirical method gives better results since at any time the function obtained takes all the various factors into account, whereas the second theoretical method must make certain assumptions which are more or less valid as the case may be. Both methods however give similar results. As will be seen below, the solution finally adopted to establish the relating function is partly theoretical and partly empirical in some instances.

8.6.5. THEORETICAL APPROACH—LAPLACE EQUATION

The theoretical analysis is carried out in two stages. Firstly g is considered constant and secondly the small effects of its variation are examined. Consider a small element of the atmosphere of unit cross-section. The variation in pressure dP is related to the variation in height dh by

$$dP = -g\rho\, dh \tag{27}$$

The sign is negative because pressure decreases with height. Now P, v, and T, the pressure, volume, and absolute temperature of a perfect gas are related to their values under standard conditions P_s, v_s, and T_s by the combination law

$$Pv/T = P_s v_s / T_s$$

i.e.
$$v = (P_s v_s / P)(T/T_s) \tag{28}$$

Vertical Control

If ρ is the density of air under the conditions (P, v, T) and ρ_s is its density under the standard conditions, equating the mass of air

$$\rho v = \rho_s v_s$$

$$\therefore v = \rho_s v_s / \rho \qquad (29)$$

Equating v from (28) and (29)

$$\frac{1}{\rho} = \frac{P_s}{\rho_s} \frac{T}{T_s} \frac{1}{P} \qquad (30)$$

Substituting (30) in (27)

$$dh = -\frac{dP}{g\rho} = \frac{P_s}{g\rho_s} \frac{T}{T_s} \frac{dP}{P}$$

$$\therefore h_2 - h_1 = \int_{h_1}^{h_2} dh = \int_{P_2}^{P_1} \frac{P_s}{g\rho_s} \frac{T}{T_s} \frac{dP}{P}$$

Putting $g = g_s$

$$\frac{P_s}{g_s \rho_s} = h_s \simeq 26\,000 \text{ ft}$$

$$\therefore h_2 - h_1 = +\int_{P_2}^{P_1} \left(\frac{h_s T}{T_s}\right) \frac{dP}{P} = \frac{h_s T}{T_s}(\log_e P_1 - \log_e P_2)\dagger$$

$$= h_s \log_e 10 \, T/T_s (\log_{10} P_1 - \log_{10} P_2)$$

And finally

$$h_2 - h_1 = 60\,370 \, \frac{T}{T_s} \log_{10}\left(\frac{P_1}{P_2}\right) \qquad (31)$$

It will be noted that the logarithms are used as functions in their own right and not merely as aids to computation. To allow for the variation of g with latitude ϕ and the mean height above sea level of the barometers, and to allow for the fact that air is not a perfect gas but contains water vapour, the full version of formula (31) is

$$h_2 - h_1 = 60\,370 \left(1 + 0.002\,64 \cos 2\phi + \frac{h_1 + h_2}{R} + \frac{3e}{8P_m}\right) \frac{T}{T_s} \log\left(\frac{P_1}{P_2}\right) \qquad (32)$$

where R is the radius of the Earth, e the vapour pressure, and P_m the mean pressure $\frac{1}{2}(P_1 + P_2)$. $h_2 - h_1$ in feet, and T is the absolute temperature.

This is the classical *Laplace formula* for the reduction of aneroid heights, and is based on the *isothermal assumption*. It is possible to take into account the drop of temperature with height (the lapse rate) but as actual conditions will still differ from the assumed lapse rate nothing is gained, and a correction for temperature still has to be applied. The lapse rate formula is therefore omitted from this book. (See C. A. Biddle, The Reduction of Aneroid Readings, E.S.R., No. 116 April 1960, p. 291 for discussions on this matter.)

† It is important to note that this integration was effected by considering the temperature T constant over the range of the integration, i.e. this formula results from the isothermal assumption.

8.6.6. CALIBRATION OF AN ANEROID IN HEIGHT UNITS

For ease of working in the field, it is convenient to calibrate a barometer in terms of height instead of pressure. Such an instrument reading directly in height is called an *altimeter*. The humidity, latitude and temperature terms of equation (32) are denoted by $1 + c$ i.e. equation (32) becomes

$$h_2 - h_1 = K(1 + c) \log (P_1/P_2) \qquad (33)$$

where $K = 60\,370$. For standard conditions of humidity, latitude, and temperature

$$h_2 - h_1 = \alpha \log (P_1/P_2) \qquad (34)$$

(α a constant.)

If $h_1 = 0$ and P_1 is an accepted standard value for pressure, say 29·92 in Hg the height values h corresponding to various pressures P can be calculated from (34). Hence a *barometer* reading *pressure* may be calibrated as an *altimeter* reading *height* according to formula (34). A height difference obtained from altimeter readings will approximate to the correct value only, since corrections are required for the latitude, temperature and humidity variations from standard. The variation with latitude is normally omitted, and the temperature and humidity correction factor obtained from a table such as table 8.8. A factor F is obtained by entering this table with the wet and dry bulb thermometer readings taken on a psychrometer. The final height difference between station (1) and (2) is then $(h_2 - h_1)F$, where h_1 and h_2 are the altimeter readings, i.e. if H_1 and H_2 are the corrected heights of the two stations

$$H_2 - H_1 = F(h_2 - h_1) \qquad (35)$$

This is the equation used in practical field work with an altimeter.

8.6.7. METEOROLOGICAL FACTORS

8.6.7.1. *Wet and Dry Bulb Thermometers*

The reader is referred to §13.4.8 in which meteorological readings are discussed.

8.6.7.2. *Empirical Method*

As an alternative to obtaining the factor F from tables, it may be derived from readings taken at two stations whose heights are already known, i.e. if the heights H_2 and H_1 of equation (35) are known, F is calculated from

$$F = (H_2 - H_1)/(h_2 - h_1) \qquad (36)$$

Where possible this empirical method should be employed, though met. readings should also be taken as a check, or as a safeguard against faulty readings at a base station due perhaps to an unreliable observer. Since F is obtained from simultaneous readings at two stations of known height, a third altimeter is required for the field readings at new stations.

Table 8.8. Altimeter Heights. Air Temperature and Humidity Factors

The table is based on the isothermal assumption, standard temperature 50° F.

True height difference = Observed height difference × Factor

$$\Delta H = \Delta h \times F$$

Dry bulb temp. (°F)	Wet bulb temperature °F									Dry bulb temp. (°F)
	40 and under	50	60	70	75	80	85	90	95	
20	0·943									20
2	0·947									2
4	0·951									4
6	0·955									6
8	0·959									8
30	0·963									30
2	0·967									2
4	0·971									4
6	0·975									6
8	0·979									8
40	0·984									40
2	0·987									2
4	0·991									4
6	0·995									6
8	0·998									8
50	1·002	1·005								50
2	1·005	1·008								2
4	1·009	1·012								4
6	1·013	1·016								6
8	1·017	1·019								8
60	1·020	1·023	1·026							60
2	1·024	1·027	1·030							2
4	1·027	1·030	1·034							4
6	1·031	1·034	1·037							6
8	1·035	1·038	1·041							8
70	1·038	1·041	1·045	1·049						70
2	1·042	1·045	1·048	1·052						2
4	1·046	1·049	1·052	1·056						4
6	1·049	1·052	1·056	1·060	1·062					6
8	1·053	1·056	1·059	1·063	1·065					8
80	1·057	1·060	1·063	1·067	1·070	1·072				80
2		1·063	1·067	1·071	1·074	1·076				2
4		1·067	1·070	1·074	1·077	1·080				4
6		1·071	1·074	1·078	1·081	1·083	1·086			6
8		1·074	1·078	1·082	1·085	1·087	1·089			8
90		1·078	1·081	1·085	1·088	1·091	1·094	1·097		90
2		1·081	1·085	1·089	1·092	1·094	1·098	1·101		2
4		1·085	1·089	1·093	1·096	1·098	1·101	1·104		4
6		1·089	1·092	1·096	1·099	1·102	1·105	1·111		6
8		1·093	1·096	1·100	1·103	1·105	1·109	1·112	1·115	8
100		1·096	1·100	1·104	1·107	1·109	1·112	1·115	1·119	100
	40	50	60	70	75	80	85	90	95	
	Wet bulb temperature °F									

Published by permission of Mr. T. Bassett, Directorate of Military Survey—England

8.7. PRACTICAL ALTIMETRY

In all practical methods of altimetry two assumptions are made to simplify the problem:

(i) That the air mass in which the work is carried out is homogeneous throughout.
(ii) That the air temperature varies linearly with height.

Both these factors imply the existence of stable air conditions throughout the survey area. Consequently, the area covered should not exceed about 5 miles square and 500 ft in altitude if the best results are to be obtained. All height determinations by altimetry should be relative to one or more base points of known height, and each day's work should close between such points whenever possible. No attempt should be made to obtain the heights of bases by repeated altimetric heighting of these bases; a high degree of precision may be obtained but the accuracy will very often be poor. It is essential that all altimeters are calibrated with respect to each other and that watches are synchronized. All methods depend on obtaining readings taken at two places *at the same instant of time* and in effect *by identical instruments*. The atmospheric pressure at one place varies throughout the day, falling to a minimum about 14.00 hrs. Figure 8.28 indicates the manner in which the height readings on an altimeter vary

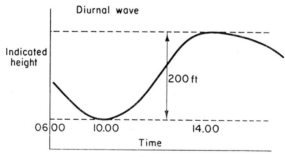

Figure 8.28

throughout the day under stable weather conditions. The diurnal variation or *diurnal wave* may be as much as 400 ft within the tropics, through 200 ft or less is more common. Hence if we trusted our altimeter to give an absolute height, our inference from a set of daily readings would be that the ground surface was rising and falling, which is clearly not so. Allowance therefore has to be taken for diurnal variation in all methods of altimetry.

Outside the tropics, the diurnal wave varies a great deal from day to day in an unpredictable manner. Because several altimeters will normally be used in the field, it is essential that their serial numbers are carefully recorded in field books and computation sheets. Since a great many points are heighted in the course of an altimetric survey, the field work must be organized and documented carefully, particularly regarding the numbers assigned to points, if ambiguity is to be avoided. If the heights are used to control aerial photography, as is generally the case, care must be exercised to relate the altimetric field books to

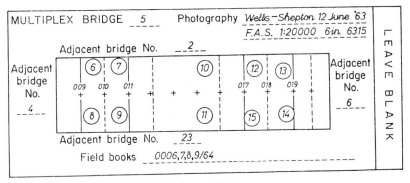

Figure 8.29(a)

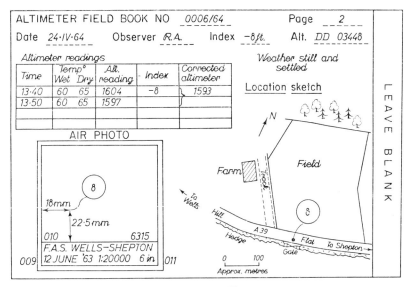

Figure 8.29(b)

the air photos. Figures 8.29 show a typical field book and its related multiplex height bridging diagram, assuming eight projector bridges. As a general rule each day's work will include one or two points heighted on a previous occasion, to check the internal consistency of the work. The results of each day's work are computed each evening and checked before proceeding with new work.

8.7.1. CALIBRATION OF ALTIMETERS

At the beginning of each day the altimeters are compared together over a period of about fifteen minutes. One is chosen as the standard instrument to which the others will be corrected. It is usual to accept a small difference between each altimeter and apply its correction, rather than to attempt to set each instrument to an identical reading. Table 8.9 shows the comparison readings taken on three altimeters before and after field work. Altimeter No. 1 is chosen as the standard to which the others are related. If the index corrections for the

Table 8.9. Base Comparison Readings

	(1)			Altimeters	(2)		
Time	I	II	III	Time	I	II	III
08·00	931	936	927	16·00	984	992	978
08·05	930	936	926	16·05	984	991	977
08·10	930	935	926	16·10	983	991	979
08·15	929	935	925	16·15	982	989	978
Mean	930	936	926	Mean	983	991	978
Index	0	−6	+4	Index	0	−8	+5
Accept		−7	+4	for day			

beginning and end of the day are different, the correction is interpolated for time throughout the day. If however the variation is small (under 2 feet) a mean correction may be used for the day.

With the Wallace and Tiernan altimeters it is possible to simulate different pressures with the aid of a pump linked to the air intake tubes of the various instruments. In this way the behaviour of the altimeters at different altitudes may be examined and a relative calibration obtained for the complete range of the instruments. For example, a typical set of such readings is given in table 8.10.

Table 8.10. Wallace and Tiernan Altimeters Relative Calibration at Various Heights.

Altimeter	Readings taken at same pressure							
I	700	800	900	1000	1100	1200	1300	1400	1500
II	705	805	906	1007	1108	1209	1310	1410	1510
III	693	794	896	996	1097	1198	1300	1401	1502

This table indicates the extent to which the index correction varies with the readings themselves. Normally this variation is small and may be neglected, though a test of this kind will indicate a faulty altimeter. Some times the calibration readings from this table are applied to the field work, using the daily index readings to check the validity of the calibration. If the daily index differs from the calibration, it is usually assumed that the whole calibration has altered by this difference.

8.7.2. ALTIMETRIC METHODS

There are five methods of heighting with an altimeter as follows:

(i) Diurnal wave.
(ii) Single base.
(iii) Leap frog.
(iv) Double base.
(v) Multiple base.

Vertical Control

One known height is required with methods (i), (ii), and (iii); two for (iv), and three or more for (v). Defining a *set of equipment* and personnel as: One surveyor, one altimeter, one watch, one wet bulb thermometer and one dry bulb thermometer; method (i) requires one set of equipment, methods (ii) and (iii) require two sets, method (iv) requires three, and method (v) requires at least four sets. When more than one person is required only one need be a surveyor, the others being assistants who need only know how to read an altimeter and be trusted to do so correctly. Where an additional known height exists a check should be made on it whenever possible.

In all methods watches must be synchronized, and the altimeters calibrated relative to each other. The correction factor F will be obtained from table 8.8 or by calculation using equation (36). Sometimes both may be used as a check.

8.7.3. DIURNAL WAVE METHOD

Prior to the establishment of new height points, the diurnal wave is established by taking readings at intervals of fifteen minutes throughout each of several days. These readings will indicate if the method is at all possible by the regular agreement of each wave. The actual times of the daily peaks, etc. will vary slightly and allowance may have to be made for this on the field days.

Having established that the method is feasible, the altimeter is taken into the field after making some readings at the base. The heights of new points are taken, and at the end of the day, base readings are taken once more. On the assumption that the shape of the diurnal curve is as previously determined, the curve for the field day is constructed by fitting the established diurnal wave to the small sections plotted from the readings on the particular field day, i.e. in the morning and evening. The heights of the field stations are then computed assuming the readings at the base were those of the wave at the particular time of the field reading.

Example: Points A and B are to heighted from a base whose height is 910 ft. The diurnal wave established over several days was as follows:

Time	09·00	10·00	11·00	12·00	13·00	14·00	15·00	16·00
Altimeter	923 ft	915	917	928	948	975	990	985

(To save space only a few of the readings have been given.) On the field day readings were taken as follows:

Station	Base	Base	Field A	Field B	Base	Base
Time	09·00	10·00	12·00	14·00	15·00	16·00
Alt. read	922 ft	918	1095	1380	992	988

The readings at base on the field day indicate that the assumption concerning the diurnal wave is reasonable and the mean wave is accepted. The wave correction from 10·00 to 12·00 on day 1 is $928 - 915 = 13$ ft. Hence the reading at A is corrected for wave to $1\,095 - 13 = 1\,082$. Similarly the corrected reading at B is $1\,380 - (975 - 915) = 1\,320$. If the dry bulb readings at A and B are 80° and 81°F respectively and the wet bulb readings both 75°F, we calculate the final heights of A and B as follows:

Stn.	C^d Field alt	Base	Δh	Temp°F Dry	Wet	Factor	ΔH	Height H
A	1082	918	+64	80	75	1·07	+68	978
B	1320	918	+402	81	75	1·07	+430	1340

Within the tropics a standard error in heights produced by this method is 10 ft, though great care must be exercised to look out for a breakdown in the weather.

8.7.4. SINGLE BASE

This method is essentially similar to the diurnal wave method except that a second altimeter is kept at the base station throughout the entire series of measurements, i.e. the actual diurnal wave is recorded for each day. At the start of the day both altimeters are read together at the base station to establish the index correction. Watches are synchronized and the field altimeter taken to various new stations. About ten minutes should be allowed at each station for the instrument to settle down before taking a few readings and recording the temperatures. Temperatures may also be recorded at the base station about every half hour. The mean temperature at base and field is used in computation for the instant of the field measurement. At the end of the day a second index comparison is made between field and base altimeters, and one corrected to the other before computation computation is as follows: field correction +4, base height 112 ft. Base and field readings are for the same instant of time.

Stn.	Field alt.	C^n	C^d field	Base alt.	Δh	Mean temp (°F) Dry	Wet	Factor	ΔH	H
17	818	+4	822	919	−97	73	60	1·050	−102	10
18	886		890	924	−34	75	62	1·056	−36	76
19	825		829	917	−88	76	62	1·057	−93	19
C/9	914		918	916	+2	77	63	1·058	+2	114
20	839		843	938	−95	79	65	1·062	−101	11
21	878		882	985	−103	82	70	1·071	−110	2

Point C/9 was a bench mark whose height was 115·3 ft. Provided the two height points are not more than five miles apart and do not differ by more than 500 ft in height, a standard error of 5 ft can be consistently achieved in stable weather conditions by this method.

8.7.5. DOUBLE BASE METHOD

In this method all three altimeters are calibrated at the beginning of the day and watches synchronized. One altimeter remains at a base whose height is typical of the lower part of the area to be heighted i.e. at the *lower base*, one altimeter is situated at a higher point—the *upper base*, whilst the third altimeter is used in the field to height new points. Temperatures are read as a safeguard.

Before computing the final heights, two of the altimeters are corrected for the mean index error for the day. The index variation with height may also be applied if significant.

Typical set of readings for double base altimetry:

Time	Lower base c^n −7 ft corrected		Upper base c^n 0	Field c^n +4 corrected		Mean temps (°F) Dry	Wet	Station
9·40	829	822	919	818	822	73	60	17
10·00	827	820	924	886	890	75	62	18
10·40	826	819	917	825	829	76	62	19
11·00	825	818	916	914	918	77	63	C/9
12·30	849	842	938	839	843	79	65	20
15·30	889	882	985	878	882	82	70	21

Upper base altimeter used as standard.

The heights of the bases are:

$$\text{Upper base} \quad 112 \text{ ft}$$
$$\text{Lower base} \quad 5 \text{ ft}$$
$$\Delta H \quad 107 \text{ ft}$$

Whence the factor F is given by $F = \Delta H/\Delta h$. The heights may then be computed as follows:

Point	(1) U. Base	(2) Field	(3) L. Base	(1)−(3) Δh	$\frac{\Delta H}{\Delta h}$ F	(2)−(1)	(2)−(3)	H_1	H_2
17	919	822	822	97	1·10	−97	0	5	5
18	924	890	820	104	1·03	−34	+70	77	77
19	917	829	819	98	1·09	−88	+10	16	16
C/9	916	918	818	98	1·09	+2	+100	114	114
20	938	843	842	96	1·12	−95	+1	6	6
21	985	882	882	103	1·04	−103	0	5	5

$$H_1 = \text{Upper base height} + F[(2) - (1)]$$
$$H_2 = \text{Lower base height} + F[(2) - (3)]$$

And H_1 should equal H_2 as a check on the arithmetic only. This method is more accurate than the single base method since an empirical assessment is continuously made of the factor F.

8.7.6. LEAP FROG METHOD

This method consists of a traverse run over some distance by a succession of height determinations using single base methods, each forward point in turn becoming the base. If the traverse is run between two fixed points, or in a loop to the same point, it may be computed by double base methods as a check.

Example: A leap frog altimeter traverse is run between two bases A and B whose heights are 29 ft and 43 ft respectively. The following table shows a synopsis of the readings, times, etc.

Time	Altimeter I At. Reading		Altimeter II At. Reading		Mean temperature (°F) Dry	Wet
08·00	Base A	1380	Base A	1384		
08·30	Base A	1374	Stn. 1	1568	70	60
09·30	Stn. 2	1333	Stn. 1	1564	74	62
11·00	Stn. 2	1329	Stn. 3	1360	76	64
14·00	Base B	1162	Stn. 3	1395	80	75
15·00	Base B	1160	Base B	1164		

The altimeters are read together at base A, altimeter II moves to station 1 whilst altimeter I remains at the base, and at the same instant both are read together. This involves taking readings at a prearranged time and that both watches are carefully synchronized, or that a system of signalling from one to the other is available. On short distances, visual signalling is practicable, whilst walkie-talkie radios have been employed over longer distances for communication. For the next stage, altimeter II acts as a base whilst altimeter I moves forward to station 2. and so on until the traverse is closed on base B. Temperatures are recorded throughout the measurement. The computation of the above readings is as follows: Altimeter I is chosen as standard, i.e. there is an index correction of −4 ft to be applied to altimeter II. The factor F is obtained from table 8.8 and each small height difference computed by single base methods.

Stations	Forward stn Corrected reading	Back stn Corrected reading	Δh	Factor F	ΔH	H
A to 1	1564	1374	+190	1·045	+199	495
1 to 2	1333	1560	−227	1·053	−239	256
2 to 3	1356	1329	+27	1·058	−29	285
3 to B	1162	1391	−229	1·070	−245	40

The misclosure on base B is 3 ft. The leap frog method invariable gives more accurate results since each individual distance covered is relatively short, with the result that the air mass covering the stations is likely to be stable. This example may also be computed by the double base method obtaining the factor F from the altimeter height difference carried through the traverse without any corrections, i.e. the uncorrected height difference is

$$+190 - 227 + 27 - 229 = -239 = \Delta h$$

The correct height difference is $43 - 29 = -253 = \Delta H$ whence

$$\text{Factor } F = 253/239 = 1·059$$

from which we obtain the individual ΔH's and heights

Stn 1   +201   497
Stn 2   −240   257
Stn 3   +29   286

8.7.7. MULTIPLE BASE METHOD

As the name suggests, this is an extension of the double base method. At least three bases are set out at the apices of a triangle covering the survey area. Two of the bases should be typical of the lower and upper heights of the region respectively. If there is a tilt in the isobaric surface to which the heights are referred and which has tacitly been assumed horizontal, multiple base readings will detect this to some extent. One base is chosen as the reference station both for height and for plan position, i.e. as the coordinate origin. The 'North–South' axis is chosen to run through one other base for simplicity. Coordinates of the other bases and the field stations are read from a map. The correction factor F for a point relative to A will then be calculated according to its plan position using the data for the three bases. For example, the following table shows data for five stations of which A, B, and C are the bases:

Stn	Height (ft)	Approx. E	coords N	Altimeter reading corrected for time and index (ft)
A	628·4	0	0	1730
B	378·0	0	+5·22	1483
C	11·4	+4·79	+2·77	1080
D	Not known	−0·16	+7·13	1444
E	Not known	−1·36	+2·66	1500

From A and B, $F_{AB} = 250·4/247 = 1·014$
From A and C, $F_{AC} = 617·0/650 = 0·949$

The temperatures at station A were 35°F, whence from table 8.8 $F_A = 0·973$. Let the factor F be expressed as:

$$F = \alpha E + \beta N + \gamma$$

where α, β, and γ are coefficients to be evaluated from the readings taken at the bases, as follows:

$$\gamma = F_A = 0·973$$
$$\therefore \quad _AF_B = \alpha(0) + \beta(5·22) + 0·973 = 1·014$$

whence $\quad \beta = +0·00785$
and $\quad _AF_C = \alpha(4·79) + \beta(5·22) + 0·973 = 0·949$
whence $\quad \alpha = -0·009\ 60$

The equation with which to calculate the factor F for any point is then

$$F = 0·973 - 0·009\ 60E + 0·007\ 85N \tag{37}$$

Applying this equation to the points D, and E in turn

$$_AF_D = 1·030\ 5 \quad \text{and} \quad _AF_E = 1·007$$

giving corrected height difference of −295 ft and −232 ft. Whence the final heights of D and E are respectively 333 ft and 396 ft.

Since the results obtained from single base and double base methods are normally sufficiently accurate for topographical mapping, multiple base has not been applied widely. It has the drawback of requiring at least three base altimeters and one field instrument, whereas the same number of persons and equipment can produce adequate results with two bases and two field instruments at a faster rate.

9 Detail Survey

9.1. INTRODUCTION

It is intended in this chapter to describe the methods and equipment used for the making of large scale plans from measurements made on the ground. Today most maps at scales of 1:10 000 and smaller are made almost entirely by air survey methods but as the scales become larger the amount of detail to be surveyed on the ground becomes greater. Furthermore small or isolated revision surveys can rarely be executed economically by air survey.

The methods described in this chapter are used to make plans at large scales given tertiary control, to revise existing large scale plans and to complete plans made in part by air survey methods.

9.2. DEFINITIONS

Detail.
 (i) Buildings, structures and features, natural or artificial whose extent and shape can be shown on the plan to scale.
 (ii) Objects and features, natural and artificial whose extent cannot be shown on the plan to scale but are sufficiently prominent to be an important feature of the plan.

Indefinite Detail. Detail as defined above which has an outline incapable of exact definition or is liable to change.

Overhead Detail. Detail which constitutes no obstruction at ground level, this is only surveyed if it is an important feature of the area being surveyed. For instance an overhead gantry running over open country.

Interior Detail. The internal features of a building, these are only surveyed if they represent an important feature such as the division between two houses.

Building Line. Where a building enters the ground is the line that is surveyed. This is termed here the 'building line'. Large scale plans show detail existing at ground level.

9.3. REPRESENTATION OF DETAIL

9.3.1. PRINCIPLES

Firm lines are used on the plan to mark the outer limits of all detail and obstructions sufficiently high to impede progress, for instance over 30 cm high.

Pecked lines represent the edge of a detail feature where no impediment to progress exists. These pecks are drawn at approximately 8 pecks to the centimetre.

Vegetation will be sketched as a conventional sign unless the feature is isolated and prominent.

9.3.2. BUILDINGS

(a) All permanent buildings appearing as 8 square millimetres or more at plan scale are shown.

(b) Permanent buildings if between 2 and 8 mm² at plan scale are shown if detached and forming an important ground feature.

(c) Temporary buildings are not shown. A building that is likely to remain in position for more than six months is regarded as permanent.

(d) In the case of an open sided building the outer line of the permanent floor is shown with a pecked line.

(e) The exact limits of roofs are not surveyed, however where a roof has been erected to protect a ground area its edge is shown marked with a pecked line. All roofs are hatched.

(f) Glazed roofs are crossed hatched.

(g) An archway or passage is shown with firm lines and a brace

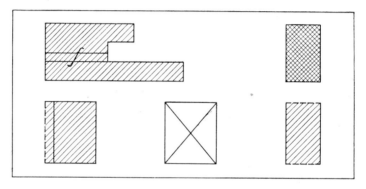

Figure 9.1

(h) Juts and recesses are only shown if they appear as greater than 1 mm at plan scale or if they abut onto a public right of way.

(i) Overhead detail if surveyed is shown with pecked lines and annotated.

(j) A building which is wholly or partly deroofed is a ruin and its extent is indicated by firm lines with firm diagonal lines to the corners.

9.3.3. FENCES, HEDGES, AND WALLS

(a) If in height greater than 30 cm a firm line is used to indicate the feature. An annotation, F, H, or W, may be used to distinguish the exact nature of the feature.

(b) If a wall appears as less than 1 mm at plan scale it is shown by a single firm line representing its centre line.

(c) Vegetation symbols are added to hedges appearing as greater than 10 mm in thickness at plan scale.

(d) A permanent step between fields is indicated by suitable annotation. If no hedge, wall, or fence lies on top of the step a pecked line may be used to mark the division.

(*e*) When two or more features, usually represented by a firm line, run parallel to each other and so close as to be incapable of being separately plotted only one feature, the most permanent, is shown in its exact position. All other features are omitted.

9.3.4. STREAMS, RIVERS, ROADS, TRACKS AND PATHS

(*a*) The width of the feature is only shown if it is in excess of 1 mm at plan scale.

(*b*) The direction of flow of a stream or river is indicated with an arrow.

(*c*) The limits of the feature are taken as the adjoining hedge or fence, sides or banks of streams are shown only if there is no adjoining hedge or fence. The edge of a metalled vehicular way is shown with a pecked line.

(*d*) Unmetalled vehicular ways are annotated track or Tk.

(*e*) Kerb lines are shown with pecked lines unless over 30 cm in height.

(*f*) Such features which are so common as to have no significance to the map user are omitted. For example traffic warning signs. Some large scale plans do however require this information particularly when made for engineering purposes.

9.3.5. VEGETATION

The character of the ground surface should be implied by the presence or absence of symbols or annotation on the map or plan. It is usual for open land, pasture or arable to be shown without annotation but where the area is predominantly covered with, for instance, scrub the open areas may be shown by symbol or annotation. Assuming open areas are shown without symbol or annotation the following apply;

(i) Principal types of vegetation are distinguished by symbols (see figure 9.2).
(ii) Isolated trees which are prominent landmarks are surveyed and the appropriate symbol used.
(iii) Other isolated trees unless specifically required are not surveyed and the appropriate symbol used to indicate the character of the ground surface.

9.3.6. NAMES

(*a*) The recognized names of all features appear on the plan.

(*b*) Buildings and features of interest to the map user are annotated, e.g. public house.

(*c*) Administrative areas are distinguished by lightly printed titles.

(*d*) Street names and house numbers are shown in sufficient detail to enable all property to be identified. Isolated houses are named provided this does not overshadow other detail.

9.3.7. OTHER FEATURES

The notes given above can be regarded as defining generally accepted methods of representing detail at scales between 1 in 1 000 and 1 in 5 000. It is not the intention to lay down specific instructions for every type of detail feature, the

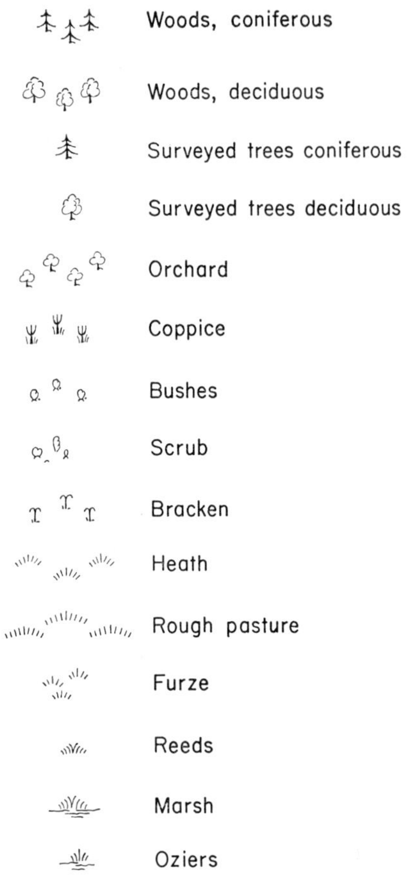

🌲🌲🌲	Woods, coniferous
🌳🌳🌳	Woods, deciduous
🌲	Surveyed trees coniferous
🌳	Surveyed trees deciduous
♧ ♧ ♧	Orchard
ψ ψ ψ	Coppice
℘ ℘ ℘	Bushes
℘ ℘ ℘	Scrub
ϒ ϒ ϒ	Bracken
‚ıli‚ ‚ıli‚	Heath
‚ııll‚ ‚ııll‚ ‚ııll‚	Rough pasture
‚ıl‚ ‚ıl‚	Furze
⋎	Reeds
⋎	Marsh
⋎	Oziers

Figure 9.2

surveyor must use his own judgement acting in line with the principles given above and following by analogy the detailed examples given.

For plans at scales larger than 1 in 1 000 the principles given above still apply and many of the notes are relevant but the specific needs of the map user should be carefully considered in selecting detail to be shown or omitted.

9.3.8. METHODS OF DETAIL SURVEY

The following principles underlie all methods:

(i) Every survey is based on a rigid framework which is regarded as correct until proved to be in error .
(ii) The object of the detail survey is to produce a plan on which no plottable inaccuracies shall appear.
(iii) The relative positions of features must be correct as regards alignment, distance apart, and orientation.
(iv) The field work is checked at all stages by independent processes.
(v) The work is planned to avoid repeated covering of the area.

Fundamentally there are only two different methods for location of detail. The first method is that given a fixed station and a fixed direction a point of detail has certain rectangular coordinates in which the fixed station is origin and the fixed direction an axis. Figure 9.3 illustrates this concept, the y ordinate

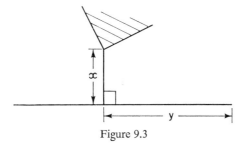

Figure 9.3

being the distance along the line and the x ordinate being a perpendicular offset. This is the basic concept in chain survey methods.

The second method is that again given a fixed station as origin and a fixed direction as axis the point of detail has polar coordinates. Figure 9.4 illustrates

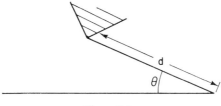

Figure 9.4

this concept, the distance d and the angle θ being the measured quantities. This is the basic concept in radiation methods.

9.4. CHAIN SURVEYING

9.4.1. INTRODUCTION

Consider a field with four irregular sides as shown in figure 9.5. A framework can be established by laying down triangles ABD and BDC; lines AB, BC, CD, DA, and DB being measured and the triangles plotted by intersection of arcs. Line AC is measured as an independent check and any error is readily isolated if the distances AE, BE, DE, and CE are noted in passing. The irregular hedge line between A' and B' is surveyed, at the same time as the line AB is measured, by measurement of perpendicular distances (offsets) at noted distances from A (chainages) to any significant changes in direction of the hedge. The hedge line can now be plotted by erecting perpendiculars to the plotted chain line.

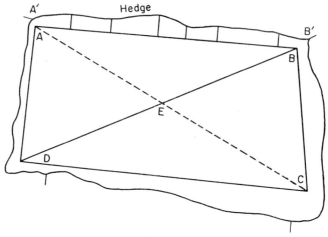

Figure 9.5

9.4.2. FIELD WORK

In planning the work the surveyor endeavours to select a framework in which:
 (i) A single line passes through the entire area.
 (ii) The lines provide triangles with apex angles of around 90° (see §13.11.3 on strength of figures in trilateration.)
 (iii) The selected lines provide each portion of the survey with adequate checks.
 (iv) The location of the lines provide short offsets to the detail.
 (v) The lines run along the longer sides of buildings.
 (vi) The lines run over level ground and avoid obstacles to ranging and chaining.

In execution of the work the following points apply:
 (i) The various stations, (ends of the lines) are pegged or suitably marked on the ground.
 (ii) A description is prepared for each station.
 (iii) A framework diagram is drawn in the field book and each station given a letter or number.
 (iv) The length of each line is measured and objects fixed by reference to the starting point by noting the chain or tape reading when either (a) a single offset is taken to the point of detail at right angles to the chain line or (b) two linear measurements from different points on the chain line are taken to the point of detail.
 (v) The maximum length of perpendicular offset depends upon the scale of the plan to be drawn from the survey and upon the error in setting out a right angle. Assuming a 2° error in setting out a right angle and a limiting plottable error of 1/200th of an inch. Then for a scale of 1 in 500, 0·24 feet can just be plotted.

  At 5·9 feet 2° subtends 0·24 feet and this represents the limit of offset. For a 1 in 2 500 plan the limit of offset is 30 feet.

Detail Survey

The perpendicular offset is most easily established without instrumental aid by holding the zero of the tape on the point of detail and swinging the tape about this as centre. The shortest distance from the point to the chain line establishes the perpendicular and where this line meets the chain line gives the chainage.

Alternatively an optical square may be used with which it is possible to set out the perpendicular to the chained line instrumentally. The optical square is a small hand instrument the size of a large pocket watch. It has three narrow openings in the sides as illustrated in figure 9.6. The mirror at F is half silvered

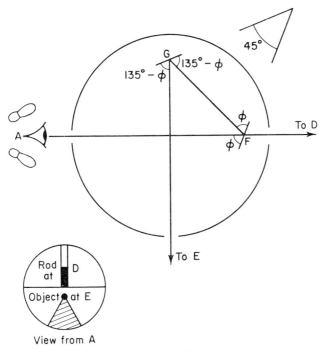

Figure 9.6

and makes an angle of 45° with the mirror at G. As a result the image in the mirror F is of a point perpendicular to the line AD and this can be brought by movement along the line into coincidence with a rod erected at D ahead on the chain line. It is usual to find that the mirror at G can be rotated enabling the instrument to be brought into adjustment. In adjustment points are established at right angles to a line on both the left- and right-hand sides of the line from a fixed point. The two points so established should be colinear with the fixed point on the line.

(vi) Dimensions of buildings are taken and prefixed by a plus symbol. These are designated plus measurements.
(vii) The field measurements are recorded in a manner as illustrated by figure 9.7. The principle adopted is that as far as possible the field record will represent by diagram or symbol detail met during taping along the line.

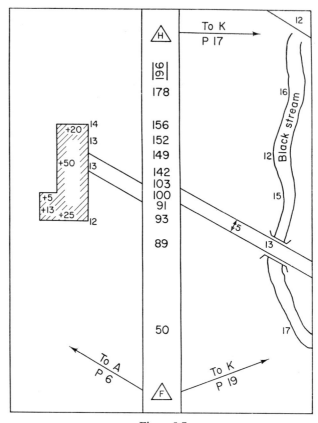

Figure 9.7

It should be noted that entries are made from the bottom of the page upwards, so detail right appears on the right-hand side of the page. Care in booking is essential and a system of transference of arrows is necessary to ensure that whole chain or tape lengths are not overlooked.

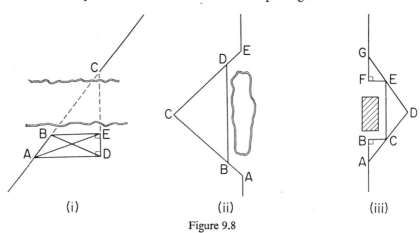

Figure 9.8

(viii) Various obstacles may obscure chaining, figure 9.8 illustrates how these may be overcome.

(i) Make BE 100 ft and ADE, BED, and BEC right angles, then:

$$BC = 100 \, AB/(AD - BE)(ft)$$

(ii) Make nBC = AC and nDC = EC, then: nDB = AE.
(iii) Make ABC and GFE right angles and BC = FE, then; BF = EC.

9.4.3. PLOTTING

A skeleton framework is plotted using the lengths of the principal lines. The longest line is plotted first as a base and the various triangles erected from it by use of beam compasses.

The check lines ensure that there is no error in the skeleton framework, however small random errors may accumulate and there may not be exact agreement between the framework and the check lines. If the skeleton has been well selected an agreement of 1 in 1 000 is to be expected.

Once the framework is plotted the points of detail relative to each line are plotted. This is achieved by marking off along each plotted line the chainages at which each offset was taken and at these points erecting perpendiculars of length equivalent at the plotting scale to the length of the offset. A special offset scale may be used to ease the work of plotting.

From the points of detail the remaining detail is filled in. Where much detail has to be plotted it is advisable to fill in the detail around each point as it is plotted. When all the detail on a page has been plotted the draughtsman or surveyor notes this on the page of the field book adding his initials and the date.

The common mistakes in plotting detail are:

(i) Marking off chainages from the wrong end of the line.
(ii) Erecting perpendiculars on the wrong side of the line.
(iii) Scaling the wrong offset distances along a perpendicular.
(iv) In filling in the detail joining the wrong points.
(v) Confusing plus measurements and offset measurements.

9.5. SIMPLE TACHEOMETRY

9.5.1. RADIATION METHODS

With a modern theodolite or level fitted with a graduated horizontal circle the two components of the radiation method can be determined; θ the angle being measured on the circle and d the distance being determined by tacheometry.

9.5.2. GENERAL CONCEPTS

This is the most commonly used method of determining distance and difference in height by optical means. The principle is that if the length of a base and the size of the apex angle are known, the length of the perpendicular to the base from the apex can be calculated. In tacheometry the apical angle is fixed in the instrument and the length of the base is measured.

A distinction must be drawn between tacheometric methods involving a vertical or slightly inclined staff and those involving a horizontal staff. The latter methods are inherently more accurate but involve the use of specialized equipment (see §9.8).

The methods involving the vertical or slightly inclined staff use stadia marks etched on the reticule of the survey telescope and thereby a fixed angle is made at or near the mechanical centre on the instrument. The intercept of this fixed angle is measured on an ordinary levelling staff or on special staves with clearly distinguishable markings.

9.5.3. OPTICAL PRINCIPLES

Considering figure 9.9 which illustrates ray paths in the external focusing

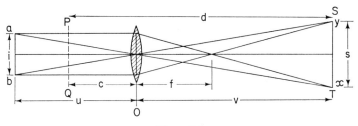

Figure 9.9

telescope. The stadia marks are at a and b, O is the optical centre of the objective, ST represents a staff held at a distance d from the vertical axis of the instrument PQ, x and y are the readings on the staff, $y - x$ being the intercept s.

Now $d = v + c$; $i/s = u/v$; and $1/v + 1/u = 1/f$ where f is the focal length of the objective.

From the equations above

$$\frac{1}{v}\left(1 + \frac{s}{i}\right) = \frac{1}{f}$$

$$v = fs/i + f$$

hence $\qquad d = fs/i + (f + c) \qquad (1)$

For a particular instrument f/i is a constant, $f + c$ will vary when focusing is achieved by movement of the objective but the variation in c is small compared with the errors that arise in reading s and $f + c$ is usually treated as a constant.

Hence we may say $\qquad d = k_1 s + k_2 \qquad (2)$

in which stadia marks are set so that k_1 is 100.

To determine k_1 and k_2 tape several distances from the instrument and set up and measure s in each case. For each pair of observations one solution for the values of k_1 and k_2 can be obtained. If more than two are taken mean values can be determined or alternatively a series of observation equations can be solved by the method of least squares.

Example 1:

Staff intercept (ft)	1·00	2·00	3·00	4·00	5·00
Measured distance (ft)	81·7	161·9	242·5	321·6	403·0

Observation equations

$$1·00 \, k_1 + k_2 = 81·7$$
$$2·00 \, k_1 + k_2 = 161·9$$
etc.

Forming the normal equations as in §5.6.4 we obtain

$$55 k_1 + 15 k_2 - 4\,434·4 = 0$$
$$15 k_1 + 5 k_2 - 1\,210·7 = 0$$

which are easily solved to give

$$k_1 = 80·23$$
$$k_2 = 1·45$$

9.5.4. TACHEOMETRY AND THE INTERNAL FOCUSING TELESCOPE

The above formulae apply only to the external focusing telescope, the modern telescope is internal focusing and its focal length depends upon the separation of the two lenses. This distance is in turn dependent upon the distance of the object from the telescope. Consequently the multiplying constant has become a variable. However by choice of lenses of suitable powers it is possible to create a telescope in which the angle subtended by the staff intercept at the mechanical centre of the instrument is very nearly a constant. In effect the anallactic point is made to coincide with the mechanical axis. It should be appreciated that the internal focusing telescope is rarely perfectly anallactic but it is so constructed that the additive constant can be neglected and any variations in the multiplying constant are less significant than the errors in the field methods used. Within the range 20 to 1000 feet all internal focusing telescopes of modern design have a multiplying constant correct to 1 part in a thousand.

It is outside the scope of this book to consider the variations in the internal focusing telescope made to achieve a near perfect anallactic design. The following points based on Henrici's paper in *Trans. Opt. Soc.*, Vol. 22, p. 20 illustrates how this is achieved.

If d is the distance between the instrument's mechanical axis and the staff then:

$$d = \frac{f}{i}\left(\frac{f' + A - D_\infty}{f'}\right) s + c + f - \frac{f}{i}\left(\frac{D - D_\infty}{f'}\right) s \tag{3}$$

where f is the focal length of the objective, f' the focal length of the internal lens, i the distance between the stadia marks, A the distance between the stadia marks and the objective, D the distance between the objective and the internal lens, D_∞ is the value for D when the telescope is at an infinite focus, s the staff intercept, and c the distance between the objective and the mechanical axis. $D - D_\infty$ decreases in (3) above as s increases until when $D - D_\infty = 0$, s is infinitely large.

If in designing the telescope it is arranged that for values of s within the range 0·2 to 10·0 ($d = 20$ to 1 000 ft)

$$f + c = \frac{f}{i}\left(\frac{D - D_\infty}{f'}\right)s$$

to within a few inches.

Then for all practical purposes

$$d = Ks \tag{4}$$

9.5.5. PORRO'S ANALLACTIC TELESCOPE

Prior to the development of the internal focusing telescope Porro designed a modified external focusing telescope (see figure 9.10) which was completely

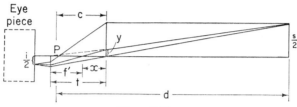

Figure 9.10

anallactic. The first telescope of this type was produced around 1830, focusing is achieved by movement of the eyepiece system including the reticule, the objective remaining at a fixed distance (c) from the mechanical axis of the instrument (p). In the telescope an additional convex lens termed the anallactic lens (focal length f') is placed at a constant, calculated, distance from the objective (t).

Considering figure 9.10 for an objective of focal length f

$$\frac{1}{x} - \frac{1}{c} = \frac{1}{f} \quad \text{or} \quad \frac{1}{f} = \frac{c-x}{cx}$$

$$\frac{d}{\frac{1}{2}s} = \frac{c}{y} \quad \text{and} \quad \frac{f'}{\frac{1}{2}i} = \frac{x}{y}$$

Also $x = t - f'$ which substituted in the above gives

$$c(t - f') = fc - f(t - f')$$

or

$$c = \frac{f(t - f')}{f' + f - t}$$

hence

$$\frac{2d}{s} = \frac{f(t - f')}{(f' + f - t)y}$$

but

$$y = \frac{ix}{2f'} = \frac{i(t - f')}{2f'}$$

$$\frac{2d}{s} = \frac{f(t - f')2f'}{(f' + f - t)i(t - f')}$$

which reduces to

$$d = \left\{\frac{ff'}{(f' + f - t)i}\right\}s$$

in which d is proportional to s.

9.5.6. INCLINED SIGHTS

It is usual to find that a telescope that is anallactic or nearly so will have a multiplying constant K that is 100 and so for all practical purposes the distance to the staff is 100 times the staff intercept when the line of sight is horizontal. When a theodolite is used inclined tacheometric shots can be taken. In such cases the difference in height can also be determined.

Two cases exist:

(i) When the staff is held vertical.
(ii) When the staff is held normal to the line of sight.

Consider case (i) illustrated by figure 9.11.

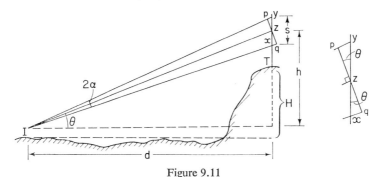

Figure 9.11

For the elevated sight in the anallactic telescope (both by Porro and internal focusing)

$$Iz = Kpq \quad \text{and} \quad d = Iz \cos \theta$$

On the vertical staff yx or s is measured.

To find the relationship between pq and yx let pq $= t$, yzI $= 90° + \theta$, yzp $= \theta$, and yIx $= 2\alpha$.

then
$$\frac{yz}{\cos \alpha} = \frac{pz}{\cos (\theta + \alpha)}$$

and
$$\frac{zx}{\cos \alpha} = \frac{qz}{\cos (\theta - \alpha)}$$

By addition

$$\frac{s}{\cos \alpha} = \tfrac{1}{2}t\{\sec (\theta + \alpha) + \sec (\theta - \alpha)\}$$

$$= \frac{t\{\cos (\theta - \alpha) + \cos (\theta + \alpha)\}}{2 \cos (\theta + \alpha) \cos (\theta - \alpha)}$$

$$= \frac{2t \cos \theta \cos \alpha}{2 (\cos^2 \theta \cos^2 \alpha - \sin^2 \theta \sin^2 \alpha)}$$

$$\therefore \quad t = s\left\{\frac{\cos^2 \theta \cos^2 \alpha - \sin^2 \theta \sin^2 \alpha}{\cos \theta \cos^2 \alpha}\right\}$$

$$= s \cos \theta (1 - \tan^2 \theta \tan^2 \alpha)$$

Provided θ does not exceed 45°, which it is unlikely to do, the term $\tan^2 \theta \tan^2 \alpha$ is very small and can be neglected for all tacheometric work where s is only measured to the nearest 0·01, and $\tan \alpha$ is about 1/200.

The horizontal distance is now given by

$$d = Ks \cos^2 \theta \tag{5}$$

or in the case of the external focusing telescope

$$Iz = k_1 pq + k_2 = k_1 s \cos \theta + k_2$$

and

$$d = k_1 s \cos^2 \theta + k_2 \cos \theta \tag{5a}$$

Considering the difference in height h

$$h = Iz \sin \theta = Ks \cos \theta \sin \theta \tag{6}$$

or in the case of the external focusing telescope

$$h = k_1 s \cos \theta \sin \theta + k_2 \sin \theta \tag{6a}$$

If i is the height of instrument and Tz the centre hair reading on the staff, the difference in level, H, between the ground at instrument and staff is for the anallactic telescope

$$H = Ks \cos \theta \sin \theta + i - Tz \tag{7}$$

A similar expression for the external focusing telescope can be developed.

For the depressed sight the distance is determined from expressions (5) or (5a) but

$$H = Ks \cos \theta \sin \theta - i + Tz$$

If θ is regarded as positive for an angle of elevation and negative for an angle of depression a general expression for H may be used which will overcome ambiguity when θ is small.

Then expression (7) is used in all cases. It will give a value for H which is negative or positive according to the sign of the difference in height.

Consider case (ii) illustrated by figure 9.12. Here the staff is held perpendicular

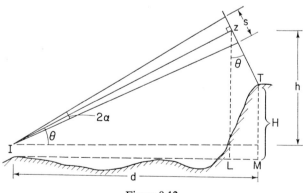

Figure 9.12

to the line of sight by means of a small sighting tube (a diopter sight) attached at right angles to the staff.

In the anallactic telescope

$$Iz = Ks \quad \text{and} \quad d = IL + LM$$
$$IL = Iz \cos \theta \quad \text{and} \quad LM = Tz \sin \theta$$
$$d = Ks \cos \theta + Tz \sin \theta \tag{8}$$

or in the case of the external focusing telescope

$$d = k_1 s \cos \theta + k_2 \cos \theta + Tz \sin \theta \tag{8a}$$

For the difference in height h

$$h = Iz \sin \theta = Ks \sin \theta \tag{9}$$

or in the case of the external focusing telescope

$$h = k_1 s \sin \theta + k_2 \sin \theta \tag{9a}$$

With nomenclature as in (7) above

$$H = Ks \sin \theta + i - Tz \cos \theta \tag{10}$$

A similar expression for the external focusing telescope can be developed.

For the depressed sight LM lies beyond IM then

$$d = Ks \cos \theta - Tz \sin \theta \tag{11}$$

or for the external focusing telescope

$$d = k_1 s \cos \theta + k_s \cos \theta - Tz \sin \theta \tag{11a}$$

For the difference in height expression (10) can be used noting that for an angle of depression $\sin \theta$ is negative and $\cos \theta$ positive.

9.5.7. RELATIVE MERITS OF THE VERTICAL AND NORMAL STAFF HOLDING POSITIONS

It is found that in:

Ease of reduction. There is little to choose between the two methods as $Tz \sin \theta$ is frequently assumed to be zero and $Tz \cos \theta$ to be Tz.

Facility of staff holding. The vertical position is more easily maintained with a spirit level or plumb bob attached to the staff. This requires less skill on the part of the staff man who is performing a largely repetitive operation. To hold the staff normal to the line of sight involves the use of a diopter sight by the staff man.

Effect of careless staff holding. Errors are generally greater in the case of the vertical staff. A further examination of this point follows.

Let V be the distance determined with the vertical staff and N be the distance determined with the staff held normal to the line of sight. In figure 9.13

$$V = (Ks \cos \theta') \cos \theta$$

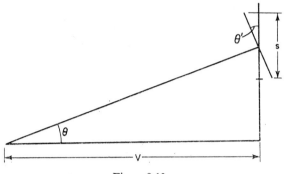

Figure 9.13

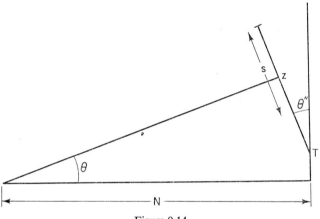

Figure 9.14

and in figure 9.14 $\quad N = Ks \cos\theta + Tz \sin\theta''$

Usually $\theta = \theta' = \theta''$ but if the staff is carelessly held this is not so. For purposes of comparison let $\theta' = \theta''$ and let $\theta - \theta' = \theta - \theta'' = d\theta$.

If the error in V be dV and in N be dN

$$dV = (Ks \sin\theta' \, d\theta') \cos\theta$$
$$dN = Tz \cos\theta' \, d\theta'$$

As $d\theta'$ is small it can be taken that $\cos\theta' = \cos\theta$ and $\sin\theta' = \sin\theta$.

Then $\quad\quad\quad\quad\quad\quad dV = V \tan\theta \, d\theta'$

and $\quad\quad\quad\quad\quad\quad dN = Tz \cos\theta \, d\theta'$

Considering these expressions dV increases as θ increases and the greater V the greater the error, if $d\theta' = 2°$, $V = 400$ feet and $\theta = 4°$,

$$dV = 0\cdot 93 \text{ feet.}$$

This is just perceptible.

Detail Survey   453

dN is a maximum when $\theta = 0°$, as then dN = Tz dθ', in practice Tz will rarely exceed 10 feet if d$\theta' = 2°$ d$N = 0·33$ feet.

Summarizing, it has been shown that up to $\theta = 4°$ the errors from careless staff holding are imperceptible in either method provided reasonable care has been exercised. After this the vertical method is inferior. In spite of this the vertical method is usually preferred due to the simplicity of the field method and custom.

9.5.8. FIELD PROCEDURE FOR TACHEOMETRY WITH A VERTICAL STAFF

Using the procedure outlined below it is possible to run a traverse or locate in distance and direction spot heights or points of detail. A theodolite is commonly used but a level with a horizontal circle can be used on level or near level ground, in such a case the reduction of the observations is considerably simpler.

Assuming a theodolite is being used the procedure is as follows:

(i) Set the instrument over the station mark carrying out the usual station adjustments. The instrument should be in good adjustment as it is customary to measure vertical and horizontal angles on only one face. If the levelling is completed by use of the altitude bubble its setting before reading the vertical circle is not required.

(ii) Measure and record the height of the instrument axis.

(iii) By use of the lower plate clamp and slow motion screw or by rotation of the horizontal circle set the theodolite so that it is directed on the orienting station with a horizontal circle reading of zero. The staff is only held at this station when the distance to it is required, or the difference in height, or both. If the orienting station is not well defined a ranging rod is left.

(iv) Place the staff at a previously agreed starting point. Direct the telescope at it, and bring the upper stadia hair to a whole foot mark. This will be the lowest of the three hair readings.

(v) Read and record the three hairs. If all three hairs cannot be read two will suffice as the middle hair reading is effectively the mean of the other two. The use of all three hairs will act as a check which should be applied in the field.

(vi) Direct the staff man to proceed to the next staff station. It is convenient to swing round in a clockwise direction making one or more complete horizons at each station.

(vii) Whilst the staff man is moving, level the altitude bubble and read and record the vertical and horizontal angles. In both cases the reading is taken to the nearest minute.

(viii) Repeat the procedure for subsequent staff stations until on completion of the observations the telescope is redirected at the orienting station to check that the horizontal circle has not been disturbed.

The features of a typical booking sheet are illustrated in figure 9.15. The sketch is a vital part of the record and besides showing the location of the staff

Location of survey...Somerton...		Observer...Spray...			Date...22-7-64.			Page...1...of...17...		
Instrument...DKM.2..No.650.29		Booker...White...								

At Ht. of inst. (i) Redcd level (H)	Sketch	To	Horizontal reading	Vertical reading Angle	Staff U c L	's' U−L	'D' $100s\cos^2\alpha$	Δh $50s\sin 2\alpha$	$\Delta h - c$	Reduced level
$L/_3$ (i) 4·5 ft (H) 148·94	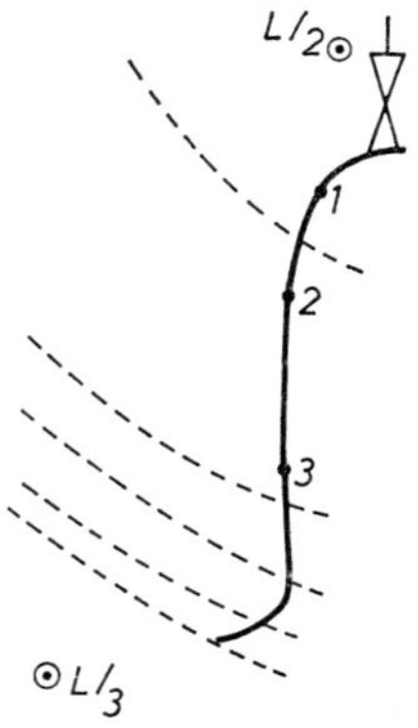	$L/_2$	00° 00'	94° 45' +4° 45'	4·08 2·42 0·78	3·30	328 ft	+27·22	+24·80	178·24 (Data)
		1	02° 19'	94° 58' +4° 58'	5·00 3·58 2·16	2·84	282 ft	+24·48	+20·90	174·3
		2	04° 21'	95° 58' +5° 58'	8·00 6·73 5·47	2·53	250 ft	+26·16	+19·43	172·9
		3	11° 46'	97° 38' +7° 38'	8·00 7·01 6·02	1·98	194 ft	+26·06	+19·05	172·5

Figure 9.15

Detail Survey

stations it should show the nature of the detail, form lines and miscellaneous information such as width of ditches, gateways.

9.5.8.1. *Tacheometric Traverses*

The traverse is designed to provide control for tacheometric detailing, each station being selected for its suitability for the survey of surrounding detail. During the traverse, detail and traverse observations are taken simultaneously the procedure being:

9.5.8.2. *Traverse Angles*

A face left pointing is made to the staff held in turn at the back and forward stations before the detailing commences. Readings are recorded to the nearest minute. At both back and forward stations a ranging rod is left and after the detailing is completed the telescope is directed at the back station to check that the horizontal circle has not been disturbed. The traverse angle is now measured a second time with a new zero on face right.

9.5.8.3. *Traverse Distances*

These are determined identically to detail distances, but each line is measured twice, back and forward. These distances are reduced in the field and should agree to within 1 foot.

9.5.9. REDUCTION OF TACHEOMETRIC OBSERVATIONS

When observations are made to a vertical staff with an anallactic or internal focusing telescope, distance to the staff is given by $Ks \cos^2 \theta$ and the difference in height by $Ks \cos \theta \sin \theta$ (see expressions (5) and (6)). The reduction of these quantities without special tables can prove tedious particularly as hundreds of observations may be taken in a relatively small survey.

If no special tables are available ordinary tables of natural trigonometrical functions can be used with the expressions (12) and (13) below.

As
$$\cos^2 \theta = \tfrac{1}{2}(1 + \cos 2\theta)$$
$$Ks \cos^2 \theta = Ks - \tfrac{1}{2}Ks(1 - \cos 2\theta) \tag{12}$$

Also
$$Ks \sin \theta \cos \theta = \tfrac{1}{2}Ks \sin 2\theta \tag{13}$$

With a calculating machine, reduction using these expressions is simple and quick, particularly if natural tables tabulating sine and cosine on the same page are used. For slopes of up to 2° with distances under 400 feet the corrective term in expression (12), $\tfrac{1}{2}Ks(1 - \cos 2\theta)$, is less than $\tfrac{1}{2}$ a foot and can be ignored.

Quite commonly tables of $100 \cos^2 \theta$ and $100 \sin \theta \cos \theta$ are available; here distance and height difference is directly obtained by multiplying by s. Again however a calculating machine is required.

Jordans, *tacheometric tables*, which are unfortunately out of print, give values of $100s \cos^2 \theta$ and $100s \sin \theta \cos \theta$ for values of s at 0·01 intervals from 0·10 to 2·00; being tabulated in single minutes. F. A. Redmond's *Tacheometric Tables*,

(Simpkin, Marshall and Co., Ltd., London: 1931) give values for $100s \cos^2 \theta$ and $100s \sin \theta \cos \theta$ for integers of $100s$ from 50 to 850; θ being at 20' intervals from 00° 20' to 20° 00'. If these tables are used the vertical angle must be brought to an exact 20'; then the lower hair is set on a whole foot. The method suggested in Redmond's tables of moving the telescope, after reading the centre hair, to a whole foot graduation is prone to error, no check on the staff intercept remaining.

Graphs of $50 s(1 - \cos 2\theta)$ can be prepared for various values of s and serve as a useful check on reduction or for preliminary reductions. Similarly graphs of $50 s \sin 2\theta$ can be prepared at the same time and used to determine the height differences. More convenient are tacheometric slide rules with scales in $\cos^2 \theta$ and $\sin \theta \cos \theta$; these are sufficiently accurate for most work carried out with the vertical staff.

Tacheometric traverses should always be reduced by use of tables and checked by a less exact method.

9.5.10. PLOTTING

The framework is first plotted and checked, the misclosure of a tacheometric traverse should always be within the plottable error of the plan and no adjustment is permitted. If the traverse is lengthy it may be computed and plotted by rectangular coordinates.

In plotting the detail and spot heights, the bearings are first marked off with a circular protractor oriented with reference to the back station or orienting station. Opposite each mark is written the reference number used in the field book. The distances are scaled in the direction of the appropriate mark after the full horizon has been completed and each staff station is located on the plan with a dot and small circle against which is written the elevation. Intervening detail is filled in from the sketches and measurements in the field book and contours interpolated with the aid of the form lines shown therein.

9.5.11. ACCURACIES ATTAINABLE AND SOURCES OF ERROR

The degree of accuracy attained depends upon the length of the line, the power of the telescope, the type of staff, the graticule in the telescope, the conditions of visibility and the inclination of the line of sight.

Considering the determination of distance, where $d = Ks \cos^2 \theta$,

$$\log d = \log K + \log s + \log \cos \theta + \log \cos \phi$$

θ being the vertical angle and ϕ being the angle the staff makes to the normal to the line of sight. Usually $\phi = 0$.

By differentiation

$$\frac{dd}{d} = \frac{dK}{K} + \frac{ds}{s} - \frac{d\theta \sin \theta}{\cos \theta} - \frac{d\phi \sin \phi}{\cos \phi}$$

Let E_K, E_s, E_θ, and E_ϕ be the errors in K, s, θ, and ϕ respectively, then

$$\left(\frac{E_d}{d}\right)^2 = \left(\frac{E_K}{K}\right)^2 + \left(\frac{E_s}{s}\right)^2 + (E_\theta \tan \theta)^2 + (E_\phi \tan \phi)^2 \tag{14}$$

where E_d is the error produced in d.

Detail Survey

It has been said earlier that K may be in error by 1 part in 1 000, this is a maximum value, here E_K is taken as $\pm 1/3\ 000$.

The intercept s, being the difference of two staff readings each taken to the nearest 0·01 feet, will be correct under most circumstances to $\pm 0·007$ feet. This may be regarded as the maximum error and here E_s is taken as $\frac{1}{3}$ of this $\pm 0·002\ 3$.

The vertical angle θ is read to the nearest minute and the maximum error is $\pm 30''$ here E_θ is taken as $\pm 10''$.

The staff will not necessarily be held vertical, with a competent staff man it should be within $2°$ taking this as a maximum then here E_ϕ is taken as $40'$.

Using expression (14) and taking s as 4·00 and θ as $4°$ then $d = 398·04$ ft, and

$$\left(\frac{E_d}{398·04}\right)^2 = \left(\frac{0·33 \times 10^{-3}}{10^2}\right)^2 + \left(\frac{0·23 \times 10^{-2}}{4}\right)^2 + \left(\frac{10 \times 0·069\ 9}{206\ 265}\right)^2$$

$$+ \left(\frac{40 \times 0·699}{3\ 438}\right)^2$$

$$\therefore \frac{E_d}{398} = (0·012 + 3\ 306·25 + 0·123 + 6\ 638·99)^{\frac{1}{2}} 10^{-5}$$

$$E_d = \pm 0·39\ \text{ft}$$

The following conclusions can be drawn:

(i) An accuracy of about 1 in 1 000 can normally be expected,
(ii) The serious sources of error are in the measurement of the intercept and in the verticality of the staff,
(iii) Errors arising from the measurement of the vertical angle and the multiplying constant are negligible.

Considering the determination of differences in height, where $h = Ks \cos \theta \sin \theta$, then with terminology as in (14)

$$\left(\frac{E_h}{h}\right)^2 = \left(\frac{E_K}{K}\right)^2 + \left(\frac{E_s}{s}\right)^2 + (E_\theta \cot \theta)^2 + (E_\phi \tan \phi)^2 \qquad (15)$$

where E_h is the error produced in h.

Taking s as 4·00 and θ, as $4°$

$$h = 27·83\ \text{ft} \quad \text{and} \quad \frac{E_h}{27.8} = (14\ 752·4)\ 10^{-5}$$

$$E_h = 0·033\ 8\ \text{ft}$$

The following conclusions can be drawn:

(i) The difference in height can only be determined to within 0·05 feet and spot heights should not be shown to more than one decimal place.
(ii) Although more care in measurement of the vertical angle will increase the accuracy a large residual will always be left due to the errors in the determination of the distance.

9.6. SELF-REDUCING TACHEOMETERS WITH VERTICAL STAFF

9.6.1. INTRODUCTION

Based on the principles of normal tacheometry these instruments provide for a reduction of the distance between the stadia lines with elevation of the telescope. The solution of Hammer of around 1910 provides for a pair of curves etched on a glass plate to be reflected into the telescope; the portion of the plate that is viewed depends upon the elevation of the telescope. The design provides that the separation between the curves gives an intercept on a vertical staff that is 1/100th of the horizontal distance between instrument and staff.

9.6.2. THE WILD RDS SELF-REDUCING TACHEOMETER

This is essentially a normal theodolite of the Wild T 16 type modified to be a self-reducing tacheometer (see figure 9.16). Considering the modifications it

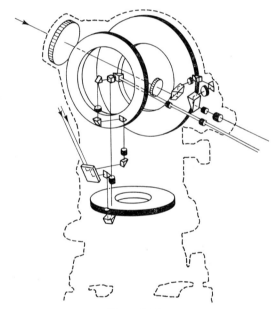

Figure 9.16

is found that the telescope gives an upright image, the graticule consists of a vertical line with a short cross line at right line at right angles in the middle of the field of view, and besides there also appears in the field of view an image of parts of curves being reflected into the telescope. Figures 9.17 and 9.18 illustrate the nature of the image seen in the field of view. In the terminology of self-reducing tacheometry the reflected curves are termed the diagram and the circle of glass upon which they are etched the diagram circle.

The diagram circle is mounted adjacent to the telescope on the side opposite to the vertical circle and concentric with the transit axis. It is enclosed in a housing cast in one piece with the telescope body. On this glass circle are etched

Detail Survey

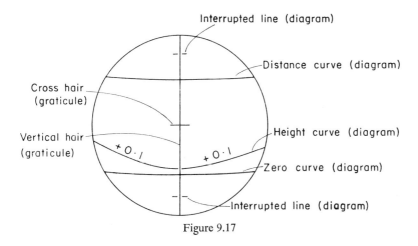

Figure 9.17

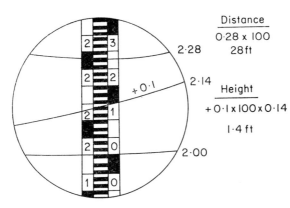

Figure 9.18

the distance curve, the zero circle, and the eight height curves. The zero circle forms a reference for both distance and height measurements, it is the curve which appears lowest in the field of view and in angular measure is 17′.2 below the short cross line in the graticule. When the vertical circle reading is 89° 42′.8 the zero curve is defining a horizontal line of sight and in this situation the distance curve lies 17′.2 above the short cross line. The intercept on a vertical staff of the zero and distance curves gives 1/100 of the distance to the staff and as the telescope is elevated or depressed different portions of the distance curve come into the field of view maintaining this relationship.

Between the distance curve and the zero curve there appears in the field of view the height curve. On the diagram circle there are eight height curves which for convenience of mental calculation are designated by multiplication constants 0·1, 0·2, ½, and 1 and are prefixed by plus or minus. The plus or minus indicates whether the line of sight as defined by the zero circle is elevated or depressed. When the zero circle is defining a horizontal line of sight the height curves are symmetrical with respect to the vertical line on the graticule.

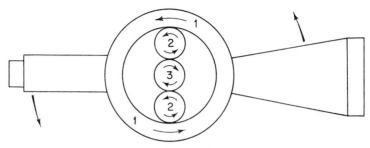

Figure 9.19

Location of the exact position of this horizontal line of sight is facilitated by two short interrupted horizontal lines on the diagram circle located in the uppermost and lowest portion of the field of view.

The instrument is operative as a tacheometer over a range from 45° in elevation to 45° in depression. This is due to the gearing of the glass circle on which the diagram is etched. A movement of the telescope in a vertical direction is magnified three times in the movement of the diagram circle in the opposite direction. This is achieved by a gearing as illustrated in simplified form by figure 9.19. Ring 1 is rigidly connected to the telescope, and teeth on the inside of this ring cause the pinion 2 to rotate at three times the speed of ring 1; the result is that gear 3 will turn at three times the speed and in the opposite direction to ring 1 and the telescope axis. In figure 9.20 consider a spot A, with elevation of the telescope through an angle α, point B on the diagram, 4α above A, will now appear in the field of view.

Figure 9.21 illustrates the optical train through the telescope and diagram circle. Rays from the staff pass through the objective and internal focusing

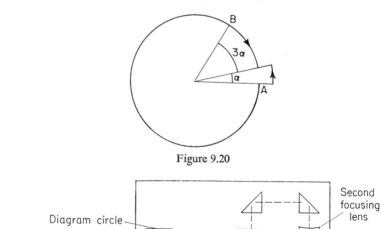

Figure 9.20

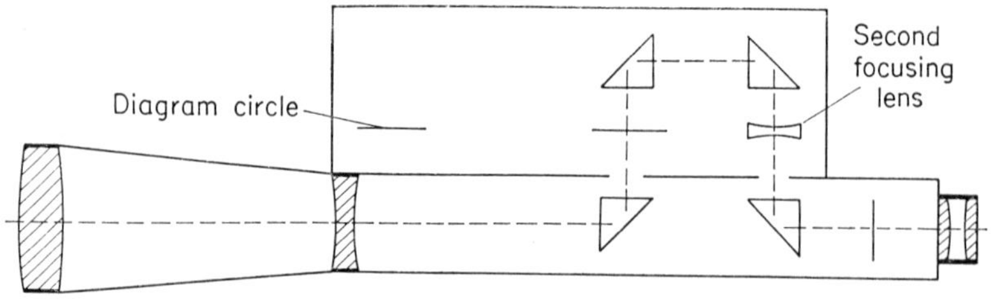

Figure 9.21

Detail Survey

lens to impinge on a prism which reflects them at right angles to the line of the telescope onto a plane at right angles to the plane of the diagram circle. An image is now formed on the further side of the diagram circle coincidental with the etched lines. Further reflection of this image now takes place through three prisms to bring the rays back into the line of sight of the telescope. A lens in this part of the train provides that a new image is formed in the plane of the reticule. Thereby the observer sees in the plane of the reticule an image of the staff, the etched lines on the reticule and the etched lines on the diagram circle relevant to the particular elevation of the telescope.

9.6.3. USE IN THE FIELD

Normal angular observations are carried out in the usual way using the graticule only. If vertical angles are measured care should be taken to use the short cross line and not the line of the height curve.

Tacheometry is carried out on face left, that is with the vertical circle on the left of the telescope and the diagram circle and focusing knob on the right. The zero curve now falls in the lower field of view.

For measurement of distance the vertical hair is placed such that the vertical staff is equally bisected and the zero curve brought, by means of the vertical slow motion screw, to fall on an even foot graduation. The intercept between the zero and distance curves gives 1/100 of the horizontal distance to the staff.

For measurement of differences in height the staff intercept between the zero and height curves gives as follows:

1/10 the difference in height when the multiplication constant adjacent to the height curve is 0·1,
1/20 the difference in height when the constant is 0·2
1/50 the difference in height when the constant is $\frac{1}{2}$
1/100 the difference in height when the constant is 1

The sign of the constant indicates whether there is a rise or fall in the line of sight defined by the zero circle.

If instead of placing the zero curve on a whole foot graduation it is placed on the staff at the height of instrument the ground station differences in height are directly obtained. Some observers prefer this practice but reduction of distances is made more tedious.

9.6.4. ADJUSTMENT

The adjustment of the plate level, horizontal collimation and index error is carried out in the usual manner (see §3.8). The diagram circle is correctly adjusted when the height curves +0·1 and −0·1 are symmetrical with respect to the vertical hair and the line of sight defined by the zero curve is horizontal. The interrupted short horizontal lines on the diagram circle are now bisected by this vertical hair.

9.6.4.1. *Procedure*

(*a*) A staff reading for a horizontal line of sight is obtained by use of the vertical circle and the horizontal hair in the graticule. Provided the altitude bubble has been adjusted for index error this is defined by a central bubble and a vertical circle reading of 90° 00′0.

(b) Elevate the telescope until the zero curve falls on the same point on the staff. The vertical circle should now read 89° 42′8, the height curves be symmetrical about the vertical hair, and the short interrupted lines accurately bisected by the vertical hair.

(c) If the conditions described in (ii) above do not pertain, the diagram circle is moved after slackening the retaining screws. Care must be exercised in executing this adjustment.

9.6.5. ACCURACY

It must be stressed that the accuracy of vertical staff self-reducing tacheometers is no better than tacheometers based on normal stadia lines. Their advantage lies in the speed of reduction particularly in areas where a great number of points must be surveyed. With these instruments the possibilities of gross errors are considerably reduced.

9.7. LARGE SCALE PLANE TABLE EQUIPMENT

9.7.1. THE TELESCOPIC ALIDADE

This instrument will provide distances and differences in height with little reduction. As the directions can be plotted directly on a sheet mounted or clipped to a plane table board, detail can be filled in by the surveyor with both the ground and the plan simultaneously visible. The telescopic alidade consists of a telescope fitted with cross and stadia hairs and a vertical graduated arc with the usual level tube. The whole instrument is set on a heavy base which rests on the plane table and can be used as a ruler, a parallel attachment is usually fitted and a circular bubble for rough levelling.

The vertical arc is usually graduated in degrees and can be read to the nearest minute by means of an attached vernier. It is usual to fit in addition a Beaman stadia arc as described below.

9.7.2. THE BEAMAN STADIA ARC

Graduations are set on the vertical arc at the points where $\cos \theta \sin \theta = 0.01, 0.02, 0.03$, etc. which correspond to values of θ equal to $00° 34′ 23″$, $1° 08′ 46″$, $1° 43′ 12″$, etc. As the telescope is anallactic and the numbering of these graduations is $100 \cos \theta \sin \theta$, viz. 1, 2, 3, etc. then the difference in height is given by

$$h = rs \qquad (16)$$

where r is the reading on the Beaman stadia arc and s is the staff intercept.

To avoid possible errors in determination of the signs of the graduation the arc may be constructed so that the zero reading is 50. In this case 50 is subtracted from the scale reading to give r in expression (16) above, a positive value being an elevation and a negative value being a depression.

A second scale is provided to give a correction to reduce $100s$ to the horizontal distance. This is of the form $50 (1 - \cos 2\theta)$ as indicated by expression (12)

Detail Survey

in §9.5.9., then the distance is given by

$$100s - ps \tag{17}$$

where p is the reading on the Beaman stadia arc.

The scale divisions on the Beaman stadia arc are unequal in length, hence verniers or micrometers cannot be used. To overcome this feature the telescope is elevated or depressed after a preliminary pointing at the staff until the Beaman stadia reading r is on a whole number, the p or distance correction reading is estimated. After setting the telescope the two stadia marks and the centre hair are read in the usual way.

9.7.2.1. *Use in the Field*

With the telescopic alidade set on a plane table rays are taken to a staff held at points of detail or changes of slope. The distance is obtained in the field, the point plotted and after determination of the difference in height the elevation of the staff station obtained. This value is written on the plan and when sufficient points have been located the contours are interpolated in the field.

The telescopic alidade is most suited for scales between 1 in 2 500 and 1 in 25 000. It is limited in value in the United Kingdom by the unsuitable weather. Control can be established by traversing on the board but for a survey covering several sheets it is better to establish control by conventional taped traversing or traversing by horizontal bar tacheometry.

The reduction is simple and can be performed with the aid of a simple slide rule, for example:

Let $\qquad r' = 56, \qquad p = 0.46$

and the three staff readings be 7·59, 8·23, and 8·87.

The station where the board is set is of height 374·8 feet A.O.D. and the board is 3·5 feet above ground level.

Now $\qquad s = 1·28, \qquad r = r' - 50 = +6$

$\qquad\qquad h = +7·68 \quad$ and $\quad H = +3·0 \ (7·68 + 3·5 - 8·23)$

Also $\qquad d = 100 \times 1·28 - 1·28 \times 0·46 = 127·4$

Hence the height of the staff station is 377·8 and the distance 127 feet.

9.7.3. KERN PLANE TABLE EQUIPMENT RK

More refined is the Kern plane table equipment with the self-reducing RK alidade. This alidade differs from the type described above in several respects. The differences enable the work to be speeded up, especially in reduction of height and distance. The four main differences are: a self-reducing scale, an eyepiece fixed at a convenient angle, a precision parallel rule, and a fine pointing device. The instrument is illustrated by figure 9.22.

The reticule carrying the stadia marks is geared to the horizontal axis of the objective prism. As the objective prism is tilted various stadia patterns move across the field of view as in the self-reducing tacheometer described in §9.6.

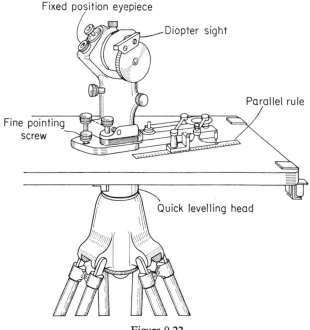

Figure 9.22

Figure 9.23 illustrates the pattern viewed for about +5°. The intercept between the outer lines gives 1/100th of the distance and the intercept of the second set of curves which lie within the distance curves gives 1/20, 1/50, or 1/100 the difference in height according to which pair of curves are seen. The factor to use is indicated by short ticks, a factor 1/20 by 2 ticks, a factor of 1/50 by 5 ticks, and 1/100 by 1 tick; in the figure the factor is 1/20. For slopes in excess of 40° there are no self-reducing stadia curves and the fixed lines in the left of the field of view are used, such slopes are rarely encountered. The difference in height obtained is from the axis of the objective prism to the staff reading of the small cross central in the field of view. The procedure is to read

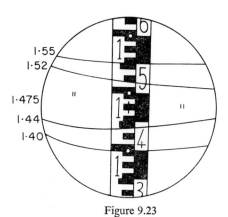

Figure 9.23

Detail Survey

the two outer curves and multiply the intercept by 100 to give the distance, to read the two inner curves and multiply the intercept by 20, 50 or 100 as indicated by the ticks to give the difference in height from which (for angles of elevation) is subtracted the staff reading of the central cross and to which is added the objective prism height. For example:

Lower line 4·00 feet
Upper line 5·85 feet
Distance 185 feet
Lower elevation line 4·71 feet
Cross 4·93 feet
Upper elevation line 5·13 feet
Ticks elevation (2)
Height of ground station 148·7 feet
Height of axis above ground 3·5 feet

Space between elevation curves increasing from left to right indicating depression of line of sight.

Difference in height is

$$148 \cdot 7 - (20 \times 0 \cdot 42) - 4 \cdot 93 + 3 \cdot 5 = 138 \cdot 9 \text{ feet}$$

It will be noted that the calculation of height could have been simplified by bringing the lower elevation curve to a whole foot mark on the staff after reading the distance curves.

The telescope is viewed through an eyepiece fixed in position at 30° to the horizontal this enables the surveyor to set the board at an elevation comfortable for plotting. The image is erect and the line of sight is inclined by rotating an objective prism which is geared to the moving reticule.

The plotting device is similar to a draughting mechanism, the alidade arm being connected to the alidade body by a carefully constructed parallelogram enabling exact plotting from the station point without tedious setting. To this arm is clipped a plotting scale, a plotting needle is located at the zero of the scale and the distance is read at the station point. The scale can be changed as required and studs on the arm ensure exact positioning of the zero on the needle.

The telescope and the plotting arm can be moved without movement of the alidade by means of the horizontal slow motion screw. Although this movement is relatively small it does permit very exact pointing. A diopter sight enables a preliminary pointing to be made in azimuth and elevation.

The staff used with this equipment may be of the conventional type but full advantage of the equipment is obtained when a staff of a type similar to figure 9.23 is used. Exact bisection of the staff is then achieved ensuring that the correct portions of the curves are used.

9.8. HORIZONTAL BAR TACHEOMETRY

9.8.1. INTRODUCTION

Typical of the various types of tacheometer using the horizontal staff is the Wild model which will be described here in detail. Other models in fairly common use are made by Kern and Zeiss. Self-reducing tacheometers of the Wild type are based on an invention of R. Bosshardt, patented in 1923. The instrument telescope is so designed as to give a split field of view, some of the incident rays reaching the objective lens without refraction while others are refracted by achromatic wedges mounted ahead of the normal telescope objective. Thereby a parallactic or anallactic angle is made at the instrument and by use of a special staff distances can be determined. Mechanical gearing causes rotation of the wedges with change in elevation of the telescope and the parallactic angle is reduced to correct the slant distance to the horizontal.

9.8.2. WILD RDH REDUCTION TACHEOMETER WITH HORIZONTAL STAFF

Figures 9.24 and 9.26, illustrate the optical train in the Wild RDH, it is seen that two achromatic wedges W1 and W2 cover the whole of the objective 0 of the normal telescope; however half the light is cut off by prism A which lets

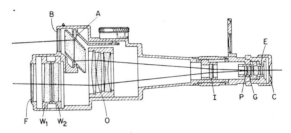

Figure 9.24

in undeviated light through window B. The two wedges produce a maximum deflection of $34'\ 22''\!.6$ when the telescope is horizontal. The light through window B is from the vernier on the centre portion of the staff, see figure 9.25, and the light through the wedges is from a point on the staff which varies with the distance of the staff. The prism P, placed in the focal plane of the two images, superimposes the half circles of the exit pupils and the slit diaphragm C cuts off the rays which would produce double images. The image as viewed is illustrated by figure 9.27, the two halves being separated by a horizontal line which is the edge of prism A. When required the eyepiece sleeve can be rotated thereby opening the diaphragm C and passing a shutter across the lower deviated field. This brings into the field of view only the image viewed through B and in these circumstances the instrument can be used as a theodolite.

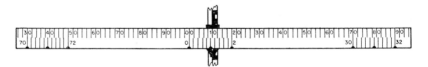

Figure 9.25

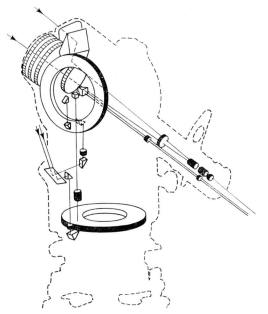

Figure 9.26

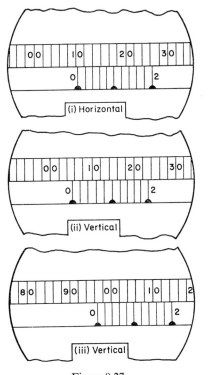

Figure 9.27

Figure 9.28 illustrates in diagrammatic form the action of the two wedges. In the horizontal position for the telescope the maximum deflection occurs and the distance is given by Ks where K is the cotangent of the parallactic angle, viz. 100, and s is the intercept on the horizontal staff. By elevation of the telescope the two wedges which are geared to its rotation turn in opposite directions at an identical rate. The axis of rotation is the line of sight through the telescope and consequently the lateral deflections are additive and the vertical deflections cancel out. When the wedges are each turned through an angle α, the vertical angle to the staff, given by the angle of elevation or depression of the telescope, the total deflection of the wedges is reduced by $\cos \alpha$. Consequently the portion of staff viewed is reduced to $s \cos \alpha$ where s is that portion which would have been viewed had rotation not taken place. It is by this reduction of the intercept that the true horizontal distance is deduced. The slant distance is given by $100s$ and the horizontal distance by $100\ s \cos \alpha$, viz. 100 times the staff intercept.

When the two wedges are rotated in opposite directions through 90° the staff intercept has become $s \sin \alpha$ and $100\ s \sin \alpha$ is the difference in elevation between the staff and the apex of the parallactic angle. This contrarotation is effected by a change over switch.

Prism A, see figure 9.24, can be rotated by a small amount about a vertical axis. The effect of this rotation is to laterally displace the line of sight through B and provided the axis of the rotation of the prism is truly vertical this displacement will be restricted to the horizontal plane. By these means the line of sight through B can be deviated and a line on the vernier portion of the staff made to coincide with a graduation on the main scale. The movement of the prism is read on a graduated drum which is rotated to effect this movement. It should be noted that when horizontal angles are measured with the drum not around the 10 position there will be a certain amount of displacement of the line of sight from its true position. The error in the horizontal angle which results is discussed in §3.10.4; this error is effectively cancelled out if angles are measured on both faces and provided the setting of the micrometer drum is not changed. The Wild RDH and similar instruments can be used as normal theodolites provided this point is noted.

In horizontal bar tacheometry a zero setting of the circle is convenient and is achieved as follows:

With the circle clamp in the 'up' position the circle is held fixed and the alidade may be turned until the image of the zero mark of the horizontal circle is brought into coincidence with the zero mark of the scale.

The pointing onto the reference mark is then made with the circle clamp in the 'down' or horizontal position.

The standard strut type tripod consists of a 2 m rod which is plumbed by means of an attached circular bubble. The attachment for the horizontal staff provides for its exact location in a position at right angles to the rod. The verticality of the rod is achieved after centring of the rod by use of the extending rings on the struts.

9.8.3. ANALLACTIC DESIGN

With regard to the anallactic requirements of an instrument of this type, it must be appreciated that the anallactic, or strictly speaking parallactic, point is mid-way between the wedges 74 mm ahead of the intersection of the main axis and the transit axis. On examining the staff and rod it will be found that the graduated surface of the staff is not vertically over the point at the lower end of the support rod for the staff, but is 40 mm closer to the instrument. Thus the distance between the apex of the parallactic angle and the staff is 114 mm less than the distance from the instrument station to the point of the rod. This difference is compensated by shifting the three staff verniers by 0·2 mm thereby achieving a correction of 20 mm in line length and by making the graduated drum read 9·4 when the prism A is not deviating light from the staff.

With horizontal sighting, the true horizontal distance is obtained. When there is a vertical angle, say α, the parallactic point is now 74 cos α mm horizontally in front of the intersection of the axes and accordingly a small correction of magnitude 74 (1 − cos α) mm is necessary. This correction, which is always

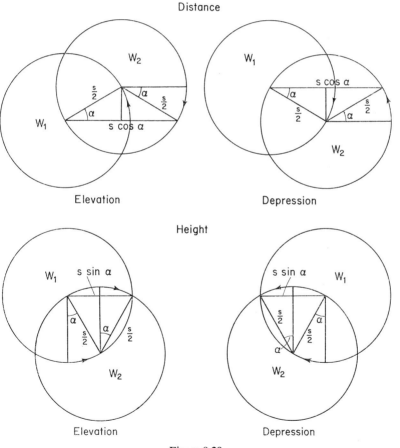

Figure 9.28

negative, is read from the graduated ring adjacent to the change-over switch. When differences in height are required the change-over switch is moved to bring the wedges into position. The mechanical arrangements are the same as for distance measurement and accordingly the reading with the telescope horizontal is not zero but 114 mm. Also since the prism A shifts the optical axis vertically, a 23 mm wide strip appears cut out of the horizontal staff. Thus the deflected line of sight strikes the staff at a point 11·5 mm below the actual centre. To correct for these two constants the reference marks are set 102·5 mm above the true centre of the horizontal staff. When the vertical angle is α a correction of magnitude $74 \sin \alpha$ mm is required. This correction has the same sign as the relative elevation reading and is read off the same graduated ring as used for the correction to the distance measurement.

9.8.4. OPERATION OF THE INSTRUMENT: DISTANCE MEASUREMENT

(a) Carefully level the instrument. This is vital for accurate measurement especially if the line of sight is considerably elevated. The altitude fine motion screw is provided with an adjustable index which when lined up with a fixed index will indicate the verticality of the main axis in the plane of the telescope axis. This serves as a quick check on the levelling of the instrument; the instrument may be taken to be level in this direction if the ends of the split bubble are less than half a bubble width apart when the indices are lined up.

(b) Check the position of the change-over switch, the horizontal black line should be visible to the observer and the auxiliary scale should be pressed under tension against its stop.

(c) Rotate in an anticlockwise direction the inverter ring (the coarse milled ring) to produce a narrow slit and a split image. The separation line of the images is carefully focused by means of the graticule focusing ring.

(d) Aim the telescope at the staff so that the line of separation lies just on the lower edge of the main scale. Carefully focus the image of the staff by means of the telescope focusing screw. Check that there is no parallax.

(e) Bring a vernier (apparent lower scale) graduation into coincidence with a main scale graduation by means of a micrometer knob.

(f) See figure 9.27 (i) which represents a typical field of view, the distance is given by

Main scale	units of 2 m	10·00
Vernier scale	units of 0·2 m	0·40
Micrometer scale	units of 0·02 m	0·06
	Horizontal distance	10·46 m

For distances over 70 metres the image of the 0 mark is beyond the end of the staff, the telescope is now directed at a vernier, graduated from 70 to 72, on the left-hand edge of the staff. The readings are taken in the same manner as indicated above but 70 metres is added to the reading.

(g) When the vertical angle is large a correction must be applied from the auxiliary scale. The figures on this scale are in centimetres and the value should be subtracted from the distance reading.

Detail Survey

(*h*) When the part of the staff being observed is covered by some obstruction the staff may be shifted laterally. In shifting the staff it may become excessively inclined, this is less likely to be serious if the shift is always made to the left on looking at the staff.

9.8.5. OPERATION OF THE INSTRUMENT: HEIGHT MEASUREMENT

(*a*) Check the levelling of the instrument, if necessary relevel the instrument using the altitude bubble with the indices coinciding.

(*b*) Rotate the change-over switch so that the red vertical line is visible. The auxiliary scale is now pressed under tension against its stop.

(*c*) Make an accurate bisection of the staff with the separation line of the two images. Complete the setting by a clockwise turn of the vertical slow motion screw.

(*d*) See figure 9.27 which represents typical fields of view, the difference in height is given by 9.27(ii). This is an elevation, the black vernier and the black figures being used.

Main scale	4·00
Vernier scale	1·60
Micrometer scale	0·16
Difference in height	+4·76 m

9.27(iii) This is a depression, the red main scale graduations being used.

Main scale (red)	96·00
Vernier scale	0·60
Micrometer scale	0·14
Difference in height	96·74 − 100·00
	−3·26 m

If the difference in height exceeds 70 metres the end verniers must be used, if both the red scale and the red vernier is used 200 metres must be subtracted from the reading.

(*e*) All height measurements are from the instrument transit axis to the two reference marks on the staff.

(*f*) When the vertical angle is large the red auxiliary scale is used. The figuring is in centimetres which are added to the absolute magnitude of the difference, that is to say they always increase the difference in height irrespective if it is an elevation or depression.

9.8.6. ADJUSTMENTS

The adjustments for the plate bubble and for horizontal collimation are carried out in the usual way (see §3.8) on theodolite adjustments. It should however be noted that to adjust for horizontal collimation the micrometer drum should be set to read 10.

9.8.6.1. *Index Error*

The adjustment for index error is complicated by the shifting of the line of sight upward by 23 mm parallel to itself when the instrument is in the normal distance measuring position, i.e. face left. In the face right position the line of sight is shifted downward by the same amount. When testing for index error it is necessary either to take two aiming points 46 mm vertically apart or to observe to an aiming point at least 2 m away.

Procedure

(i) On a white post or wall make two marks 46 mm apart and set the instrument at about 150 metres from the post.
(ii) Level the instrument carefully and with the instrument in the normal position, viz. face left, sight onto the apparent lower mark since the telescope image is inverted. Level the altitude bubble and note the vertical circle reading, say 87° 26'8.
(iii) Change face, the micrometer drum is now below the telescope, and sight onto the apparent upper mark. Again note the vertical circle reading after levelling the altitude bubble, say 272° 32'4.
The required face right reading is 272° 32'8.
(iv) Leave the instrument in the face right position and bring the vertical circle to this reading with the altitude bubble setting screw. This will displace the bubble from its correct central position to which it is returned by use of its adjusting screw.
(v) The test is repeated until the readings when reduced are in almost exact agreement.

It should be noted that maladjustment of the index does not effect horizontal or vertical measurements made using the horizontal staff.

9.8.6.2. *Setting of Bubble Setting Screw Index*

The value of this index in determining the correctness of the levelling of the instrument has been indicated earlier. The index is quite valueless if it is not in adjustment.

Procedure

(i) Level the instrument using the altitude setting screw as indicated in §3.8.5.(ii).
(ii) The main axis of rotation is now vertical in all directions and as the two ends of the altitude bubble are in coincidence the indices should also be coincident.
(iii) If the indices are not coinciding, loosen the three screws positioned at 120° intervals around the knurled knob of the setting screw and rotate the collar to bring the index thereon into coincidence with the fixed index.
(iv) Before tightening the screws on the knurled knob check that the bubble is central when the indices coincide.

Detail Survey

9.8.6.3. Distance Adjustment

Perfection in the grinding of prism A cannot be achieved and although the precision is very high small errors are compensated for by rotation of the wedge-shaped front glass F. The accurate grinding of the wedges W_1 and W_2 is equally important and errors here are compensated for by a forward or backward tilting of the second wedge; this adjustment is only carried out by the manufacturer or his agent.

With constant use the relationship between the prism A and the micrometer drum may be slightly disturbed and this can also be compensated for by rotation of the wedge-shaped front glass F. It is this adjustment which can easily be performed in the field which will be considered here.

Procedure

(i) Centre and level the instrument with great care at one end of a base the lenght of which is known to within a few millimetres.
(ii) Read the distance in the usual way; if this does not agree with the measured length of the base, adjust as follows:
(iii) Slightly loosen the three locking screws on the frame of the glass wedge mounted in front of the rotating wedges.
(iv) If the reading is short turn the roughened frame clockwise by one division and read the staff to calculate the effect. An anticlockwise rotation is made if the reading is too great.
(v) Knowing the effect of one graduation an adequate rotation is made to obtain a distance reading correct to within 5 mm.
(vi) Retighten the lock screws and check over a different control length.

9.8.6.4. Change-Over Switch Stops

As the change-over switch is rotated a gear ring in the right-hand standard rotates between two adjustable stops and rotates each of the two wedges W_1 and W_2 through 90°. An error in the adjustment of a stop will cause an over or under rotation of the wedges which will have the same effect on the observations as displacement of the main axis in the line of the telescope.

(a) Height stop adjustment

Procedure

(i) Level the instrument with great care and place the staff about 20 metres away.
(ii) Set the change-over switch to the height measurement position.
(iii) With the telescope in the normal measuring position make an accurate bisection of the staff and take the reading, say h_1.
(iv) Transit the telescope and take a second reading, say h_2.
If the stop is correctly positioned

$$h_1 = (h_2 - 0.023)$$

it is not desirable to attempt this adjustment if

$$h_1 - h_2 + 0.023 < 0.005$$

(v) The adjustment is carried out by rotating the forward adjusting screw on the right standard until the correct relationship is obtained.

(b) *Distance stop adjustment*
Procedure

   (i) With the same arrangement as for the height stop adjustment and the change-over switch in the distance position take a reading on the staff with the telescope in the normal position. Let this reading be q_1.
   (ii) Transit the telescope and repeat the reading on the staff, say q_2. The wedges are now deflecting in the reverse direction and combined with the built in corrections for the anallactic constant.

   Let $$q_1 - (100 \cdot 228 - q_2) = C$$

   (iii) Repeat the readings over a steep sight, say from a rooftop to ground level where the vertical angle is about 30° or more. Let the readings be r_1 and r_2 then
   $$r_1 - (100 \cdot 228 - r_2) = Cr$$

   If the stop is wrongly positioned it will be apparent as $C \neq Cr$. A large height difference is required to determine that there is error in the positioning of the distance stop; in §9.8.7 on errors in horizontal bar tacheometry it will be shown that main axis obliquity or mispositioning of the distance stop produces errors that are proportional to the difference in height of instrument and staff.
   (iv) The adjustment is carried out by rotating the rear adjusting screw above the vertical slow motion screw on the right standard, Cr being made to equal C.

9.8.6.5. *Adjustment of Staff Stand*

(a) *Diopter sight*
This should be perpendicular to the staff face.
Procedure

   (i) Erect the staff as if for measurement and direct the diopter sight at a distant clearly visible object.
   (ii) Above the diopter hold an optical square so that the distant object is viewed at the same time as the staff face. These should coincide.
   (iii) If adjustment is required move the rod supporting the staff so that the staff is in the correct position. This will displace the diopter sight which is adjusted by lateral movement of the slit sight.

(b) *Circular bubble*
It is essential that the rod supporting the staff should be perpendicular.
Procedure

   (i) Erect the staff and rod with the aid of a plumb line.

Detail Survey

(ii) If the circular bubble is not central make it so by means of the three adjusting screws beneath the bubble housing.

9.8.7. SOURCES OF ERROR IN HORIZONTAL STAFF TACHEOMETRY

9.8.7.1. *Maladjustment of Instrument and Staff*

If the distance adjustment (§9.8.6.3 above) has not been carried out a constant scale error will be present in each measurement. The resultant errors in distance and height are proportional to differences in distance and height respectively. If p is the distance and h the difference in height,

$$p = K \cos \alpha \, s \quad \text{and} \quad h = K \sin \alpha \, s$$

then
$$dp = dK \cos \alpha \, s \quad \text{and} \quad dh = dK \sin \alpha \, s$$

or
$$dp/p = dK/K \quad \text{and} \quad dh/h = dK/K$$

If the change-over stops are wrongly positioned or the instrument has not been correctly levelled in the direction of the telescope the effect is to create an error $d\alpha$ in α the vertical angle.

The
$$dp = -K \sin \alpha \, d\alpha = -h \, d\alpha$$
and
$$dh = K \cos \alpha \, d\alpha = p \, d\alpha$$

It will be seen that the error in distance is proportional to the difference in height and the error in height is proportional to the horizontal distance.

If the diopter sight is not correctly positioned a systematic error is present. If the staff is not exactly at right angles to the line by an amount ϵ then the apparent intercept $s' = s \cos \epsilon$ where s is the correct intercept, hence

$$ds = s(1 - \cos \epsilon) = \tfrac{1}{2} s \, \epsilon^2$$

If a precision of 1/10 000 is sought ϵ should not exceed 48′·5.

If the circular bubble is not in adjustment the rod support will not be vertical and measurement will be made to a point not plumb over the ground mark.

9.8.7.2. *Defects in Instrument and Staff*

Regular checking of the equipment over control bases will ensure that the relative positioning of the wedges has not been disturbed. Minor changes are compensated for by adjustment of the wedge shaped front glass F. If this adjustment is excessively large the instrument should be returned to the manufacturer or his agent.

For correct measurement highly accurate graduation of the staff is essential. The staff should be checked with a beam compass before it is first used.

A further condition is that the staff must be straight. If the face of the staff is concave this amounts to a shortening of the intercept and for a precision of 1:10 000 a deviation of 1 cm at the ends of the staff may be tolerated. If the staff is convex the length of the line will be overmeasured as the coincidence point is beyond the ground mark. A figure in this case of $\tfrac{1}{2}$ cm may be taken as the maximum deviation for the end of the staff.

9.8.7.3. *Atmospheric Conditions*

Shimmer leading to an apparent 'dancing' of the images can make reading difficult. Usually both images will appear to move together and several co-incidences should be made with the micrometer. Experience has proved that the loss of accuracy in distance measurement arising from shimmer is smaller than might be expected. In heighting the loss of accuracy is comparable to that which might occur in levelling under comparable conditions.

Rays passing close to walls heated by the sun should be avoided as lateral refraction will lengthen the apparent length of the line.

9.8.7.4. *Errors by the Observer*

Misreading of the staff by counting the 2 metre graduation as 1 metre or the 2 decimetre graduation as 1 decimetre should be guarded against.

In double image instruments, such as the horizontal staff tacheometers, observer errors arise when the two separate images pass through different parts of the eye lens which may be deformed and cause a displacement of one image relative to the other. This possible personal error is overcome by causing the two light pencils to pass through the same spot in the exit pupil. It is for this purpose that the prism P is introduced in the optical train.

9.8.8. APPLICATION OF HORIZONTAL STAFF TACHEOMETRY

Horizontal staff tacheometry is a standard ground method used by the Ordnance Survey of Great Britain to provide control for large scale plans in resurvey areas and in development areas of resurveyed plans under continuous revision. The Kern DKRT equipment is used for both control traversing and detailing. The survey is executed to provide an instrumentally surveyed framework of detail from which the remaining detail can be surveyed by graphic and short taped methods (see §9.9).

A network of tacheometric traverses is established, tied to minor control points established by triangulation or major traverses. These points are about 1 km apart. The traverses are plotted by rectangular coordinatograph and the detail observed from the traverses is plotted by polar coordinatograph. Connecting detail is then penned-in by reference to a diagram drawn up as the survey is made.

The tacheometric traverse stations are sited and usually marked by permanent ground features, these stations serve as permanent reference points for future revision. The whole area to be surveyed is broken down into a number of blocks of about 100 hectares. Each block is bounded by a tacheometric traverse and within each block all work is carried out with the same instrument and pair of staves. During each traverse detail and traverse observations are taken simultaneously. The following rules govern the execution of the traverses:

  (i) The terminals of the traverses must be either minor control points or tacheometric stations previously fixed. Sufficient angular observations must be taken to orient the traverse.
  (ii) The lengths of traverses should not exceed 1 500 metres in the case of

traverses between minor control points and 750 metres between other fixed points.

(iii) A traverse should be as straight as possible and should not deviate from the direct line by more than half the distance between the terminals.

(iv) Acute changes of direction and large differences in lengths of consecutive lines are to be avoided; where a short line cannot be avoided a deviation (see §7.8.5.) will be observed.

(v) Traverses will not be looped outside of the terminals and parallel traverses of over 800 metres in length and within 100 metres of each other will be tied near their mid-points.

(vi) Crossing traverses will be tied to each other.

(vii) Subsidiary points may be established during the traverse by triangulation or by a single distance and direction.

(viii) Each traverse station is selected for its suitability for the survey of surrounding detail. Detail surveyed from each station is selected so as to enable the subsequent detail survey to be executed. This detail must be readily recognizable to the detail surveyor and must be spread over the whole area regardless of any road network. The best results are obtained if the points of detail are junctions of firm straight detail which can be connected at the plotting stage. Pylons, flagstaffs, etc., can be intersected from three or more traverse stations; such points are co-ordinated and plotted with the traverse stations.

At each station the following procedure is adopted;

Traverse angles

One pointing is made on each face, recorded to the nearest decimal of a minute. Included in this round are observations to any intersected points and to a local R.O. which is used to check that no movement of the circle has taken place during the observations to the detail points.

Traverse distances

For each traverse line 3 forward and 3 backward observations are made. The distances are recorded to the nearest millimetre and the mean of the forward and backward readings should agree to 1 part in 4 000. The same horizontal staff is used in the forward position throughout the traverse.

Detail directions and distances

One measure of each is made. Angles are recorded to the nearest minute and distances to the nearest centimetre. If necessary distances can be taped, or where the staff cannot be set over the point of detail it may be hand held, so that the back of the staff is held against the point. The staff holder must ensure by means of an attached diopter sight that the staff is perpendicular to line of sight and the observer must ensure the horizontality of the staff. The staff may be slightly offset or held short of the point and the excess distance taped.

Cul-de-sac stations

Where selected points of detail cannot be fixed from stations on the main traverse a single spur or cul-de-sac station may be used. Such a point is fixed by observations from a traverse station on both faces and by a forward and backward measure of the distance as if it were a traverse station. To check against an error of orientation a point of detail is fixed by observation from both a main traverse station and from the cul-de-sac station.

9.9. COMPLETION AND FIELD EXAMINATION

9.9.1. GENERAL PRINCIPLES

Given a well-selected distribution of instrumentally fixed points, say with self-reducing horizontal bar tacheometer, it is the usual practice to complete the plan by graphical methods. That is to say methods which involve little or no measurement and are closely akin to plane tabling (see §1.1.).

Graphical methods involve the use of few instruments, the surveyor works on his own in the field with the plan in front of him. The surveyor uses one or more of the following techniques to breakdown the existing framework, which is represented by points of surveyed detail, until the whole of the plottable detail has been fitted into position.

9.9.2. TECHNIQUES

Three basic operations are involved, namely:

(i) The shot—this is when a sight with the unaided eye from one point to another is extended to a third point.
(ii) The straight—this is when a sight is formed by extending the alignment of any straight detail feature, e.g. a wall or fence.
(iii) The tie—this is a short linear measurement made with a linen tape.

Points of detail may now be fixed by one of the following techniques:

(i) Intersection—when two shots or straights meet at a point of detail.
(ii) Production—when a shot or straight together with a tie locates a point of detail. This point is then 'tied out' by measurement to another fixed point along the shot or straight.
(iii) Resection—when by use of a line ranger the surveyor places himself on line between two fixed points of detail so establishing a shot and then repeats the process to two different points, preferably nearly at right angles to the first pair. Once the surveyor has located his exact position on the plan any detail coinciding with that point is located and nearby detail is located by ties.
(iv) Pull out—when the surveyor positions himself on the prolongation of a straight not plotted and then fixes his position by one of the methods above. Given one other point fixed on the unplotted straight its line is established. Ties complete the location.
(v) Shoot a straight—when the point where the forward prologation of a straight is noted and used to locate a point of detail, possibly with a short tie.

9.9.3. EQUIPMENT

Very little equipment is required other than a stable base for the original plan which is taken into the field. Graphical methods of completion have been carried to a high degree of perfection by the Ordnance Survey whose terminology has been used in part in this section. The Ordnance Survey prepare original plans for use in the field on butt jointed aluminium plates surfaced with an off-white enamel. Each butt jointed plate is exactly 20 cm × 20 cm and a set of 4 make up a 500 metre square sheet at a scale of 1:1 250. The 4 plates are mounted in a special frame and as the surveyor reaches the edge of a plate he can work straight across onto the next plate. Edge comparisons are thereby not required. The butt jointed plates are exactly machined so that any 4 will fit together, exact positioning of adjacent plates is determined by fine registration marks cut on the plate at the time of plotting the control.

Without butt jointed plates graphical methods of completion are particularly difficult around the edges of each sheet. Other than plotting about 100 metres of control around each sheet, no satisfactory alternative exists.

A silver point or a 9H pencil is used to draw on the butt jointed plates, scales and straight edges are additional items of equipment required together with a line ranger and linen tape. The surveyor works without assistance.

9.9.4. FIELD EXAMINATION

The purpose of a field examination is to prove that the surveyors work has been satisfactorily carried out. The techniques of completion are extensively used, the field examination particularly checking that blocks of work are correctly positioned and oriented with respect to each other. This is usually achieved by taking shots and straights across the sheet.

Where an error is found a correction is made in a distinctive colour. If there are several errors over a sheet an extensive check is carried out, otherwise only part of the work is checked. If an error can be regarded as a difference of opinion between the surveyor and the examiner the latter corrects the sheet but writes against the correction M.O. (matter of opinion) so that no blame attaches to the surveyor.

10 Curve Design and Setting Out

10.1. SETTING OUT OF WORKS—GENERAL

Setting out engineering works involves the siting on the ground of the various elements of the works in accordance with the dimensioned plans and drawings supplied by the designer.

Since it is often not possible to set out the whole of the works before construction commences, the accurate positioning of each element independently is highly important and errors or mistakes can be very expensive.

It is important to remember that inaccurate surveys are not unknown, nor are dimensions scaled from distorted drawings or prints, so that it is advisable to check all leading dimensions on the site before commencing any setting out whatsoever.

Normally the dimensions of individual elements (buildings, roads, bridges, etc.) will be fully-figured on the drawings but for their relative position reliance may have to be made on scaled dimensions. In such cases it is advisable to look for any controlling factors that will influence the actual positioning of the element, e.g. if space is to be left between two houses for two pre-fabricated garages, each 8′ 3″ wide, then these houses must be set out 16′ 6″ apart against a possible scaled dimension of 16′ 2″.

The usual practice is to mark key points (corners of buildings, centre-lines, or kerbs of roads, etc.) with wooden pegs, with a pencil cross or a small nail in the top if greater precision is required. A pipe nail can be driven into tarmac or asphalt, whilst it may be necessary to cut a cross with a cold chisel on stone or concrete.

Much wasted labour can be avoided by establishing permanent reference points (by a peg surrounded with concrete or by driving a short length of small diameter steel tube) adjacent to the works, but secure from damage by excavating machinery or construction traffic from which the pegs can be re-established quickly and easily (possibly by a charge hand) if they are lost.

The general procedure in setting out is to establish a main control framework (a single base line will often suffice), usually by triangulation or traversing, from which the detail can be set out by means of off-setting, tie lines, radiation, intersection, etc. In the case of 'route' works (railways, roads, waterways, pipe-lines, etc.) the control framework is usually a traverse and it is convenient to utilize the main intersecting straights for this purpose, leaving only the curves to be established by other means.

For small sites, or when a high precision is not required, chain and tape methods will often be adequate, the 3:4:5 triangle for setting out right-angles, and the principle of equality of diagonals for checking the 'squareness' of rectangles, can also be employed. Where greater precision is required, instrumental methods are to be preferred, and of course, will be essential if use is to be made of radiation and intersection.

In setting out, a system of coordinates is obviously called for, e.g. rectangular coordinates for off-setting methods, and polar coordinates for radiation methods. For setting out curves, although both these systems have their uses, a third system, using chords and deflection angles (see §§10.3.2 and 10.3.2.3) is often employed.

Since in this system the location of each point is dependent on the location of the preceding point, the method is known as curve 'ranging'.

Setting out elements consisting of straight lines requires no further explanation, but the setting out of curves needs amplification, and the bulk of this chapter relates to this work.

In setting out road and railway works it is usual to establish pegs at 100 ft intervals along the actual centre-line of the route, these pegs being consecutively numbered from the commencement of the route. Thus any point on the centre-line may be identified by its 'chainage', i.e. its distance along the centre-line from the commencement. For example a point 327 feet along the route would be described as being at 'chainage $(3 + 27)$', i.e. 27 ft beyond the third 100 ft peg.

In setting out curves it is convenient to have some of the pegs defining the curve coincident with these 100 ft pegs, this procedure being known as 'preserving running chainage'. In all examples used here the chainage will be assumed to be increasing from left to right.

Proceding in the direction of increasing chainage the first tangent to a curve is known as the 'running-on' tangent, and the second as the 'running-off' tangent. The junction of a straight and a curve is known as a 'tangent point' (TP), the junction of two portions of a compound or combined curve as a 'junction point' (JP), the point at which the running-on and running-off tangents meet as the 'intersection point' (I), and the external angle between the tangents as the 'intersection angle' (β). (The internal angle is known as the 'apex angle' but is rarely used in curve calculations.)

These terms are illustrated in figure 10.1.

10.2. TYPES OF HORIZONTAL CURVE

The types of horizontal curve which the surveyor normally has to deal with in connection with building and civil engineering works are as follows (see figure 10.1(*a*) to (*e*)):

(i) *Simple curves*—circular curves of constant radius.
(ii) *Compound curves*—two or more consecutive simple curves of different radii.
(iii) *Reverse curves*—two or more consecutive simple curves of the same or different radii with their centres on opposite sides of the common tangent.
(iv) *Transition curves*—curves with a gradually varying radius (often referred to as 'spirals').

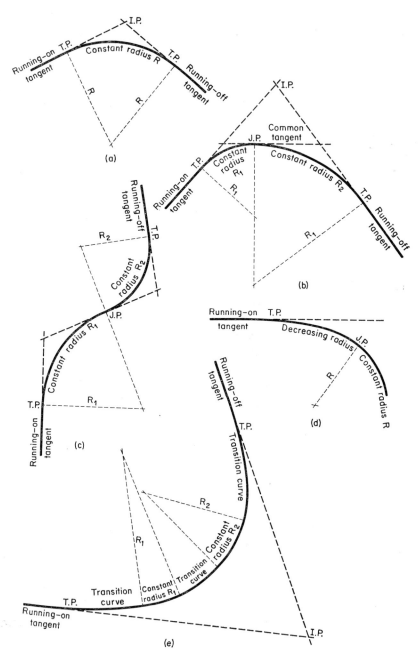

Figure 10.1

(v) *Combined curves*—consisting of consecutive transition and simple circular curves. This is the usual manner in which transition curves are used in road and railway practice, to link a straight and a circular curve, or two branches of a compound or reverse curve.

The problems connected with horizontal curves are twofold, first the design and fitting of curves of suitable radius, or rate of change of radius to suit site conditions, vehicle speeds, etc., and secondly the setting out of the designed curves in the field.

The first of these problems is often the concern of the engineer-designer, the computation of setting out data and the field setting out only falling to the surveyor. For this reason, these last two portions of the work will be dealt with first, but design and fitting will also be considered briefly in §10.6.

10.3. BASIC CURVE GEOMETRY

10.3.1. CIRCULAR CURVES

Circular curves can be defined by 'radius' which is self-explanatory, or by 'degree'. The 'degree' D of a circular curve is defined as being the angle in degrees, subtended at the centre by a chord 100 ft long so that (figure 10.2)

$$D° = 2 \sin^{-1} 50/R \quad (R \text{ in ft}) \qquad (1)$$

or
$$R = 50/\sin \tfrac{1}{2} D°$$

(If R is large $D° \simeq 5\,729.6/R$.)

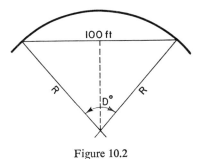

Figure 10.2

(Note: Some authorities define the 'degree' of a curve as being the angle subtended at the centre by an arc 100 ft long, the difference being negligible in the range that $D^c \simeq 2 \sin \tfrac{1}{2} D$)

Intersection angle = Angle subtended at centre

= Twice the angle between the chord and the tangent (see figure 10.3).

(i) Since A = C = 90°, ABCO is a cyclic quadrilateral, therefore

$$\text{CBD} = \text{AOC} = 2\theta \qquad (2)$$

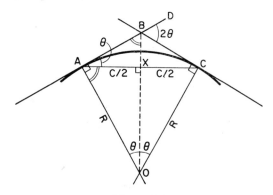

Figure 10.3

(ii) In the right-angled triangles, ABX and OBA, angle ABO is common, therefore the triangles are similar, and hence,

$$\text{AOB} = \text{BAX} = \theta \qquad (3)$$

(N.B. for 100 ft chord, $\theta = \tfrac{1}{2}D$)

(iii) From triangle AOX

$$\sin \theta = c/2R \qquad (4)$$

Angle between adjacent chords = Angle subtended at the centre. ABO = CBO = $(90 - \theta)$, therefore (see figure 10.4)

$$\phi = 180 - (180 - 2\theta) = 2\theta \qquad (5)$$

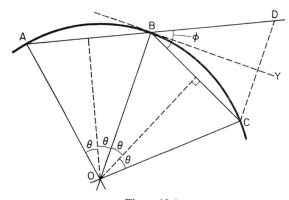

Figure 10.4

10.3.2. TRANSITION CURVES

A transition curve is one in which the curvature varies uniformly with respect to arc, in order to allow a gradual change from one radius to another (a straight being merely a circular curve of infinite radius) and to permit a gradual change in the superelevation. It must, of course, have the same radius of curvature at its ends, as the circular curves that it links.

Curve Design and Setting Out

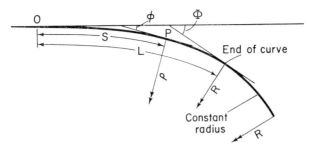

Figure 10.5

By definition, the transition curve must have a constant rate of change of curvature with respect to arc, i.e. if ϕ is the tangential angle and s is the arc, $d^2\phi/ds^2 = K$, where K is a constant.

Considering, for the time being, only the case of a transition curve linking a straight and a circular curve of (constant) radius R (see figure 10.5)

$$d^2\phi/ds^2 = K \tag{1}$$

whence, $\qquad d\phi/ds = \int K\, ds = Ks + K_1$

but, since the curvature at the origin 0, is zero, $d\phi/ds = 0$ when $s = 0$, and so $K_1 = 0$, and so

$$d\phi/ds = Ks \tag{2}$$

Integrating again,

$$\phi = \int Ks\, ds = \tfrac{1}{2}Ks^2 + K_2$$

but, again, since $\phi = 0$ when $s = 0$, $K_2 = 0$, and the fundamental equation to the transition curve becomes that of the mathematical curve, the clothoid

$$\phi = \tfrac{1}{2}Ks^2 \tag{3}$$

This may also be written, $\qquad s = C\phi^{\frac{1}{2}}$

where $C = 2/K$.

If L is the total length of the curve, then when $s = L$, $d\phi/ds = 1/R = KL$, therefore $K = 1/LR$, and the equation to the curve becomes,

$$\phi = s^2/2LR \tag{4}$$

or $\qquad s = (2LR\phi)^{\frac{1}{2}} \tag{5}$

Note: That since $d\phi/ds = Ks = s/LR = 1/\rho$, where ρ is the radius of curvature corresponding to the arc s then

$$s\rho = LR = \text{constant} \tag{6}$$

This is an important, and very useful, property of the clothoid.

When $s = L$ (i.e. at the end of the curve) the total tangential, or 'spiral' angle for the curve Φ is given by

$$\Phi = L^2/2LR = L/2R \tag{7}$$

This is half the tangential angle turned through by a circular curve of the same length, and of radius R.

The above equations do not lend themselves readily to the computation of curve components or to the field setting out of the curves and it is necessary to arrange them in a more convenient form.

Three systems of representing the curve are commonly used for these purposes, namely

(i) A system of rectangular coordinates.
(ii) A system of polar coordinates.
(iii) A system utilizing a deflection angle corresponding to the polar amplitude and the length of the arc, (represented in the field by a series of chords).

All these systems involve using approximations, the extent of the approximation being dependent on the degree of accuracy finally required. In this connection it must be borne in mind that even the maximum curvature of a practical transition curve is comparatively small and that the curvature shown in the diagrams is grossly exaggerated for the sake of clarity.

10.3.2.1. *Rectangular Coordinates*

It is customary to use the tangent point as the origin, the continuation of the tangent as the X axis, and offsets from this continuation as the Y axis.

Considering a small element of arc δs at a distance s from the origin, it will be clear from figure 10.6 that

$$x = \int \cos \phi \, ds \tag{1}$$

$$y = \int \sin \phi \, ds \tag{2}$$

In order to perform these integrals it will be necessary to substitute for ϕ in terms of s or *vice versa*. The choice of the most expeditious substitution is dependent on the relationship it is required to establish and one illustration should suffice.

From equation (4) §10.3.2

$$\phi = s^2/2LR$$

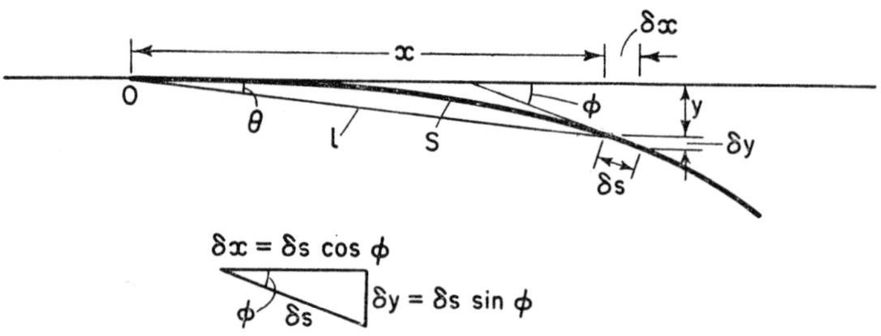

$\delta x = \delta s \cos \phi$

$\delta y = \delta s \sin \phi$

Figure 10.6

whence
$$x = \int \cos(s^2/2LR)\,ds$$
$$= \int \left\{1 - \frac{(s^2/2LR)^2}{2} + \frac{(s^2/2LR)^4}{24} \text{ etc.}\right\} ds$$

whence, since $x = 0$ when $s = 0$,
$$x = s - s^5/40(LR)^2 + s^9/3\,456(LR)^4 \text{ etc.} \tag{3}$$

Similarly,
$$y = \int \sin(s^2/2LR)\,ds$$
$$= \int \left\{(s^2/2LR) - \frac{(s^2/2LR)^3}{6} + \frac{(s^2/2LR)^5}{120} \text{ etc.}\right\} ds$$

whence, since $y = 0$ when $s = 0$,
$$y = s^3/6LR - s^7/336(LR)^3 + s^{11}/42\,240(LR)^5 \text{ etc.} \tag{4}$$

By substituting $s = (2LR\phi)^{\frac{1}{2}}$ expressions can be found for x and y in terms of ϕ, alternatively since $ds/d\phi = (2LR)^{\frac{1}{2}}/2\phi^{\frac{1}{2}}$ the expressions

$$x = \int \frac{\cos\phi(2LR)^{\frac{1}{2}}}{2\phi^{\frac{1}{2}}}\,d\phi$$

and
$$y = \int \frac{\sin\phi(2LR)^{\frac{1}{2}}}{2\phi^{\frac{1}{2}}}\,d\phi$$

can be expanded and integrated, leading to

$$x = (2LR)^{\frac{1}{2}}\left\{\phi^{\frac{1}{2}} - \frac{\phi^{\frac{5}{2}}}{5/2} + \frac{\phi^{\frac{9}{2}}}{9/4} - \frac{\phi^{\frac{13}{2}}}{13/6} + \text{etc.}\right\} \tag{5}$$

$$y = (2LR)^{\frac{1}{2}}\left\{\frac{\phi^{\frac{3}{2}}}{3} - \frac{\phi^{\frac{7}{2}}}{7/3} + \frac{\phi^{\frac{11}{2}}}{11/5} - \text{etc.}\right\} \tag{6}$$

The expressions for x and y in terms of s ((3) and (4) above) can be rewritten

$$x = s\left\{1 - \frac{1}{40}\frac{s^4}{(LR)^2} + \frac{1}{3\,456}\cdot\frac{s^8}{(LR)^4} \text{ etc.}\right\} \tag{7}$$

$$y = \frac{s^3}{6LR}\left\{1 - \frac{1}{56}\cdot\frac{s^4}{(LR)^2} + \frac{1}{7\,040}\cdot\frac{s^8}{(LR)^4} \text{ etc.}\right\} \tag{8}$$

which are both power series in $s^4/(LR)^2$ and since s is less than L, this term will be less than $L^4/(LR)^2 = (L/R)^2$, so that if, as is normally the case, L/R is small, all the terms except the first, in these series, will be small and may be regarded as 'correction' terms.

If all these 'tail' terms are neglected we have the simple relationship, $x = s$ and $y = s^3/6LR$ whence

$$y = x^3/6LR \tag{9}$$

This is the cubic parabola and is commonly used, without correction, as a transition curve where the final curvature is small. It should be noted however that it has a very important limitation. When the ratio y/x exceeds about 0·15 (equivalent to a deflection angle of 8° 34′) the curvature ceases to increase as the distance round the curve increases, and begins to decrease again, so that the cubic parabola is useless as a transition curve outside this range.

$$y = \frac{x^3}{6LR}, \quad \frac{dy}{dx} = \frac{x^2}{2LR}, \quad \frac{d^2y}{dx^2} = \frac{x}{LR}$$

$$\rho = \frac{[1 + (dy/dx)^2]^{\frac{3}{2}}}{d^2y/dx^2} = \frac{LR}{x}\left[1 + \left(\frac{x^2}{2LR}\right)^2\right]^{\frac{3}{2}}$$

$$\frac{d\rho}{dx} = \frac{LR}{x^2}\left\{\frac{3}{2}\left[1 + \left(\frac{x^2}{2LR}\right)^2\right]^{\frac{1}{2}}\frac{x^4}{L^2R^2} - \left[1 + \left(\frac{x^2}{2LR}\right)^2\right]^{\frac{3}{2}}\right\}$$

$$= \frac{LR}{x^2}\left[1 + \left(\frac{x^2}{2LR}\right)^2\right]^{\frac{1}{2}}\left(\frac{3x^4}{2L^2R^2} - 1 - \frac{x^4}{4L^2R^2}\right)$$

For minimum radius of curvature $d\rho/dx = 0$, whence, neglecting infinite and negative solutions, $5x^4 = 4L^2R^2$ i.e. $x^2 = 2LR/5^{\frac{1}{2}}$, whence $y/x = x^2/6LR = 1/(3.5^{\frac{1}{2}})$ i.e. $y/x \simeq 0\cdot 15$ and $\tan \theta \simeq 8° 34'$.

10.3.2.2. *Polar Coordinates*

It is convenient to take the tangent point as the origin, and the continuation of the tangent as the initial line, The amplitude θ is then referred to as the 'deflection angle', and the radius vector l as the 'long chord'.

It will then be seen from figure 10.6 that

$$\tan \theta = y/x \tag{1}$$

and that $l = (x^2 - y^2)^{\frac{1}{2}}$ or preferably

$$l = x/\cos \theta \tag{2}$$

In practical problems θ and l can be calculated from x and y where these have already been found, alternatively a general expression can be found.

From equations (5) and (6) §10.3.2.1

$$\tan \theta = \frac{y}{x} = \frac{(2LR)^{\frac{1}{2}}(\phi^{\frac{3}{2}}/3 - \phi^{\frac{7}{2}}/42 + \phi^{\frac{11}{2}}/1\,320 - \text{etc.})}{(2LR)^{\frac{1}{2}}(\phi^{\frac{1}{2}} - \phi^{\frac{5}{2}}/10 + \phi^{\frac{9}{2}}/216 - \text{etc.})}$$

$$= \frac{\phi/3 - \phi^3/42 + \phi^5/1\,320}{1 - \phi^2/10 + \phi^4/216} \quad \begin{array}{l}\text{neglecting powers}\\ \text{of } \phi \text{ above 5}\end{array}$$

which, by the binomial expansion equals,

$$\left\{\frac{\phi}{3} - \frac{\phi^3}{42} + \frac{\phi^5}{1\,320}\right\}\left\{1 + \frac{\phi^2}{10} - \frac{\phi^4}{216} + \frac{\phi^4}{100}\right\}$$

$$= \frac{\phi}{3} + \phi^3\left(\frac{1}{30} - \frac{1}{42}\right) + \phi^5\left(\frac{1}{1\,320} - \frac{1}{420} + \frac{29}{16\,200}\right)$$

whence, $\tan \theta = \phi/3 + \phi^3/105 + 26\phi^5/155\,925 + \text{etc.}$ \hfill (3)

Curve Design and Setting Out 489

but, $\theta = \tan\theta - \tfrac{1}{3}\tan^3\theta + \tfrac{1}{5}\tan^5\theta - $ etc.

$$= \frac{\phi}{3} + \frac{\phi^3}{105} + \frac{26\phi^5}{155\,925} - \frac{1}{3}\left(\frac{\phi^3}{27} + \frac{\phi^5}{315} + \cdots\right) + \frac{1}{5}\left(\frac{\phi^5}{243} + \cdots\right)$$

or $$\theta = \frac{\phi}{3} - \frac{8\phi^3}{2\,835} - \frac{32\phi^5}{467\,775} \text{ etc.} \tag{4}$$

Note that, to the first order of small quantities, $\theta \simeq \tfrac{1}{3}\phi$, and that the correction terms, being independent of any other variables, may be used for any clothoid.

So as to utilize this first approximation in the correction terms, we write

$$\theta \simeq \frac{\phi}{3} - \frac{8 \times 27}{2\,835}\left(\frac{\phi}{3}\right)^3 - \frac{32 \times 243}{467\,775}\left(\frac{\phi}{3}\right)^5 - \cdots$$

$$= \frac{\phi}{3} - \frac{216}{2\,835}\left(\frac{\phi}{3}\right)^3 - \frac{7\,776}{467\,775}\left(\frac{\phi}{3}\right)^5 - \cdots$$

$$= \frac{\phi}{3} - \left\{0\cdot0762\left(\frac{\phi}{3}\right)^3 + 0\cdot0166\left(\frac{\phi}{3}\right)^5 + \cdots\right\} \tag{5}$$

All the angles used in the above descussion are, of course, in circular measure. In practice, the angles will be in degrees, and it is convenient to have the correction terms in seconds. To this end, the expression can be modified, as follows

$$\theta° = \phi°/3 - 206\,265\left(\frac{0\cdot076\,2(\phi°/3)^3}{(57\cdot3)^3} + \frac{9\cdot016\,6(\phi°/3)^5}{(57\cdot3)^5}\right)''$$

or, $$\theta° = \phi°/3 - [0\cdot083\,6(\phi°/3)^3 + 0\cdot000\,005\,5(\phi°/3)^5]'' \tag{6}$$

e.g. if ϕ is 30°, then θ is approximately 10°, and the correction is

$$(83\cdot6 + 0\cdot55)'' = 84\cdot2'' = 1'\,24\cdot2'',$$

and the corrected value of θ is 9° 58' 36".

An expression can now be developed for the long chord l.

$$l = \frac{x}{\cos\theta} = \frac{(2LR)^{\frac{1}{2}}(\phi^{\frac{1}{2}} - \phi^{\frac{5}{2}}/10 + \phi^{\frac{9}{2}}/216\text{—etc.})}{1 - \tfrac{1}{2}(\phi/3 - 8\phi^3/2\,835)^2 + \tfrac{1}{24}(\phi/3 \text{ etc.})^4}$$

$$= \frac{(2LR)^{\frac{1}{2}}(\phi^{\frac{1}{2}} - \phi^{\frac{5}{2}}/10 + \phi^{\frac{9}{2}}/216)}{1 - \tfrac{1}{2}(\phi^2/9 - 16\phi^4/8\,505) + \phi^4/1\,944}$$

$$= (2LR\phi)^{\frac{1}{2}}\left(1 - \frac{\phi^2}{10} + \frac{\phi^4}{216}\right)\left(1 - \frac{\phi^2}{18} + \frac{\phi^4}{687}\right)^{-1}$$

$$= (2LR\phi)^{\frac{1}{2}}\left(1 - \frac{\phi^2}{10} + \frac{\phi^4}{216}\right)\left(1 + \frac{\phi^2}{18} - \frac{\phi^4}{687} + \frac{\phi^4}{324}\right)$$

$$= (2LR\phi)^{\frac{1}{2}}\left(1 - \frac{\phi^2}{19} + \frac{\phi^4}{216} + \frac{\phi^2}{18} - \frac{\phi^4}{180} + \frac{\phi^4}{613}\right)$$

$$= (2LR\phi)^{\frac{1}{2}}\left(1 - \frac{2\phi^2}{45} + \frac{2\phi^4}{2\,835}\right)$$

Substituting $\phi = s^2/2LR$, this becomes,

$$l = s\left(1 - \frac{s^4}{90(LR)^2} + \frac{s^8}{22\,680(LR)^4}\right)$$

or,
$$l = s - \frac{s^5}{90(LR)^2} + \frac{s^9}{22\,680(LR)^4} \text{ etc.} \quad (7)$$

N.B. In an interesting article† to which the reader is referred, G. J. Thornton-Smith describes a method by which, for a wide range of values of ϕ, by means of a 'substitute' series, a 'one term' correction can be obtained for all the important elements of the clothoid, for which series have been derived above.

10.3.2.3. *Chords and Deflection Angles*

As the system of coordinates here utilizes the distance round the curve from the origin s, and the deflection angle θ which is the same as that found in the previous section, no further calculations are required at this stage.

10.4. COMPUTATION OF CURVE COMPONENTS

The initial requisites for setting out any curve are

(i) The location of the straights and their intersection points.
(ii) The determination of the intersection angles.

This information may be supplied by the planning engineer, determined in the field by direct measurement, or determined indirectly from field measurements (e.g. a traverse).

It will also be necessary for some of the curve components (e.g. radius or tangent length, etc.) to be fixed, and these again will normally be supplied by the engineer or must be determined from traffic considerations or by site controls (e.g. property boundaries, etc.). This aspect of curve design will be dealt with in §10.6.

The present section assumes that the above data has been obtained, and will deal with the determination of the remaining components and the location of tangent points, junction points and, where necessary, subsidiary intersection points.

10.4.1. SIMPLE CIRCULAR CURVES

If the intersection angle β is fixed, the only 'free' components here are the radius R, the length of the curve L, and the two equal tangent lengths T_1I and IT_2.

It will be seen at once from figure 10.7 that $L = R\beta^c$ and that $T_1I = IT_2 = R\tan\frac{1}{2}\beta$ and the following examples should suffice to illustrate the necessary computations.

† G. J. Thornton Smith, *Survey Review*, Vol. 17, Jan 1963.

Curve Design and Setting Out 491

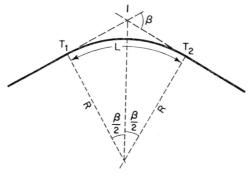

Figure 10.7

Example: Given: $\beta = 75°\ (= 1·309\ 00^c)$, $R = 1\ 000$ ft, chainage $I = 28 + 63·2$ (figure 10.7). Determine tangent length, curve length, and chainage of tangent points.

$$T_1I = IT_2 = R\tan 37\tfrac{1}{2}° = 1\ 000 \times 0·608\ 76 = 608·8 \text{ ft}$$
$$L = R\beta^c \simeq 1·000 \times 1·309\ 00 = 1\ 309·0 \text{ ft}$$

$$\begin{aligned}
\text{Chainage } I &= 28 + 63·2 \\
T_1I &= 6 + 8·8 \\
\hline
\text{Chainage } T_1 &= 22 + 54·4 \\
L &= 13 + 9·0 \\
\hline
\text{Chainage } T_2 &= 35 + 63·4 \text{ ft}
\end{aligned}$$

Example: Given (figure 10.7): $\beta = 75°\ (= 1·309\ 00^c)$, chainage $I = 28 + 63·2$ ft and chainage $T_1 = 24 + 36·6$ ft. Determine R, the length of curve, and the chainage of T_2.

$$\begin{aligned}
\text{Chainage } I &= 28 + 63·2 \\
\text{Chainage } T_1 &= 24 + 36·6 \\
\hline
T_1I &= 4 + 26·6
\end{aligned}$$

$R = 426·6/\tan 37\tfrac{1}{2}°$ $\text{Log } T_1I = 2·630\ 02$
$ = 556·0$ ft $\text{Log }\tan 37\tfrac{1}{2}° = 9·884\ 98$

$L = R\beta^c = 556 \times 1·309\ 00$ $\text{Log } R = 2·745\ 04$
$ = 727·8$ ft $\text{Log }\beta^c = 0·116\ 94$
         $\text{Log } L = 2·861\ 98$

$$\begin{aligned}
\text{Chainage } T_1 &= 24 + 36·6 \\
L &= 7 + 27·8 \\
\hline
\text{Chainage } T_2 &= 31 + 64·4
\end{aligned}$$

Example: Given (figure 10.7): $ABC = 149°\ 44'\ 40''$, $BCD = 137°\ 43'\ 40''$, $BC = 1\ 023·3$ ft. A circular curve of radius 850 ft is to be inserted, tangential to AB and CD. If the chainage of B is 30 + 15·6 ft, find the chainages of the initial and final tangent points.

$$\text{IBC} = 180° - 149° \ 44' \ 40'' = 30° \ 15' \ 20''$$
$$\text{ICB} = 180° - 137° \ 43' \ 40'' = 42° \ 16' \ 20''$$
$$\beta = 72° \ 31' \ 40''$$
$$= 1\cdot265 \ 85^c$$

$$\text{BI} = \frac{\text{BC} \sin \text{ICB}}{\sin \beta} \qquad \text{IC} = \frac{\text{BC} \sin \text{IBC}}{\sin \beta}$$

Log BC = 3·010 01
Log sin β = 9·979 49

	3·030 52		3·030 52
Log ICB =	9·827 79	Log IBC =	9·702 31
Log BI =	2·858 31	Log IC =	2·732 83
BI =	721·6 ft	IC =	540·5 ft

$$L = R\beta^c \qquad T_1I = IT_2 = R \tan \tfrac{1}{2}\beta$$

Log L   = 3·031 80   L = 1 076·0 ft
Log   = 0·102 38
Log R   = 2·929 42
Log tan $\tfrac{1}{2}\beta$ = 9·865 46
Log IT   = 2·794 88   IT = 623·6 ft

Chainage B = 30 + 15·6
BT = 7 + 21·6
Chainage I = 37 + 37·2
T₁I = 6 + 23·6
Chainage T₁ = 31 + 13·6
L = 10 + 76·0
Chainage T₂ = 41 + 89·6

10.4.2. COMPOUND CIRCULAR CURVES

As will be seen from figure 10.8 the centres, O_1 and O_2 lie on a straight line O_2J which is perpendicular to the common tangent PJQ. Note that in general O_2J does not pass through I.

Assuming the main intersection angle β to be fixed, there are eight remaining components, R_1, R_2, T_1I, T_2I, β_1, β_2, L_1 and L_2, and three of these must be fixed for the curve to be uniquely defined. Five equations must therefore be established in order to obtain the remaining components.

From triangle IPQ:

$$\beta = \beta_1 + \beta_2 \tag{1}$$
$$L_1 = R_1\beta_1^c \tag{2}$$
$$L_2 = R_2\beta_2^c \tag{3}$$

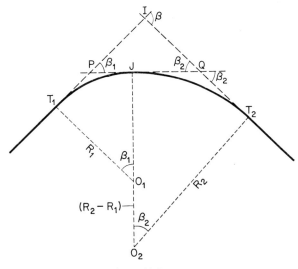

Figure 10.8

In the polygon $O_1O_2T_2IT_1O_1$, the algebraic sum of the projections of the sides on to any one side must be zero.

Considering first projection on to O_2T_2:

$$R_2 + 0 - IT_1 \sin \beta - R_1 \cos \beta - (R_2 - R_1) \cos \beta_2 = 0$$

whence $\quad IT_1 \sin \beta = R_2 - R_1 \cos \beta - (R_2 - R_1) \cos \beta_2$

(Note: That if β is greater than 90° its cosine will be negative so that the above equation is consistent.)

For the purposes of computation the above equation can be rearranged as follows:

$$\begin{aligned} IT_1 \sin \beta &= R_1(\cos \beta_2 - \cos \beta) + R_2(1 - \cos \beta_2) \\ &= R_1(1 - \cos \beta) - (1 - \cos \beta_2) + R_2(1 - \cos \beta_2) \\ &= R_1 \text{ versin } \beta + (R_2 - R_1) \text{ versin } \beta_2 \end{aligned} \quad (4)$$

Similarly, projecting on to O_1T_1:

$$R_1 - IT_2 \sin \beta - R_2 \cos \beta + (R_2 - R_1) \cos \beta_1 = 0$$

whence

$$\begin{aligned} IT_2 \sin \beta &= R_1 - R_2 \cos \beta + (R_2 - R_1) \cos \beta_1 \\ &= R_2 (\cos \beta_1 - \cos \beta) + R_1(1 - \cos \beta_1) \\ &= R_2[(1 - \cos \beta) - (1 - \cos \beta_1)] + R_1(1 - \cos \beta_1) \\ &= R_2 \text{ versin } \beta + (R_1 - R_2) \text{ versin } \beta_1 \end{aligned} \quad (5)$$

(Note: This equation can be obtained from equation (4) by transposing R_1 and R_2, and β_1 and β_2.)

These equations are not always convenient to solve, e.g. when the unknowns are R_1 and R_2, and the following approach produces equations which are more amenable under these circumstances.

In figure 10.9 O is the centre of an 'equivalent' circle of radius R, which is tangential to all three lines T_1P, PQ, and QT_2, the tangent points being K, M, and N.

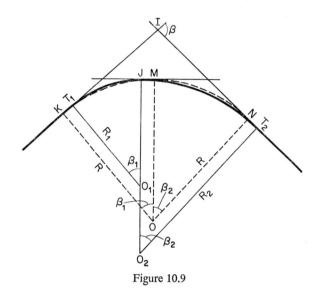

Figure 10.9

Then
$$KT_1 = JM = (R - R_1) \tan \tfrac{1}{2} \beta_1 \qquad (6)$$
$$NT_2 = JM = (R_2 - R) \tan \tfrac{1}{2} \beta_2 \qquad (7)$$

Whence
$$R(\tan \tfrac{1}{2}\beta_1 + \tan \tfrac{1}{2}\beta_2) = R_1 \tan \tfrac{1}{2}\beta_1 + R_2 \tan \tfrac{1}{2}\beta_2$$

or
$$R = \frac{R_1 \tan \tfrac{1}{2}\beta_1 + R_2 \tan \tfrac{1}{2}\beta_2}{\tan \tfrac{1}{2}\beta_1 + \tan \tfrac{1}{2}\beta_2} \qquad (8)$$

Also
$$T_1 I = IK - JM \qquad (9)$$
$$IT_2 = IN + JM \qquad (10)$$

where
$$IK = IN = R \tan \tfrac{1}{2}\beta$$

Example: Given figure 10.8: $\beta = 75°$, $\beta_1 = 30°$, $R_1 = 800$ ft, $R_2 = 1\,000$ ft, Determine β_2, L_1, L_2, $T_1 I$, and IT_2.

From (1) $\beta_2 = 75° - 30° = 45°$

From (2) and (3) $L_1 = \tfrac{1}{6}\pi \times 800 = 481 \cdot 2$ ft
$L_2 = \tfrac{1}{4}\pi \times 1\,000 = 785 \cdot 4$ ft

Curve Design and Setting Out   495

Using (4) and (5)

$$T_1I \sin 75° = 800 \text{ vers } 75° + 200 \text{ vers } 45°$$

$$\begin{aligned} 800 \times 0\cdot741\ 18 &= 592\cdot94 \\ 200 \times 0\cdot292\ 89 &= \underline{58\cdot58} \\ T_1I \sin 75° &= 651\cdot52 \end{aligned}$$

$$T_1I = \frac{651\cdot52}{0\cdot965\ 93} = 674\cdot5 \text{ ft}$$

$$IT_2 \sin 75° = 1\ 000 \text{ vers } 75° - 200 \text{ vers } 30°$$

$$\begin{aligned} 1\ 000 \times 0\cdot741\ 18 &= 741\cdot18 \\ 200 \times 0\cdot133\ 98 &= \underline{26\cdot38} \\ IT_2 \sin 75° &= 714\cdot38 \end{aligned}$$

$$IT_2 = \frac{714\cdot38}{0\cdot965\ 93} = 739\cdot6 \text{ ft}$$

Alternatively (figure 10.9)

From (8)

$$R = \frac{800 \tan 15° + 1\ 000 \tan 22\tfrac{1}{2}°}{\tan 15° + \tan 22\tfrac{1}{2}°}$$

$$\begin{array}{ll} 800 \times 0\cdot267\ 95 = 214\cdot36 & \quad 0\cdot26795 \\ 1\ 000 \times 0\cdot414\ 21 = \underline{414\cdot21} & \quad \underline{0\cdot41421} \\ 628\cdot57 & \quad 0\cdot68216 \end{array}$$

$$R = \frac{628\cdot57}{0\cdot682\ 16} = 921\cdot47 \text{ ft}$$

$$IK = IN = R \tan \tfrac{1}{2}\beta = 921\cdot47 \tan 37\tfrac{1}{2}° = 921\cdot47 \times 0\cdot767\ 33 = 707\cdot07 \text{ ft}$$
$$JM = (R - R_1) \tan \tfrac{1}{2}\beta_1 = (R_2 - R) \tan \tfrac{1}{2}\beta_2$$

$$\begin{array}{ll} 121\cdot47 \times 0\cdot267\ 95 & \quad 78\cdot53 \times 0\cdot414\ 21 \\ JM = 32\cdot53 & \quad JM = 32\cdot53 \end{array}$$

$$\begin{aligned} \underline{739\cdot6} &= IT_2 \\ 707\cdot07 & \\ \underline{32\cdot53} & \\ T_1I &= 674\cdot5 \end{aligned}$$

Example: Given: $\beta = 75°$, $\beta_1 = 30°$, $\beta_2 = 45°$, $T_1I = 600$ ft, $IT_2 = 700$ ft. Determine R_1 and R_2.

$$\overline{\begin{array}{l} 2IN = 1\cdot300 \\ IN + JM = 700 \\ IN - JM = 600 \\ \hline 2JM = 100 \end{array}} \quad \begin{array}{l} IN = 650 \text{ ft} \\ JM = 50 \text{ ft} \end{array}$$

$$R \tan 37\tfrac{1}{2}° = 650 \quad \therefore R = \frac{650}{0\cdot767\,33} = 847\cdot09 \text{ ft}$$

$$R - R_1 = \frac{JM}{\tan 15°}$$

$$\therefore R_1 = 847\cdot09 - \frac{50}{0\cdot267\,95} = 847\cdot09 - 186\cdot60 = 660\cdot5 \text{ ft}$$

$$R_2 = 847\cdot09 + \frac{50}{\tan 22\tfrac{1}{2}°} = 847\cdot09 + \frac{50}{0\cdot414\,21} = 847\cdot09 + 120\cdot71 = 967\cdot8 \text{ ft}$$

10.4.3. REVERSE CIRCULAR CURVES

From traffic considerations it is desirable to introduce a section of straight and transition curves between the branches of a reverse curve, so that the pure reverse curve is rarely used in practice. It must however, be considered, the treatment being the same in principle as that applied to compound curves.

It will be seen from figure 10.10 that the centres O_1 and O_2 lie on the straight line O_1JO_2 which is perpendicular to the common tangent PJQ, and O_1JO_2 does not in general, pass through the main intersection point I of the running-on and running-off tangents T_1P, and QT_2. Note that, I may be at some considerable distance (infinite if T_1P and QT_2 are parallel), from the curves, and will commonly be inaccessible. The value of the main intersection angle however will be known, being the difference in bearing between the lines PT_1 and QT_2.

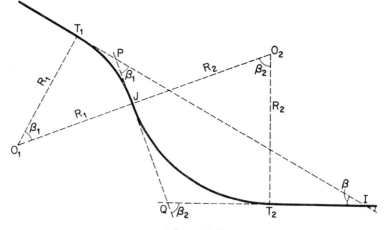

Figure 10.10

Curve Design and Setting Out

As for compound curves there are 8 remaining components, R_1, R_2, T_1I, T_2I, β_1, β_2, L_1, and L_2, of which three must be fixed to define a unique curve system, and again, five equations must be established to obtain the remaining components.

From triangle IPQ (figure 10.10), $\beta_2 = \beta_1 - \beta$ and in all cases β will be the difference (+ve) between β_1 and β_2, i.e.

$$\beta = (\beta_1 \sim \beta_2) \tag{1}$$

As before, also

$$L_1 = R_1 \beta_1^c \tag{2}$$

$$L_2 = R_2 \beta_2^c \tag{3}$$

As with compound curves, the sum of the projections of the sides of the polygon $O_1O_2T_1IT_2O_1$ on to any one side, will be zero, leading by an exactly similar process as that used in §10.4.2, to the equations

$$IT_1 \sin \beta = (R_1 + R_2) \operatorname{versin} \beta_2 - R_1 \operatorname{versin} \beta \tag{4}$$

and

$$IT_2 \sin \beta = (R_1 + R_2) \operatorname{versin} \beta_1 - R_2 \operatorname{versin} \beta \tag{5}$$

The alternative approach, utilizing the 'equivalent' circle, can also be used here, but the extra steps required to accommodate the method invalidate its usefulness.

One particular case of reverse curves, when the running-on and running-off tangents are parallel, requires special treatment as the standard equations are not applicable.

From figure 10.11 all the following relationships will be perfectly clear.

$$\beta_1 = \beta_2 = \beta \text{ say}$$
$$AJ = R_1 \operatorname{versin} \beta \qquad JB = R_2 \operatorname{versin} \beta$$
$$AB = (R_1 + R_2) \operatorname{versin} \beta$$
$$T_1A = R_1 \sin \beta \qquad BT_2 = R_2 \sin \beta$$
$$T_1C = T_2D = (R_1 + R_2) \sin \beta$$
$$T_1J = 2R_1 \sin \tfrac{1}{2}\beta \qquad JT_2 = 2R_2 \sin \tfrac{1}{2}\beta$$
$$T_1T_2 = 2(R_1 + R_2) \sin \tfrac{1}{2}\beta$$

and since $\sin \tfrac{1}{2}\beta = AB/T_1T_1$,

$$T_1T_2 = [2AB(R_1 - R_2)]^{\tfrac{1}{2}}$$

10.4.4. TRANSITION AND COMBINED CURVES

Once the rate of change of curvature of a transition curve (hereafter referred to as a 'spiral') has been fixed (by the methods discussed in §10.6), the only component remaining to be found to define the curve uniquely, is its length, consequently the components of combined curves will be investigated in this section.

In the discussion which follows and for which it will be necessary to refer to figure 10.12, the following notation will be used.

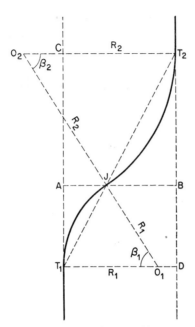

Figure 10.11

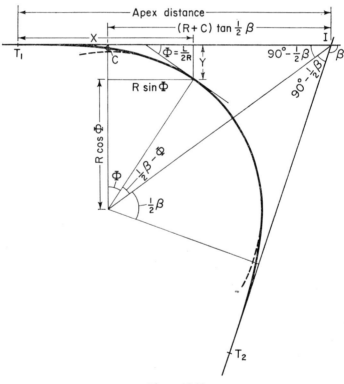

Figure 10.12

Curve Design and Setting Out

L is the total length of each spiral, R the radius of central circular arc (and, of course, the radius of curvature of the spiral at its junction with the circular arc), and c the shift (see §10.4.4.1). T denotes springing point of spiral from straight. J denotes junction point between spiral and circular arc. s, ϕ, x, y, θ, and l, all have the meanings assigned to them in §10.3.2 and the upper case letters, Φ, X, Y, and Θ refer to the particular values of ϕ, x, y, and θ at the point J, (i.e. when $s = L$).

To avoid confusion with the tangent lengths for purely circular curves, the distances IT_1, IT_2, etc. will be referred to as 'apex distances'.

In this section, the 'spiral' used will be the true transition curve, the clothoid, which, if the correction terms are ignored, becomes the cubic parabola. Other curves, which approximate to a transition curve, will be treated separately in §10.7. Some of the relationships developed below hold good whatever curve is used, and where this is so it will be indicated in the text.

10.4.4.1. Shift

For it to be possible to insert a spiral between a straight and a circular curve it is necessary to move the circular curve away from the straight by an amount known as the 'shift'. Similarly in order to insert a spiral between two branches of a compound curve it is necessary to move the circular curve with the smaller radius inwards, or the circular curve with the larger radius outwards, i.e. to separate the two curves. It should be noted that in this latter case the effect of inserting a spiral is to change the position of the common tangent. In the unusual circumstances that this is not practicable it will be necessary to substitute a completely new curve system.

For the single circular curve it will be seen at once from figure 10.12 that $R + c = R \cos \Phi + Y$, whence

$$c = Y - R(1 - \cos \Phi) \tag{1}$$

This holds good whatever spiral is used.

For the clothoid
From equation (4) §10.3.2.1

$$y = \frac{s^3}{6LR} - \frac{s^7}{336(LR)^3} + \frac{s^{11}}{42\,240(LR)^5} \text{ etc.}$$

so that, when $s = L$

$$Y = \frac{L^2}{6R} - \frac{L^4}{336R^3} + \frac{L^6}{42\,240R^5} \text{ etc.} \tag{2}$$

From equation (4) §10.3.2 $\phi = s^2/2LR$ so that $\Phi \simeq L/2R$. By expansion

$$1 - \cos \Phi = 1 - \left(1 - \frac{\phi^2}{2} + \frac{\phi^4}{24} - \frac{\phi^6}{720} \text{ etc.}\right)$$

whence

$$(1 - \cos \Phi) = \frac{L^2}{8R^2} - \frac{L^4}{384R^4} + \frac{L^6}{46\,080} \text{ etc.} \tag{3}$$

so that from (1), (2), and (3)

$$c = \left[\frac{L^2(8-6)}{R\ 48}\right] - \left[\frac{L^4(384-336)}{R^3\ 129\ 024}\right] + \left[\frac{L^6(46\ 080-42\ 240)}{R^5\ 1{\cdot}95\times 10^6}\right]$$

i.e.
$$c = \frac{L^2}{24R} - \frac{L^4}{2\ 688R^3} \quad \text{etc.} \qquad (4)$$

(Note. As LR is a fixed value for the curve this last expression is often more conveniently written:

$$c = L^3/24(LR) - L^7/2\ 668(LR)^3 \qquad (4a))$$

For shift between two branches of a compound curve see §10.4.5.1 post.

10.4.4.2. Apex Distance

For the single circular curve it will be seen at once from figure 10.12 that
$T_1 I = X + (R+c)\tan \tfrac{1}{2}\beta - R\sin\Phi$

i.e.
$$T_1 I = IT_2 \simeq (X - R\sin\Phi) + (R+c)\tan\tfrac{1}{2}\beta \qquad (1)$$

This holds good whatever spiral is used. For the clothoid
From equation (3) §10.3.2.1

$$x = s - s^5/40(LR)^2 + s^9/3\ 456(LR)^4 \quad \text{etc.}$$

so that when $s = L$,

$$X = L - L^3/40R^2 + L^5/3\ 456R^4 \quad \text{etc.} \qquad (2)$$

also
$$\sin\Phi = \Phi - \Phi^3/6 + \Phi^5/120 \quad \text{etc.}$$

and since $\Phi = L/2R$

$$\sin\Phi = L/2R - L^3/48R^3 + L^5/384R^5 \quad \text{etc.} \qquad (3)$$

so that from (2) and (3):

$$X - R\sin\Phi = L(1 - \tfrac{1}{2}) - \frac{L^3(48-40)}{R^3\ 1\ 920} + \frac{L^5(3\ 840 - 3\ 456)}{R^5\ 13\ 271\ 040}$$

$$\simeq \frac{L}{2} - \frac{L^3}{240R^2} + \frac{L^5}{34\ 560R^4} \quad \text{etc.} \qquad (4)$$

whence

$$T_1 I = IT_2 = \frac{L}{2} + (R+c)\tan\tfrac{1}{2}\beta - \frac{L^5}{240(LR)^2} + \frac{L^9}{34\ 560(LR)^4} \quad \text{etc.} \qquad (5)$$

10.4.4.3. Length of Combined Curve

It has been shown that the angle consumed by one (clothoid) spiral $= L/2R$ so that the angle consumed by two identical spirals $= 2\Phi = L/R$ leaving the angle available for the central circular curve as $\beta - L/R$. The length of the central circular arc is therefore

$$R(\beta - L/R)^c = R\beta^c - L \qquad (1)$$

Curve Design and Setting Out

and the total length of combined curve is

$$R\beta^c + L \qquad (2)$$

Example Illustrating Use of §§10.4.4.1–10.4.4.3: Referring to figure 10.12 and using the notation of the previous sections, assume $\beta = 75°$ ($= 1\cdot309\ 00^c$), $R = 500$ ft, $L = 400$ ft ($LR = 200\ 000$) and chainage $I = (28 + 63\cdot2)$ ft.

Required to find the position and chainage of the tangent points, and the chainage of the junction points.

$$\text{Shift } c = \frac{L^2}{24R} - \frac{L^7}{2\ 688(LR)^3}$$

$$= \frac{160\ 000}{24 \times 500} - \frac{16\ 384 \times 10^{14}}{2\ 688 \times 8 \times 10^{15}}$$

$$= 13\cdot333 - 0\cdot076 = 13\cdot26 \text{ ft}$$

(Note: Slide rule will suffice for the correction terms.)

$$\text{Apex distance} = \tfrac{1}{2}L + (R + c)\tan\tfrac{1}{2}\beta - \frac{L^5}{240(LR)^2}$$

$$= 200 + 513\cdot26 \tan 37\tfrac{1}{2}° - \frac{1\ 024 \times 10^{10}}{240 \times 4 \times 10^{10}}$$

$$= 200 + 393\cdot84 - 1\cdot07 = 592\cdot77 \text{ ft}$$

Length of central circular arc $= 500 \times 1\cdot309\ 00 - 400$

$$\simeq 654\cdot50 - 400 = 254\cdot50 \text{ ft}$$

$$\begin{aligned}
\text{Ch. I} &= 28 + 63\cdot2 \\
\text{IT}_1 &= \underline{5 + 92\cdot8} \\
\text{Ch. T}_1 &= 22 + 70\cdot4 \\
\text{T}_1\text{J}_1 &= \underline{4 + 00\cdot0} \\
\text{Ch. J}_1 &= 26 + 70\cdot4 \\
\text{Arc J}_1\text{J}_2 &= \underline{2 + 54\cdot5} \\
\text{Ch. J}_2 &= 29 + 24\cdot9 \\
\text{J}_2\text{T}_2 &= \underline{4 + 00\cdot0} \\
\text{Ch. T}_2 &= 33 + 24\cdot9
\end{aligned}$$

10.4.5. COMBINED CURVES INCORPORATING COMPOUND CURVES

10.4.5.1. *Shift*

It will be seen from the foregoing sections that the shift represents the movement of the centre of the circular curve away from the tangent. So, here the relative shift between the two branches represents the relative displacement of the two centres.

From figure 10.13 it is apparent that the distance between O_1 and O_2 in the x

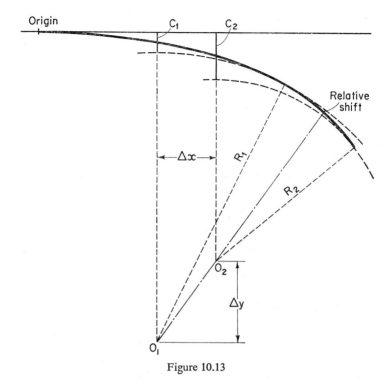

Figure 10.13

direction Δx is $(\tfrac{1}{2}L_2 - L_2^3/240R_2^2) - (\tfrac{1}{2}L_1 - L_1^3/240R_1^2)$ and that the distance between O_1 and O_2 in the y direction Δy is $(R_1 + c_1) - (R_2 - c_2)$ and therefore the shift is given by

$$c_3 = (\Delta x^2 + \Delta y^2)^{\frac{1}{2}} - (R_1 - R_2) \qquad (1)$$

However the length of spiral between two branches of a compound curve is usually short, and the shift small, consequently a higher degree of approximation can safely be used.

From equation (6) §10.5.8.2 the displacement of the spiral from the circular curve at distance $s \simeq s^3/6LR$ so that referring to figure 10.14, if l is the length of

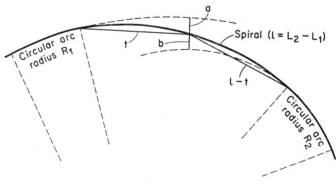

Figure 10.14

Curve Design and Setting Out 503

spiral to be inserted,
$$a = t^3/6LR \quad \text{and} \quad b = (l - t)^3/6LR \quad (1)\ \&\ (2)$$
whence
$$a + b = (l^3 - 3l^2t + 3lt^2)/6LR \quad (3)$$
But when a and b lie on O_1O_2 produced $a + b$ will be a minimum

i.e.
$$d(a + b)/dt = (-3l^2 + 6lt)/6LR = 0$$
whence, ignoring impracticable solutions, $t = l/2$

thus
$$a = b = l^3/48LR \quad (4)$$
and shift
$$c_3 = l^3/24LR \quad (5)$$

Although the LR value will normally be the same for all three spirals in the combined curve and so will have already been obtained, it is sometimes convenient to express it in a different form,

$$l = L_2 - L_1 = LR\left(\frac{1}{R_2} - \frac{1}{R_1}\right) = \frac{LR(R_1 - R_2)}{R_1R_2}$$

i.e.
$$LR = lR_1R_1/(R_1 - R_2)$$
whence
$$\text{Shift } c_3 = l^2(R_1 - R_2)/24R_1R_2 \quad (6)$$
(Cf. $L^2/24R$ where L is replaced by l and R by $R_1R_2/(R_1 - R_2)$.)

10.4.5.2. *Apex Distance*

By reference to figure 10.15 it will be seen that
$$IT_1 = IV + L_1/2 - L_1^3/240R_1^2 \quad (1)$$
and
$$IT_2 = IW + L_2/2 - L_2^3/240R_2^2 \quad (2)$$
and by direct comparison with §10.4.2 that
$$IV \sin \beta = (R_2 + c_2) - (R_1 + c_1) \cos \beta + (R_1 - R_2 - c_3) \cos \beta_2 \quad (4)$$
and $\quad IW \sin \beta = (R_1 + c_1) - (R_2 + c_2) \cos \beta - (R_1 - R_2 - c_3) \cos \beta_1 \quad (5)$

These formulae do not however lend themselves to transposition into versine form, or to the use of the equivalent circle method.

10.4.5.3. *Length of Combined Curve*

$$\text{Angle consumed by the first spiral} = L_1/2R_1 \quad (1)$$
and $\quad$ Angle consumed by the first half of the central spiral $= l/2R_1 \quad (2)$
$$\text{Angle consumed by the second spiral} = L_2/2R_2 \quad (3)$$
Angle consumed by the second half of the central spiral $= l/2R_2 \quad (4)$
whence $\quad$ Arc $J_1J_2 = (R_1B_1 - L_1/2 - l/2)^c \quad (5)$
and $\quad$ Arc $J_3J_4 = (R_2B_2 - L_2/2 - l/2)^c \quad (6)$
and Total length of the combined curve $= \frac{1}{2}(L_1 + L_2) + R_1\beta_1 + R_2\beta_2 \quad (7)$

Example Illustrating Use of §§10.4.5.1–10.4.5.3: Referring to figure 10.15 and using the notation of the foregoing sections, assume $\beta = 75°$, $\beta_1 = 30°$, $\beta_2 = 45°$. $R_1 = 800$ ft $L_1 = 500$ ft, $R_2 \simeq 1\ 000$ ft and $L_2 = 400$ ft. $L_1R_1 = L_2R_2 = 400\ 000$) and that the chainage of $I = 28 + 63·2$ ft.

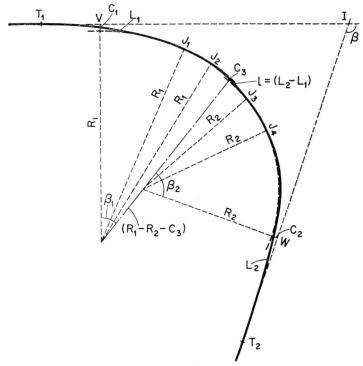

Figure 10.15

Required to find the positions and chainage of the tangent points and the chainage of the junction points.

$$c_1 = \frac{500^2}{24 \times 800} - \frac{625 \times 10^8}{2\,688 \times 512 \times 10^6}$$
$$= 13.02 - 0.05 \simeq \underline{12.97 \text{ ft}}$$

$$c_2 = \frac{400^2}{24 \times 1\,000} - \frac{256 \times 10^8}{2\,688 \times 10^9}$$
$$= 6.67 - 0.01 = \underline{6.66 \text{ ft}}$$

$$c_3 = \frac{200^3}{24 \times 400\,000} \simeq \frac{10}{96} = \underline{0.10 \text{ ft}}$$

$(R_1 + c_1) = 812.97 \qquad (R_2 + c_2) = 1\,006.66$

$R_2 - R_1 - c_3 = 1\,000 - 800 - 0.10 = 199.90$

$\sin 75° = 0.965\,93, \qquad \cos 75° = 0.258\,82, \qquad \cos 30° = 0.866\,03,$

$\cos 45° = 0.707\,11 \qquad 30° = 0.523\,60°, \qquad 45° = 0.785\,40°$

$0.965\,93 \text{ IV} = 1\,006.66 - 812.97 \times 0.258\,82 - 199.9 \times 0.707\,11$
$= 1\,006.66 - 210.41 - 141.35 = 654.90$

$\text{IV} = 654.90/0.965\,93 = \underline{678.00 \text{ ft}}$

Curve Design and Setting Out

$$0.965\ 93\ \text{IW} = 812.97 - 1\ 006.66 \times 0.258\ 82 + 199.90 \times 0.866\ 03$$
$$= 812.97 - 260.54 + 173.12 = 725.55$$
$$\text{IW} = 725.55/0.965\ 93 = \underline{751.14}\ \text{ft}$$

$$T_1V = 250 - 3\ 125 \times 10^{10}/(240 \times 16 \times 10^{10}) = 250 - 0.81$$
$$= \underline{249.19}\ \text{ft}$$

$$T_2W = 200 - 64 \times 16 \times 10^{10}/(240 \times 16 \times 10^{10}) = 200 - 0.27$$
$$= \underline{199.73}\ \text{ft}$$

$$T_1I = 678.00 + 249.19 = \underline{927.19}\ \text{ft}$$
$$IT_2 = 751.14 + 199.73 = \underline{950.87}\ \text{ft}$$

$$\text{Arc } J_1J_2 = 800 \times 0.523\ 60 - 250 - 50 = 418.88 - 300.0$$
$$= \underline{118.88}\ \text{ft}$$

$$\text{Arc } J_3J_4 = 1\ 000 \times 0.785\ 40 - 200 - 50 = 785.40 - 250$$
$$= \underline{535.40}\ \text{ft}$$

$$\begin{aligned}
\text{Ch. I} &= 28 + 63.2 \\
IT_1 &= \underline{9 + 27.2} \\
\text{Ch. } T_1 &= 19 + 36.0 \\
L_1 &= \underline{5 + 00.0} \\
\text{Ch. } J_1 &= 24 + 36.0 \\
J_1J_2 &= \underline{1 + 18.9} \\
\text{Ch. } J_2 &= 25 + 54.9 \\
l &= \underline{1 + 00.0} \\
\text{Ch. } J_3 &= 26 + 54.9 \\
J_3J_4 &= \underline{5 + 35.4} \\
\text{Ch. } J_4 &= 31 + 90.3 \\
L_2 &= \underline{4 + 00.0} \\
\text{Ch. } T_2 &= \underline{\underline{35 + 90.3}}
\end{aligned}$$

10.4.6. WIDE ROADS AND DUAL CARRIAGEWAYS

For roads having a large formation width, and in particular for dual carriageways, it will be necessary to set out the individual kerbs, embankment edges, etc. independently, the method of off-setting from the centre-line only being applicable to narrow roads. This makes no difference to the setting out of the curves themselves, but does affect the calculation of the curve components.

Two cases will be considered, that where the carriageway retains constant width throughout, and also the case where, to meet traffic engineering considerations, the carriageway is widened gradually along the spiral but maintains constant (increased) width round the circular portion.

In both cases, only the condition that the circular arcs remain truly concentric, and the normal arrangement for each curve of spiral–circular arc–spiral, will be considered. (It should be noted that the methods here employed have the effect that spirals used for each kerb utilize different speed (LR) values and that methods which use the same speed (LR) values are possible but do not, however satisfy the other conditions set out above.)

In the argument that follows the suffices 0, 1, and 2 relates to the centre-line, the inner kerb, and the outer kerb respectively, b is the carriageway width i.e. distance between inner and outer kerbs) and w the increase in width.

To illustrate the methods a simple numerical example will be carried through simultaneously.

10.4.6.1. Carriageway Constant Width

Let the centre-line radius R_0 be 800 ft, the centre-line spiral length L_0 be 500 ft (i.e. $L_0 R_0 = 400\ 000$) and the carriageway width b be 80 ft.

$$R_1 = R_0 - \tfrac{1}{2}b = 760 \text{ ft} \qquad (1)$$

$$R_2 = R_0 + \tfrac{1}{2}b = 840 \text{ ft} \qquad (2)$$

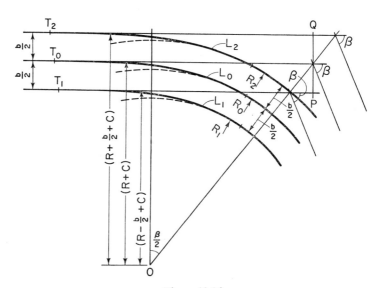

Figure 10.16

From figure 10.16 it will be clear that since the three circles are concentric and the difference in their radii is equal to the perpendicular distance between their tangents, the shifts for the three spirals must all be equal.

i.e. $\qquad\qquad c_1 = c_0 = c_2 = c$ (say) $\qquad\qquad (3)$

$$c = L_0^2/24R_0 - L_0^4/2688R_0^3$$

$$\simeq \frac{25 \times 10^4}{24 \times 800} - \frac{625 \times 10^8}{2\ 688 \times 512 \times 10^6} = 13\cdot02 - 0\cdot05 = 12\cdot97 \text{ ft} \qquad (4)$$

similarly

$$c = \frac{L_1^2}{24R_1} - \frac{L_1^4}{2\,688R_1^3} = 12\cdot 97 \text{ ft.}$$

This equation could of course be solved as a quadratic in L_1^2 but since the correction term is very small and only the *difference* in correction terms is involved it will be sufficiently accurate to say

$$L_1^2/24R_1 = L_0^2/24R_0 = 13\cdot 02$$

i.e.
$$L_1 = L_0(R_1/R_0)^{\frac{1}{2}} = 500(760/800)^{\frac{1}{2}} = 500 \times 0\cdot 95^{\frac{1}{2}}$$
$$= 500 \times 0\cdot 974\,68 = \underline{487\cdot 34 \text{ ft}} \qquad (5)$$

and
$$L_2 = L_0(R_2/R_0)^{\frac{1}{2}} = 500(840/800)^{\frac{1}{2}} = 500 \times 1\cdot 05^{\frac{1}{2}}$$
$$= 500 \times 1\cdot 024\,70 = \underline{512\cdot 35 \text{ ft}} \qquad (6)$$

It will also be seen from figure 10.16 that the distance

$$I_1P = QI_2 = \tfrac{1}{2}b \tan \tfrac{1}{2}\beta \qquad (7)$$

All the rest of the components are, of course calculated exactly as in the previous sections.

10.4.6.2. Carriageway Progressively Widened Along Spiral

Let the numerical data be the same as in the last section and the increase in width on the circular curve w be 4 ft, all the widening being applied on the inside of the curve.

The calculations for R_2 and L_2 will, of course, be identical to those in the last section, but now

$$R_1 = R_0 - (\tfrac{1}{2}b + w) = 800 - 44 = 756 \text{ ft} \qquad (1)$$

It will be seen from figure 10.17 that the centre line and inner circles are

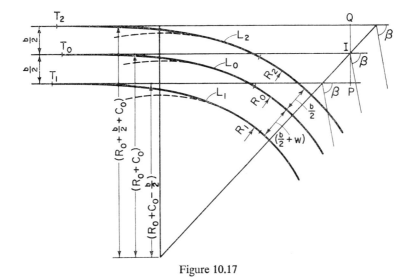

Figure 10.17

concentric, but that the difference in their radii is 4 ft greater than the perpendicular distance between their tangents, so that the shift for the inner circle is 4 ft greater than that for the centre-line circle, i.e.

$$c_1 = c_0 + w \qquad (2)$$

Once again the *difference* in the correction terms is involved, and so can be neglected, whence

$$\frac{L_1^2}{24R_1} = \frac{L_0^2}{24R_0} + w \sim (13\cdot02 + 4)$$

i.e. $\qquad L_1 = (24 \times 756 \times 17\cdot02)^{\frac{1}{2}} = (30\cdot9 \times 10^4)^{\frac{1}{2}} = 555\cdot9 \text{ ft} \qquad (3)$

and still $\qquad$ Distance $I_1 P = \frac{1}{2} b \tan \frac{1}{2} \beta \qquad (4)$

The widening (or part of it) could of course be applied on the outside of the curve in which case the shift for the outer curve would be given by

$$c_2 = c_0 - w \qquad (5)$$

but this will produce a very short spiral

$$L_2 = (24 \times 844 \times 9\cdot02)^{\frac{1}{2}} = (18\cdot64 \times 10^4)^{\frac{1}{2}} = 431\cdot7 \text{ ft} \qquad (6)$$

and a correspondingly small LR (speed) value (364 435 compared with 400 000) which is undesirable from the traffic engineering point of view.

In the extreme case when the widening is equal to the shift of the centre-line curve no spiral at all is possible (although it may still be highly desirable from traffic considerations), and if the widening exceeds the centre-line shift, some form of reverse curve will be required, which is ridiculous.

10.5. FIELD SETTING OUT

10.5.1. GENERAL

As has been explained at the beginning of the chapter, a system of coordinates is required for all setting out, the three most convenient being: rectangular coordinates, polar coordinates, and chords and deflection angles, and each of these will now be considered in detail.

In reading the ensuing sections, sight should not be lost of the fact, that for a circular curve of radius 100 ft or less, whose centre is accessible, no improvement can be made on swinging the chain about the centre.

Although in practice 'route' curves are nearly always 'combined' curves (see §10.2), in the following sections the setting out procedure for circular curves and spirals will be kept separate as they are in fact set out separately, and the procedure at the junctions between them will be dealt with in the sections relating to spirals.

10.5.2. CIRCULAR CURVES—RECTANGULAR COORDINATES

These methods are normally used for short curves, and are particularly useful for such situations as urban road intersections where the centres are inaccessible. They suffer from the disadvantage that the chords are not equal in length.

10.5.2.1. *Off-sets from the Tangent*

From figure 10.18, using the tangent point as the origin, the distance along the tangent as the X coordinate and the rectangular off-set as the Y coordinate,

$$(R - y)^2 = R^2 - x^2$$

whence
$$(R - y) = (R^2 - x^2)^{\frac{1}{2}}$$

and
$$y = R - (R^2 - x^2)^{\frac{1}{2}} \tag{1}$$

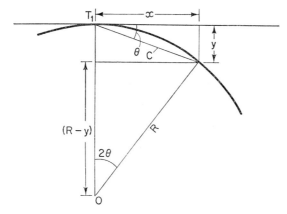

Figure 10.18

Also $\sin \theta = y/C = \frac{1}{2}C/R$, whence

$$y = C^2/2R \tag{2}$$

and if y/x is small

$$y \simeq x^2/2R \tag{3}$$

10.5.2.2. *Off-Sets from the Chord*

In addition to the uses mentioned above, this method is particularly useful for in-filling additional chord points on any circular curve once the main chord points have been established by curve ranging methods.

Referring to figure 10.19 and using the mid-point of the chord as origin, distances along the chord as the X coordinate, and rectangular off-sets as the Y coordinate.

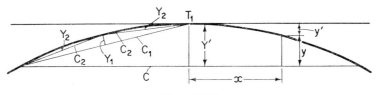

Figure 10.19

From equations (1) and (3), §10.5.2.1 $y' = R - (R^2 - x^2)^{\frac{1}{2}} \simeq x^2/2R$

whence $\qquad Y' = R - [R^2 - (\tfrac{1}{2}C)^2]^{\frac{1}{2}} \simeq C^2/8R$

and $\qquad y = Y' - y' = (R^2 - x^2)^{\frac{1}{2}} - [R^2 - (\tfrac{1}{2}C)^2]^{\frac{1}{2}} \simeq (C^2 - 4x^2)/8R \qquad (1)$

As a further approximation, commonly used by roadmen, $C_1 \simeq \tfrac{1}{2}C$ and

$$Y_1 \simeq (\tfrac{1}{2}C)^2/8R \simeq Y'/4 \qquad (2)$$

Similarly, $Y_2 \simeq Y_1/4$, $Y_3 \simeq Y_2/4$, etc., this process being referred to (obviously) as 'quartering'.

10.5.3.1. Circular Curve Ranging Using Chain and Tape Only

Establishing the tangent points

Referring to figure 10.20 peg any two points P and Q on IT_1 and IT_2 so that $IP = IQ$, measure PQ, bisect it, peg its mid-point M and measure IM.

$$IT_1(= IT_2) = R \tan \tfrac{1}{2}\beta = R.IM/MP \qquad (1)$$

and $\qquad PT_1(= QT_2) = (IT_1 \sim IP) \qquad (2)$

Measure off PT_1 and QT_2 and peg T_1 and T_2.

Ranging the curve

Referring to figure 10.21(a) where the chords are all equal to C, from equation (2) §10.5.2.1

$$AA' = C^2/2R \simeq (T_1A')^2/2R \qquad (1)$$

and since from equation (2) §10.3.1 $B'AB = T_1OA = 2\theta$, the isosceles triangles

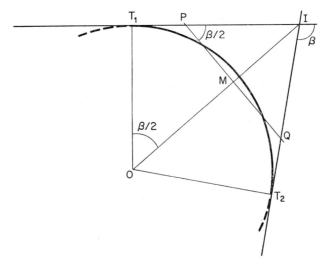

Figure 10.20

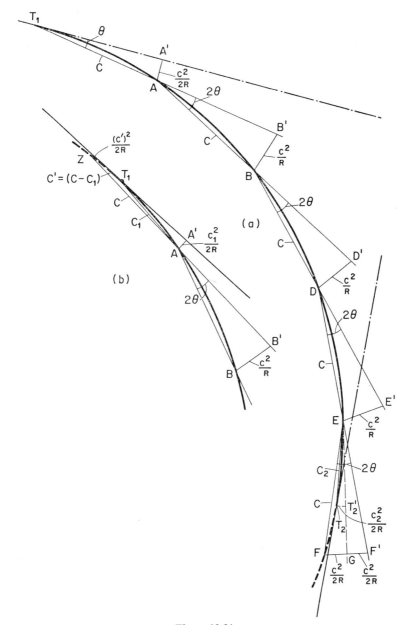

Figure 10.21

B'AB and T_1OA are similar so that

$$BB'/AB = T_1A/T_1O \quad \text{and} \quad BB' \simeq (T_1A \cdot AB)/T_1O = C^2/R \qquad (2)$$

similarly $C^2/R = BB' = DD' = EE' =$ etc.

To range the curve (preferably using a chord length chain), swing the stretched chain about T_1 until the perpendicular off-set A'A is equal to $C^2/2R$, and then peg A. Produce T_1A an additional chord length to B', mark B', and swing the stretched chain about A until AA' is equal to C^2/R.

Repeat this process for each chord until one chord point *beyond* T_2 is reached. (in this case F). Bisect FF' (at G), measure the final sub-chord $ET_2 = C_2$, stretch the chain along EG, which will be the tangent to the curve at E, and measure the perpendicular off-set $T_2'T_2$. If this off-set $T_2'T_2$ is equal to $C_2^2/2R$, the closure of the curve is checked.

This method is rapid and its accuracy is only limited by the accuracy of linear measurements although, of course, any errors will tend to be cumulative. A suitable chord length is dictated by the frequency with which pegs are required for the job in hand, 100 ft or 50 ft being commonly used. If intermediate points are required, e.g. for the mason or timberman's road pins, they can be inserted by off-sets from the main chords (see §10.5.2.2).

If it is required to maintain running chainage (see §10.1) an initial sub-chord C_1 will be required to make the chainage of the first chord point up to a round number and this calls for a modified procedure for establishing this point only, thereafter the procedure being as above.

If chainage is being preserved, and the curve is a long one, it must be remembered that the distance between each chord point will be in error by the difference between the chord and the arc, and either the chords must be reduced by the chord/arc correction, or the chord length kept sufficiently short for the correction to be insignificant. To introduce the initial sub-chord the following procedure is usually adopted.

From figure 10.21(b) if Z represents one full chord point *before* T_1, it will be seen that if it is sufficiently accurate to assume that the chord and the arc are equal then it will also be sufficiently accurate to assume that $C' = (C - C_1)$. A and Z can then be set out by the same procedure used before for A, but now using the chords C_1 and C' and the perpendicular off-sets $C_1^2/2R$ and $(C')^2/2R$ respectively. If ZA is now produced one full chord length to B' everything can now proceed as before.

10.5.3.2. *Circular Curve Ranging Using a Theodolite*

Establishing the tangent points

The first step in setting out the curve must be to establish the tangent points, which will involve measuring or deriving the intersection angle, calculating the tangent distances and pegging the tangent points, and determining their chainage. This has been fully covered in §10.4.

Ranging the curve

Throughout this section it will be assumed that running chainage is to be maintained, and that the theodolite is to be set up at one or other of the two tangent points. Cases where this latter is not possible or desirable are dealt with in §10.5.8 under the heading of 'obstructed curves'.

In figure 10.22 C_1 and C_2 are the initial and final subchords and all the remaining chords are equal to C. The deflection angles δ are referred to the running-on tangent T_1I, and P_1, P_2, etc. are the chord points.

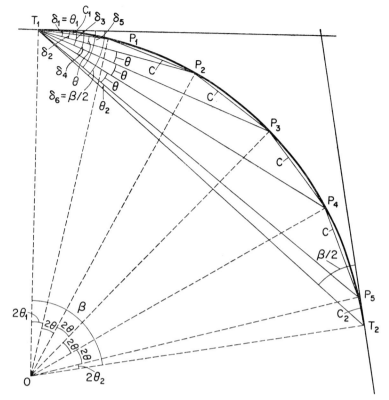

Figure 10.22

From equation (4) §10.3.1,

$$T_1OP_1 = \theta_1 = \sin^{-1} c_1/2R \tag{1}$$

$$P_1OP_2 = P_2OP_3 = P_3OP_4 = \text{etc.} = \theta = \sin^{-1} c/2R \tag{2}$$

$$P_nOT_2 = \theta_2 = \sin^{-1} c_2/2R \tag{3}$$

Note that if θ is small, then

$$\theta' \simeq 3\,437 \cdot 747 c/2R \simeq 1\,718 \cdot 9 c/R \tag{2a}$$

$$\theta_1 \simeq \theta c_1/c \tag{1a}$$

$$\theta_2 \simeq \theta c_2/c \tag{3a}$$

From equation (3) §10.3.1

$$\delta_1 = IT_1P_1 = \tfrac{1}{2}T_1OP_1 = \theta_1 \tag{4}$$

$$\delta_2 = IT_1P_2 = \tfrac{1}{2}T_1OP_2 = \theta_1 + \theta \tag{5}$$

$$\delta_3 = IT_1P_3 = \tfrac{1}{2}T_1OP_3 = \theta_1 + 2\theta \tag{6}$$

$$\delta_n = IT_1P_n = \tfrac{1}{2}T_1OP_n = \theta_1 + (n-1)\theta \tag{7}$$

$$\delta_{n+1} = IT_1T_2 = \tfrac{1}{2}T_1OT_2 = \theta_1 + (n-1)\theta + \theta_2 = \tfrac{1}{2}\beta \tag{8}$$

The fact that the total deflection angles is equal to half the intersection angle is useful as a check on the arithmetic of the calculations.

If the circular curve is defined by its 'degree' instead of its radius, and $c = 100$ ft (which is common) then as previously shown $\theta° = \frac{1}{2}D°$ and the calculation of the deflection angles is considerably simplified. This is the main advantage of the 'degree' method of definition.

From the above equations a table of deflection angles for setting out the curve can be prepared as in the following example.

Example: Two straights making a deflection angle of 75°, intersect at I (chainage 28 + 63·2 ft). A circular curve of 1 000 ft radius, deflecting left, is to be set out in 100 ft chords. Tabulate the setting-out data assuming that the instrument is set up at T_1 and that the initial reading of the horizontal circle when the instrument is bisecting I, is 214° 35′ 20″.

Tangent length = 1 000 tan $37\frac{1}{2}$ = 608·8 ft
Length of arc = 1 309·0 ft
Chainage T_1 = 22 + 54·4 ft
Chainage T_2 = 35 + 63·4 ft

Example of §10.4.1 refers.

$$\sin \theta = 100/2\ 000 = 0·05 \qquad \theta = 2° 51′ 58″$$

$$(\text{or } \theta \simeq (1\ 718·9/10) = 171·89′ = 2° 51′ 89′ = 2° 51′ 53″)$$

$$c_1 = 100 - 54·4 = 45·6 \text{ ft} \qquad c_2 = 63·4 \text{ ft}$$

$$\theta_1 = 171·89′ \times 45·6/100 = 78·38′ = 1° 18′ 23″$$

$$\theta_2 = 171·89′ \times 63·4/100 = 108·98′ = 1° 48′ 59″$$

When the tabulation (table 10.1) has been completed, the curve can be set out.

Once the instrument has been set up over the tangent point and the intersection point bisected, it is advisable to bisect the other tangent point, and measure the total deflection angle to check that it is in fact equal to half the intersection angle before proceeding any further. This having been done, lay-off the first deflection angle, stretch the chain a distance c_1 from T_1, line in the end of the chain and drive in a peg at P_1. Lay-off the second deflection angle, stretch the chain a full chord's length c from P_1, line in the end of the chain and drive in a peg at P_2. Proceed in this manner until peg P_n has been driven, and then measure the final sub-chord c_2. The amount by which this differs from the value of c_2 as calculated from $c_2 = 2R \sin \theta_2$ will be a measure of the accuracy with which the curve has been set out.

The accuracy of the method is obviously dependent on the care which is taken in chaining the chords and in lining in the pegs. If the ground is unsuitable for chaining, it may be preferable to use two theodolites, one at each tangent

Curve Design and Setting Out

Table 10.1

Chainage	Increase in deflection angle	Deflection angle	Angle to be set on instrument
22 + 54·4		—	214° 35' 20"
23 + 00·0	1° 18' 23"	1° 18' 23"	213° 17' 00"
24 + 00·0	2° 51' 53"	4° 10' 16"	210° 25' 00"
25 + 00·0	2° 51' 53"	7° 02' 09"	207° 33' 00"
26 + 00·0	2° 51' 53"	9° 54' 02"	204° 41' 20"
27 + 00·0	2° 51' 53"	12° 45' 55"	201° 49' 25"
28 + 00·0	2° 51' 53"	15° 37' 48"	198° 57' 30"
29 + 00·0	2° 51' 53"	18° 29' 41"	196° 05' 40"
30 + 00·0	2° 51' 53"	21° 21' 34"	193° 13' 40"
31 + 00·0	2° 51' 53"	24° 13' 27"	190° 22' 00"
32 + 00·0	2° 51' 53"	27° 05' 20"	187° 30' 00"
33 + 00·0	2° 51' 53"	29° 57' 13"	184° 38' 00"
34 + 00·0	2° 51' 53"	32° 49' 06"	181° 46' 20"
35 + 00·0	2° 51' 53"	35° 40' 59"	178° 54' 20"
35 + 63·4	1° 49' 01"	37° 30' 00"	177° 05' 20"

By subtraction $O_2 = 1° 49' 01"$, cf. $1° 48' 58"$

point, lay-off the deflection angles from both tangent points simultaneously, and locate the chord points by intersection. This of course, presupposes that the chord points are all visible from both tangent points. Where the ground is undulating and accurate lining in is difficult (whether using one theodolite or two), it will be preferable to change the position of the instrument as in the procedure for obstructed curves (see §10.5.8.2). This procedure is also necessary when the intersection angle between the ray and the chain becomes small. (i.e. a poor 'fix' (see figure 10.23)).

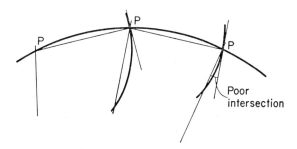

Figure 10.23

When the theodolite is set up at T_2 a fresh tabulation of deflection angles must be made, using, of course, the same values of θ, θ_1, and θ_2.

Chainage	Increase in deflection angle	Deflection angle
35 + 63·4		
	1° 48′ 58″	
35 + 00·0		1° 48′ 58″
	2° 51′ 53″	
34 + 00·0		4° 37′ 51″
	2° 51′ 53″	
33 + 00·0		7° 29′ 44″
etc.	etc.	etc.

10.5.4. CIRCULAR CURVE—POLAR COORDINATES

Circular curves can of course be set out using polar coordinates, with the tangent point as the origin, the tangent as the reference direction, the deflection angles as calculated in the last section as the amplitude, and the long chord (calculated from $l = 2R \sin \delta$) as the radius vector. This is obviously an uneconomical method because of the excessive amount of chaining involved, nevertheless it has its uses when normal curve ranging is impossible on account of an obstructed chord point and will be referred to again in §10.5.8.1—Obstructed curves.

10.5.5. TRANSITION CURVES—RECTANGULAR COORDINATES

The calculation of the rectangular coordinates for a spiral has been fully dealt with in §10.3.2.1, from which (to sufficient accuracy for most normal purposes), the distance along the tangent from $T_1 = x = s - s^5/40(LR)^2$ and the perpendicular off-set $= y = s^3/6LR - s^7/336(LR)^3$ and the field method of setting out (once the tangent points have been established) requires no further explanation.

As soon as the off-sets become sufficiently long for aids to be required to ensure its perpendicularity, the method becomes uneconomical and the method of chords and deflection angles becomes preferable. The method (even resorting if necessary to instrumental methods for ensuring the right-angle) may still be of value in dealing with obstructed curves.

Assuming, however that chain and tape only are being used, and that the circular curve is also to be set out with the same equipment, the procedure at the junction between the spiral and the circular curve must be considered.

It should be noted that the usual procedure is to set out each spiral independently from its own tangent point and then to run the circular curve in between, starting from the end point of one or other of the two spirals. It is of course possible to set out the spiral from the end point of the circular arc, the procedure for this being dealt with in §10.5.8.2—Obstructed curves.

Referring to figure 10.24, it will be seen that to set out the circular curve from J_1, the end of the spiral, it will be necessary to establish the common

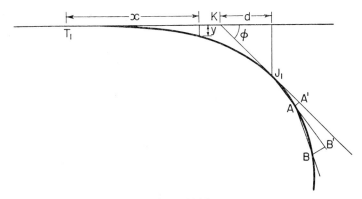

Figure 10.24

tangent to the two curves at J_1. This can be done by pegging the point K, the circular curve can then be set out in the normal manner (see §§10.5.2.1 and 10.5.3.1) from its tangent point J_1 and its tangent KJ_1 produced.

The X coordinate of K is $X - d$, where

$$d = Y \cot \Phi \qquad (1)$$

For the clothoid, $Y = L^3/6LR - L^7/336(LR)^3$, $\Phi = L/2R$, and (by expansion) $\cot \Phi = \Phi^{-1} - \Phi/3 - \Phi^3/45$, whence

$$d = \left(\frac{L^2}{6R} - \frac{L^4}{336R^3}\right)\left(\frac{2R}{L} - \frac{L}{6R} - \frac{L^3}{360R^3}\right)$$

$$d = \frac{L}{3} - \frac{17L^3}{504R^2} = \frac{L}{3} - \frac{17L^5}{504(LR)^2} \qquad (2)$$

i.e. for most practical cases where this method would be used, $d \simeq L/3$.

10.5.6. TRANSITION CURVES—POLAR COORDINATES

The calculation of the polar coordinates for the spiral has been dealt with in §10.3.2.2, but the same comments on its unsuitability as a field method of setting out as made for the circular curve in §10.5.4 apply. If it is used for obstructed curves, the procedure at the junction between the spiral and the circular arc is identical with that employed when the chord and deflection angle method of setting out the spiral is used as described in the next section.

10.5.7. TRANSITION CURVE RANGING USING A THEODOLITE

The method employed here, using chords and deflection angles is exactly the same as that described in section §10.5.3.2 for setting out a circular curve, except of course, for the calculation of the table of deflection angles. A few minor points of procedural detail also arise.

In the case of circular curves, the only limit to the length of chord employed is the effect of the chord–arc error on the chainage, and intermediate points on the curve can always be in-filled by off-sets from the chord.

The calculation of off-sets from the chord for a spiral is not a practicable proposition, as of course, they will be different for each chord, and so the chord length employed in setting out a spiral will depend on the frequency with which peg points are required on the curve. Chord lengths of 10, 15, 20, and at the most 25 ft, are commonly employed. This obviously has an effect on the procedure for preserving running chainage.

It will be seen later that inserting an initial sub-chord into a spiral makes the calculation of the table of deflection angles more laborious (although, of course quite possible) and so it is more common practice to set out the spiral in equal chords, inserting the round 100 ft chainage pegs, either by calculating additional deflection angles (only $\frac{1}{10}$ to $\frac{1}{4}$ of the work of keeping running chainage all through) or by inserting the additional points by one of the procedures described for obstructed curves. It is also possible to set out the curve from the first chord point (after the initial sub-chord has been set out), instead of from the tangent point. This procedure would, of course involve relying on one very short backsight for the whole of the curve which is likely to lead to considerably inaccuracy.

Once again it is usual practice to set out the two spirals independently, and then insert the circular arc in the centre from the end of one of the spirals. Normal chord length and normal procedure for keeping running chainage can of course be reverted to for the central circular arc. Also again, the procedure for setting out a spiral backwards from the end of the circular arc is described in §10.5.8.2—Obstructed curves.

10.5.7.2 Calculation of Deflection Angles

It has been shown in §10.3.2.2 that the deflection angle θ in minutes is given by the following expression

$$\theta' = (\phi/3)' - [0{\cdot}083\ 6(\phi°/3)^3 + 0{\cdot}000\ 005\ 5(\phi°/3)^5]'' \tag{1}$$

where ϕ in the correction term is in degrees and the correction term itself is in seconds, and where $\phi^c = s^2/2LR$.

It will be seen that if the deflection angles are required to the nearest 10" which is sufficient for most purposes, that the correction term does not become significant until θ is almost 5° which will embrace the whole of a number of practical spirals and certainly the early part of all spirals. It is obvious therefore that the deflection angles will be calculated from $\theta^c = s^2/6LR$, with a small correction being applied where necessary. As all the early deflection angles are small it is also most convenient to calculate θ in minutes, converting to degrees when it becomes sufficiently large, so that, for small angles,

$$\theta' = 3\ 437{\cdot}7s^2/6LR = 572{\cdot}96s^2/LR \tag{2}$$

The distance round the curve, s going to be set out as a series of equal chords, so that s can be replaced by nc, where c is the chord length and n is the number of chords, from which equation (2) becomes

$$\theta' = 572{\cdot}96n^2c^2/LR \tag{3}$$

This can be rearranged as $\qquad \theta' = n^2 572{\cdot}96c^2/LR \tag{4}$

Curve Design and Setting Out

and if c is referred to as the 'unit chord' and the first deflection angle (when $n = 1$) as the 'unit deflection angle' (θ_u), the equation to the curve for small angles becomes

$$\theta' = n^2 \theta_u' \tag{5}$$

and the equation for larger angles,

$$\theta = (n^2 \theta_u')' - [0.083\ 6(\tfrac{1}{3}\phi°/3)^3 + 0.000\ 005\ 5(\tfrac{1}{3}\phi°/3)^5]'' \tag{6}$$

Note that the main term is in *minutes* and the correction in *seconds*, whereas θ_u is in *minutes* in the main term, and $\tfrac{1}{3}\phi/3$ is in *degrees* in the correction terms.
e.g.
    If $LR = 100\ 000$, $R = 400$ ft, and $c = 25$ ft,

$$\theta_u = (572.96 \times 625/100\ 000)' = 3.581'$$

and for the eight deflection angle ($n = 8$, $s = 200$) as a first approximation

$$\theta = (64 \times 3.581)' = 229.184'$$
$$= 3.819\ 7° = 3° 49' 11''$$

and the correction is $(0.083\ 6 \times 3.819\ 7^3)'' = 4.7''$ so that $\theta = 3° 49' 06''$, or $3° 49' 10''$ to the nearest $10''$.

It will be obvious that, although straight forward, these calculations will be rather laborious if a large number have to be done, but it often happens in practice that there is no special reason, from an engineering standpoint, why a particular radius for the circular curve, or length of spiral must be used, and provided the radius and the LR value are not reduced below those required for the design speed (see §10.6), either L or R or both, may be modified to produce a 'round number' unit deflection angle, i.e. in the example above, if the LR value is increased to 119 367 then θ_u becomes exactly 3 minutes and the calculation for θ_8 becomes, as a first approximation,

$$\theta_8 \simeq (64 \times 3)' = 192' = 3° 12' = 3.2°, \quad 3.2^3 = 32.77$$

and the correction $0.083\ 6 \times 32.77 = 2.7''$ (note that in this range the slide-rule or even mental calculation $0.08 \times 33 = 2.6$ is more than adequate), giving θ_8 as $3° 12' 00''$ to the nearest $10''$.

It should be observed that decreasing the unit deflection angle will give a larger (and therefore safer) LR value, and also, of course, it must not be forgotten that any alterations to the LR value must be made *before* any calculations for curve components (apex distance, curve length, chainages, etc.) are made.

For the example taken, assuming that the radius remains at 400 ft, but that the curve length is increased to $119\ 367/400 = 298.42$ ft, and that the chainage of T_1 is $16 + 11.3$, the full tabulation would appear as in table 10.2.

Table 10.2 also illustrates how additional angles can be inserted in order to set out the even chainage pegs, using in each case a sub-chord of 13.7 ft.

The same table will of course serve for setting out the second spiral, backwards from T_2, except that the deflection angles for the even chainage pegs will have to be recalculated.

Table 10.2.

Chainage	Dist round curve	n	n^2	$n^2\theta_u$ (mins)	$n^2\theta_u$	θ^3	CN	Corrected angle
16 + 11.3	0	0	0	0	0	0	—	0' 00"
	25	1	1	3	3' 00"	—	—	3' 00"
	50	2	4	12	12' 00"	—	—	12' 00"
	75	3	9	27	27' 00"	—	—	27' 00"
	88.7	3.55	12.60	37.80	37' 48"	—	—	37' 50"
17 + 00	100	4	16	48	48' 00"	—	—	48' 00"
	125	5	25	75	1° 15' 00"	—	—	1° 15' 00"
	150	6	36	108	1° 48' 00"	—	—	1° 48' 00"
	175	7	49	147	2° 27' 00"	—	—	2° 27' 00"
	188.7	7.55	57.00	171.00	2° 51' 00"	23	2	2° 51' 00"
18 + 00	200	8	64	192	3° 12' 00"	33	3	3° 12' 00"
	225	9	81	243	4° 03' 00"	36	6	4° 02' 50"
	250	10	100	300	5° 00' 00"	125	10	4° 59' 50"
	275	11	121	363	6° 03' 00"	221	19	6° 02' 40"
19 + 00	288.7	11.55	133.40	400.20	6° 40' 12"	297	25	6° 39' 50"
19 + 09.7	298.4	11.94	142.56	427.68	7° 07' 41"	345	29	7° 07' 10"

10.5.7.3. Junction Between Spiral and Circular Curve

Referring to figure 10.25, it will be seen that to set out the circular curve from J_1, the end of the spiral, it will again be necessary to establish the common tangent to the two curves at J_1. This can be done by transferring the instrument to J_1, bisecting T_1, laying off the angle $\alpha = (\Phi - \Theta)$ and transiting the telescope, (or by laying off $(180° + \alpha)$), which will leave the instrument sighted along the common tangent. The deflection angles for the circular curve can then be added to the circle reading in the usual manner.

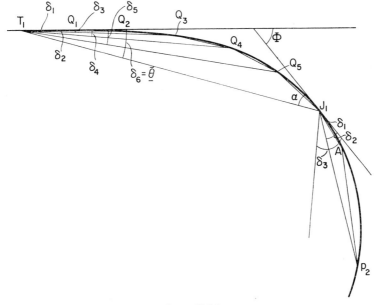

Figure 10.25

For the example used in the previous section, Θ from the table $= 7° 07' 10"$

$$\Phi = L/2R = 298 \cdot 42/800 = 0 \cdot 373\ 02° = 21° 22' 20".$$

so that $(\Phi - \Theta) = 14° 15' 10"$. (Compare this with $2\Theta = 14° 14' 20"$.)

10.5.8. OBSTRUCTED CURVES

Obstructions to curve ranging fall into two main categories.

(i) Obstructions on the line of sight of a deflection ray.
(ii) Obstructions on, or inaccessibility of the curve itself.

Both of these prevent the use of the setting out methods described so far.

In most cases the setting out can be achieved by the following general methods, used singly or in combination.

(i) Setting out a visible and accessible point beyond the obstruction by means of the deflection angle and long chord.

(ii) Setting out by normal chord and deflection angle methods from a point on the curve which has already been established.
(iii) Setting out by off-sets from subsidiary tangents or long chords which have already been established.
(iv) Setting out by radiation or intersection from known points either on the curve and its tangents or in extreme cases from known points remote from the curve.

The problems that can arise in ranging obstructed curves will often require individual solutions calling for ingenuity on the part of the surveyor and the use of his whole repertoire of setting out procedures.

The following examples illustrate only some of the methods that can be employed in the more common cases.

10.5.8.1. *Obstructions on the Curve, Negotiated by Use of Long Chord*

The curve has been set out from T_1 as far as possible (in figure 10.26, as far as chord point (2)). Chord points (3) and (4) are inaccessible.

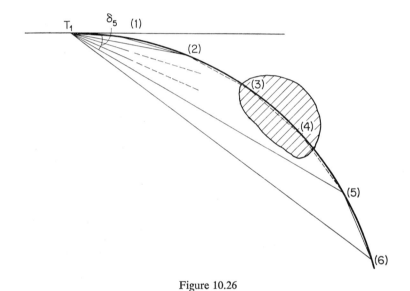

Figure 10.26

With instrument still at T_1, lay off deflection angle δ_5, set out distance $T_1(5)$, peg (5) and continue ranging curve in normal manner.

For circular curve
From equation (4) §10.3.1,
$$T_1(5) = 2R \sin \delta_5 \qquad (1)$$

For spiral
From equation (8) §10.3.2.2,
$$T_1(5) = s - s^5/90(LR)^2 + \text{etc.} \qquad (2)$$

Where $s = nc$, in this case, $5c$.

Curve Design and Setting Out

When the points (3) and (4) eventually become accessible, they can be filled in, either by off-sets from the chord (2)(5), or by setting out by chords and deflection angles from (2) or (5) (see §10.5.8.2 post).

10.5.8.2. Obstruction to the Line of Sight, Negotiated by Setting out from Point on the Curve

The curve has been set out from T_1 as far as possible (in figure 10.27 and 10.28, as far as chord point (3)). The rays to chord points (4), (5), (6), etc. are obstructed.

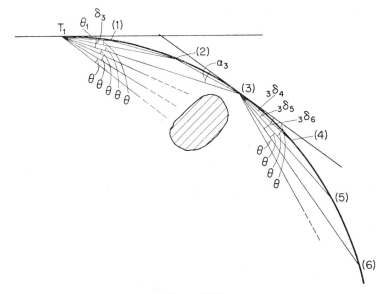

Figure 10.27

Transfer the instrument to chord point (3), bisect T_1, transit, and lay off angles $(\alpha_{3+3}\delta_4)$, $(\alpha_{3+3}\delta_5)$ to chord points (4), (5), etc, where α_3 is the back angle between the chord T_1 (3), and the tangent to the curve at (3), and $_3\delta_n$ is the deflection angle from the tangent at chord point (3) to the nth chord point.

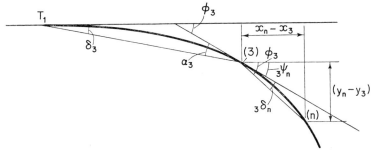

Figure 10.28

Provided the new deflection angles are used, the curve can now be ranged beyond chord point (3) in the normal manner.

Note that this procedure can be used whenever it is desired to move the instrument forward, whether obstructions are involved or not, and that if chord point (3) was the only chord point to be visible from T_1 it could have been established by means of the long chord $T_1(3)$, as in §5.8, and then the remainder of the curve (foreward and backward), set out from (3).

(N.B. For setting out backwards, bisect T_1, and lay off angles $(\alpha_3 - {}_3\delta_n)$).

Calculation of $(\alpha_3 + {}_3\delta_n)$ for Circular Curves: From figure 10.27 it will be clear that, since the lengths of the tangents to a circular curve are equal, $\alpha_3 = \delta_3 = (\theta_1 + 2\theta)$, and that ${}_3\delta_n = (n - 3)\theta$, so that

$$(\alpha_3 + {}_3\delta_n) = (\theta_1 + (n - 1))\theta = \delta_n \tag{1}$$

i.e. the angle to lay off to $(4) = \delta_3 + \theta = \delta_4$, the angle to lay off to $(5) = \delta_3 + 2\theta = \delta_5$, etc. so that in effect no calculation at all is required as the deflection angles to the various chord points will have been already tabulated. Similarly for setting out backwards from (3), the angle to lay off from (3) T_1 to chord point (2) is $\delta_3 - \theta = \delta_2$, etc.

Calculation of $(\alpha_3 + {}_3\delta_n)$ for Spiral: Unlike the previous section the calculation of the deflection angles requires a fuller investigation.

Back angles

It will be seen from figure 10.28 that

$$\alpha_3 = \phi_3 - \delta_3 \tag{1}$$

where $\phi_3 = (s_3)^2/2LR$ and $\delta_3 = [(s_3)^2/6LR]$ − Correction. In both of these expressions, if equal chords are being used, $s_3 = 3c$.

Deflection angles

It will be seen from figure 10.28 that

$${}_3\delta_n = {}_3\psi_n - \phi_3 \tag{2}$$

where ${}_3\psi_n$ is the angle between the chord $(3)(n)$ and the tangent to the spiral at the origin.

As before $\phi_3 = (s_3)^2/2LR$ and

$$\tan {}_3\psi_n = (y_n - y_3)/(x_n - x_3) \tag{3}$$

and from equations (7) and (8) §10.3.2.1

$$y = s^3/6LR - s^7/336(LR)^3 + \text{etc.}$$

and

$$x = s - s^5/40(LR)^2 + \text{etc.}$$

By substituting $s_3 = 3c$ and $s_n = nc$ in the above a table of deflection angles can be drawn up and the curve set out.

If, however, as is commonly the case, the angles ${}_3\delta_n$ are small, an approximation can be used, which considerably simplifies the calculations.

Denoting the arcual distance from chord point (3) to chord point (n) by nc, and assuming small angles,

$$\tan {}_3\psi_n \simeq {}_3\psi_n = \frac{y_n - y_3}{x_n - x_3}$$

$$= \frac{(s_3 + nc)^3/6LR - s_3^3/6LR}{(s_3 + nc) - s_3} \quad (4)$$

$$= \frac{3s_3^2 nc + 3s_3(nc)^2 + (nc)^3}{6LRnc}$$

$$= \frac{s_3^2}{2LR} + \frac{s_3(nc)}{2LR} + \frac{(nc)^2}{6LR}$$

but, as the product, length × radius is constant for the clothoid, $LR = s_3 R_3$ where R_3 is the radius of curvature of the spiral at the chord point (3), i.e. the radius of the osculating ('kissing') circle at (3), so that

$$_3\delta_n = {}_3\psi_n - \phi_3$$

$$= \frac{s_3^2}{2LR} + \frac{s_3 nc}{2R_3 s_3} + \frac{(nc)^2}{6LR} - \frac{s_3^2}{2LR} \quad (5)$$

$$_3\delta_n = \frac{nc}{2R_3} + \frac{(nc)^2}{6LR}$$

But $nc/2R_3 \simeq \sin^{-1}(nc/2R_3)$ = Deflection angle for setting circle of radius R_3 from its tangent, and $(nc)^2/6LR$ = Deflection angle for setting out spiral from the tangent at its origin, so that equation (5) can be expressed in words as follows:

'The deflection angle for a spiral from the tangent to the spiral is approximately equal to the sum of the deflection angles for the osculating circle and the deflection angle for a spiral having its origin at the instrument point'.

By a precisely similar analysis it can be shown that the deflection angle for setting the curve out backwards is given by the difference between the deflection angle for the osculating circle and that for a spiral having its origin at the instrument point.

A simpler way of expressing both of these relationships would be to say that the spiral deflects away from the osculating circle at the same rate as same spiral deflects away from the tangent at its origin.

From this it is clear that the off-set from the osculating circle is the same as the off-set from the tangent at the origin at the same distance from the origin, i.e.

$$y \simeq s^3/6LR \quad (6)$$

A numerical example, based on a spiral of length 298·42 ft and minimum radius 400 ft, for which the deflection angles for 25 ft chords are tabulated in §10.5.7.2 will serve to illustrate the use of equation (5).

526   *Practical Field Surveying and Computations*

The instrument is to be transferred to the peg at a distance of 200 ft round the curve, i.e. chord point (8), and the chord points (6) and (7) and from (9) to the end of the curve are to be set out from there.

The angles in column 6 (table 10.3) are the deflection angles from the tangent at (8), and those in column 7 are those to be laid off from chord (8) T_1 for chord points (6) and (7), and from $T_1(8)$ produced for the remainder.

Back angle $\alpha_8 = \phi_8 - \delta_8 \simeq 2\delta_8 = 2 \times 3° 12' = 6° 24' 00''$
Radius of osculating circle at (8) $= LR/200 = 596·84$ ft
Deflection angle for 25 ft chord on osculating circle
$= \sin^{-1}(25/2 \times 596·84) \simeq 25 \times 1\ 718·9/596·84 = 72$ mins
Deflection angle for 13·7 ft chord $= 0·55 \times 72 = 39·6$ mins
Deflection angle for 23·4 ft chord $= 0·94 \times 72 = 67·7$ mins

Table 10.3

1	2	3	4	5	6	7
Dist. round curve	Dist. from 8	n	Defln. for circle	Defln. for spiral	Total defln.	Angle from $T_1 8$
150	50	2	2° 24' 00"	12' 00"	2° 12' 00"	4° 12' 00"
175	25	1	1° 12' 00"	3' 00"	1° 09' 00"	5° 15' 00"
200	0	0	0	0	0	0
225	25	1	1° 12' 00"	3' 00"	1° 15' 00"	7° 39' 00"
250	50	2	2° 24' 00"	12' 00"	2° 36' 00"	9° 00' 00"
275	75	3	3° 36' 00"	27' 00"	4° 03' 00"	10° 27' 00"
288·7	88·7	3·55	4° 15' 36"	37' 48"	4° 53' 24"	11° 17' 20"
298·4	98·4	3·94	4° 43' 42"	46' 34"	5° 30' 16"	11° 54' 20"

The angles in column 5 are taken directly from the table in §10.5.7.2 except the last one which must be calculated, as follows,

$$n = 3·94, \quad n^2 = 15·52, \quad n^2\theta_u = 15·52 \times 3 = 46·56 \text{ mins}$$
$$= 46' 34''.$$

N.B. If (as in this case) the curve is to be set out in equal chords, and *there is no initial short chord*, the calculation of the deflection angle for the osculating circle can be even further simplified, so that the whole of the calculations for the new table can be made by mental arithmetic.

From equation (4) the expression for the deflection angle for the osculating circle is $s_3(nc)/2LR$, but if equal chords are used, $s_3 = 3c$, so that the angle becomes $3nc^2/2LR = 9nc^2/6LR = 9n\theta_u$ and equation (5) simplifies to

$$_3\delta_n = n9\delta_1 + \delta_n \tag{7}$$

For the numerical example above, where the curve is being set out from chord point (8), the deflection angle for the osculating circle becomes $8nc^2/2LR = 24nc^2/6LR = 24n\delta_1 = 72n$ mins $= n(1° 12' 00'')$.

10.5.8.3. Obstruction to Line of Sight, Negotiated by Moving Instrument along the Curve

This method is only an extension of the method described in the last section, with the difference that the backsight, instead of being made to T_1, is made to one of the previously established chord points. The only difference that this makes to the calculations is in the calculation of the back angle α.

Calculation of back angle for a circular curve

See figure 10.29. Instrument at chord point (5), back sight to chord point (3)

$$_5\alpha_3 = 2\theta \quad (= 5\theta - 3\theta)$$

The angle to lay off from (3)(5) produced for chord point $(n) = 2\theta + (n-5)\theta = n\theta - 3\theta$.

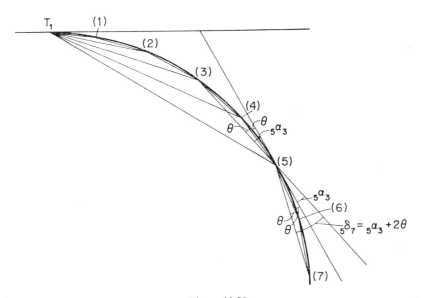

Figure 10.29

But $\delta_n = \theta_1 + (n-1)\theta$ and $\delta_3 = \theta_1 + (3-1)\theta$ so that $n\theta - 3\theta = \delta_n - \delta_3$, i.e. the angle to lay off from (3)(5) produced for the chord point (n), is the deflection angle from T_1 for chord point (n), minus the deflection angle from T_1 for the chord point to which the backsight is taken.

Calculation of back angle for spiral

See figure 10.30. Instrument at chord point (5), back sight to chord point (3). Back angle $_5\alpha_3$ is the deflection angle for setting out chord point (3) from the tangent to the curve at chord point (5) which from §10.5.8.2 is

$$(5-3)c/2R_5 - (5-3)^2c^2/6LR = (5-3)c/2R_5 - (5-3)^2\theta_u$$

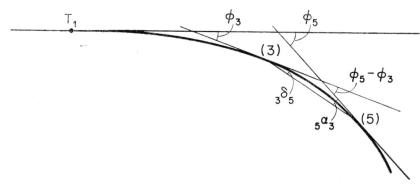

Figure 10.30

If there is no initial short chord, then

$$R_5 = LR/5c \quad \text{and} \quad (5-3)c/2R_5 = (5-3)5c^2/2LR$$
$$= (5-3)5 \cdot 3 \cdot c^2/6LR = (5-3) \cdot 5 \cdot 3 \cdot \theta_u$$

and $\quad _5\alpha_3 = [(5-3) \cdot 5 \cdot 3 - (5-3)^2]\delta_1$

For the numerical example of §10.5.8.2, with the instrument at (8) and the backsight to (2)

$$_8\alpha_2 = [(8-2) \cdot 8 \cdot 3 - (8-2)^2]\delta_1 - (144 - 36)\delta_1 = 108 \times 3 \text{ mins}$$
$$= 324' = 5°\ 24'\ 00''$$

Note that if the back sight is to T_1

$$_8\alpha_{T_1} = [(8-0) \cdot 8 \cdot 3 - (8-0)^2]\delta_1 = (192 - 64) \times 3 = 384' = 6°\ 24'\ 00''$$

which checks with the original calculation.

10.5.8.4. *Inaccessible or Obstructed Tangent Points*

When only one of the tangent points is obstructed the curve can be set out from the other end (or, of course, by one of the methods given below), but the problems of maintaining the chainage and the closure of the curve remain, and they will be dealt with first.

One tangent point only obstructed

The line of the initial tangent AI (figure 10.31) must be extended beyond the obstruction by any of the standard methods of chaining round an obstacle

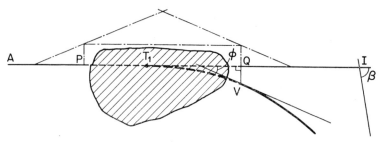

Figure 10.31

(see Chapter 9) and the chainages of I and P and Q (two accessible points on the tangent) recorded. When the intersection angle β has been measured and the curve data fixed, the tangent length (or apex distance) IT_1 can be calculated and hence the chainage of T_1 ($=$ Ch.I $-$ IT_1) and T_2 ($=$ Ch.T_1 $+$ Total length of curve) found. The distance T_1Q ($=$ Ch.Q $-$ Ch.T_1) can also be determined. When the curve has been set out (from T_2 or some other point on the curve, the point V can also be set out and the length of the perpendicular off-set QV measured. For closure of the curve this measured distance QV must agree with the value calculated for it from the length of T_1Q and the curve data. (For circular curve $QV \simeq (T_1Q)^2/2R$ and for spiral $QV \simeq (T_1Q)^3/6LR$.)

If it is the final tangent point that is obstructed a similar procedure is used except that in this case it will be the chainage of P that is required ($=$ Ch.T $+$ TP).

Both tangent points obstructed

Many different methods are possible for dealing with this situation, only three of which are now described.

(1) Determine the chainage of T_1, and the distances T_1Q and QV (figure 10.31) as in the previous section. The curve is then set out from V with VQ as the backsight, $90° + \phi$ being added to the deflection angles from the tangent at V.

For circle $\qquad \phi \simeq 2 \times$ Deflection angle from T_1 to Q

For spiral $\qquad \phi \simeq 3 \times$ Deflection angle from T_1 to Q

The method is very inaccurate if VQ is short.

(2) Determine the chainage of T_1 as before, and hence the chainage of K the mid-point of the curve (figure 10.32) ($=$Ch.T_1 $+$ $L/2$).

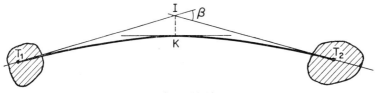

Figure 10.32

Set up at I, bisect the angle T_1IT_2 measure off IK and peg K.

For circle $\qquad IK = R(\sec \tfrac{1}{2}\beta - 1)$ (1)

For spiral $\qquad IK = (R + \text{Shift})(\sec \tfrac{1}{2}\beta - 1)$ (2)

The curve is set out backwards and forwards from K, using KI as the backsight and adding 90° to the deflection angles from the tangent at K.

Again the method is inaccurate if IK is short.

(3) Determine the chainage of T_1 and the distance PT_1 as before (figure 10.33). Set up at P and establish point M by means of the deflection angle γ

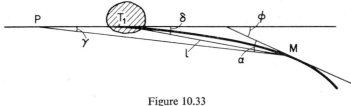

Figure 10.33

and the long chord PM. The chord T_1M $(= l)$ and the angle IT_1M $(= \delta)$ can be calculated from the curve data. Then

$$PM \sin \gamma = l \sin \delta$$

$$PM \cos \gamma = l \cos \delta + PT_1$$

whence
$$\tan \gamma = l \sin \delta / (1 \cos \delta + PT_1) \tag{3}$$

$$PM = l \sin \delta / \sin \gamma = (l \cos \delta + PT_1)/\cos \gamma \tag{4}$$

The curve is set out forwards and backwards if necessary, from M using MP as the backsight, and adding the back angle α $(= \phi - \gamma)$ to the deflection angles from the tangent at M.

If the angles γ for all the chord points are tabulated, the whole curve could of course be set out from P, once M has been pegged.

10.5.8.5. *Setting out from Points Remote from the Curve*

Where numerous obstructions exist, or where earthworks or construction plant prevent the use of any permanent ground marks on the curve itself, it may be more expedient to set out the curve from permanent stations remote from the curve. In such cases, once the calculations have been made and tabulated, the curve can be quickly re-established whenever required.

Although only two such permanent stations are shown in figure 10.34, in practice several will probably be required, not only to obtain visibility of the whole curve, but to obtain good 'fixes'.

The rectangular coordinates of all the chord points on the curve must be calculated, and the coordinates of the permanent stations (A, B, etc.) on the same axes obtained, usually by traversing. The curve can then be set out either by intersection, calculating the angles PAB and PBA for each chord point P,

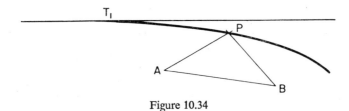

Figure 10.34

Curve Design and Setting Out 531

by radiation, calculating ray AP and angle PAB for each chord point, or, if the permanent stations are numerous and close to the curve, by tie lines, calculating rays AP and BP for each chord point.

10.6. DESIGN AND FITTING OF HORIZONTAL CURVES

It has already been pointed out that the design of road and railway layouts is the province of the specialist engineer, and space only permits general principles to be discussed here. It has also been shown that both circular curves and clothoid spirals are each defined uniquely if any two of their properties are fixed. For circular curves these two will usually be selected from the following: radius (or degree), length of arc, intersection angle, or tangent length. For spirals they will be selected from: minimum radius, length of curve, maximum spiral angle, shift. (Definition of spirals by minimum radius and shift has been demonstrated in §10.4.6 dealing with calculation of curve components for wide roads and dual carriageways.)

The combined curve, consisting of two identical spirals, and a central circular arc, is the case most commonly met in practice, and as purely circular curves are very much simpler to deal with, such a combined curve only will be assumed in general in this section.

On severely restricted sites curves may have to be designed from purely geometrical considerations, but in the main, design is based on traffic requirements for safety and comfort, possibly modified by aesthetic considerations or site restrictions.

In most practical cases, the overall intersection angle is predetermined by the general layout, and the problem finally resolves into determining a suitable radius for the circular arc, and length for the spiral, from the traffic viewpoint.

It will be obvious that from traffic considerations the largest possible radius, and longest possible spiral are desirable, but that some restriction will always arise from site conditions or cost. The first problems to tackle therefore, are the determination of suitable *minima* for the central radius and the length of the spiral for given traffic speeds.

As the speed of vehicles using a particular road or railway is a variable quantity, a 'design speed' must be defined, which is normally taken as the 85 percentile speed, i.e. the speed that will not be exceeded by 85% of the vehicles using, or expected to use, the road.

10.6.1. MINIMUM SAFE RADIUS FOR CIRCULAR CURVES

It will be appreciated that vehicles travelling round a curve will be subject to a centrifugal acceleration, which must be combated by the superelevation of the track, combined with the reaction of the rails on the wheel flanges, or the frictional force between the tyres and the road surface.

As in general, the track will have to accommodate vehicles travelling at a wide range of speeds, the superelevation (tan θ) cannot be so large that it will be uncomfortable for slow travelling or stationary vehicles. For railways, where the vehicles have a low centre of gravity, a maximum value of $\frac{1}{10}$ is often adopted, and for roads, 1/14·5, (= 0·069).

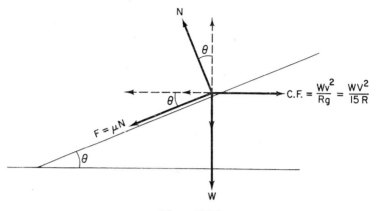

Figure 10.35

Referring to figure 10.35, resolving forces horizontally and vertically, and replacing v^2/Rg by $V^2/15R$ (where V is the speed in miles/hr),

$$WV^2/15R = N \sin \theta + F \cos \theta \tag{1}$$

$$W = N \cos \theta - F \sin \theta \tag{2}$$

Replacing F by $\mu N = N \tan \lambda$ (where μ is the coefficient of friction and λ the angle of friction), and dividing (1) by (2),

$$\frac{V^2}{15R} = \frac{\sin \theta + \cos \theta \tan \lambda}{\cos \theta - \sin \theta \tan \lambda}$$

and dividing top and bottom of the right-hand side by $\cos \theta$,

$$\frac{V^2}{15R} = \frac{\tan \theta + \tan \lambda}{1 - \tan \theta \tan \lambda}$$

or
$$\frac{V^2}{15R} = \tan(\theta + \lambda) \tag{3}$$

An ideal situation arises if for a given speed, $F = 0$, i.e. $\mu = 0$, and $\lambda = 0$, when $\tan \theta = V^2/15R$. e.g. for a speed of 40 miles/hr, and maximum superelevation of 1/14·5, $\quad 1/14\cdot 5 = 1\,600/15R$,

i.e.
$$R = 1\,600 \times 14\cdot 5/15$$
$$= 1\,547 \text{ ft}$$

In practice, some assistance from the frictional force can be expected, and in the U.K. at present (1967) use is made of the M.O.T. empirical formula,

$$\tan \theta = V^2/15R - 0\cdot 15 \tag{4}$$

e.g. for design speed 40 miles/hr, and maximum superelevation 1/14·5 ($= 0\cdot069$),

$$0\cdot 069 = (1\,600/15R) - 0\cdot 15$$

and minimum safe radius $R_m \simeq 1\,600/15 \times 0\cdot 22 \simeq 485$ ft

10.6.2. MINIMUM LENGTH OF SPIRAL

Various theories have been put forward for finding a suitable length of spiral, but the most widely accepted is one that limits the rate of change (with respect to time) of centrifugal acceleration to a value lying between $\frac{1}{2}$ ft/sec³ and 2 ft/sec³, depending on the degree of comfort required, a value of 1 ft/sec³ being most commonly used.

For the clothoid spiral the rate of change of curvature $(1/\rho)$ is constant, so that for constant speed, the rate of change of centrifugal acceleration $a = (v^2/\rho)$ will be constant, and equal to the change in acceleration divided by the time in which the change takes place,

i.e. $\qquad a = (v^2/R - 0)/t$

where R is the radius at the end of the curve, and t is the time taken by a vehicle travelling round the curve at constant speed v.

But, $t = L/v$, so that $a = v^3/LR$, or, if V is the speed in miles/hr,

$$LR = 3 \cdot 155 V^3/a \qquad (1)$$

It has already been shown (equation (6) §10.3.2), that the product of arc and radius LR, is constant for the spiral, and fixes the shape of the curve. From equation (1) §10.6.2, it is known as the 'speed value' of the curve.

Once the central radius R has been fixed (from §10.6.1 or otherwise), the minimum length of the curve can be found from the minimum LR value (or speed value) $3 \cdot 155 V^3/a$. e.g. Assuming $a = 1$ ft/sec³ for 40 miles/hr,

$$LR = 3 \cdot 155 \times 64\ 000 = 201\ 920 \simeq 202\ 000$$

If radius of central circular curve is 1 000 ft,

$$\text{Minimum length of spiral} = 202 \text{ ft}$$

If radius of central circular curve is the minimum for 40 miles/hr ($= 485$ ft, see §10.6.1)

$$\text{Minimum length of spiral} = 202\ 000/485 \simeq 416 \text{ ft}$$

Provided site conditions allow, it will always be preferable for both L and R to exceed the minima.

10.6.3. OTHER DESIGN CONSIDERATIONS

The intersection angle being assumed fixed, and the minimum values of R_m and (LR) having been found, the problem remaining is to find a combination of L and R to suit the particular case.

Two factors will influence the final choice of L and R.

10.6.3.1. *Where the Site is Unrestricted*

If there are no site restrictions whatsover, the ideal combined curve is one in which the length of the central circular arc is approximately equal to the length of each of the spirals. The angle consumed by one spiral $= L/2R$, the angle consumed by the circular arc is L/R, so that

$$\text{Intersection angle } \beta = L/2R + L/R + L/2R = 2L/R$$

i.e. $$L/R = \beta/2$$

e.g. Intersection angle $\beta = 30° = 0\cdot523\ 60°$, design speed $= 40$ miles/hr

From §10.6.1, $R_m = 485$ ft

From §10.6.2, Minimum value of $LR \simeq 202\ 000$

From above, $L/R = 0\cdot523\ 60/2 = 0\cdot261\ 80$

whence, $R^2 = 202\ 000/0\cdot261\ 8 \simeq 771\ 600$ and $R = 878$ ft

This is comfortably above the minimum value of 485 ft. Assuming we adopt this value for R, the minimum value for L will be $202\ 000/878 \simeq 230$ ft. Any convenient value, above 202 000, which will give a round number unit deflection angle (see §10.5.7.2) can now be chosen, $LR = 238\ 755$, giving $\theta_u = 1'\ 30''$ for a 25 ft chord would be suitable. Selecting $L = 250$ ft as giving an exact multiple of the chord length, $R = 238\ 733/250 = 954\cdot9$ ft, the angle consumed by the two spirals is $2 \times (L/2R) = L/R = 250/954\cdot9 = 0\cdot261\ 81°$, and the length of the central arc is $R(0\cdot523\ 60 - 0\cdot261\ 81) = 954\cdot9 \times 0\cdot261\ 79 = 250$ ft.

N.B. This almost exact equality of length of spiral and central arc is a coincidence in this case, for had we decided to make R the round number, e.g. 950 ft, then L would have been $238\ 733/950 = 251\cdot3$ ft, the angle consumed by two spirals, $L/R = 251\cdot3/950 = 0\cdot264\ 53°$, and the length of the central arc, $R(0\cdot523\ 60 - 0\cdot264\ 53) = 950 \times 0\cdot259\ 07 = 246\cdot1$ ft.

10.6.3.2. *Where the Site is Restricted*

The most common field restriction is inadequate apex distance. In figure 10.36 the straight BC must accomodate the apex distances for both the curves shown, with preferably at least 100 ft of straight between them.

In extreme circumstances, the central circular arc may have to be dispensed with entirely (see figure 10.37) in which case, the common radius of curvature of the two spirals at their junction point J, must of course, not fall below the minimum radius for the design speed.

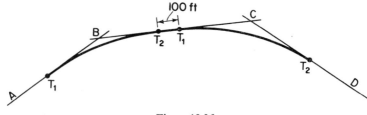

Figure 10.36

Omitting the central circular arc, and using the minimum radius and minimum value of LR will produce the shortest possible apex distance.

In this case, the angle consumed by the two spirals will exactly equal the intersection angle, i.e. $2(L/2R) = L/R = \beta^c$. A gain for maximum rate of change of centrifugal acceleration of 8 ft/sec³, design speed 40 miles/hr, and intersection angle 30°.

Making use of previous calculations, $LR = 201\,920$, and $R_m = 485$ ft. From above, $L/R = 0{\cdot}523\,60$.

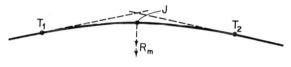

Figure 10.37

These figures are not compatible. If we omit the central arc, $L = 0{\cdot}523\,60R$, $R^2 = 201\,920/0{\cdot}523\,60 = 385\,638$, and $R \simeq 621$ ft which is above the maximum. $L = 201\,920/612 = 325{\cdot}2$ ft, the shift $\simeq L^2/24R = 325{\cdot}2^2/24 \times 621 = 7{\cdot}1$ ft. and since $\tan 15° = 0{\cdot}267\,95$, the apex distance $= 325{\cdot}2/2 + 268{\cdot}1 \times 0{\cdot}267\,95 = 162{\cdot}6 + 71{\cdot}8$ ft $= 234{\cdot}4$ ft.

If however we use the minimum radius the length of the spiral will be $201\,920/485 = 416{\cdot}3$, and the angle consumed by two spirals, $L/R = 416{\cdot}3/485 = 0{\cdot}858\,35^c = 49°$, which is greater than the intersection angle which is impossible.

If the space available for the apex distance is less than 234·4 ft, it will be necessary to rearrange the layout, or reduce the design speed, or increase the allowable rate of change of centrifugal acceleration.

10.7. OTHER MATHEMATICAL CURVES USED AS TRANSITION CURVES

Although the clothoid is the only mathematical curve which fulfills the requirements of a transition curve as defined in §10.3.2, it has already been shown that the cubic parabola approximates to the clothoid over a limited range, and has considerable advantages over the clothoid in respect of ease of calculation of components and setting out data.

It is obviously possible to find a number of other mathematical curves which will approximate to the clothoid over a limited range, and may have advantages in some respects. Various curves of this type are described by G. J. Thornton-Smith†.

The most commonly used of these alternatives, however, is the lemniscate, which has the polar equation, $l = C(\sin 2\theta)^{\frac{1}{2}}$, where l is the radius vector or long chord, θ the amplitude or deflection angle, and C a constant, the shape of the curve from $\theta = 0°$, to $\theta = 360°$ being sketched in figure 10.38. The main

---

† G. J. Thornton-Smith, *E.S.R.*, No. 117, Vol. XV, July 1960 and No. 120, Vol. XVI, April 1961.

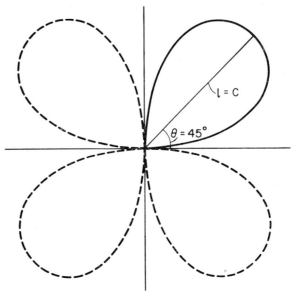

Figure 10.38

points influencing the possible choice of this curve in preference to the clothoid, are:

  (i) The rate of change of curvature whilst not being constant, decreases as the curvature increases, which is regarded by some engineers as being desirable for high speed roads.
 (ii) Once the constant C has been evaluated, polar and rectangular coordinates can be calculated easily and exactly, the determination of the corresponding distances round the curve however requiring the summation of a series which does not converge very rapidly. It does not lend itself easily to maintaining chainage round the curve, pegging equal chords, or setting out by chords and deflection angles.
(iii) As will be seen later, the spiral angle ϕ is exactly three times the deflection angle, which simplifies some of the calculations. This is not of as much help as it would seem due to the difficulty in finding chord lengths. The main advantage is that for a clover-leaf crossing, or similar circumstance where the intersection angle is 270° (i.e. the curve has to make a complete loop as in one quadrant of figure 10.38) since the spiral angle ϕ is exactly 135° when the deflection angle θ is 45°, the loop can consist of two identical spirals, back to back, without the necessity for the short piece of circular arc that would be needed if the clothoid were used.

To summarize the above, the lemniscate is of most value when the deflection angles are large (and the 'tail' terms in the various series used to calculate the component of the clothoid, would be required), and setting out can be conveniently accomplished by polar or rectangular coordinates, without the need for equal chords or running chainage.

10.7.1. CALCULATIONS FOR LEMNISCATE

The equation to the lemniscate (figure 10.39) is

$$l = C(\sin 2\theta)^{\frac{1}{2}} \tag{1}$$

For any curve expressed in polar coordinates,

$$\tan \alpha = l \, d\theta/dl \tag{2}$$

From (1), $\quad dl/d\theta = (2C \cos 2\theta)/2(\sin 2\theta)^{\frac{1}{2}}$

whence $\quad \tan \alpha = C(\sin 2\theta)^{\frac{1}{2}}(\sin 2\theta)^{\frac{1}{2}}/C \cos 2\theta = \tan 2\theta$

from which $\alpha = 2\theta$, and so $\quad \phi = 3\theta \tag{3}$

For any curve expressed in polar coordinates, the radius of curvature ρ at any point is given by

$$\rho = \frac{(l^2 + (dl/d\theta)^2)^{\frac{3}{2}}}{l^2 + 2(dl/d\theta)^2 - l \, d^2l/d\theta^2} \tag{4}$$

As before, $dl/d\theta = C \cos 2\theta/(\sin 2\theta)^{\frac{1}{2}}$, from which, $d^2l/d\theta^2 = -C(1 + \sin^2 2\theta)/(\sin 2\theta)^{\frac{3}{2}}$. Substituting these values in (4),

$$\rho = C/3(\sin 2\theta)^{\frac{1}{2}} = \tfrac{1}{3} \sin 2\theta \tag{5}$$

hence, $\quad C = 3\rho(\sin 2\theta)^{\frac{1}{2}} \tag{6}$

At the end of the curve, $\quad C = 3R(\sin 2\theta)^{\frac{1}{2}} \tag{6a}$

If x and y are the rectangular coordinates, as used throughout this chapter,

$$x = l \cos \theta \tag{7}$$
$$y = l \sin \theta \tag{8}$$

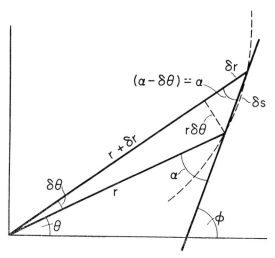

Figure 10.39

Example: Two straights, intersecting at 150° are to be connected by a combined curve consisting of a central circular curve of 900 ft radius and two identical lemniscate spirals, so that the length of the central arc is approximately the same as that of each of the spirals. $150° = 2·618 \ 0^c$. As shown in §10.6.3, if the three arcs are to be equal, the angle consumed by the circular arc will be roughly $\frac{1}{2}\beta$, and that consumed by each of the spirals, roughly $\frac{1}{4}\beta$, so that in this case we will take the maximum spiral angle Φ for one spiral to be 0·654 5^c ($= 37\frac{1}{2}°$), leaving 1·309^c ($= 75°$) for the central arc which will then have a length of $900 \times 1·309 = 1 \ 178·1$ ft.

From (6a),

$$C = 3R(\sin 2\theta)^{\frac{1}{2}} = 2 \ 700 \sin (2 \times 37\frac{1}{2}/3)^{\frac{1}{2}}$$
$$= 2 \ 700(\sin 25°)^{\frac{1}{2}} = 2 \ 700 \times 0·650 \ 1 = 1 \ 755·3 \text{ ft}$$

At the end of the curve,

$$l_m = 1 \ 755·3(\sin 25°)^{\frac{1}{2}} = 1 \ 755·3 \times 0·650 \ 1 = 1 \ 141·1 \text{ ft}$$
$$X = l_m \cos \theta = 1 \ 141·1 \cos 12\frac{1}{2}° = 1 \ 141·1 \times 0·976 \ 30 = 1 \ 114·1 \text{ ft}$$
$$Y = l_m \sin \theta = 1 \ 141·1 \sin 12\frac{1}{2}° = 1 \ 141·1 \times 0·216 \ 44 = 247·0 \text{ ft}$$

From equation (1) §10.4.4.1, the shift $= Y - R(1 - \cos \Phi)$

$$\cos \Phi = \cos 37\frac{1}{2}° = 0·793 \ 35, \ (1 - \cos \Phi) = 0·206 \ 65$$

and shift,

$$c = 247·0 - 900 \times 0·206 \ 65 = 247·0 - 186·0 = 61·0 \text{ ft}$$

From equation (1), §10.4.4.2,

$$\text{Apex distance} = (X - R \sin \Phi) + (R + c) \tan \frac{1}{2}\beta$$
$$R \sin 37\frac{1}{2}° = 900 \times 0·608 \ 76 = 547·9$$
$$\tan 75° = 3·732$$

and

$$\text{Apex distance} = (1 \ 114·1 - 547·9) + 961·0 \times 3·732 =$$
$$566·2 + 3 \ 586·4 = 4 \ 152·6 \text{ ft}$$

This enables the tangent points to be established.

The setting out data can now be tabulated, using equation (1) if polar coordinates are required, and equations (7) and (8) if rectangular coordinates are required. It is often useful to tabulate both.

Although equal chords are inconvenient to arrange, it is usually advantageous to make the pegs roughly equidistant. This can be achieved by selecting the values of θ to be tabulated, so that they increase proportionately to the square of the distance. In the example shown, the length of the curve is of the order of 1 200 ft so that 48 chords, each roughly 25 ft long could be used. The deflection angle for the first chord can then be found from $12·5 \times 3 \ 600/48^2 = 19·5$ secs, say 20 secs (it could also be found from $\sin 2\theta = (25/1 \ 755·3)^2$, or $\theta = (206 \ 265/2)(25/1 \ 755·3)^2 = 19·5$, say 20 secs.

These calculations to determine convenient values of θ, need not, of course be exact (except for the last peg), and as with the clothoid, the value of C can be

Table 10.4. Tabulation of Setting out Data for Lemniscate.

Peg No.	n^2	θ(secs)	2θ	$\sin 2\theta$	$(\sin 2\theta)^{\frac{1}{2}}$	$l = C(\sin 2\theta)^{\frac{1}{2}}$	$\cos \theta$	$\sin \theta$	$x = l\cos\theta$	$y = l\sin\theta$
0	0	0	00°00'00"	0	0	0	1·0	0	0	0
1	1	20	40"	0·00019	0·0138	24·2	1·0	0·0001	24·2	0·0
2	4	80	2'40"	0·00078	0·0279	49·0	1·0	0·0004	49·0	0·02
3	9	180	6'00"	0·00175	0·0418	73·4	1·0	0·0009	73·4	0·06
4	16	320	10'40"	0·00310	0·0557	97·8	1·0	0·0015	97·8	0·15
5	25	500	16'40"	0·00485	0·0696	122·2	1·0	0·0024	122·2	0·30
6	36	720	24'00"	0·00698	0·0835	146·7	1·0	0·0035	146·7	0·51
45	2025	40 500	22°30'00"	0·38268	0·61861	1085·8	0·98079	0·19509	1064·9	211·8
46	2116	41 320	23°00'00"	0·39073	0·62508	1097·2	0·97992	0·19937	1075·2	218·7
47	2209	44 180	24°32'00"	0·41522	0·64438	1131·1	0·97717	0·21246	1105·3	240·3
48			25°00'00"†	0·42262	0·65010	1141·1	0·97630	0·2644	1114·1	247·0

† Actual value of θ

adjusted to make θ_u a round number, provided the new value of C is used in all the calculations for curve components.

Should it be necessary, for any purpose, to find the actual length of a particular chord, it can always be calculated from the x and y coordinates.

For further information on transition curves, and in particular, on the lemniscate, the reader is referred to the following books:

F. G. Royal-Dawson, *Elements of Curve Design for Road,
Railway, and Racing Track*
F. G. Royal-Dawson, *Road Curves for Safe Modern Traffic*
F. G. Royal-Dawson, *Motorways, Flyovers and Mountain Roads*
R. B. M. Jenkins, *Curve Surveying*

10.8. VERTICAL CURVES

Whenever two gradients intersect on a road or railway, it is obviously necessary, from the traffic point of view, that they should be connected by a vertical curve, so that the vehicles can pass smoothly from one gradient to the next.

Although in practice road and rail gradients are comparatively flat and it is therefore relatively unimportant what type of curve (circular, parabolic, sinusoidal, etc.) is used, it is usually assumed that a curve having a constant rate of change of gradient (i.e. a parabola) is the most satisfactory, and it so happens that this is the easiest to calculate.

For modern high speed roads, some engineers prefer a curve in which the rate of change of gradient increases or decreases uniformly along the curve, but as roads designed to these standards have also very flat gradients, this is usually regarded as an unnecessary refinement. Such curves are dealt with briefly in §10.8.3, but otherwise throughout §10.8 the simple parabola only will be considered.

In modern road and railway practice, gradients are expressed as percentages, e.g. a $+4\%$ gradient is one in which the level rises 4 units vertically in 100 units *horizontally*, a positive gradient indicating a rising gradient, and a negative gradient indicating a falling gradient.

This method of representing gradients, not only has the advantage of simplifying the calculation of rises and falls in level (e.g. a $+4\%$ gradient 2·6 chains long rises $4 \times 2 \cdot 6 = 10 \cdot 4$ ft), but also, apart from a factor of 1/100, such gradients are identical with those used in coordinate geometry or calculus, e.g. if the X axis denotes the horizontal, and the Y axis the vertical, then a 4% gradient is a mathematical gradient of 0·04 or $dy/dx = 0 \cdot 04$. Even simpler, if all horizontal distances are measured in chains, and all vertical distances in feet, then a 4% gradient will be a mathematical gradient of 4 and $dy/dx = 4$. Although much used in practice, to prevent confusion, this notation will not be used in this text, and both vertical and horizontal distances will be assumed to be in feet, unless specifically stated otherwise.

Because of the well-known simple geometry of the parabola, many different methods are put forward for calculating the properties of vertical curves, but the approach used here is selected because it is felt that it is the simplest to understand and because it will readily yield a solution to any problem that arises. In some cases a quicker solution can be found to a particular problem by utilizing simple geometry, but the same method applied to another problem may become very involved.

10.8.1. TYPES OF VERTICAL CURVE

Vertical curves can take various different forms depending on the sign and magnitude of the intersecting gradients (figure 10.40). It will be observed that when there is a decrease in gradient, a curve convex upwards is formed, and when there is an increase in gradient a curve concave upwards is formed. The

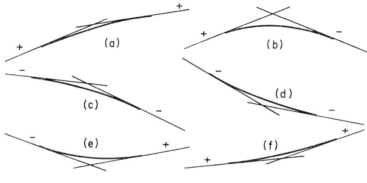

Figure 10.40

former are known as 'summit' curves, and the latter as 'valley' curves. It should be noted, however, that a true summit (highest point), or a true valley (lowest point), can only occur when there is a change of sign between the two gradients. It will be obvious that the gradient at the highest or lowest point is zero.

10.8.2. DERIVATION OF EQUATION TO CURVE AND CALCULATION OF SETTING OUT DATA—SIMPLE PARABOLA

Assume rectangular coordinates, with X axis horizontal and Y axis vertical in figure 10.41.

The basic requirement for the curve is that the rate of change of gradient

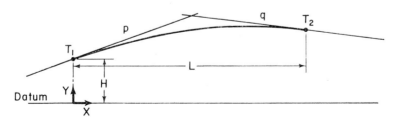

Figure 10.41

(with respect to distance) shall be constant. This can be expressed in two different ways.

(i) $$d^2y/dx^2 = K \tag{1}$$

(ii) Since it is constant, the rate of change of gradient is given by the change in gradient, divided by the distance in which it takes place. i.e. if the initial and final gradients dy/dx are denoted by p and q respectively, and the *horizontal* length of the curve by L,

$$(q - p)/L = K \tag{2}$$

The gradient at any point on the curve can be found from

$$dy/dx = \int (d^2y/dx^2)\, dx = \int K\, dx = Kx + A \tag{3}$$

If T_1 is taken as the origin for the X axis, then at T_1, $x = 0$, and $dy/dx = p$, whence $A = p$, and equation (3) becomes

$$dy/dx = Kx + p \tag{4}$$

or $$dy/dx = (q - p)x/L + p \tag{4a}$$

Note that when $x = L$, $q = KL + p$, i.e. $K = (q - p)/L$ (see equation (2) above).

The level of any point on the curve can be found from

$$y = \int (dy/dx)\, dx = \int (Kx + p) = \tfrac{1}{2}Kx^2 + px + B \tag{5}$$

If T_1 is taken as the origin for the Y axis, then at T_1, $x = 0$, $y = 0$, and so $B = 0$, but it is usually more convenient to take the level datum as the origin for the Y axis, so that when $x = 0$, $y = B = $ Reduced level of $T_1 = H$ say. In this case, equation (5) becomes

$$y = (q - p)x^2/2L + px + H \tag{5a}$$

This is an equation of the form $y = ax^2 + bx + c$ (a parabola with its axis vertical) where a is the $\tfrac{1}{2}$(Rate of change of gradient), b the initial gradient at T_1, and c the reduced level of T_1.

In evaluating these coefficients, care must be taken with their signs. The sign of b and c will be perfectly obvious in a practical problem, but the following will be of use in ensuring the correct sign for a.

The equation $y = bx + c(= px + H)$ is the equation to the straight line passing through the point T_1 and having a gradient p, i.e. the equation to the initial gradient tangent. If the curve is a summit curve it will fall away from this tangent and so the sign of a will be negative. If the curve is a valley curve, it will rise above the initial tangent and so the sign of a will be positive.

This will be demonstrated by numerical examples for each case illustrated in figure 10.40.

Figure 10.40(a) $p = +4\%$, $q = +3\%$, then $K = (3 - 4)/100L$ and
$$y = \frac{-x^2}{200L} + \frac{4x}{100} + H$$

Figure 10.40(b) $p = +4\%$, $q = -3\%$, then $K = (-3 - 4)/100L$ and
$$y = \frac{-7x^2}{200L} + \frac{4x}{100} + H$$

Figure 10.40(c) $p = -3\%$, $q = -4\%$, then $K = (-4 + 3)/100L$ and
$$y = \frac{-x^2}{200L} - \frac{3x}{100} + H$$

Figure 10.40(d) $p = -4\%$, $q = -3\%$, then $K = (-3 + 4)/100L$ and
$$y = \frac{x^2}{200L} - \frac{4x}{100} + H$$

Figure 10.40(e) $p = -4\%$, $q = +3\%$, then $K = (3 + 4)/100L$ and
$$y = \frac{7x^2}{200L} - \frac{4x}{100} + H$$

Figure 10.40(f) $p = +3\%$, $q = +4\%$, then $K = (4 - 3)/100L$ and
$$y = \frac{x^2}{200L} + \frac{3x}{100} + H$$

It will be clear from the above derivation of equations, that when the gradients have been fixed, only one other factor, either the horizontal length of the curve, or the rate of change of gradient, is required to define the curve uniquely. The choice of these factors is dealt with in §10.8.4, and for the remainder of the present section it will be assumed that the curve has been defined, and all that remains is to determine the setting out data.

Two properties of the parabolic vertical curve must be established to assist in calculating this data.

(1) In figure 10.42, let the horizontal distance from T_1 to I be l, and the equation to the curve be
$$y = (q - p)x^2/2L + px + H$$
Substituting L for x, the R.L. of T_2 becomes
$$\frac{(q - p)L^2}{2L} + pL + H = \frac{(p + q)L}{2} + H$$

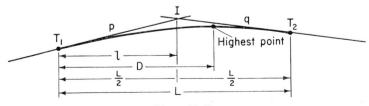

Figure 10.42

But the reduced level of T_2 is also given by

$$H + pl + q(L - l) = H + (p - q)l + qL$$

Equating these two expressions

$$\tfrac{1}{2}(p + q)L + H = H + (p - q)l + qL$$

whence
$$(p - q)l = \tfrac{1}{2}(p - q)L$$

i.e.
$$l = \tfrac{1}{2}L \qquad (6)$$

This very important relationship, that the horizontal distances from the tangent points to the intersection point are equal, is of considerable use in solving vertical curve problems, and yet is often overlooked.

(2) In figure 10.42, D denotes the distance to the highest point of the curve (or the lowest if the curve is a valley). (It should be noted that this will *not* coincide with l unless $p = -q$.)

Remembering that the gradient at the highest (or lowest) point is zero, and making use of the definition, that the rate of change of gradient (which is constant) is given by the change of gradient, divided by the distance in which the change takes place,

$$\frac{0 - p}{D} = \frac{q - 0}{L - D} = \frac{q - p}{L}$$

from which
$$D = \left(\frac{-p}{q - p}\right)L \qquad (7)$$

and
$$(L - D) = \left(\frac{q}{q - p}\right)L \qquad (8)$$

It will have been seen (figure 10.40) that a true summit (highest point) or true valley (lowest point) only occurs when there is a change of sign between p and q, so that the term $q - p$ in the above equations, will always be given by the *numerical* sum of the two gradients, and that the signs in the two equations are consistent.

e.g. Summit $\qquad p = +4\%, q = -3\%$.

$$D = \left(\frac{-4}{-3 - 4}\right)L = \left(\frac{4}{7}\right)L; \qquad L - D = \left(\frac{-3}{-3 - 4}\right) = \left(\frac{3}{7}\right)L$$

e.g. Valley $\qquad p = -4\%, q = 3\%$.

$$D = \left(\frac{4}{3 + 4}\right)L = \left(\frac{4}{7}\right)L; \qquad L - D = \left(\frac{3}{3 + 4}\right)L = \left(\frac{3}{7}\right)L$$

10.8.2.1. Location of Tangent Points

This is best illustrated by numerical examples.

Example:
A rising gradient of $+3\%$ is followed by a falling gradient of -4% (figure 10.43), and they are to be connected by a parabolic vertical curve of horizontal length 300 ft. The reduced level of a point A, at chainage $4 + 00$ on the first gradient is 96·57 ft, and the reduced level of a point B, at chainage $7 + 50$ on the

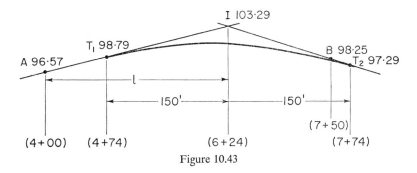

Figure 10.43

second gradient is 98·25 ft. Find the chainage and reduced levels of the tangent points.

Let the horizontal distance from A to the intersection point I be l ft, then, since distance AB = 350 ft,

$$\text{R.L.I} = 96 \cdot 57 + 3l/100 = 98 \cdot 25 + 4(350 - l)/100$$

i.e. $7l/100 = 98 \cdot 25 - 96 \cdot 57 + 14 \cdot 00 = 15 \cdot 68$

and $l = 1\,568/7 = 224$ ft and $(350 - l) = 126$ ft

$$\text{Ch.I} = (4 + 00) + 224 \text{ ft} = (6 + 24) \text{ ft}$$
$$\text{Ch.T}_1 = (6 + 24) - \tfrac{1}{2}L = (6 + 24) - 150 \text{ ft} = (4 + 74) \text{ ft}$$
$$\text{Ch.T}_2 = (6 + 24) + \tfrac{1}{2}L = (6 + 24) + 150 \text{ ft} = (7 + 74) \text{ ft}$$
$$L = 300 \text{ ft}$$

$$\text{R.L.I} = 96 \cdot 57 + 3 \times 224/100 = 96 \cdot 57 + 6 \cdot 72 = 103 \cdot 29 \text{ ft}$$
$$= 98 \cdot 25 + 4 \times 126/100 = 98 \cdot 25 + 5 \cdot 04 = 103 \cdot 29 \text{ ft}$$
$$\text{R.L.T}_1 = 103 \cdot 29 - 3 \times 150/100 = 103 \cdot 29 - 4 \cdot 50 = 98 \cdot 79 \text{ ft}$$
$$\text{R.L.T}_2 = 103 \cdot 29 - 4 \times 150/100 = 103 \cdot 29 - 6 \cdot 00 = 97 \cdot 29 \text{ ft}$$

Example: Points A and B, and C and D (figure 10.44), data for which are given below, lie on two intersecting gradients, which are to be connected by a parabolic vertical curve of horizontal length 500 ft.

Point	Chainage	Reduced level
A	12 + 00	81·83
B	14 + 00	77·33
C	17 + 00	89·33
D	20 + 00	98·93

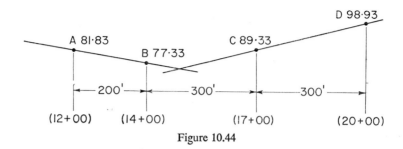
Figure 10.44

Determine the chainage and reduced level of the tangent points.
Distance AB = 200 ft, BC = 300 ft, CD = 300 ft.

$$\text{Gradient AB} = (77 \cdot 33 - 81 \cdot 83)/200 = -4 \cdot 50/200 = -2 \cdot 25\%$$
$$\text{Gradient CD} = (98 \cdot 93 - 89 \cdot 33)/300 = 9 \cdot 60/300 = 3 \cdot 2\%$$

Let distance from B to I be l, then,

$$\text{R.L.I} = 77 \cdot 33 = 2 \cdot 25 l/100 = 89 \cdot 33 = 3 \cdot 2(300 - l)/100$$

i.e.
$$5 \cdot 45 l/100 = 77 \cdot 33 - 89 \cdot 33 + 9 \cdot 60 = -2 \cdot 40$$

and
$$l = -240/5 \cdot 45 = -44 \cdot 0 \text{ ft, and } (300 - 1) = 344 \text{ ft}$$

(N.B. As l turns out to be negative, I lies to the left of B as shown in figure 10.45)

$$\text{Ch.I} = (14 + 00) - 44 \text{ ft} = (13 + 56) \text{ ft}$$
$$\text{Ch.T}_1 = (13 + 56) - 250 \text{ ft} = (11 + 06) \text{ ft}$$
$$\text{Ch.T}_2 = (13 \cdot 56) + 250 \text{ ft} = \underline{(16 + 06) \text{ ft}}$$
$$L = 500 \text{ ft}$$

$$\text{R.L.I} = 77 \cdot 33 + 2 \cdot 25 \times 44/100 = 77 \cdot 33 + 0 \cdot 99 = 78 \cdot 32$$
$$= 89 \cdot 33 - 3 \cdot 2 \times 344/100 = 89 \cdot 33 - 11 \cdot 01 = 78 \cdot 32$$
$$\text{R.L.T}_1 = 78 \cdot 32 + 2 \cdot 25 \times 250/100 = 78 \cdot 32 + 5 \cdot 62 = 83 \cdot 94 \text{ ft}$$
$$\text{R.L.T}_2 = 78 \cdot 32 + 3 \cdot 2 \times 250/100 = 78 \cdot 32 + 8 \cdot 00 = 86 \cdot 32 \text{ ft}$$

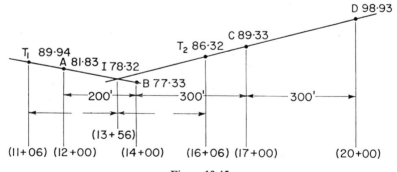
Figure 10.45

10.8.2.2. Tabulation of Reduced Levels on Curve

Once the chainage and reduced level of the tangent points have been determined, it remains to determine the values of spot levels along the curve. This is normally done for a series of equidistant points, usually referred to as 'chord' points, although the horizontal distances between them are in fact the horizontal components of the chords.

The calculations will be considerably simplified if the curve is set out in equal 'chords', and the total horizontal length of the curve is an exact multiple of the 'chord' length. In general, both these conditions can not be satisfied if running chainage is to be preserved, but a convenient compromise can be obtained if the length of the curve (subject to the design considerations discussed in §10.8.4) is chosen to make the final 'chord' only a short chord, as in the second of the examples that follow.

These two example are calculated on the basis of the formula previously developed (equation (5a) §10.8.2, but it should be noted that, for a parabola, second differences are constant, and this provides an alternative method of calculating the curve levels. For practical purposes, however, it is usually suitable to calculate levels to the nearest 0·01 ft but to achieve this by the difference method it is usually necessary to calculate the second differences to two further decimal places, so that very little time is saved, unless the second difference is exact to 0·01.

The calculation of the second differences after the calculations have been made from equation (5a) §10.8.2 acts as a valuable check, as is demonstrated in the two following examples.

Example 1: Running chainage preserved, but curve length *not* adjusted.

The levels will be calculated for the 50 ft even chainage points for the curve used in example I of §10.8.2.1 above. i.e., $p = +3\%$, $q = -4\%$, $L = 300$ ft: $Ch.T_1 = (4 + 74)$, $R.L.T_1 = 98·79$, $Ch.I = (6 + 24)$, $R.L.I = 103·29$, $Ch.T_2 = (7 + 74)$, and $R.L.T_2 = 97·29$.

Equation to curve is $y = ax^2 + bx + c$, where

$$a = (q - p)/2L = (-4 - 3)/200 \times 300 = -1·166\ 7/10\ 000$$
$$b = p = 3/100, \qquad c = H = 98·79$$

Ch.	x	x^2	ax^2	bx	ax^2+bx	R.L.	Δ^1	Δ^2
4+74	0	0	0	0	0	98·79		
5+00	26	676	−0·08	0·78	0·70	99·49	0·91	
5+50	76	5 776	−0·67	2·28	1·61	100·40	0·32	−0·59
6+00	126	15 876	−1·85	3·78	1·93	100·72	−0·26	−0·58
6+50	176	30 976	−3·61	5·28	1·67	100·46	−0·85	−0·59
7+00	226	51 076	−5·96	6·78	0·82	99·61	−1·43	−0·58
7+50	276	76 176	−8·89	8·28	−0·61	98·18		
7+74	300	90 000	−10·50	9·00	−1·50	97·29		

Example 2: Running chainage preserved, curve length adjusted to obviate initial short chord.

Same data as previous example, but curve length increased to make chainage of $T_1(4 + 50)$.

$$\text{Length of curve} = 2[(6 + 24) - (4 + 50)] = 2 \times 174 = 348 \text{ ft}$$
$$\text{Ch.}T_2 = (6 + 24) + 174 = 7 + 98$$
$$\text{R.L.}T_1 = 98 \cdot 79 - 3 \times 24/100 = 98 \cdot 79 - 0 \cdot 72 = 98 \cdot 07$$
$$\text{R.L.}T_2 = 97 \cdot 29 - 4 \times 24/100 = 97 \cdot 29 - 0 \cdot 96 = 96 \cdot 33$$

Equation to curve is

$$y = \left(\frac{-(4-3)}{200 \times 348}\right)x^2 + \frac{3x}{100} + 98 \cdot 07$$

$$y = \frac{-1 \cdot 005\ 7x^2}{10\ 000} + \frac{3x}{100} + 98 \cdot 07$$

As equal 50 ft chords are being used, x can be replaced by $50n$, where n is the number of chord lengths, and equation becomes,

$$y = -0 \cdot 251\ 4n^2 + 1 \cdot 5n + 98 \cdot 07$$

thus producing more manageable figures.

Ch.	n	n^2	an^2	bn	an^2+bn	R.L.	Δ^1	Δ^2
4+50	0	0	0	0	0	98·07		
5+00	1	1	−0·25	1·5	1·25	99·32	1·25	−0·51
5+50	2	4	−1·01	3·0	1·99	100·06	0·74	−0·49
6+00	3	9	−2·26	4·5	2·24	100·31	0·25	−0·51
6+50	4	16	−4·02	6·0	1·9	100·05	−0·26	−0·50
7+00	5	25	−6·28	7·5	1·22	99·29	−0·76	−0·51
7+50	6	36	−9·05	9·0	−0·05	98·02	−1·27	
7+98	6·96	48·44	−12·18	10·44	−1·74	96·33		

N.B. Distance to highest point = $(3/7) \cdot 348 = 149 \cdot 14$ ft
Chainage to highest point = $(5 + 99 \cdot 14)$

10.8.3. Derivation of equation to curve and calculation of setting out data—cubic parabola

See figure 10.46, and assume rectangular coordinates, with X axis horizontal and Y axis vertical. The basic requirement for the curve is that the increase (or decrease) of rate of change of gradient (with respect to distance) shall be constant, the rate of change increasing from the tangent points towards the centre of the curve. This means, of course, that a curve continuous from T_1 to T_2 cannot be used, but that two curves, back to back, and having a common tangent at their junction must be employed (similar to two transition curves

Curve Design and Setting Out   549

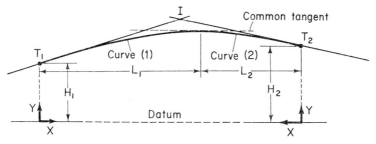

Figure 10.46

without any central circular arc). This means that when the curve has a true summit or valley (and there is no justification for using the cubic parabola unless this is the case), the common tangent occurs at the highest (or lowest) point of the curve, and is horizontal (i.e. $dy/dx = 0$).

Considering, to start with, only one curve from T_1 to the summit (or valley)

$$d^3y/dx^3 = \text{Constant} = M_1 \tag{1}$$

(It will subsequently be seen that a different constant M_2, will be required for the second curve.)

The rate of change of gradient at any point can be found from

$$d^2y/dx^2 = \int (d^3y/dx^3)\,dx = \int M_1 dx = M_1 x + A$$

but as the rate of change of gradient at T_1 is 0, $A = 0$.

$$d^2y/dx^2 = M_1 x \tag{2}$$

The rate of change of gradient reaches its maximum value K at the junction point at distance L_1 from T_1, so that

$$K = M_1 L_1 \tag{3}$$

or

$$M_1 = K/L_1 \tag{3a}$$

The gradient at any point can be found from

$$dy/dx = \int (d^2y/dx^2)\,dx = \int M_1 x\,dx = \tfrac{1}{2}M_1 x^2 + B$$

but since gradient at T_1, where $x = 0$, is p

$$dy/dx = \tfrac{1}{2}M_1 x^2 + p \tag{4}$$

At the junction point, where $x = L_1$, the gradient (for a true valley or summit) must be zero, so that, $0 = \tfrac{1}{2}M_1 L_1^2 + p = \tfrac{1}{2}KL_1 + p$, from which,

$$L_1 = -2p/K \tag{5}$$

and, from (3a),

$$M_1 = K/L_1 = -K^2/2p \tag{6}$$

The level at any point can be found from

$$y = \int (dy/dx)\,dx = \int (\tfrac{1}{2}M_1 x^2 + p)\,dx = M_1 x^3/6 + px + C \qquad (7)$$

and again, since $x = 0$ at T_1, $C = \text{R.L.T}_1 = H_1$, and, substituting for M_1,

$$y = (-K^2/12p)x^3 + px + H \qquad (7a)$$

This an equation of the form $y = ax^3 + bx + c$, i.e. a cubic parabola.

If the origin is transferred to T_2, and x measured from right to left instead of from left to right, the same equation will serve for the second curve if we substitute q for p, and M_2, L_2, and H_2 for M_1, L_1, and H_1.

(N.B. If the direction of the X axis is reversed in this manner, the gradient q in figure 10.46 is now *rising* from T_2 to I and so will be *positive*, i.e. for a true valley or summit, both gradients will have the *same* sign, unlike the case of the simple parabola where a continuous curve throughout is used.)

Note that the total length of the complete curve is $L = L_1 + L_2$, which from (5)

$$= -2(p + q)/K \qquad (8)$$

which is twice the length of the simple parabola for the same initial and final gradients and the same *maximum* rate of change of gradient (K).

It will be seen from equations (3) and (5), that, for first curve, $-K^2/M_1 = 2p$, for second curve, $-K^2/M_2 = 2q$. So that the same values of both K and M cannot be used for both curves (unless $p = q$). It is obviously preferable that K should be the same for both curves, and that it should be M that differs. Hence the use of M_1 and M_2, and L_1 and L_2 in the above equations.

10.8.3.1. Location of Tangent Points

The only way in which this differs from the procedure used for the simple parabola is in the calculation of the horizontal distance from the intersection point of the gradients to the tangent points, which in this case *unlike* the simple parabola, is *not* equal to $\tfrac{1}{2}L$.

Denoting horizontal distance $T_1 I$, by l and by direct calculation along the path, $T_1 I T_2$,

$$H_2 - H_1 = pl - (L_1 + L_2 - l)q = (p + q)l - q(L_1 + L_2)$$

whence
$$l = [(H_2 - H_1) + q(L_1 + L_2)]/(p + q)$$

By calculation round the combined curve,

$$H_2 - H_1 = (-K^2/12p)L_1^3 + pL_1 - (-K^2/12q)L_2^3 - qL_2$$

but, $L_1 = -2p/K$, and $L_2 = -2q/K$, so that,

$$H_2 - H_1 = 2p^2/3K - 2p^2/K - 2q^2/3K + 2q^2/K$$
$$= (-4/3K)(p^2 - q^2) = (-4/3K)(p - q)(p + q)$$

so that, since $L_1 + L_2 = -2(p + q)/K$,

$$l = -\tfrac{4}{3}K(p - q) - \frac{2q}{K} = -\frac{4p + 2q}{3K} \qquad (9)$$

Curve Design and Setting Out 551

and $$(L - l) = -(4q + 2p)/3K \qquad (10)$$

The use of these equations will be demonstrated in the numerical example in §10.8.3.2.

10.8.3.2. Tabulation of Reduced Levels on Curve

This can best be demonstrated, by a numerical example, using the same basic data (as far as applicable) as that used in example 1 in §10.8.2., i.e. $p = +3\%$, $q = -4\%$ (which becomes $+4\%$ as far as present example is concerned), Ch.I $= (6 + 24)$, R.L.I $= 103 \cdot 29$ ft, length of *simple parabolic curve* $= 300$ ft.

For *simple parabola* from equation (2) §10.8.2, $K = (-4 - 3)/(300 \times 100) = -7/30\,000$. This becomes the *maximum* rate of change of gradient to be used in this example. From equation (5), §10.8.3

$$L_1 = -2p/K = (-6/100)(30\,000/-7) = 1\,800/7 = 257 \cdot 1 \text{ ft}$$
$$L_2 = -2q/K = (-8/100)(30\,000/7) = 2\,400/7 = 342 \cdot 9 \text{ ft}$$
$$L = \overline{600 \text{ ft}}$$

From equations (9) and (10) §10.8.3

$$l = -(4p + 2q)/3K = [(12 + 8)/300](30\,000/7) = 2\,000/7$$
$$= 285 \cdot 7 \text{ ft}$$
$$L - l = +(4q + 2p)/3K = [(16 + 6)/300](30\,000/7) = 2\,200/7$$
$$= 314 \cdot 3 \text{ ft}$$
$$L = \overline{600 \text{ ft}}$$

$$\text{Ch.T}_1 = (6 + 24) - 285 \cdot 7 \text{ ft} = 3 + 38 \cdot 3$$
$$\text{Ch.T}_2 = (6 + 24) + 314 \cdot 3 \text{ ft} = 9 + 38 \cdot 3$$
$$L = 600 \text{ ft}$$

$$\text{R.L.T}_1 = 103 \cdot 29 - 3 \times 285 \cdot 7/100 = 103 \cdot 29 - 8 \cdot 57 = 94 \cdot 72 \text{ ft}$$
$$\text{R.L.T}_2 = 103 \cdot 29 - 4 \times 314 \cdot 3/100 = 103 \cdot 29 - 12 \cdot 57 = 90 \cdot 72 \text{ ft}$$

From equation (6) §10.8.3

$$M_1 = -K^2/2p = -(49 \times 100)/(9 \times 10^8 \times 2 \times 3) = -0 \cdot 907\,41/10^6$$
$$\text{and} \quad a_1 = M_1/6 = -0 \cdot 151\,23/10^6$$
$$M_2 = -K^2/2q = -(49 \times 100)/(9 \times 10^8 \times 2 \times 4) = -0 \cdot 689\,56/10^6$$
$$\text{and} \quad a_2 = M_2/6 = -0 \cdot 113\,43/10^6$$

From equations (7) and (8) §10.8.3, equation to curve (1) is

$$y = -0 \cdot 15\,123x^3/10^6 + 3x/100 + 94 \cdot 72$$

Equation to curve (2) is

$$y = -0 \cdot 11\,343x^3/10^6 + 4x/100 + 90 \cdot 72$$

Neglecting running chainage (for simplicity) and calculating levels for 50 ft chords, equations can be simplified by substituting $x = 50n$

$$y = -0 \cdot 018\,90n^3 + 1 \cdot 5n + 94 \cdot 72 \qquad (1)$$
$$y = -0 \cdot 014\,18n^3 + 2n + 90 \cdot 72 \qquad (2)$$

In the table that follows, the calculations are made downwards from T_1 to the junction point, and upwards from T_2 to the junction point.

Ch.	Dist.	n	n^3	a_1n^3	pn	$pn - a_1n^3$	R.L.
3+38·3	0	0	0	0	0	0	94·72
3+88·3	50	1	1	0·02	1·5	1·48	96·20
4+38·3	100	2	8	0·15	3	2·85	97·57
4+88·3	150	3	27	0·15	4·5	3·99	98·71
5+38·3	200	4	64	1·21	6	4·79	99·51
5+88·3	250	5	125	2·36	7·5	5·14	99·86
5+95·4	257·1	5·14	136·0	2·57	7·71	5·14	99·86
5+95·4	342·9	6·86	322·4	4·57	13·71	9·14	99·86
6+38·3	300	6	216	3·06	12	8·94	99·66
6+88·3	250	5	125	1·77	10	8·23	98·95
7+38·3	200	4	64	0·91	8	7·09	97·81
7+88·3	150	3	27	0·38	6	5·62	96·31
8+38·3	100	2	8	0·11	4	3·89	94·61
8+88·3	50	1	1	0·01	2	1·99	92·71
9+38·3	0	0	0	0	0	0	90·72
Ch.	Dist.	n	n^3	a_2n^3	qn	$qn - a_2n^3$	R.L.

10.8.4. CHOICE OF DESIGN CONSTANTS FOR VERTICAL CURVES

It has been shown, that once the value of the two intersecting gradients, and the position (vertically and horizontally) of their intersection point have been fixed, only one further property is required to define a vertical curve uniquely. For comparison purposes, the most convenient additional property to fix is the (horizontal) length.

The determination of a suitable length for a particular vertical curve is normally the responsibility of the designer, but it is felt that at least the principles influencing the choice have a proper place in this volume. Where actual recommendations are made, they are based on current practice in the U.K. in 1967, and may not be applicable in circumstances where different traffic conditions prevail.

It will be appreciated, that where the length of the curve is derived from traffic considerations, these will be *minimum* lengths, and the actual curves will usually be longer, even if only to obtain a whole number of chords, or to preserve running chainage, and in consequence, any calculations made to assist in the choice of length, need only be approximate.

Safety and free flow of traffic are the first considerations in traffic engineering, and with this in view comfort and prevention of stress on vehicle and driver are the main objectives for valley curves, but for summits, whilst these are still of equal importance, visibility over the crest must also be taken into account, and as in most (but not all) practical cases, it will require a longer curve, it will be the overriding criterion.

N.B. *All the arguments that follow relate to the simple parabolic vertical curve.*

10.8.4.1. Design by Limitation of Rate of Change of Gradient (Valleys and Summits)

This is an entirely empirical method, in which the maximum rate of change of gradient, $d^2y/dx^2 (= (q-p)/L)$, is limited to an arbitrarily chosen value, which in practice usually lies between 1/3 000 and 1/20 000 and is often expressed as the change of percentage gradient per 100 ft, i.e. $(q-p)/L = 1/5\,000$ is a rate of change of gradient of 2% per 100 ft, or 2% per station.

e.g. If a valley curve links a gradient of -3% and $+4\%$ and the maximum rate of change of gradient is to be 2% per 100 ft, then the minimum length of curve is found from

$$(4-3)/100L = 1/5\,000 \quad \text{and} \quad L = 350 \text{ ft}$$

or

$$(4-3)/L = 2/100 \quad \text{and} \quad L = 350 \text{ ft}$$

N.B. If the same criterion is applied to a summit curve linking a $+3\%$ and a -4% gradient, L is found from $(-4-3)/L = -7/L = 2/100$, and $L = 350$ ft, the negative sign in this instance having no significance.

10.8.4.2. Design by Minimum Radius (Valleys and Summits)

For any curve the radius of curvature R at any point is given by

$$R = \{1 + (dy/dx)^2\}^{3/2}/(d^2y/dx^2)$$

and as for a practical vertical curve dy/dx is always small $(dy/dx)^2$ becomes a second-order small quantity, and we can make use of the approximation,

$$d^2y/dx^2 \simeq 1/R$$

Since for the simple parabolic curve d^2y/dx^2 is constant it will be sufficiently accurate for present purposes, to replace the parabola by a circular curve of radius R. (This does not of course apply to the cubic parabola for which d^2y/dx^2 varies.)

The empirical method of designing the curve by means of an arbitrarily chosen minimum radius is therefore identical with the method of the previous section, merely defining the curve by saying that the minimum radius is (for example) 5 000 ft, instead of saying that the maximum rate of change of gradient is 1/5 000.

10.8.4.3. Design by Limitation of Vertical Acceleration (Valleys and Summits)

This is a much more rational method of design, taking into account the design speed of the road. (The design speed is usually taken to be the 85 percentile speed, i.e. the speed that is not exceeded by 85% of the vehicles using the road.)

It will be clear that if a vehicle traverses the curve at constant speed, that since there is a change of gradient between T_1 and T_2, it will have a change of vertical component velocity and hence be subject to a vertical acceleration f. It will further, be clear that at the lowest (or highest) point, this vertical acceleration will in fact be the centrifugal acceleration, i.e. $f = v^2/R$ (where v is the speed in ft/sec²).

For safety and comfort this vertical acceleration must be limited to a reasonable value, 1 ft/sec² being commonly accepted, with ½ ft/sec² desirable for first class roads, and 2 ft/sec² regarded as just tolerable.

From above $1/R = f/v^2$, or if V is the speed in miles/hr, L can be found from

$$1/R = 1/2 \cdot 15 V^2 = (q - p)/L$$

e.g. If $p = -3\%$, $q = +4\%$, $f = 1$ ft/sec², and $V = 40$ miles/hr

$$7/100L = 1/(2 \cdot 15 \times 1\,600) \quad \text{and} \quad L = 241 \text{ ft}$$

If p and q have the same values, but $f = \frac{1}{2}$ ft/sec² and $V = 70$ miles/hr

$$7/100L = \tfrac{1}{2}/(2 \cdot 15 \times 4\,900) \quad \text{and} \quad L = 1\,475 \text{ ft}$$

10.8.4.4. *Design by 'Vision Distance'* (*Summits only*)

For summits, in addition to applying one of the foregoing criteria, the question of visibility over the brow of the hill must also be considered.

The 'vision distance D', is defined as being the length of the sight line between two points at driver's eye level h above the road (see figure 10.47), h being normally assumed to be 3·75 ft.

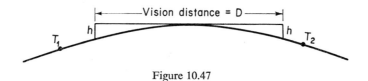

Figure 10.47

The required length of this sight line for two way roads (i.e. single carriageway) is based on the minimum distance required for overtaking, and for one-way roads, on the minimum stopping distance, both of these being dependent on the design speed.

Typical values (in feet) being given in table 10.5.

Two separate cases will have to be considered depending on whether the length of the curve, when calculated, turns out to be greater or less than the vision distance.

Table 10.5

Design speed miles/hour	Vision distance	
	Two-way	One-way
40	950	300
50	1200	425
60	1400	650
70	1650	950

Length of curve greater than vision distances

Assuming a circular curve (figure 10.47), h is the off-set from the tangent to the curve, so that, $h \simeq (\frac{1}{2}D)^2/2R$, whence,

$$(p - q)/L = 1/R = 8h/D^2 \quad \text{or} \quad L = D^2(p - q)/8h$$

If $h = 3\cdot 75$ ft, $L = D^2(p - q)/30$ ft.

e.g. (i) If $p = +3\%$, $q = -4\%$, and $D = 1\,200$ ft,

$$L = (1\,440\,000 \times 7)/(30 \times 100) = 3\,360 \text{ ft.}$$

e.g. (ii) If p and q are the same, but $D = 300$ ft,

$$L = (90\,000 \times 7)/(30 \times 100) = 210,$$

which is less than the vision distance, and the formula is invalid.

Length of curve less than vision distance

Again assuming a circular curve figure 10.48 (AT$_1$ parallel to sight line), as before, $h_1 = (\frac{1}{2}L)^2/2R$, and $1/R = (p - q)/L$, so that

$$h_1 = L(p - q)/8$$

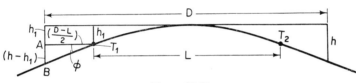

Figure 10.48

From triangle AT$_1$B, $(h - h_1)/\frac{1}{2}(D - L) = \tan \phi = \frac{1}{2}(p - q)$ and so

$$h = \frac{1}{4}(D - L)(p - q) + \frac{1}{8}L(p - q) = \frac{1}{8}(p - q)(L + 2D - 2L)$$

or $\qquad 2D - L = 8h/(p - q) \quad$ and $\quad L = 2D - (8h/(p - q))$

If $h = 3\cdot 75$ ft, $L = 2D - [30/(p - q)]$. e.g. If p and q are again $+3\%$ and -4%, respectively, and D is again 300 ft,

$$L = 600 - (30 \times 100)/7 = 600 - 429 = 171 \text{ ft}$$

By reference to the example in §10.8.4.3, this is less than the length required from vertical acceleration considerations, and so in this case it is this latter requirement that rules.

It will be noticed that for a vision distance of 300 ft, if the numerical sum of the gradients is 5%, the required length of curve is zero. This obviously satisfies the vision distance criterion (see figure 10.49) but the required length of curve

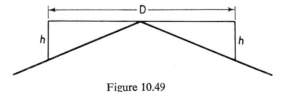

Figure 10.49

will be taken care of by applying the vertical acceleration, or other valley method of design.

10.8.4.5. Design of Curve Controlled by Fixed Levels

In certain cases none of the above criterial will apply, and the curve may have to be designed to pass above or below some point of fixed level.

One numerical example should demonstrate this.

Example: A falling gradient of 1·8% is followed by a rising gradient of 2·5%, their intersection occurring at chainage 8 + 73·2 ft and R.L. 72·56 ft. In order to provide sufficient clearance under an existing bridge, the R.L. at chainage 7 + 95·8 on the curve must not be higher than 74·20 ft. Find the greatest

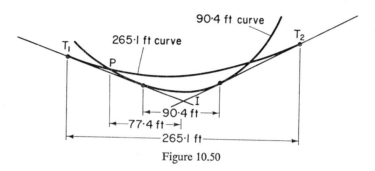

Figure 10.50

length of simple parabolic vertical curve that can be used. Let P represent control point, then distance

$$PI = (8 + 73·2) - (7 + 95·8) = 77·4 \text{ ft}$$
$$R.L.T_1 = 72·56 + 1·8L/200 = H$$

Equation to curve is

$$y = H - \frac{1·8x}{100} + \frac{(2·5 + 1·8)x^2}{200L}$$

$$R.L.P = 74·20 = 72·56 + 0·009L - 0·018(\tfrac{1}{2}L - 77·4)$$
$$+ 0·021\,5(\tfrac{1}{2}L - 77·4)^2/L$$

whence

$$(\tfrac{1}{2}L - 77·4)^2 = L(1·64 - 1·393\,2)/0·021\,5$$
$$L^2 - 309·6L + 23\,963 = 45·9L$$
$$L^2 - 355·5L + 23\,963 = 0$$

from which, $L = 265·1$ ft or 90·4 ft.

Since $\tfrac{1}{2} \times 90·4$ is less than 77·4, the second solution is impracticable.

In problems of this kind a quadratic equation will always occur producing two solutions as shown in figure 10.50. Common sense, aided where necessary by a rough sketch, will indicate which solution is applicable.

11 Volumes of Earthworks

11.1. PRELIMINARY NOTE ON PRISMOIDAL AND END-AREA RULES

As land surfaces are invariably irregular, all earthwork calculations must of necessity be approximate, the precision obtained being entirely dependent on the density of the measured ground heights.

Volumes are usually obtained by calculating the volume of the regular geometrical solid whose surfaces most nearly represent the actual ground surfaces. The geometrical solid most frequently used for this purpose is the prismoid, which is defined as a solid figure having plane parallel ends and plane sides (see figure 11.1).

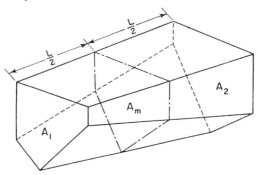

Figure 11.1

It will be clear that the area A of a cross-section, parallel to the ends, will be a function of the perpendicular distance x of the cross-section from one of the ends, and that if the cross-section area A is plotted against the distance x (see figure 11.2), the volume will be equal to the area under the curve $= \int_0^L A \, dx$.

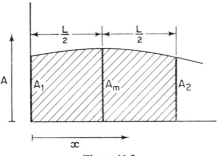

Figure 11.2

As will be seen in §12.4 (dealing with the determination of areas), an approximation to the area under the graph can be found, using Simpson's rule, by taking three ordinates, A_1, A_m, and A_2, equally spaced at distance $\frac{1}{2}L$ apart, i.e.
$V = (\frac{1}{3}) \times \frac{1}{2}L(A_1 + 4A_m + A_2) = (L/6)(A_1 + A_m + A_2)$.

(N.B. A_m is the cross-section area *mid-way* between A_1 and A_2, and *not* the mean of A_1 and A_2.)

For a true prismoid, the cross-sectional area is a quadratic function of x, the distance from one end, so that the above relationship is exact, and the equation is known as the *prismoidal rule*.

When this rule is applied to the portion of a normal road or railway cutting or embankment, between two parallel cross-sections, as shown in figure 11.3,

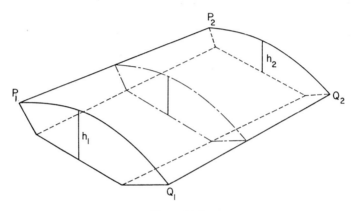

Figure 11.3

where three of the side faces are planes, then if the ground surface is also a plane, the figure is a true prismoid. Since (as also shown in §12.4) the rule is also exact when the cross-sectional area is a cubic function of x, the rule will also give the true volume, if either

(i) The slope of the ground at right angles to the centre-line is uniform (i.e. PQ is a straight line), and the central height h is a quadratic function of x (i.e. the longitudinal profile on the centre-line is parabolic).

(ii) The transverse profile PQ is parabolic, and h is directly proportional to x, (i.e. the longitudinal profile is a straight line). Thus it will be seen, that provided the ground surface curves smoothly, the rule will give a very good approximation to the true volume for this type of solid.

A less accurate estimate of the volume of a prismoid can be obtained by using the trapezoidal rule (see §12.4) to find the approximate value of the area under the curve in figure 11.2, i.e. $V \simeq \frac{1}{2}L(A_1 + A_2)$, which, when applied to volumes, is known as the *end-area rule*. The best accuracy will be obtained from this rule when A_1 is approximately equal to A_2, and the accuracy will decrease as the difference between A_1 and A_2 increases. As an extreme example, for a pyramid or cone, where one end-area is zero, and the true, or prismoidal volume is $\frac{1}{3}$ base × height, the rule gives $\frac{1}{2}$ base × height, an error of 50%.

11.2. DETERMINATION OF VOLUME FROM CONTOURS

This method, which depends largely for its accuracy on the contour vertical interval, and on the accuracy with which the contours have been determined and plotted, is extremely valuable where very large volumes are involved, e.g. reservoirs, land reclamation schemes, open-cast mining, etc., and is also very useful in the preliminary stages of route projects (roads, railways, canals, etc.) for making initial estimates of cost, comparison of alternative routes, selection of best profile, etc.

Basically the method consists of splitting the solid, along the contour planes, into a series of horizontal slabs, each slab being then regarded as a prismoid whose length is the contour vertical interval, and whose end areas are the areas enclosed by the contour lines at the height in question, the areas being taken off the map or plan by planimeter or one of the other methods described in §12.4.

If the prismoidal rule is to be applied to each slab individualy, it will be necessary to interpolate the contour lines mid-way between those already plotted, in order to obtain the mid-area of each slab. It is usually adequate however, to take the slabs in pairs and find the volume by Simpson's rule. Any portions of the solid which are not embraced by two contour planes will have to be treated separately and the volume of the nearest appropriate geometrical solid (usually a pyramid or wedge) found.

Referring to figure 11.4, (vertical interval = 10 ft) by end-area rule,

$$V = \tfrac{10}{2}[A_{120} + 2(A_{130} + A_{140} + A_{150}) + A_{160}] + \tfrac{7}{3}A_{160}$$

By prismoidal (or Simpson's) rule,

$$V = \tfrac{10}{3}[A_{120} + 4A_{130} + 2A_{140} + 4A_{150} + A_{160}] + \tfrac{7}{3}A_{160}$$

When the new profile (i.e. the ground profile after the works have been carried out) is anything other than a horizontal plane, the new contour lines for the finished work will have to be superimposed over the existing ground contours, and the end areas of the prismoids will then be the areas on the plan enclosed between the new and the old contour lines for the height in question.

In figure 11.5 the new contours for the benching shown in the cross-section have been superimposed onto the contours for a natural hillside, and the prismoidal end-areas at heights 20 ft and 25 ft about datum indicated by hatching.

In all but the simplest earthwork projects, the plotting of the contours for the finished work may require some thought, and figure 11.5 also illustrates some of the principles involved.

It will be assumed that the embankment surrounding the benching on sides AB, BC, and CD has a slope of $\tfrac{1}{3}$ (i.e. 1 vertical in 3 horizontal), and that the vertical interval is, as shown, 5 ft. Then, if at any point of known height on the plan we move away along the line of greatest slope for a distance of $3 \times 5 = 15$ ft, the level will have dropped 5 ft and a point on the next lower contour line will have been found. For example, if from points B and C (both at 20 ft above datum) we move out 15 ft in a direction normal to BC, two points on the

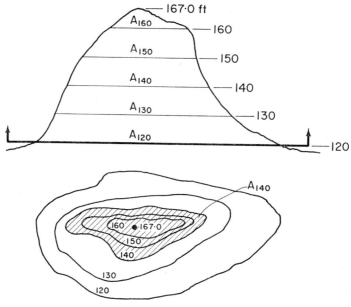

Figure 11.4

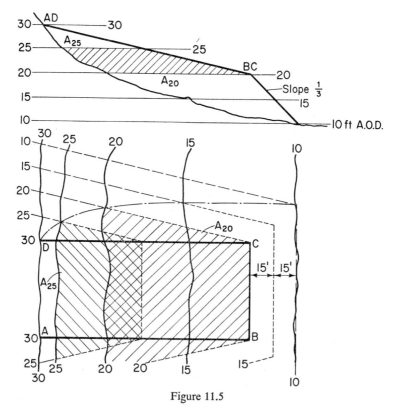

Figure 11.5

15 ft contour will have been found; if we move out 30 ft, two points on the 10 ft contour will have been found, and so on. As for a plane, the contours will be a series of equidistant parallel lines, the contours for the embankment falling away from the line BC can now be plotted.

Similarly, if we move away 15 ft from C in a direction normal to CD, and 45 ft away from D in the same direction, two further points on the 15 ft contour for the embankment will have been found. Since the normal to CD at D, in this case, lies along the 30 ft contour, the point we have just located on the 15 ft contour for the embankment lies below existing ground level, and the 15 ft contour for the finished work will only extend part of the way along the line between the two points by which its position has been established. Clearly as the point where the new and the old 15 ft contours intersect the new and existing ground levels are the same, and the actual new contour line will not extend beyond this point.

By linking up all such points where new and existing contours of the same value intersect, the position of the toe of the embankment can be established. This is also shown in the figure.

Figure 11.6 shows the same principle applied to a section of road running in cutting and on embankment, with again some of the prismoid end-areas marked to indicate the areas that will have to be taken off with the planimeter. It should be noted, that in both these cases, the cross-sectional view is not required for the purposes of the volume calculations, and is only included to illustrate the meaning of the construction lines on the plan. A very small part of a project, drawn to a large scale, has been used for the sake of clarity, but it must be realized that in the ordinary way, due to the difficulty of obtaining sufficiently accurate contours, the accuracy of the volumes obtained under these conditions would be considerably inferior to those obtained by the cross-section method described in §11.4.

In the road example illustrated in figure 11.6, the road gradient is uniform, and the plan centre-line straight. If the vertical profile of the road is laid to a vertical curve, the new contours crossing the road, will no longer be equidistant, and the side-slope contours will cease to be equidistant, parallel lines, and become instead equidistant, parallel curves, although, unless the change in gradient is very accute, the curvature of the side-slope contours is very slight.

The case where the road centre-line is curved on plan is different, as will be seen from figure 11.7. The line of greatest slope of the side-slope is usually normal to the edge of the road, so that the same technique of stepping off intervals of n times the vertical interval (where the side-slope is $1/n$) from points of known height on the edge of the road, still applies, as illustrated.

Figure 11.8 demonstrates one further application of contour geometry to the solution of earthwork volume problems. Here a tunnel 16 ft in diameter, and with its axis horizontal is to be driven into a hillside whose contours are shown. In order to be able to plot on the plan view, the position of the interpenetration curve between the tunnel and the hillside, and the shape of the end-areas of the horizontal slabs, it is necessary, in this case to draw the front elevation of the mouth of the tunnel, and from this view obtained the true length (such as *ab*)

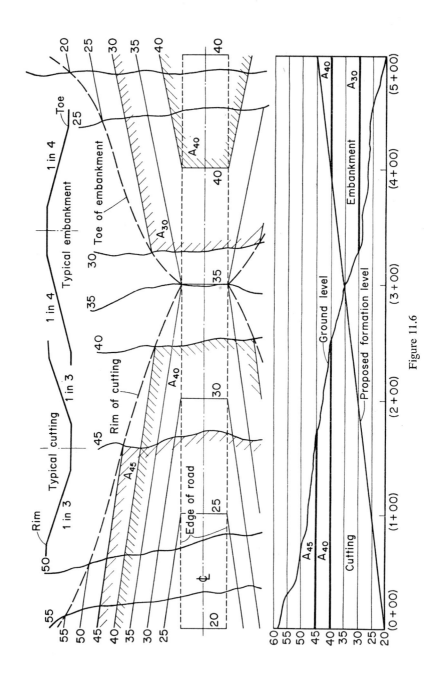

Figure 11.6

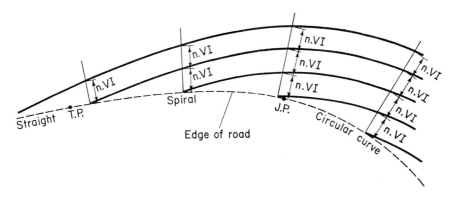

Figure 11.7

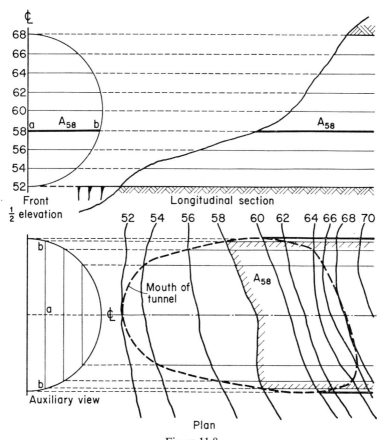

Figure 11.8

of the horizontal distances from the centre-line to the lines of intersection of the contour planes with the walls of the tunnel. In the figure this has been achieved by drawing the auxiliary view of the tunnel mouth, so that these intersection lines can be obtained by simple projection. The position of points on the plan view of the tunnel mouth denoted by the intersection of the old and new contour lines at 58 ft above datum, and the shape of the slab end-area at the same height are marked on the diagram. The embankment carrying the road up to the tunnel mouth has been omitted, as the construction for this would be similar to that shown in figure 11.6.

11.3. DETERMINATION OF VOLUME FROM SPOT HEIGHTS

This method, which depends for its accuracy only on the density of the levels taken is usually used for large open excavations such as reservoirs, for ground levelling operations such as parks or playing fields, or for building sites. The use of photogrammetric methods of heighting from air photographs, coupled with the use of electronic computers for processing the data, has opened up a field in which, by enormously increasing the number of spot heights that can be used, the method can be extended to almost any earthwork project for which photogrammetrical heights are sufficiently accurate.

In normal ground methods, the site is 'gridded' by a series of lines forming squares (or occasionally rectangles), and the ground levels are determined at the intersections of the grid lines, together with such additional points (referenced to the grid by off-sets or tie lines) as may be necessary to pick up random boundaries or exceptional ground irregularities or discontinuities. The spacing of the grid lines will depend on the nature of the ground, and should be sufficiently close for the ground surfaces between the lines to be reasonably regarded as planes.

The proposed formation levels at the corners of the grid squares are obtained from the designer's drawings, and the volume within each square is taken as being the plan area of the square, multiplied by the average of the depths of excavation (or fill) at the four corners of the square. The volumes of the portions lying between the outermost grid lines and the random boundaries of the site are taken as the plan areas of the nearest equivilent trapezia or triangles, each multiplied by the average of their corner depths (see figure 11.9).

As all the depths at the internal intersections of the grid lines are used in the calculation of the volume of more than one square, a formula on the lines of the following can be evolved

$$V = (l^2/4)(\sum h_1 + 2 \sum h_2 + 3 \sum h_3 + 4 \sum h_4) + \sum R$$

where l is the length of side of square, h_1 the depths such as a and e which are used once, h_2 the depths such as b, c, and d, which are used twice, h_3 the depth such as f which is used 3 times, h_4 the depth such as g, h, etc. which are used 4 times, and R the volume of random trapezia and triangles, which must, of course be calculated individually.

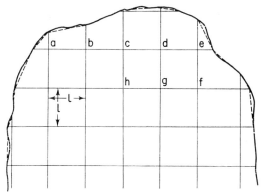

Figure 11.9

Certain difficulties will arise if there is a change over from cut to fill, or *vice versa*, within one grid square (see figure 11.10).

If $h_1 = -h_4$ and $h_2 = -h_3$, the calculation, $V = l^2(h_1 + h_2 + h_3 + h_4)/4$ will produce zero volume for the square which is patently incorrect, and whatever the values of the depths the calculation will only produce the *net* volume of cut or fill (whichever is the larger).

If the grid lines are fairly close together, the formulae

$$V_{cut} = l^2(h_1 + h_2 + 0 + 0)/4$$

and

$$V_{fill} = l^2(0 + 0 + h_3 + h_4)/4$$

will not produce excessive inaccuracy.

If, however, the grid lines are widely spaced, some calculation such as the

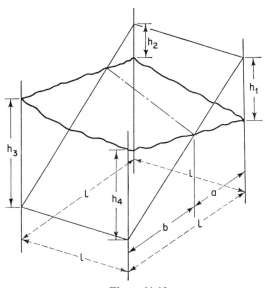

Figure 11.10

following (suitably modified if three of the corner depths have a different sign from the fourth), will have to be used.

Assuming the ground to be plane,

$$a = lh_1/(h_1 - h_4), \quad b = lh_2(h_2 - h_3), \quad \text{etc.}$$

(N.B. Since h_3 and h_4 are negative $h_1 - h_4$ will be the *numerical sum* of h_1 and h_4, and so on.) Whence, using end-area rule,

$$V_{cut} = \frac{l^2}{4}\left(\frac{h_1^2}{h_1 - h_4} + \frac{h_2^2}{h_2 - h_3}\right)$$

$$V_{fill} = \frac{l^2}{4}\left(\frac{h_4^2}{h_1 - h_4} + \frac{h_3^2}{h_2 - h_3}\right)$$

Clearly, if a large number of such squares are involved, the calculation required will be very laborious, and a closer grid may prove more economical, even though the ground slopes may be reasonably uniform.

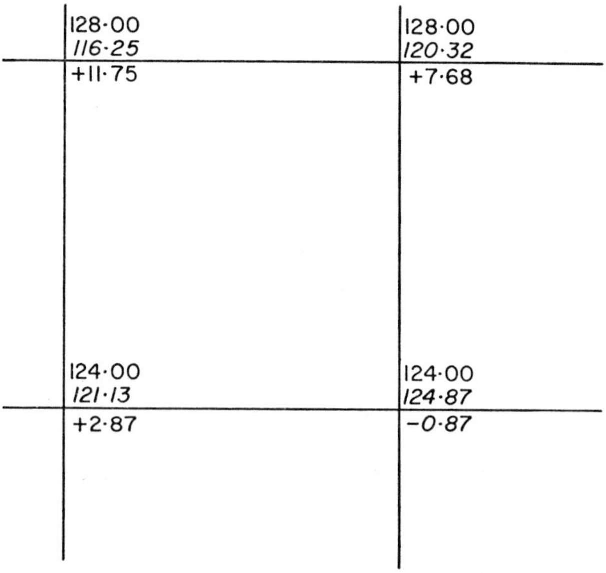

128·00 116·25 +11·75	128·00 120·32 +7·68
124·00 121·13 +2·87	124·00 124·87 −0·87

Figure 11.11

To avoid mistakes occurring, and to assist in the tabulation of the figures, it is often convenient to extract the ground levels from the field book, and the formation levels from the drawings, and to record them in coloured inks on a gridded plan of the site in the manner indicated in figure 11.11, where the first level (sepia) is the ground level, the level below it (green) the proposed formation level, and the difference where positive (red) indicates cut, and where negative (blue) indicates fill.

11.4. DETERMINATION OF VOLUME FROM CROSS-SECTIONS

11.4.1. GENERAL

This method is that most widely used for engineering projects of all types, and lends itself particularly well to route projects, roads, railways, waterways, etc.

Cross sections are taken on lines at right angles to the main job centre-line and the volume between adjacent cross-sections found for a prismoid with the areas of the cross-sections as its end and, if necessary, its mid-areas. For route surveys it is usually convenient to take the cross-sections at the round number chainage points, with additional sections at points of unusual irregularity or discontinuity.

When the ground is very irregular it is best to plot the profiles of the existing ground and the proposed formation to a convenient scale and obtain the areas of the cross-sections by planimeter or counting squares, etc. In this connection is important to note that it is customary, for clarity, to plot cross-sections to an exaggerated vertical scale, and suitable allowance must be made for this if the areas of the cross-sections are to be found by the above methods.

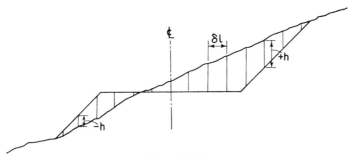

Figure 11.12

When the ground surface is reasonably regular, and the ground profiles at the cross-sections can be satisfactorily represented by uniform slopes, it is more expeditious to make use of suitable formulae for calculating the cross-sectional areas.

As referred to in the previous section, photogrammetrical heighting, coupled with computer processing, can greatly assist with this method. In such cases, the area of the cross-sections can be obtained from a series of equidistant spot heights. As will be seen from figure 11.12, if the depths are determined at sufficiently close intervals δl, the areas will be given by

$$A_{cut} = \delta l \sum (+h) \quad \text{and} \quad A_{fill} = \delta l \sum (-h)$$

and as depths taken beyond the end of the section will be zero, there is no need to determine the overall width of the cross-section independently.

When normal ground methods of survey are employed, and the ground slopes can be regarded as uniform, the following calculation methods will be found useful.

It will be noted that the general case considered here is the 'three-level' cross-section, in which the ground slopes are assumed to be uniform from A to E, and from E to D, in figure 11.13. Although such a situation, with a sudden change of slope exactly on the route centre-line will rarely, if ever, occur in practice, this assumption gives a much closer approximation to the true ground profile when this is a smooth curve, as shown in the figure, than would be given by the assumption of a straight line from A to D. Where the

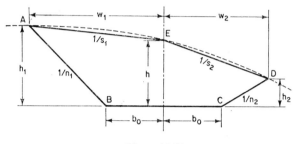

Figure 11.13

ground surface approximates to a plane, however, the 'two-level' cross-section is quite justifiable.

In figure 11.13 the slopes are expressed as 1 vertical in n or s horizontal, the rest of the symbols being self-explanatory.

Unless the field method of 'slope staking' (see §11.4.2) has been used the half breadths w_1 and w_2 and the corresponding depths h_1 and h_2 will not be known, and they will have to be determined from the other data. The ground slopes can be obtained by direct measurement with a clinometer or gradienter, or by calculation from two spot heights at known positions on the slope. The formation breadth and level, and the values of the side slopes will have been supplied by the designer.

From the figure,

$$h_1 = \frac{w_1 - b_0}{n_1} = h + \frac{w_1}{s_1}$$

whence

$$w_1\left(\frac{1}{n_1} + \frac{1}{s_1}\right) = h + \frac{b_0}{n}$$

and

$$w_1 = \frac{(h + b_0/n)s_1 n_1}{s_1 - n_1} \tag{1}$$

and similarly,

$$w_2 = \frac{(h + b_0/n)s_2 n_2}{s_2 + n_2} \tag{2}$$

equation (1) applying when the ground rises from the centre-line, and equation (2) applying when the ground falls away from the centre-line. It will be realized that, if a slope falling away from the centre line is regarded as being negative, equation (1) will suffice, e.g. for slopes as shown in the figure, if the formation breadth is 40 ft, the central depth is 7 ft, both the side slopes are $\frac{1}{5}$, and the ground slopes are both $\frac{1}{20}$;

$$w_1 = \frac{(7 + 20/5)20 \times 5}{20 - 5} = \frac{11 \times 100}{15} = 73 \cdot 3 \text{ ft}$$

$$w_2 = \frac{(7 + 20/5)20 \times 5}{20 + 5} = \frac{11 \times 100}{25} = 44 \cdot 0 \text{ ft}$$

Once the half breadths w_1 and w_2 have been found, the area of the cross-section can be found in several ways.

See figure 11.14, $FH = b_0/n_1$, $FG = b_0/n_2$.

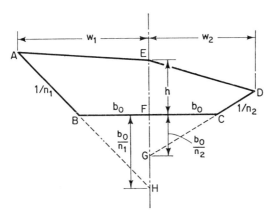

Figure 11.14

A is the area of triangles AEH + DEG minus triangles, DFH + CFG., i.e.

$$A = \tfrac{1}{2}w_1\left(h + \frac{b_0}{n_1}\right) + \tfrac{1}{2}w_2\left(h + \frac{b_0}{n_2}\right) - \frac{\tfrac{1}{2}b_0{}^2}{n_1} - \frac{\tfrac{1}{2}b_0{}^2}{n_2} \qquad (3)$$

If, as is usual, the side slopes are equal, i.e. $n_1 = n_2 = n$

$$A = \tfrac{1}{2}(w_1 + w_2)(h + b_0/n) - b_0{}^2/n \qquad (3a)$$

This is usually the most convenient way to use the formula, i.e. w_1 and w_2 are calculated first, and the numerical values substituted in the above expression, the term $b_0{}^2/n$ usually being constant for a series of cross-sections.

If however, in addition to the side slopes being equal, the section is a 'two level' cross-section, i.e. $s_1 = s_2 = s$ and AED is a straight line, the expression can be modified as follows.

From (1) and (2),

$$w_2 = w_1(s - n)/(s + n), \quad \text{and} \quad \tfrac{1}{2}(w_1 + w_2) = w_1 s/(s + n)$$

also from (2)

$$(h + b_0/n) = w_2(s + n)/sn$$

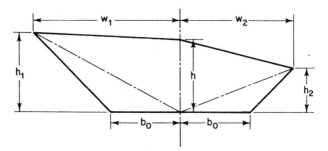

Figure 11.15

substituting these values in (3) gives,

$$A = (w_1 w_2 - b_0^2)/n \qquad (4)$$

and if it is required to dispense with the independent calculation of w_1 and w_2, equations (1) and (2) can again be substituted in (4) to give

$$A = \frac{s^2 n(h + b_0/n)^2}{s^2 - n^2} - \frac{b_0^2}{n} \qquad (5)$$

When slope staking (§11.42) has been employed in the field, and w_1, w_2, h_1, and h_2 have been determined directly, the area of the cross-section can be found much more easily (using the triangles shown in figure 11.15) from

$$A = \tfrac{1}{2}h(w_1 + w_2) + \tfrac{1}{2}b_0(h_1 + h_2) \qquad (6)$$

this equation applying equally to the 'two-level' or the 'three-level' cross-section.

It should be clear that by turning the diagrams upside down, all the above equations apply equally to embankments having the same characteristics.

In addition to the above, one further type of cross-section, that shown in figure 11.16, must be considered.

It should be noted that in this case it is more common for n_1 and n_2 to be unequal, the slope for the fill usually being flatter than that for the cut.

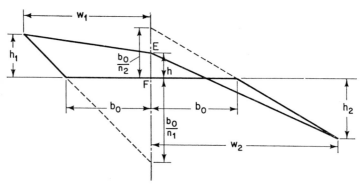

Figure 11.16

Referring to the figure, for the left-hand side, from equation (1) §11.4.1

$$w_1 = (h + b_0/n_1)s_1n_1/(s_1 - n_1), \quad \text{as before.}$$

For the right-hand side

$$h_2 = \frac{w_2 + b_0}{n_2} = \frac{w_2}{s_2} - h$$

whence

$$w_2\left(\frac{1}{n_2} - \frac{1}{s_2}\right) = \frac{b_0}{n_2} - h$$

and

$$w_2 = \frac{[(b_0/n_2) - h]s_2n_2}{s_2 - n_2} \tag{7}$$

It should be clear, that if P lies to the left of the centre-line, the above equations for w_1 and w_2 will be interchanged.

It will further be seen from figure 11.16, that the distance PF will be s_2h and the area of triangle EFP $\frac{1}{2}s_2h^2$. Then

Area of cut = Area to left of centre-line + Triangle EFP

i.e., making use of equation (3) §11.4.1

$$A_{\text{cut}} = \tfrac{1}{2}w_1\left(h + \frac{b_0}{n_1}\right) - \frac{\tfrac{1}{2}b_0^2}{n_1} + \tfrac{1}{2}s_2h^2 \tag{8}$$

or, substituting for w_1,

$$A_{\text{cut}} = \frac{\tfrac{1}{2}(h + b_0/n_1)^2 s_1 n_1}{s_1 - n_1} - \frac{\tfrac{1}{2}b_0^2}{n_1} + \tfrac{1}{2}s_2h^2 \tag{8a}$$

If a two-level cross-section is involved, and $s_1 = s_2 = s$, this expression further simplifies into

$$A_{\text{cut}} = \tfrac{1}{2}(b_0 + sh)^2/(s - n_1) \tag{9}$$

By the same principle of addition and subtraction of triangles as used in deriving equation (3) §11.4.1,

$$A_{\text{fill}} = \tfrac{1}{2}w_2\left(\frac{b_0}{n_2} - h\right) - \frac{\tfrac{1}{2}b_0^2}{n_2} + \tfrac{1}{2}s_2h^2 \tag{10}$$

or, substituting for w_2,

$$A_{\text{fill}} = \frac{\tfrac{1}{2}[(b_0/n_2) - h]^2 s_2 n_2}{s_2 - n_2} - \frac{\tfrac{1}{2}b_0^2}{n_2} + \tfrac{1}{2}s_2h^2 \tag{10a}$$

Again, if $s_2 = s_2 = s$,

$$A_{\text{fill}} = \tfrac{1}{2}(b_0 - sh)^2/(s - n_2) \tag{11}$$

It should be clear, by turning the diagram in figure 11.16 upside down, that if P lies to the left of the centre line, all these equations for cut and fill must be interchanged.

It should be observed, that wherever in these equations the term $\tfrac{1}{2}sh^2$ occurs, the slope s relating to the side of the cross-section in which the point P falls, is the one to be used.

11.4.2. SLOPE STAKING

Slope staking is a trial-and-error process for locating and staking in the field, the points at which the proposed side slopes of cuttings or embankments will intersect the existing ground surface. As it will be seen that the depths of cut and fill at these points, as well as the cross-section half breadths, are calculated as part of the procedure, not only are the limits of the earthworks defined on the ground but data for the calculation of the earthwork volumes is provided.

It will be obvious that the proposed formation levels, the formation breadth, and the values of the side slopes must have been settled before any work can be done.

11.4.2.1. Principle

In figure 11.17, any point on the side slope will be defined by the equation, $x = b_0 + ny$, consequently, at A the values of $x = w_1$ and $y = h_1$ will also satisfy this equation.

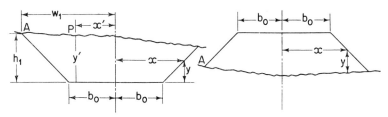

Figure 11.17

The field procedure, then, consists of placing the staff at a series of points on the cross-section such as P, determining the values of x' and y' at P, and finding out whether they satisfy the equation. When they do, the point A has been established, and w_1 and h_1 found.

This procedure may sound laborious, but in fact, after a few hours experience, very few trials are required to locate the correct point. It must be appreciated, that even on fairly smooth ground, a correspondence of levels of ±3 inches is all that can be expected, which at a ground slope of $\frac{1}{10}$ is equivalent to a variation in x of $2\frac{1}{2}$ ft. By evolving a suitable system for the fieldwork, calculation and booking, the work can be made to proceed quite rapidly.

11.4.2.2. Calculation

Once the level has been set up and the height of the horizontal place of collimation (H.P.C.) to the same datum as the predetermined formation levels, has been found, then G, known as the 'grade staff reading', can be calculated (figure 11.18).

For cuttings, $G = (\text{H.P.C.} - \text{F.L.})$
For embankments, $G = (\text{F.L.} - \text{H.P.C.})$

This is a constant for one cross-section and one instrument position.

From this, the values of y' can be found corresponding to the various values of S, the staff reading.

For cuttings, $y' = (G - S)$
For embankments, $y' = (G + S)$

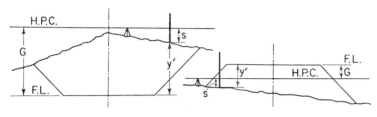

Figure 11.18

If the ground is horizontal, the half breadth will be given by, $w = b_0 + nh$ (where h is the depth at central-line), and it is convenient to calculate this for each cross-section, as a starting point for estimating w, which will be greater, or less than this value according to whether the ground slopes upwards or downwards away from the centre-line (*vice-versa* for embankments).

When the values of w_1, w_2, h, h_1, and h_2 have been found, the cross-section areas can be found from

$$A = \tfrac{1}{2}h(w_1 + w_2) + \tfrac{1}{2}b_0(h_1 + h_2) \quad \text{(see equation (6) §11.4.1)}$$

11.4.2.3. Field Work

The cross-sections are usually taken at equidistant intervals along the centre-line of the route, often at the round-number chainage points, and the work will be expedited if the ground levels on the centre-line have been determined first. (Since slope staking cannot be carried out until the formation levels have been decided, the centre-line ground levels will almost certainly have been found at some previous time.) A plentiful supply of temporary bench marks is also desirable, and the chainage pegs can be conveniently used for this purpose.

If the values of w are all expected to be less than 100 ft, the level can be set up at a convenient point for reading the staff on several cross-sections, and the distance from the centre-line measured with a chain or tape, the staff man calling out the distance x' for each point where he sets the staff.

If the values of w are greater than 100 ft, it will often be more convenient to measure x' tacheometrically, using the level stadia, although this, of course, necessitates setting the levelling instrument up on the centre-line for each cross-section.

Table 11.1. Example of Booking for Slope Staking $b_0 = 20$ ft $1/n = 1/2$

		Centre-line					Left hand side				Right hand side				Cross-section area			
1	2	3	4	5	6	7	8	9	10	11	12	13	14	15	16	17	18	19	20
Ch.	H.P.C.	F.L.	G	S_0	h_0	$b_0 + nh_0$	w_1	s_1	h_1	$b_0 + nh_1$	w_2	s_2	h_2	$b_0 + nh_2$	$w_1 + w_2$	$\frac{1}{2}h_0 \times (w_1+w_2)$	h_1+h_2	$\frac{1}{2}h_0 \times (h_1+h_2)$	$A = (17) + (19)$
4+00	106·42	92·60	13·82	7·25	6·57	33·1	45	3·14	10·68	41·4	30	11·19	2·63	25·3					
							40	3·71	10·11	40·2	27	10·80	3·02	26·0	66·4	218·1	13·21	132·1	350·2
											26·3	10·72	3·10	26·2					
5+00	106·42	94·60	11·82	5·87	5·95	31·9	40	3·91	7·91	35·8	25	8·15	3·67	27·3	62·2	185·0	11·18	111·8	296·8
							35	4·12	7·70	35·4	27	8·34	3·48	27·9					

11.4.2.4. *Booking*

Table 11.1 suggests suitable column headings for a field book, and it will be noticed that columns have been included for calculating the cross-section areas, these can of course, be completed in the office.

Column 7 is included to give a starting point for the estimation of w, as indicated under 'calculations'.

11.4.3. PRISMOIDAL CORRECTION

Once the areas of the various cross-sections have been found the volume is usually obtained by applying Simpson's rule, e.g. if A_a, A_b, A_c, A_d, etc. are a series of cross-sections all spaced a distance L apart, the volume will be given by

$$V = \tfrac{1}{3}L(A_a + 4A_b + 2A_c + 4A_d + 2A_e + 4A_f, \text{ etc.}$$

An even better estimation of the volume can be obtained by applying the prismoidal rule to the volume between any two adjacent cross-sections, e.g. $V = \tfrac{1}{6}L(A_a + 4A_M + A_b)$. This will however, involve the interpolation of the mid-area A_M. This is most conveniently achieved by developing a general expression for a 'prismoidal correction' C, which must be *subtracted* from the 'end-area volume' in order to find the 'prismoidal volume' V. Such a correction term is particularly valuable where an isolated portion for which the end-areas only are available and exhibit a large inequality.

By definition,

$$C = \tfrac{1}{2}L(A_a + A_b) - \tfrac{1}{6}L(A_a + 4A_M + A_b)$$
$$= \tfrac{1}{6}L(2A_a + 2A_b - 4A_M) = \tfrac{1}{12}L(4A_a + 4A_b - 8A_M)$$

For the general case of the three-level cross-section, with unequal side-slopes, and considering initially, only the area A', to one side of the centre-line, then, making use of equation (3) §11.4.1,

$$A_a' = \tfrac{1}{2}w_a(h_a + b_0/n) - \tfrac{1}{2}b_0^2/n$$
$$A_b' = \tfrac{1}{2}w(h_b + b_0/n) - \tfrac{1}{2}b_0^2/n$$
$$A_M' = \tfrac{1}{2} \times \tfrac{1}{2}(w_a + w_b)[\tfrac{1}{2}(h_a + h_b) + b_0/n] - \tfrac{1}{2}b_0^2/n$$

whence

$$C' = \tfrac{1}{12}L\bigg[2w_ah_a + \frac{2w_ab_0}{n} - \frac{2b_0^2}{n} + 2w_bb_b + \frac{2w_bb_0}{n} - \frac{2b_0^2}{n}$$
$$- w_ah_a - w_ah_b - \frac{2w_ab_0}{n} - w_bh_a - w_bh_b - \frac{2w_bb_0}{n} + \frac{4b^2}{n}\bigg]$$
$$= \tfrac{1}{12}L(w_ah_a + w_bh_b - w_ah_b - w_bh_a)$$

or

$$C' = \tfrac{1}{12}L(w_a - w_b)(h_a - h_b) \tag{1}$$

which can be put into words,

$$C' = \tfrac{1}{12}L \text{ (Difference half-breadths)(Difference central depths)}$$

The prismoidal correction for any complete two-level or three-level cross-section whose half-breadths are w_{1a} and w_{2a} at cross-section a, and w_{1b} and w_{2b} at cross-section b, will be

$$C = \tfrac{1}{12}L(w_{1a} + w_{2a} - w_{1b} - w_{2b})(h - h) \qquad (2)$$
$$C = \tfrac{1}{12}L(\text{Difference over-all breadths})(\text{Difference central depths})$$

In the above analysis, if $b_0 = 0$, the equation (1) holds good for a triangle with a horizontal base and height h, so that for the part cut-part fill section shown in figure 11.16,

$$C_{\text{cut}} = \tfrac{1}{12}L(w_{1a} + s_{2a}h_a - w_{1b} - s_{2b}h_b)(h_a - h_b) \qquad (3)$$
and
$$C_{\text{fill}} = \tfrac{1}{12}L(w_{2a} - s_{2a}h_a - w_{2b} + s_{2b}h_b)(h_a - h_b) \qquad (4)$$

11.4.4. CURVATURE CORRECTION

11.4.4.1. *General*

When the centre-line of a road or railway, etc., is curved on plan, the portion between two adjacent cross-sections is no longer a true prismoid, and a further correction is required.

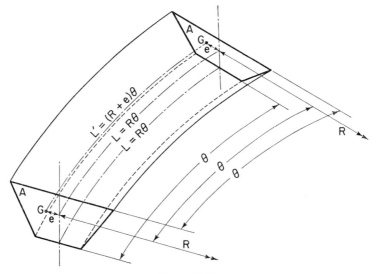

Figure 11.19

Considering first a portion of cutting or embankment of constant (but unsymmetrical) cross-section area A, having an (arcual) length L, and whose centre-line has a constant horizontal radius R, as illustrated in figure 11.19.

If the cross-section is unsymmetrical, its centroid G will be situated at a horizontal distance e, from the centre-line, where e (referred to as the 'eccentricity') is regarded as being positive or negative, according to whether G lies on the opposite, or same side, respectively, of the centre-line as the centre of curvature.

Volumes of Earthworks

If θ is the angle subtended at the vertical axis at the centre of curvature, by the two vertical planes containing the identical cross-sections at the ends of the solid, then the volume of the solid will be that generated by rotating the cross-section A through an angle θ.

By the theorem of Pappus, the volume of this solid will be the product of the area A, and the length of the path of the centroid, i.e. $V = A(R + e)\theta$. Since $\theta = L/R$, this becomes

$$V = AL + AeL/R \tag{1}$$

In this expression, AL is the volume of a prismoid of length L, and the curvature correction is AeL/R, being positive or negative according to whether e is positive or negative.

As equation (1) can also be written,

$$V = L(A + Ae/R) \tag{2}$$

the term Ae/R can be regarded as the correction to be made to the cross-sectional area, before calculating the volume as that of a normal prismoid.

The corrected area can be expressed as $A' = A(1 + e/R)$, from which it can be seen, that if e is small compared with R, the correction will be negligible.

It should be noted, that if the cross-section is symmetrical, G will lie in the centre-line, and e, and hence the correction, will both be zero.

In the general case, the shape of the cross-section will not be constant, so that neither A nor e will be constant, and if the horizontal curve is a spiral, the radius R will not be constant.

As in practice, the ratio e/R will always be comparatively small, it is usually sufficient to calculate the correction for each cross-section and then use either Simpson's rule, the prismoidal rule, or the end-area rule, as appropriate, to determine the volume in the normal manner.

In calculating the value of the correction, it is useful to remember that for the clothoid spiral (see Chapter 10), the radius of curvature is inversely proportional to the distance round the spiral from the origin, and that the product of these two quantities is equal to the LR value for the curve.

11.4.4.2. *Calculation of Eccentricity*

N.B. As the correction term to be applied to the cross-sectional area A, is Ae/R, it is the product Ae that is required for calculation purposes, so that equations for this will be derived in this section. If the actual eccentricity itself is required for any purpose, it can always be obtained numerically by dividing Ae by A which will have already been found.

Three cases only will be considered, the three-level cross-section with unequal side slopes, the same with equal side slopes, and the part cut–part fill cross-section, as all other cross-sections can be very simply derived from these.

In each of these cases the expressions will be found in terms of the half-breadths, which themselves can be found from equations (1), (2), and (7) §11.4.1 as before, and the same method of division of the cross-section into triangles as was employed in the derivation of equations (3), (8), and (10), §11.4.1 will be used.

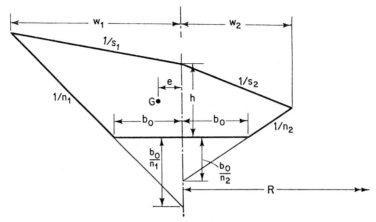

Figure 11.20

It must be remembered that the perpendicular distance of the centroid of a triangle from its base, is $\frac{1}{3}$ of its height.

From figure 11.20 FH $= b_0/n_1$ and FG $= b_0/n_2$ as before.

Assuming e to be positive when G lies to the right of the centre-line, and taking moments about the centre-line

$$Ae = \frac{1}{2}\left(h + \frac{b_0}{n_1}\right)\frac{w_1 w_1}{3} - \frac{1}{2}\frac{b_0^2}{n_1}\frac{b_0}{3} - \frac{1}{2}\left(h + \frac{b_0}{n_2}\right)\frac{w_2 w_2}{3} + \frac{1}{2}\frac{b_0^2}{n_2}\frac{b_0}{3}$$

$$Ae = \left(h - \frac{b_0}{n_1}\right)\frac{w_1^2}{6} - \frac{b_0^3}{6n_1} + \left(h + \frac{b_0}{n_2}\right)\frac{w_2^2}{6} + \frac{b_0^3}{6n_2} \tag{1}$$

If $n_1 = n_2 = n$, this simplifies to

$$\frac{1}{6}\left(h - \frac{b_0}{n}\right)(w_1^2 - w_2^2) \tag{2}$$

This can be rearranged as,

$$Ae = \tfrac{1}{2}(h - b_0/n)(w_1 + w_2)\tfrac{1}{3}(w_1 - w_2)$$

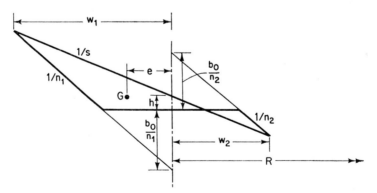

Figure 11.21

and as from (3a) §11.4.1,
$$A = \tfrac{1}{2}(h - b_0/n)(w_1 + w_2) - b_0^2/n$$
$$Ae = (A + b_0^2/n)\tfrac{1}{3}(w_1 - w_2) \qquad (3)$$

b_0^2/n will normally be constant for a whole series of cross-sections.

For the part cut-part fill section in figure 11.21, by the same process as used above,
$$Ae_{\text{cut}} = \tfrac{1}{6}[(h - b_0/n_1)w_1^2 - b_0^3/n_1 - s_2^2 h^3] \qquad (4)$$
and
$$Ae_{\text{fill}} = -\tfrac{1}{6}[(b_0/n_2 - h)w_2^2 - b_0^3/n_2 + s_2^2 h^3] \qquad (5)$$

11.4.5. NUMERICAL EXAMPLE OF CALCULATION OF EARTHWORK VOLUMES BY METHOD OF CROSS-SECTIONS

Table 11.2 overleaf gives the calculations for a 467 ft length of cutting, a typical three-level cross-section of which is given in figure 11.22.

Columns 1–5 give the field data, i.e. centre-line depth, natural ground slopes, (−ve when falling away from the centre-line), and radius of curvature of the centre-line, for the various chainages.

Columns 6–10, and 11–15, give the calculations for the half breadths, and columns 16–20, that for the cross-sectional area.

Column 23 is the 'Simpson' or trapezoidal coefficient, as appropriate, and column 26 the corresponding multiplier.

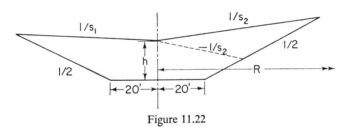

Figure 11.22

11.5. MASS-HAUL DIAGRAMS

Mass-haul diagrams are used for route earthwork projects to assist in designing the best profile and in organizing the actual work in the most economical manner. Such diagrams relate only to the longitudinal movement of earth along the route, and take no account of transverse movement of material.

Fundamentally the mass-haul diagram is a graph showing the *cumulative* volume of excavation (as ordinate) plotted against the centre-line chainage (as abscissa), and is usually (although not necessarily) plotted to the same horizontal scale as, and projected up from, the longitudinal section (see figure 11.23).

The vertical ordinate (aggregate volume) can be regarded as a measure of the volume of earth contained in the bowl of a hypothetical scraper of infinite capacity, as it moves in the direction of increasing chainage, along the route centre-line.

Table 11.2

$b_0 = 20\,ft$
$1/n = 1/2$
$b_0/n = 10\,ft$
$b_0^2/n = 200\,sq\,ft$

1	2	3	4	5	6	7	8	9	10	11	12	13	14	15
ch.	h	s_1	s_2	R	$s_1 n$	$s_1 - n$	$\frac{s_1 n}{s_1 - n}$	$h + \frac{b_0}{n}$	w_1	$s_2 n$	$s_2 - n$	$\frac{s_2 n}{s_2 - n}$	$h - \frac{b_0}{n}$	w_2
0	6·62	+20·7	+62·0	∞	41·4	18·7	2·21	16·62	36·7	+124	+60·0	2·07	16·62	34·4
1+00	7·27	18·4	+105	2000	36·8	16·4	2·24	17·27	38·7	+210	+103	2·04	17·27	35·2
2+00	8·98	+12·6	−108	1000	25·2	10·6	2·38	18·98	45·2	−210	−110	1·96	18·98	37·2
3+00	10·04	+10·4	−42·8	667	20·8	8·4	2·48	20·04	49·7	−85·6	−44·8	1·91	20·04	38·3
4+00	11·62	+8·5	−20·5	500	17·0	6·5	2·62	21·62	56·6	−41·0	−22·5	1·82	21·62	39·3
5+67	9·45	−9·7	−10·6	500	19·4	7·7	2·52	19·45	49·0	−21·2	−12·6	1·68	19·45	32·7

16	17	18	19	20	21	22	23	24	25	26	27
$w_1 + w_2$	$\frac{1}{2}(w_1+w_2)(h+\frac{b_0}{n})=A'$	$w_1 - w_2$	$\frac{w_1-w_2}{3R}=f$	$A' \times f = c$	$A' + c = A_c$	$A_c - \frac{b_0^2}{n} = A$	k	kA	Σ	x	Volume
71·1	590·8	2·3	0	0	590·8	390·8	1	390·8			
73·9	638·1	3·5	0·0006	0·4	638·5	438·5	4	1754·0			
82·4	782·0	8·0	0·0027	2·1	784·1	584·1	2	1168·2			
88·0	881·8	11·4	0·0057	5·0	886·8	686·8	4	2747·2			
95·9	1036·7	17·3	0·0115	11·5	1048·6	848·6	1	848·6	6908·8	$\frac{100}{3 \times 27}$	8529
81·7	794·5	16·3	0·0108	8·6	603·1	603·1	1	603·1	1451·7	$\frac{67}{2 \times 27}$	1801

Total 10 330 cu yd

When this imaginary machine is cutting, the volume of earth in the bowl increases, and the greater the depth and/or width of the cut, the greater will be the rate of increase of volume of material in the bowl, and the steeper will be the gradient of the plotted curve. When the machine stops cutting, and starts spreading, the ordinate will cease to increase, and commence to decrease, and a local maximum (zero gradient) will occur on the curve. Similarly, when the machine changes over from filling to cutting, a local minimum will occur on the curve.

From the above, it will be apparent that a positive gradient to the curve indicates cut, and a negative gradient indicates fill, and in fact, since $V = \int A \, dx$, the gradient of the mass-haul curve at any point will be equal in sign and magnitude (subject to a possible scale factor) to the cross-sectional area of the cutting or embankment.

The shape of the curve for a particular project, i.e. a horizontal line when there is no cut or fill, an inclined straight line when the cross-sectional area is constant, and a curve with a changing gradient when the cross-sectional area is varying, will soon become apparent from the table of cross-sectional areas from which the volumes, and hence the cumulative volume is calculated.

Although certain indications, i.e. whether it is cut or fill, and hence the sign of the mass-haul curve gradient, the position of changes from cut to fill, and hence the position of local maxima on the curve, etc., may be obtained from the longitudinal section, care must be taken in trying to relate gradients on the curve with the depths of cut and fill on the section. If cuttings and embankments with side slopes are used, the areas of the cross-sections will not be directly proportional to the depths of cut and fill, and if the formation width on the value of the side slopes vary, there will be even less correspondence between

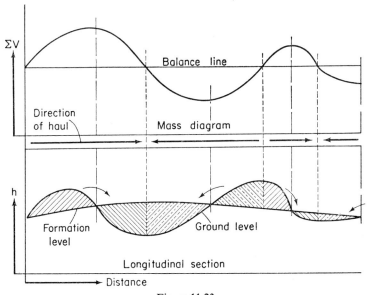

Figure 11.23

the gradients of the mass-haul diagram and the ordinates of the longitudinal section.

Since the ordinates on the mass-haul diagram represent cumulative volumes, it will be further apparent that for any two points on the curve having the same ordinate, the volume in the bowl of the hypothetical machine will be the same, and therefore, the volume that has been dug, and the volume that has been spread between those two chainages will be equal. Two points on a continuous curve that have the same ordinate must of course, embrace at least one local maximum or minimum, and the vertical distance between the points and this highest or lowest point will represent this volume which has been dug and then deposited again, i.e. is equal to the volume of the cutting, which is equal to the volume of the embankment between the two chainages. Two points on the curve having the same ordinate will, of course, be indicated by the points at which a horizontal line drawn on the diagram, cuts the curve. Any such horizontal line is referred to as a *balance line* since there is a balance of cut and fill between the chainages of all the points where it cuts the curve (see figure 11.23).

Since a positive gradient indicates cut, and a negative gradient indicates fill, it should be clear that when the curve lies above the balance line the direction of movement of the earth is forwards (i.e. increasing chainage) and when the curve lies below the balance line, the direction of movement is backwards. Thus for a continuous balance line, as in figure 11.23, the points where it cuts the curve also indicate the chainages at which the direction of haul changes.

If two (or more) balance lines are drawn on the diagram, as shown in figure 11.2.4, the earth from the cutting A is moved forwards and just fills the embankment B, and similarly the spoil from cutting C is moved backwards and just

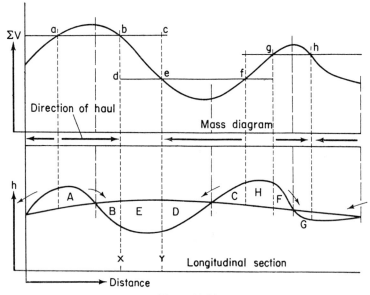

Figure 11.24

fills embankment D. If this scheme is adopted, there is no material available to fill the embankment E between chainages X and Y, and filling will have to be obtained from a borrow pit alongside the route in the vicinity of E or otherwise imported onto the job. Similarly, as the spoil from cutting F when run forwards, will be sufficient to fill G, the earth that is excavated from cutting H will have to be dumped alongside the route adjacent to H, or otherwise removed from the job.

It is customary to refer to these procedures as borrowing, when material is brought on to the job, and wasting when material is removed from the job, and it will be seen that when the balance line jumps down (as from ab to ef) borrowing must take place, and when it jumps up (as from ef to gh) wasting must occur.

If the borrowed material is brought into the embankment E, transversely, all along the route between the chainages X and Y, no balance line need be drawn for this portion. If, however the material is imported at one point the appropriate lines can be drawn. The continuation of the balance line ab to c, and then jumping down to e, would indicate that a volume equal to the ordinate ec was being imported at Y, and since the curve between X and Y lies below the balance line ac the direction of haul would be backwards. If however the line ab jumps down to d, and then continues to e and f, the material is being imported at X, and since the curve now lies above the balance line, is being moved forwards. The material could, of course be brought in at any single point between X and Y, which would be represented by a balance line somewhere between ac and df.

It has been seen that drawing more than one balance line on the diagram introduces the necessity to waste or borrow, which, apart from that required at one end of the job to accommodate any overall imbalance in the cut and fill, will involve an additional amount of excavation and filling, and before this can be justified some further factors will have to be investigated.

The area under the mass-haul curve ($= \int V \, dx$) will be the product of volume and distance hauled, and this product is known as the *haul*, the units, used in practice being usually either 'cubic yards, yards' (i.e. 1 cu yd moved through a horizontal distance of 1 yd), or 'station yards' (a dimensionally incorrectly named unit, denoting 1 *cubic* yard, moved through one 'station', i.e. moved through a horizontal distance of 100 feet).

Two simple examples will demonstrate how the mass-haul curve can be used to find out whether intermediate borrowing or wasting is desirable, and if so, the best positions for it to be carried out.

Referring to the material that has to be imported to fill embankment E in figure 11.24, the total volume required is given by the ordinates db or ec. If it is brought in at X and run forwards, the 'haul' will be given by the area bde, enclosed between the curve and the balance line de. If it is brought in at Y and run backwards, the 'haul' will be given by the area bce, enclosed between the curve and the balance line bc. As the second of these two areas (due to the shape of the curve) is smaller, importing at Y and running backwards is the more economical procedure. It should be noted that the *average haul distance* for this work can be found (if required) by dividing the 'haul' by the volume, i.e. dividing the area bce by the ordinate ec.

Similarly, for the whole job represented by the curve in figure 11.25, which has an overall excess volume of cut over fill equal to the ordinate qs which will have to be wasted somewhere. If the single balance line pq is used the direction of haul will be forwards, the whole of the excess will be wasted at the end of the job, and the 'haul' will be given by the area enclosed between the curve and the

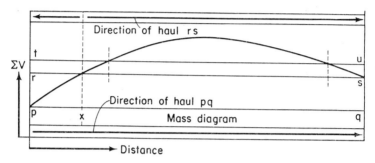

Figure 11.25

balance line pq. If however, the balance line rs is used the directions of haul will be both backwards and forwards from X, the whole of the excess volume will be wasted at the beginning of the job, and the 'haul' will be given by the sum of the two areas enclosed between the curve and the balance line rs, which is considerably less expensive than if the balance line pq was used. It will be realized that the area of the rectangle prsq represents the 'haul' involved if the whole of the excess volume was moved from one end of the job to the other.

In this project the most economical scheme from the point of view of haul, is one employing the balance line tu, in which the sum of the areas enclosed between the curve and the balance line is the minimum possible. This however involves increasing the volume to be wasted by the ordinate rt, and introducing an equal volume (su) to be borrowed at the end of the job. To decide whether this will in fact produce a cheaper scheme than that employing the balance line rs, will require a knowledge of the actual costs of borrowing, wasting, and hauling so that the cost of an increase in borrow and waste can be compared with the saving arising from a decrease in haul.

In looking for the most economical scheme a further factor must be taken into account.

When the excavated material is loaded into lorries, dumpers, etc. for transport, the cost is not directly proportional to the 'haul' (i.e. product, volume × distance), since the cost of loosening, getting out, loading, dumping, spreading and consolidating will be constant per cubic yard, regardless of the distance run by the vehicles. When the earth only has to be moved short distances, it may not be loaded into vehicles at all, or only into 'short-run' vehicles.

For these reasons it is common practice to divide the material into two categories.

Volumes of Earthworks

(i) Material that is to be moved through distances less than an agreed amount known as the *free-haul distance* (fixed by the type of plant envisaged for the job), such material being charged at a unit price per cubic yard, regardless of the distance moved (provided of course it is less than the free-haul distance), and this volume of material is known as the *free-haul volume*.

(ii) Material that is to be moved through distances greater than the free-haul distance, and which is charged for at the same rate per cubic yard as the free-haul volume, *plus* an extra charge at a unit rate per cu yd yd of *haul*, for the distances it is moved *in excess* of the free-haul distance. This volume of material is known as the *overhaul volume* and the product of the volume and the excess distance through which it is moved as the *overhaul*.

A simple example should make this clear.

Referring to figure 11.26, on which an additional balance line equal in length to the free-haul distance (say 100 yd) has been drawn. Between chainages X and Y the total volume of material to be moved is given by the ordinate ac (= 1 000 cu yd,) at *x* shillings per cu yd. Of this volume bc (= 300 cu yd) is the free-haul volume and ab (= 700 cu yd) is the overhaul volume.

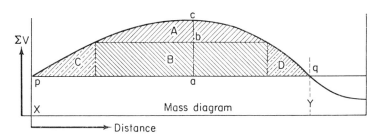

Figure 11.26

The total haul between X and Y is given by the whole area between the curve and the balance line pq. Of this, the area A (= 21 000 cu yd yd) represents the free-haul for the free-haul volume bc, all of which is moved a distance less than 100 yd, the *average* haul distance for this material being 21 000/300 = 70 yd. The area B represents the free-haul for the overhaul volume ab, being the haul for moving it the first 100 yd of its journey, and so included for in the rate of x shillings per cu yd. The areas C and D (= together, 43 000 cu yd yd) represent the haul for the overhaul volume ab, being moved through a distance beyond the first hundred yards, and so the overhaul, which is charged at y shillings per cu yd yd.

The total cost of this part of the job is therefore made up as follows:

Free-haul volume + overhaul volume = 1 000 cu yd at x s/cu yd
= 1 000 x shillings

Overhaul = 43 000 cu yd yd at y s/cu yd yd = 43 000 y shillings
Total cost = (1 000 x + 43 000 y) shillings

If required, the *average* distance moved by the overhaul volume is (43 000 + 70 000)/700 $\simeq$ 161 yd. The *average* overhaul distance = 43 000/700 $\simeq$ 61 yd, i.e. 161 yd minus F.H.D. = 161 − 100 = 61 yd, and the average distance moved by all the material is

$$(43\,000 + 70\,000 + 21\,000)/1\,000 = 134 \text{ yd}$$

In practical earthwork projects, one further factor has to be taken into account, which however, does not in any way affect the foregoing arguments.

If 1 cu yd of solid rock is excavated, and subsequently used as filling, however carefully the consolidation is carried out it will occupy a larger volume, perhaps 1·3 cu yd in its new position. This is known as 'bulking', and the material is said to have a 'bulking factor' of 1·3.

If, on the other hand, 1 cu yd of clay is excavated, and is consolidated carefully at its optimum moisture content, when used as filling, it may only occupy say 0·9 cu yd, in which case it is said to have a consolidation factor of 0·9.

All material, of course, when loosely loaded into a vehicle for transport, will occupy a greater volume than it did in the ground, and this is also referred to as 'bulking', the bulking factor in this case being somewhat larger than that referred to above. This bulking of materials in transit, however, although it affects the constructor in deciding his haulage costs, does not affect the mass-haul curve, for which only volumes in the ground are considered.

The bulking and consolidation factors first referred to, do, of course, affect the mass-haul diagram, which aims at balancing cut and fill. All that is necessary to allow for this phenomonen is, prior to calculating the cumulative volumes from which the curve is plotted, to convert all volumes of fill into 'equivalent cut'. This is most conveniently done by dividing all the cross-section areas for fill by the bulking or consolidation factor, since to fill a void of 1 cu yd with rock fill having a bulking factor of 1·3 will only necessitate 1/1·3 = 0·77 cu yd of virgin rock being excavated. Similarly to fill a void of 1 cu yd with earth having a consolidation factor of 0·9, it will be necessary to excavate 1/0·9 = 1·11 cu yd of virgin soil.

From the foregoing, it can be seen that the mass-haul diagram can be used in two ways,

(i) To assist the designer in determining the most suitable profile and horizontal lay-out for the job—In this case a diagram is drawn for a trial profile and lay-out, the curve is analysed, trial balance lines, etc. drawn and modifications made to the profile and/or lay-out in order (as far as controlling heights, ruling gradients, etc. allow) to obtain balances of cut and fill, keep the total volume of earth moved to a minimum, and in particular, keep the overhaul as small as possible. It must be realized that any modification of the profile or lay-out will necessitate the recalculation of the individual and cumulative volumes, and replotting the curve.

(ii) To assist the constructor in planning and organizing the actual work—In this case the profile has been settled, and the volumes calculated and plotted. Trial balance lines and the resultant calculation of haul volumes, hauls, haul distances, directions of haul, etc. are then carried out in order to find the most economical sequences of construction, determine the requirements for plant and transport, locate the best positions for borrow pits, spoil heaps, and so on.

12 *Cadastral Surveying*

12.1. INTRODUCTION

12.1.1. THE CADASTRAL PLAN

Any definition of cadastral surveying must be related to the meaning of the term 'cadastral plan' as this is the usual end product of a cadastral survey.

A cadastral plan may be defined as a map or plan which purports to identify a particular parcel of land for purposes of ownership and registration and thereby shows the boundaries with a degree of accuracy as defined in the legislation governing the transfer and holding of land. It is normally required that the information on the plan shall enable the boundaries to be redemarcated on the ground with a degree of tolerance as defined in the legislation. It is usual to find that such a plan can only be produced by a survey executed by a duly authorized surveyor. The term licenced surveyor is frequently used for this authorized surveyor. The licence is issued by a government department, government surveyors working in an official capacity are almost always regarded as duly authorized surveyors in the same sense as licenced surveyors.

Plans made specially for a court action, such as a boundary dispute, are not usually regarded as cadastral plans even though they may require certification by a duly authorized surveyor before they can be tendered in court as evidence. These plans can be contrasted to the cadastral plan which only shows boundaries that are legally settled and specifically defined.

All cadastral plans must carry a certificate, for example:

> I . . . . . , licenced surveyor acting in a duly authorized capacity certify that this plan is faithfully and correctly executed and accurately shows the land within the limits of the description given to me by . . . . . (here is inserted the legal instrument or other authority)
>
> Signed . . . . .   Date . . . . .

It is usual for the cadastral plan to require a check by an official survey organization, in these circumstances the plans will carry an official mark or seal. False certification of plans, and surveys found to be in error are both dealt with in the legislation. Usually the former is regarded as a criminal offence whereas the latter requires rectification by the surveyor concerned.

Other features of the legislation cover rights of entry, notices requiring attendance of interested parties to point out contiguous boundaries or rights to the land being surveyed. Technical aspects of the survey are usually dealt with under rules or regulations made under the authority of the act by some official mentioned in the act. These technical details cover the nature of the boundary posts, clearing of boundaries, execution of the survey, standards of accuracy and keeping of records. Fees, penalties, and compensation for damage may be covered in either the act or the rules.

12.1.2. DEMARCATION OF BOUNDARIES

In general the requirements for the demarcation of a boundary are as follows.

All boundaries shall be clearly defined in their entirety, it is usually sufficient for the boundary line to be cleared to a width of 6 or 8 feet, i.e. wide enough to allow passage along the boundary. Such a requirement is essential to ensure that accidental encroachment does not take place. Boundary posts must be erected at frequent intervals along the boundary, a distance of 3 000 ft is reasonable, and also at every change of direction. They must also be erected where roads, paths, and streams enter the area of the survey. The object of these boundary posts is to indicate to the public the limits of the land. Whereas the position of the boundary posts is established by the surveyor it is usual for the landowner to be required to clear and fence his boundaries within a specific period after taking occupancy of the land. The positions of all boundary posts must be coordinated and a locality sketch of the station prepared.

The problem of demarcation becomes more complex when a natural feature such as a river forms one of the boundaries. When one bank of the river forms the boundary, the boundary posts shall be placed above flood level. The direction of the boundary and the distance from these posts to water level during the dry season is indicated on the plan. The whole length of the boundary, i.e. the river bank, must be detail surveyed and will be shown on the cadastral plan. In the case of other curvilinear boundaries such as woods, tracks or paths, boundary posts are placed adjacent to the boundary and the boundary is detail surveyed (see Chapter 9). These posts are only indicatory in that the boundary itself follows a natural feature. It is therefore desirable that they should differ from a true boundary post in some essential respect. For instance they may be smaller or not carry enamel plates carried by a true boundary post. It is most essential that an indicatory post shall carry a different inscription to a true boundary post. For instance I.P. No. . . . instead of B.P. No. . . . To avoid confusion it is essential that B.P.s and I.P.s be numbered in the same series, quite obviously if there were posts marked I.P. 135 4 and B.P. 135 4 one might be confused with the other in the survey records.

The following are used for demarcation when no natural feature exists:

(i) Concrete pillars—various sizes are used, usually they are cemented into position and carry identification marks possibly moulded into the wet concrete.

(ii) Iron or steel rods—driven deep into the ground and carrying an identification mark on a bolted or welded plate. These are expensive.

(iii) Wooden pegs—driven deep into the ground, no identification mark is carried as the pegs are only suitable as temporary marks.

(iv) Dressed stone—various sizes and materials locally available are used, identification marks must be cut by hand. This feature makes the mark expensive although it may be particularly durable.

(v) Heavy nails—these can be left in the road or in nearby buildings, no identification is possible and their use is usually restricted to indicatory purposes.

(vi) Rock cairns—these are suitable for large holdings, can frequently be identified on air photographs. Beneath the cairn a fine mark is left to identify the boundary exactly.

12.1.3. CONTROL FOR CADASTRAL SURVEYS

Control points for cadastral surveys should be carefully placed so that any new survey can be readily executed without excessive field work. There should be a considerable density of points particulary where there is a private practising profession. With this in mind it is generally found that in township work traverses are preferred to triangulation (see §7.5 on the cadastral traverse). In rural areas permanently marked triangulation is often more suitable particularly if the ground is open and a breakdown network can readily be established.

12.1.4. EXECUTION OF CADASTRAL SURVEYS

Considering the case where instructions are given for the survey for title purposes of an area of definite acreage in a particular vicinity. Such instructions are usually accompanied by a rough site plan upon which little reliance should be placed. Details of adjoining cadastral surveys should be obtained.

In theory demarcation should be carried out before survey is undertaken; this is not always practicable, particularly when an exact acreage is required and possibly old boundary lines have to be re-established. If the site is open a preliminary plane table survey of the area can be made and approximate positions for the boundary posts established. In forest country a compass or rough theodolite traverse may have to be run but care should be taken not to clear lines in such a way that the owner may later mistake such clearing for the actual line of the boundary.

Once the approximate location of the boundary posts (B.P.s) has been established, temporary marks are left and a rigorous survey executed to coordinate these points. The final positions of the B.P.s by reference to the temporary marks can now be obtained and they are exactly located from their nearest temporary mark on a computed bearings and distances (corrected for scale, slope, and temperature in the reverse direction). As scale, slope, and temperature in the reverse direction).As the B.P.s are placed in position they must be checked by an independent method. Figure 12.1 illustrates a piece of land which has to be of a specific size. The traverse through the temporary marks A', B' $\cdots$ G'. is based on the existing B.P.s, BP_1, BP_2, BP_3, BP_4, whose positions have been checked by measurement of the included angles and distances between. The B.P.s are located from new marks A', B', ... on computed bearings and distances. These are then checked as follows:

A must lie on line between BP_2 and BP_3, as it has been located from A' a check is provided by a direct sighting from BP_2 to BP_3, the exact location is checked by measurement to A from either BP_2 or BP_3. The experienced surveyor

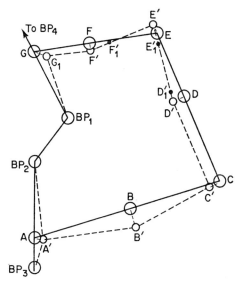

Figure 12.1

will have left a temporary mark on this line at an exact chainage during the measurements made to check the stability of the existing B.P.s.

B must lie on line between A and C, this can be checked by observation at A which will check the location of both B and C.

D and E are checked by leaving two points D_1' and E_1' on line between D' and E'. The angles $D'D_1'D$ and $E'E_1'E$ are measured and the distances $D_1'D$ and $E_1'E$.

F is similarly checked by observations from F_1'.

G must lie on the line BP_1 to BP_4 and on the line EF produced this is checked as is the location of D on line between E and C.

It is essential to prove every point placed on line and how this has been done should be mentioned in the field book. In checking the location of a B.P. good cuts are required.

When instructions are given for the survey of an already demarcated plot or when a definite acreage is not required the problem is much simpler. Care must be taken to ensure that the existing boundaries are maintained and points on line are proved.

12.1.5. FIELD BOOKS, COMPUTATIONS AND PLAN

These are legal documents and regulations are usually quite specific and clear as to how they shall be prepared. Usually it is required that in the field books:

(i) All entries shall be in ink.
(ii) All alterations shall be initialled.
(iii) All equipment used shall be specified.

(iv) All field measurements shall be recorded and dated. Cross references and an index are required.

Computations are required to be checked by suitable alternative processes. Areas must be computed and curvilinear portions determined by planimeter.

The plan is required to be at such a scale that the area to be depicted will cover more than 1 square inch of paper. It should show:

(i) True north, a scale bar, grid lines, title, reference number and date.
(ii) The location and numbers of all boundary and indicatory posts.
(iii) Survey data as required in the legislation. This may take the form of coordinates or bearings and distances between B.P.s.
(iv) Significant detail to identify the plot and the boundary lines.
(v) The area of the plot. Rectilinear and curvilinear areas are shown separately.
(vi) The actual limits of the area indicated by a uniform pink wash $\frac{1}{4}$ inch wide along the internal border of the plot. Where the boundary is curvilinear the wash follows such boundary the rectilinear limits being indicated by a pecked line.

12.2. THE DESCRIPTION

12.2.1. GENERAL CONSIDERATIONS

This is usually required when a system of deeds, registered or unregistered, is used. Only rarely is a description required under a system of registration of title. The purpose of the deed to real property is to evidence ownership, usually the deed takes the form of a conveyance. The description which goes with the deed must supply information to enable the boundaries to be identified on the ground both when the conveyance is made and at a later date.

Three principal types of description exist:

(i) By natural features, without numerical data.
(ii) Metes and bounds (numerical data).
(iii) Systematic referencing (this may be applicable in a systematic system of registration of title).

In type (i) trees, centres of roads, streams are features typically used; and also, in England, hedges, but see note below under type (iii).

In type (ii) a careful description of the starting point is required, from then on linear and angular dimensions are given.

In type (iii) a good and complete series of cadastral maps is required, provided these maps are kept up to date a description of the type—Township ...... Block ...... Plot ...... is more than adequate. The Ordnance Survey system of numbering fields on its large scale plan is typical of such a system.

The essential requirements of any description are clarity, brevity and accuracy.

Clarity. Words used must have an exact meaning, for instance north may mean true north, magnetic north, grid north; similarly bearings must be defined, true, magnetic or grid, either in the preamble to the description or in the deed itself. Terms such as parallel, right angle, etc. should be avoided unless that is exactly what is meant. Punctuation marks can cause ambiguity and care in sentence construction is important.

Accuracy. As far as practicable field observations should be used for the description rather than dimensions scaled from the plan. For this reason it is frequently required that all sides of a plot be physically measured regardless of how the survey was executed. A description should in the case of a mutation (sub-division) describe what has happened rather than what may be deduced to have happened from the original dimensions. For example (see figure 12.2) a

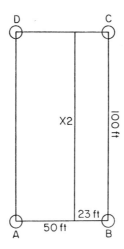

Figure 12.2

plot described as 50 ft by 100 ft from which a strip 23 ft wide is sold. The new description for the plot should read, Plot X2 less a strip 23 feet wide along the eastern boundary. Frequently in old plans the dimensions of plots is in error, in this case the plot X2 may have been 49·5 feet by 100 feet.

Brevity. For example numerical data is best given in a numerical form (36 feet instead of thirty-six feet). Frequently a tabular presentation is not acceptable.

12.2.2. RE-ESTABLISHMENT FROM OLD DESCRIPTIONS

Where doubt exists the following rules apply:

(i) Where an obvious error exists the attempt is made to render the description, and thereby the deed, valid. For instance if may be obvious that a whole tape length has been missed in one side.

(ii) The order of precedence when conflicting evidence is present is (*a*) natural features as against artifical (tree as against post), (*b*) artificial feature as against linear or angular measure, (*c*) adjoining boundaries if definitely identified carry precedence, (*d*) where conflict between area and dimensions exist the dimensions carry precedence.

(iii) The interpretation which appears from all the evidence to give effect to the intention of the original contracting parties should be taken.
(iv) Where two interpretations are possible the one that favours the purchaser should be taken.

12.2.3. EXCESSES AND DEFICIENCIES

In the case of a mutation it is not uncommon, as indicated above, to find that the dimensions given in an original description do not agree with those on the ground. Generally speaking the rule is that the excess or deficiency should be equally apportioned except if:

(i) All plots in a block have been given dimensions except one which is a remnant.
(ii) The plots are being set out (developed) one by one, then the last to be developed receives the excess or deficiency.
(iii) Erection of walls, hedges or buildings indicate an acceptance of the *status quo*.

When a plot is found to be in excess or deficient it may be that the above rules have been applied.

12.3. URBAN LAYOUTS

12.3.1. GENERAL PRINCIPLES

The unit in most cadastral surveys for urban purposes is the inch, such accuracy is costly, and can only be justified if the cost of the survey in an undeveloped state does not exceed a figure of around 20% of the value of the land. The methods used must be as economical as possible bearing in mind that:

(i) Adequate checks must be made in the field whenever a property pillar is placed.
(ii) All computations for setting out must be checked by independent means.

12.3.2. FIELD OPERATIONS

The following stages exist.

(1) *Reconnaissance.* A great deal of time is saved if during the reconnaissance the surveyor notes points which command the area and the degree of intervisibility possible between points in the layout. At this stage a working plan is prepared.

(2) *The paper traverse.* The working plan takes the form of the proposed layout superimposed upon a large scale plan of the area upon which all existing control should be plotted. Existing roads and buildings must be accurately shown.

The paper traverse is taken from the working plan. Suitable intervisible points are selected and scaled off, the bearings and distances between these points and the existing control are computed and laid out in the form of a traverse. Only a few points in the whole layout are so treated but these are points from which the rest of the layout can be set out.

Cadastral Surveying 595

(3) *Setting out the paper traverse.* Starting from existing control points the successive points in the traverse are set out. It is easier to set the bearing of a fixed line on the circle of the theodolite with the telescope correctly pointed and then turn until the required forward bearing comes up rather than to set off an angle. To lay out an exact distance, set out the approximate length, measure the slope, note the temperature, and compute in the field the length set out. This will differ from the required value and a small correction is applied with a short tape. The points so established are pillared and the setting out traverse is closed onto a fixed point. The misclosure, which will be represented by the difference between the located position for this fixed point and its actual position, should not exceed 1/3 000 of the total length of the paper traverse.

(4) *Control points.* The points set out during the paper traverse are now rigidly surveyed and their coordinates compared with those scaled from the working plan. If an appreciable difference exists the point is replotted.

(5) *Block points or block corners.* From the points of the paper traverse the block corners or block points are located. This is most suitably done by intersection, linear measurements being only used as an independent check. Intersections should be well conditioned, the apex angle should never be less than 30°. Simple string lining with two pegs on each line is the quickest and most efficient way of establishing the exact point (see figure 12.3). The coordinates of the block points must be scaled from the working plan and after pillaring the points must be coordinated, usually this can be done by 2 ray intersection checked by taping between nearby block points.

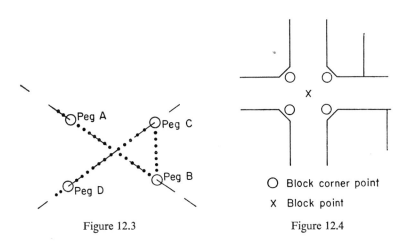

Figure 12.3   Figure 12.4

At each cross road there will be either four block corner points or one block point (see figure 12.4). With the later plot corners are located by offset from a series of points established along the centre of the road, with the former each plot corner is located by lining in between its block corners. The choice of which system to adopt depends upon the complexity of the layout. For a simple layout with small plots and narrow roads the block point is preferred.

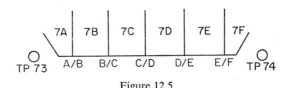

Figure 12.5

(6) *Plot points.* This a pure setting out operation by on line linear measurement only. If a line is to be produced a theodolite is used, transiting on both faces.

The distance between block points or corner points is computed from coordinates; the sum of the scaled distances of each plot frontage should equal this figure; if not adjust the frontages so that it does.

For example (see figure 12.5), block corner TP 73 to block corner TP 74 = 300·61 ft.

```
        Line  TP73-A : A-A/B : A/B-B/C : B/C-C/D : C/D-D/E
        Plan   25·0     25·0     50·0      50·0      50·0
        Use    25·05    25·05    50·10     50·10     50·10
              Line D/E-E/F : E/F-F : F-TP/4   Totals
              Plan  50·0        25·0  25·0    300·0
              Use   50·10       25·05 25·05   300·60
```

The check in setting out is the closure onto TP74 which should be correct to within 1 inch or say 0·10 feet. If there is a small but perceptible error each plot corner is adjusted. Plot corners are not surveyed, the setting out data is used to coordinate them if required. Re-establishment is always from the block or corner points. Internal points are located by a process similar to the frontage points except that the working line is the division between two sets of plots fronting on different roads. Where curves are involved the work becomes more complex, often it may appear to be as quick to set out by eye and (subsequently) survey. Alternatively block points are set at the end of all straights and plot points located by offset from a long chord (see §10.5.2.2.)

12.3.3. DESCRIPTIONS

For plots in an urban layout systematic referencing is to be preferred but if necessary plot data can be derived as follows: Bearing of frontage use bearing between block corner points, length of frontage use length used in setting out, similarly for the rear of the plot, side dimensions to correspond to side dimensions of block.

The areas of plots within a block should total to the area of the block, usually areas of individual plots can be determined by simple mensuration or by use of the planimeter; these values are then adjusted to equal the total area of the block.

12.4. DETERMINATION OF AREA

It is an integral part of a cadastral survey to determine areas. It is customary to express areas in acres and decimals of an acre or if the metric system is used, in hectares and decimals of a hectare.

Cadastral Surveying

12.4.1 GRAPHICAL METHODS

12.4.1.1. *Equalizing triangles*

If the figure is divided into triangles equalizing curvilinear boundaries and the base and perpendicular of each triangle measured.

$$\text{Twice area} = b_1 p_1 + b_2 p_2 + b_3 p_3 +$$

12.4.1.2. *Squared tracing paper*

If a sheet of tracing paper ruled in squares is placed over the plan, a count of squares and partial squares will give the area and provided the relationship between square size and an acre or hectare is known the area of the plot in these units is directly obtained.

12.4.1.3. *Intercepted Parallelograms*

If the figure is divided by a series of equidistant lines, the length of each line intercepted between the boundary is scaled and the sum of these lengths multiplied by the constant breadth will give the area. A device known as a computing scale adds by means of a cursor the length of each line to the previous sum.

12.4.1.4. *Trapezoidal and Simpson's Rules*

In the case of a very long narrow strip, such as a road or railway reservation a line can be drawn axially through the strip and perpendicular measured at regular intervals. (These may in fact have been measured in the field.) From the ordinates O and their common distance apart d the area can be calculated by the trapezoidal rule or Simpson's rule.

In the trapezoidal rule the area is assumed to be broken into a number of trapezoids and the area is given by:

$$(\tfrac{1}{2}O_0 + O_1 + O_2 + \cdots + \tfrac{1}{2}O_n)d$$

where $nd = L$ if the total length is L.

If there is an apex at one or both of the ends it is not disregarded but is treated as a zero ordinate and the resultant expression becomes:

$$(O_1 + O_2 + O_3 + \cdots + O_{n-1})d$$

Simpson's rule assumes that the boundaries being curved may be regarded as portions of a function of the third degree. Consider the curve represented in figure 12.6, it may be represented by the expression

$$ax^3 + bx^2 + cx + d = y$$

The area under the curve $A = \int_{-h}^{+h} y \, dx$

$$
\begin{aligned}
A &= \int_{-h}^{+h} (ax^3 + bx^2 + cx + d) \, dx \\
&= [\tfrac{1}{4}ax^4 + \tfrac{1}{3}bx^3 + \tfrac{1}{2}cx^2 + dx]_{-h}^{+h} \\
&= \tfrac{2}{3}bh^3 + 2dh = \tfrac{2}{3}h(bh^2 + 3d) \\
&= \tfrac{1}{3}h[(bh^2 + d) + 4d + (bh^2 + d)]
\end{aligned}
$$

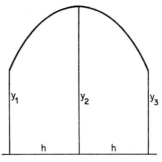

Figure 12.6

but
$$y_1 = -ah^3 + bh^2 - ch + d$$
$$y_2 = d$$
$$y_3 = ah^3 + bh^2 + ch + d$$

as
$$A = \tfrac{1}{3}h[(bh^2 + d) + 4d + (bh^2 + d)]$$
$$A = \tfrac{1}{3}h(y_1 + 4y_2 + y_3)$$

By extending this result a rule can be obtained for the area under a curve of any length, provided an odd number of equidistant ordinates be chosen.

Hence using Simpson's rule the area between alternate ordinates is given by

$$A = (O_0 + 4O_1 + O_2)\tfrac{1}{3}d$$

and considering five ordinates the area is given by

$$(O_0 + 4O_1 + O_2)\tfrac{1}{3}d + (O_2 + 4O_3 + O_4)\tfrac{1}{3}d$$
$$= (O_0 + 4O_1 + 2O_2 + 4O_3 + O_4)\tfrac{1}{3}d$$

The rule and its developed formula requires that there shall be an odd number of ordinates. If this proviso is not adhered to the extra portion must be calculated separately.

12.4.1.5. *Mechanical Integration*

(5) The planimeter is a mechanical integrator used for the measurement of plotted areas. The common feature of all planimeters is that a pointer is guided around the boundary which causes a displacement of another part of the mechanism such as to enable the area to be recorded.

12.4.2 THEORY OF THE PLANIMETER

The standard reference is H. Lamb's *Infinitesimal Calculus* (Cambridge University Press). Consider a rod AB of length b, suppose this is displaced to a new position A'B', then B'A' will meet BA at O not necessarily at a finite distance (see figure 12.7). The area AA'B'B will be equal to the difference between the two triangles OAA' and OBB', i.e.

$$\tfrac{1}{2}OB \cdot OB' \sin \delta\theta - \tfrac{1}{2}OA \cdot OA' \sin \delta\theta$$

but if the motion is small OB = OB', OA = OA' and sin $\delta\theta = \delta\theta$. Therefore

$$\text{Area swept out} = \tfrac{1}{2}\delta\theta(\text{OB}^2 - \text{OA}^2)$$
$$= \tfrac{1}{2}\delta\theta b(\text{OB} + \text{OA})$$

but $\tfrac{1}{2}\delta\theta(\text{OB} + \text{OA})$ is the small amount that the centre of the rod has moved, CC' in the figure. Let this be represented by dc. Then the area swept out is $b\, dc$ and the total area swept out during a finite movement of the rod is $b\int dc$. It is important to realize that the area does not depend upon the actual displacement of C the mid-point of the rod but upon the displacement of C perpendicular to each position of the rod.

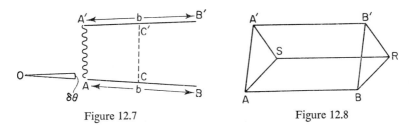

Figure 12.7　　　　　　　　Figure 12.8

When the rod AB is moved back towards its original position via the position SR as shown in figure 12.8 the quantity $b\int dc$ will represent the difference between the area of triangle AA'S and the area of triangle BB'R. Alternatively this may be expressed as the difference of the areas marked out by A and B. Provided that during the rotation of AB neither end describes a complete circle about the other before returning to its original position, it can be shown that $\int dc$ for the centre point C is equal to $\int dc_1$ for any other point A_1 on AB or AB produced (see figure 12.9).

$$dc = dc_1 + \text{A}'\text{C}'\, \delta\theta$$

and
$$\int dc = \int dc_1 + \text{A}'\text{C}' \int \delta\theta$$

If the mid-point of the bar returns to its original position without describing a complete circle

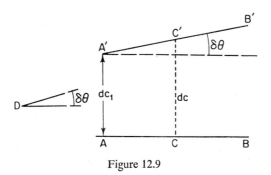

Figure 12.9

$$\int \delta\theta = 0 \quad \text{then} \quad \int dc = \int dc_1$$

12.4.2.1. *Construction of a typical planimeter* (See figure 12.10)

Two bars DA and AB are hinged at A, the point D being held fixed by means of a weight or needle point at some convenient point outside of the area to be measured. At B is a fine pointer which is passed along the boundary of the plot whose area is required. It is essential that B be brought back to its starting point. The motion of A will be along an arc of radius DA and provided a complete rotation about D is not made the area marked out by A will be zero as A passes along an arc and back along the same arc to its starting point. In this case $b\int dc_1$ is the area traced out by the pointer at B which is the required area.

The value of $\int dc_1$ is measured by a small roller the axis of which is parallel to AB and as a result the components of movements perpendicular to AB can be recorded by a dial attached to the roller and the component of any movement parallel to AB produces only a sideways slip in the roller and is not recorded.

The dial attachment to the roller may take the following form: the edge of the roller is graduated into ten divisions and provided with a vernier reading to 1/100 of a main division. There is a horizontal dial also graduated into ten divisions each being equal to a complete revolution of the roller with which it is geared. By use of the three scales it is possible to read the instrument to four significant figures. The area so determined may be expressed in square inches, square centimetres, or other units by suitable adjustment of the bar length,

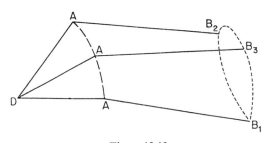

Figure 12.10

i.e. AB (b), graduation marks near the hinge at A being provided for this purpose.

If the diameter of the roller is known the correct setting can be determined as follows

Circumference of roller $= \pi d$

Area traced by one revolution $= b\pi d$

If it required that one revolution shall represent 10 square inches and b and d are in inches then b will be made to equal $10/\pi d$.

A large area can be determined by subdivision into smaller parts or by placing D, the fixed point, within the area. In the latter case a constant will have to be added to the area determined from the roller. This is an instrumental constant which varies in part with the value of AB (b) and may be established as follows (see figure 12.9).

$$b \int dc = \text{Area traced by B} - \text{Area traced by A} \qquad (1)$$

but
$$\int dc = \int dc_1 + A'C' \int \delta\theta \qquad (2)$$

hence
$$b \int dc = b \left(\int dc_1 + A'C' \int \delta\theta \right) \qquad (3)$$

from (3) and (1) $b(\int dc_1 + A'C' \int d\theta) = \text{Area traced by B} - \pi DA^2$
when $A'C' = \tfrac{1}{2}b$

$$\text{Area traced by B} = b \int dc_1 + \pi(b^2 - DA^2)$$

in all cases

Area traced by $B = b \int dc_1 +$ Constant with various values for b.

In determining areas with a planimeter the true scale of the plan or drawing must be taken into account; this includes shrinkage and any scale of projection and sea level factors applied to measured distances used in compiling the plan.

12.4.3. AREAS OF RECTILINEAR FIGURES FROM COORDINATES

The various methods of determining areas from coordinates all utilize the same general formula which by virtue of its inherent simplicity is strongly recommended.

Consider a triangle ABC with coordinates of the vertices A, B, C, $(E_1 N_1)$, $(E_2 N_2)$, $(E_3 N_3)$ figure 12.11.

Then Area of ABC = BCKL − BAML − ACKM

$$2\text{BCKL} = (N_2 + N_3)(E_3 - E_2)$$
$$2\text{BAML} = (N_1 + N_2)(E_1 - E_2)$$
$$2\text{ACKM} = (N_3 + N_1)(E_3 - E_1)$$

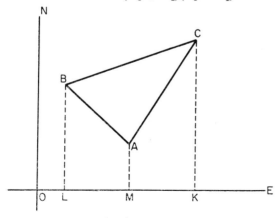

Figure 12.11

Regrouping terms

$$2ABC = (N_1E_2 + N_2E_3 + N_3E_1) - (N_1E_3 + N_2E_1 + N_3E_2)$$

Consider a figure whose vertices have coordinates as shown in figure 12.12

$$\text{Area triangle A} = \tfrac{1}{2}(N_1E_2 + N_2E_n + N_nE_1)$$
$$-\tfrac{1}{2}(N_1E_n + N_2E_1 + N_nE_2)$$
$$\text{Area triangle B} = \tfrac{1}{2}(N_2E_3 + N_3E_n + N_nE_2)$$
$$-\tfrac{1}{2}(N_2E_n + N_3E_2 + N_nE_3)$$

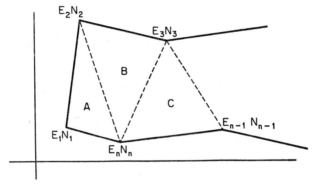

Figure 12.12

Similarly for each succeeding triangle and the area of the whole polygon will be given by A where

$$A = \tfrac{1}{2}(N_1E_2 + N_2E_3 + N_3E_4 + \cdots + N_{n-1}E_n + N_nE_1)$$
$$-\tfrac{1}{2}(N_1E_n + N_2E_1 + N_3E_2 + \cdots + N_{n-1}E_{n-2} + N_nE_{n-1})$$

In tabular form for a triangle this may be written:

$$\begin{array}{cc} N_1 & E_1 \\ N_2 & E_2 \\ N_3 & E_3 \\ N_1 & E_1 \end{array} = 2A$$

The oblique full lines should be multiplied and the products added, the difference between this quantity and the sum of the products of the oblique pecked lines will give twice the area.

This tabular form of treatment holds good with any polygon, the essential points to note are that the first station is repeated in the tabular layout and that twice the area is given.

A sometimes convenient variant of this treatment is to multiply the algebraic differences of the E's by the algebraic sum of the N's of each adjacent pair of coordinates and sum the products algebraically. This gives twice the area and as a check the process is reversed.

Cadastral Surveying

In tabular form for a triangle the method is expressed as:

$$\begin{array}{cc} \text{Differences} & \text{Sums} \\ (E_2 - E_1) \times & (N_2 + N_1) \\ (E_3 - E_2) \times & (N_3 + N_2) \\ (E_1 - E_3) \times & (N_1 + N_3) = 2A \end{array}$$

If expanded this forms an expression identical to that above. An alternative method very suitable for use with a calculating machine is as follows.

Consider the triangle ABC with coordinates of vertices as before, then it will be seen that twice the area is given by

$$(E_3 - E_1) \times N_2 + (E_1 - E_2) \times N_3 + (E_2 - E_3) \times N_1 = 2A$$

The procedure on a hand calculating machine is as follows:

(i) Set E_1 in multiplier register with zero in setting register.
(ii) Set N_2 in setting register.
(iii) Change E_1 to E_3 by successive positive and negative turns of the handle and use of the shift keys.
(iv) Set N_1 in setting register.
(v) Change E_3 to E_2.
(vi) Set N_3 in setting register.
(vii) Change E_2 to E_1.

This may be represented by the tabular presentation below where the curved arrow represents the changing of the number in the multiplier register with the right-hand number in the setting register.

$$\begin{array}{l} E_1 \\ \quad \curvearrowright \leftarrow N_2 \\ E_3 \\ \quad \curvearrowright \leftarrow N_1 \\ E_2 \\ \quad \curvearrowright \leftarrow N_3 \\ E_1 \end{array}$$

When more than three stations are involved the method is similar except that if there are an even number of stations the initial station is repeated, e.g.

(1) An odd number of stations

$$\begin{array}{l} E_1 \\ \quad \curvearrowright \leftarrow N_2 \\ E_3 \\ \quad \curvearrowright \leftarrow N_4 \\ E_5 \\ \quad \curvearrowright \leftarrow N_1 \\ E_2 \\ \quad \curvearrowright \leftarrow N_3 \\ E_4 \\ \quad \curvearrowright \leftarrow N_5 \\ E_1 \end{array}$$

(2) An even number of stations

$$\begin{array}{l} E_1 \\ \quad \curvearrowright \leftarrow N_2 \\ E_3 \\ \quad \curvearrowright \leftarrow N_4 \\ E_1 \\ \quad \curvearrowright \leftarrow N_1 \\ E_2 \\ \quad \curvearrowright \leftarrow N_3 \\ E_4 \\ \quad \curvearrowright \leftarrow N_1 \\ E_1 \end{array}$$

With a little experience the whole computation of an area can be carried out with only a list of coordinates, and a check computation can be carried out by carrying the E coordinates in the setting register instead of the N coordinates.

In all of these methods it is convenient to reduce the size of the figures being used by subtracting a constant. Given a list of coordinates, suitably reduced, the method of oblique multiplication is as convenient as any other method regardless of whether a calculating machine or logarithms are being used. If the latitudes and departures are available, as in the case of a traversed boundary, the use of the method of sums and differences may be preferred. When a large area is involved and a calculating machine is available the method given above is preferable as the sums of the products increases more slowly.

12.4.4. SUBDIVISION

A frequently occurring problem in cadastral surveys is the subdivision of a plot into given parts either by a line of given bearing or from a given point on the boundary.

12.4.4.1. *By a line of given bearing* (Figure 12.13 Example 1)

Calculate from coordinates the area of the whole plot.

On an accurately drawn plan of the whole plot draw a line from one station, or any other convenient point on the boundary, on the given bearing to cut off approximately the required area. For purposes of this explanation let the required area be Z.

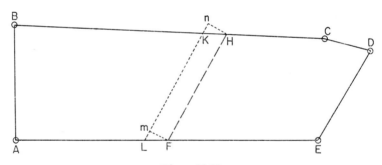

Figure 12.13

Knowing the starting point of the line of known bearing determine the coordinates of where it cuts the boundary. In the example station F is the starting point, the mid-point of the line AE, and H is where the line of known bearing cuts the boundary. The most convenient formula knowing bearings FH and BH and the coordinates of B and F is

$$N_p = \frac{N_A \tan AP - N_B \tan BP - E_A + E_B}{\tan AP - \tan BP}$$

$$E_p = E_A + (N_p - N_A) \tan AP = E_B + (N_p - N_B) \tan BP$$

Example. 1: Subdivision of Plot ABCDE (Figure 12.13) into Two Equal Parts by a line Parallel to DE

Example 1

Data

Station	Eastings (ft)	Northings (ft)
A	270 446.83	1132 246.44
B	270 445.71	1132 497.25
C	271 126.43	1132 461.56
D	271 227.89	1132 440.31
E	271 117.64	1132 249.77

(1) Bearing from D to E

E_D 271 227.89 N_D 1132 440.31
E_E 271 117.64 N_E 1132 249.77
      −110.25   −190.54

tan α   0.578 619
Bearing 210° 03′ 16″

Bearing from B to C

E_B 270 445.71 N_B 1132 497.25
E_C 271 126.43 N_C 1132 461.56
      +680.72   −35.69

cot α   0.052 430
Bearing 93° 00′ 05″

(2) Coordinates of F (Mid-point AE)

E_A 270 446.83 N_A 1132 246.44
E_E 271 117.64 N_E 1132 249.77
  541 564.47   2264 496.21

F   270 782.24   1132 248.10

(3) Coordinates of H (Intersection of line parallel to ED from F onto line BC)

N_B 1132 497.25   tan BH −19 073.129 E_B 270 445.71
N_F 1132 248.10   tan FH +0.578 619 E_F 270 782.24
  (tan−tan) −19 651 748

N_H 1132 472.79 E_H 270 912.24
                Check 270 912.25 ✓

(4) Area of West portion

Station	Eastings (ft)	Northings (ft)
A	270 446.83	1132 246.44
B	270 445.71	1132 497.25
H	270 912.24	1132 472.79
F	270 782.24	1132 248.10
2 Area	192 130.4670 sq.ft	
Area	96 065.2335 sq.ft	

Area of East portion

Station	Eastings (ft)	Northings (ft)
F	270 782.24	1132 248.10
H	270 912.24	1132 472.79
C	271 126.43	1132 461.56
D	271 227.89	1132 440.31
E	271 117.64	1132 249.77
2 Area	142 280.9387 sq.ft	
Area	71 140.4694 sq.ft	

(5) Area of whole plot (Check)

from data   2 Area   334 407.9905
          Area   167 203.9952

from (4)   2 Area West   192 130.4670
          2 Area East   142 280.9387
          2 Area   334 411.4057
          Area   167 205.7028
          Difference   1.7 sq.ft

A difference of this magnitude is to be expected as F and H are not exactly on line, as coordinates only correct to second place of decimals

Required half area $ABHF$ 96 065.23
Area required for $FHKL$ 12 462.38

(6) Bearing from A to E

E_A 270 446.83   1132 246.44
E_E 271 117.64   1132 249.77
  +670.81   +3.33

cot α   0.004 964
Bearing 89° 42′ 57″

(7) Bearings of H_n and F_m

α = 300° 03′ 16″ − 273° 00′ 05″ = 27° 03′ 11″
β = 300° 03′ 16″ − 269° 42′ 57″ = 30° 20′ 19″

Length F.H.

F   270 782.24   1132 248.10
H   270 912.24   1132 472.79
   259.59   +224.69

Equation

$12,462.38 = \frac{1}{2} x \{519.10 + x(0.585\,25 - 0.510\,69)\}$

$0.03720 x^2 + 259.59 x - 12,462.38 = 0$

$\therefore x = \frac{-259.59 \pm (67,386.97 + 1,856.39)^{\frac{1}{2}}}{0.07456}$

$= \frac{3.555}{0.07456} = 47.68$

$FL = x \sec β = 47.68 (0.86305) = 55.24$
$HK = x \sec α = 47.68 (0.89059) = 53.54$

(8) Coordination of K and L.

K Bearing from (1) 273° 00′ 05″

Distance	53.54		Cosine	0.052 35
Sine	0.999 63			+2.40
	−53.47		N_H	1132 472.79
E_H	270 912.24		N_K	1132 475.59
E_K	270 858.77			

L Bearing from (6) 269° 42′ 57″

Distance	55.24		Cosine	0.004 95
Sine	0.999 99			−0.27
	−55.24		N_F	1132 248.10
E_F	270 782.24		N_L	1132 247.83
E_L	270 727.00			

(9) Area West Portion

Station	Eastings ft	Northings ft
A	270 446.83	1132 246.44
B	270 445.71	1132 497.25
K	270 858.77	1132 475.59
L	270 727.00	1132 247.83
2 Area	167 203.6743	
Area	83 601.84	

Using the determined coordinates of H calculate the area of one of the excised portions of the main plot. Let this calculated area be Z'. Calculate the distance HF.

If line LK in the example is the line required to cut off the required area then:

$$Z - Z' = \text{Area of FHKL}$$

If Hn and Fm are perpendicular to FH, n and m lying on the line LK which is parallel to FH, then

$$\text{Hn} = \text{Fm} = x$$

Angles HKn and LFm can be determined from the known bearings of the sides and let these be denoted as α and β. Then in the example KL HF $- x \tan \alpha + x \tan \beta$.

N.B. In a different example this relationship may be different, hence the importance of an accurately drawn figure.

$$\text{Area of FHKL} = Z - Z' = \tfrac{1}{2}x\{2\text{HF} + x(\tan \beta - \tan \alpha)\}$$

this can be solved as a quadratic equation to find x.

$$x = \frac{-\text{HF} \pm \sqrt{\text{HF}^2 + 2(\tan \beta - \tan \alpha)(Z - Z')}}{\tan \beta - \tan \alpha}$$

Having found x, $\qquad$ FL $= x \sec \beta$

and $\qquad$ HK $= x \sec \alpha.$

These distances can be set out on the ground and the coordinates of L and K calculated. The area of the subdivided part is recalculated as a check on the whole process.

12.4.4.2. *Into given parts from a fixed point on the boundary*

It is required to subdivide the area shown in figure 12.14 from a definite point H mid-way between F and E so that a specific area Z is excized.

(i) Calculate from coordinates the area of the whole plot.
(ii) On an accurately drawn plan of the whole plot find the turning point on the boundary where the area bounded by a line from it to H is approximately as required. Point A in the figure.
(iii) Calculate the area AHFG. Let this be Z'.
(iv) Calculate the bearings and distances HA and AG.

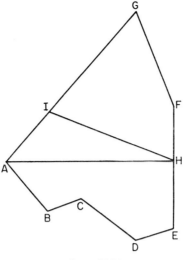

Figure 12.14

(v) The area of triangle AHI where I is the required point is $Z - Z'$ and

$$Z - Z' = \tfrac{1}{2} AH \cdot AI \sin HAG$$

in which only AI is unknown and can easily be found.

(vi) With AI known the coordinates of I can be calculated.

This distance can be set out on the ground and the area of the subdivided portion recalculated as a check on the whole process.

13 *Electromagnetic Distance Measurement*

13.1. INTRODUCTION

Since the introduction of Bergstrand's Geodimeter in 1946, and Wadley's Tellurometer in 1956 many manufacturers throughout the world have produced similar instruments based on the same principles as these originals. Discussion in this book is restricted to two typical instruments, (1) *the Geodimeter*, and (2) the *Tellurometer*, with a brief mention of others.

Electromagnetic measurement of terrestrial distances has not merely supplanted other forms of distance measurement; it has also caused considerable change of emphasis in survey technique. Major national control frameworks are being surveyed by *trilateration*, and *traverse* methods associated with steel towers even when a triangulation is possible. The use of *radiation* methods in the survey of control for aerial photography and tacheometric surveying is widespread, while many congested engineering works have found these instruments exceedingly useful. The problem of compensation to landowners for the damage done to crops in the course of taping lines has been greatly reduced as a result of these 'invisible' measurements. The duplex telephone system of the Tellurometer has improved communications in survey parties and has made practicable trigonometrical heighting by truly simultaneous vertical angle observations. At the end of this chapter we shall return to the field application of these instruments.

13.1.1. BASIC PRINCIPLES OF ELECTROMAGNETIC DISTANCE MEASUREMENT— E.D.M.

Electromagnetic measurement of distance depends on the fact that electromagnetic waves, of which radio and visible light are examples, travel through air with a velocity v of about 300×10^6 metres per second, and since by definition

$$\text{Distance } D = \text{Velocity } v \times \text{Time } t \qquad (1)$$

an unknown distance may be found by measuring the travel or *transit time* if the velocity is known. This simple principle has long been understood, but hitherto, time measurement to a sufficient degree of accuracy, i.e. $1/10^9$ second, has been impossible with portable equipment. The development of electronics has provided the technological means of achieving such accuracy.

13.1.2. MEASUREMENT OF TRANSIT TIME

Although direct readings of time are not made with the Geodimeter and the latest Tellurometers, but readings are converted to linear units for convenience, time units are non-the-less being used. It is therefore proper to consider both the Geodimeter and Tellurometer as *time measuring* instruments.

Any periodic phenomenon which oscillates regularly between maximum and minimum values may be analysed as a simple harmonic motion (s.h.m.) or as compounded of several s.h.m.'s. Consider figures 13.1(a) and (b). Point P

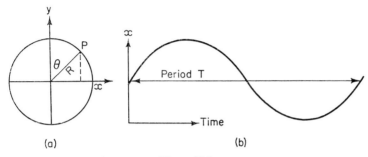

Figure 13.1

moves in a circle with constant angular velocity ω. The radius vector R will take up various positions represented by the phase angle θ. A graph of $x = R \sin \theta$ takes the customary sine curve shape of figure 13.1(b), in which the values of x are plotted against time. The process is reversible, i.e. any phenomenon which has a periodic variation represented by graph figure 13.1(b) can be thought of in terms of the circular motion illustrated in figure 13.1(a). Thus electrical currents, impulses of radio or light waves can be analysed in this way. The time taken for R to make one complete cycle or revolution is the period T, and the number of cycles occurring per second is the frequency F. Frequency is the reciprocal of the period, i.e.

$$F = 1/T \quad (2)$$

Since R moves with constant angular velocity ω, the *phase angle* θ can be used to measure fractions of the period T, just as the hands of a clock are used to measure time.

Consider an impulse x_1 phase θ_1 being generated at A figure 13.2(a) and

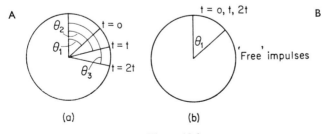

Figure 13.2

transmitted towards B. If the radiation at B is analysed in the reverse sense, the information of figure 13.2(b) will be obtained, i.e. it will illustrate the situation at A when the impulse was originally transmitted. Thus at time $t = t$ an analysis of the electromagnetic impulses at B exhibits the phase condition existing at A at time $t = 0$. Meantime the generating vector at A has moved on to position θ_2. If the impulse at B is instantaneously retransmitted back to A and sampled again it still exhibits a vector position θ_1, while the original vector has moved to position θ_3. Hence the phase difference $\theta_3 - \theta_1$ is a measure of the double transit time over the distance AB in units of the period T. Most electromagnetic distance measuring systems use this *phase comparison* principle.

Since the transit time over most survey lines exceeds one period T, phase comparison using only one period will give the amount by which the transit time exceeds an integral number of complete periods but not the integral number itself. For example, a watch with only a minute hand can only count the fractions of an hour. Although it would be possible in theory to provide a second period of say a hundred times the duration of the first to count the number of smaller periods, it is impracticable electronically to have a 'gear ratio' of more than 1/10. Instead, the number of whole periods in the transit time is counted by use of a time vernier system, similar to the rhythmic time signal used in astronomy. Several periods of slightly different duration are used to measure the transit time by a subtraction process.

13.1.3. MODULATION

In §13.1.2 the electromagnetic waves were considered as a simple wave trace. In practice, the wave pattern used for the measurement of transit time is complicated, and is derived from other wave patterns of much larger frequencies. For example, if the amplitude R of the harmonic motion in figure 13.1 is varied cyclicly between maximum and minimum values, the net effect of both oscillations can be depicted as a waveform shown in Figure 13.3(a). The larger

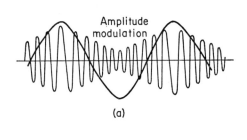

(a)

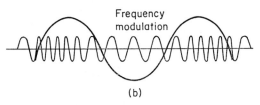

(b)

Figure 13.3

frequency curve is thus used to carry one of smaller frequency by a variation in amplitude, illustrated by the heavy line. This technique of using one frequency to carry the pattern of another is known as *modulation*; the basic wave being the *carrier*, the superimposed pattern being the *modulation*. This particular type of modulation is *amplitude modulation* (A.M.). Likewise, if the basic *frequency* is varied cyclically between limits, a *frequency modulation* takes place, as illustrated in figure 13.3(*b*).

In the Geodimeter system, the visible light carrier has a frequency of about 60×10^{11} cycles per second modulated to 60×10^6 (60 megacycles); and in the Tellurometer MRA/1 and 2 systems the carrier is 3 000 megs modulated to 10 megs. Since a radio signal cannot be limited to a narrow beam, part of the signal used in the measurement system is reflected from the ground and intervening objects, the amount depending on the reflection characteristics of the surface concerned, and a drop in the accuracy of the measurement occurs. Generally speaking, the greater the frequency of the carrier, the less is the reflection effect, though the modulation frequency is also a factor. The increase frequency of the Tellurometer MRA 3 has reduced the reflection effect somewhat. Little reflection occurs with visible light whose beam can be made very nearly parallel. For a detailed discussion of this and other scientific aspects of E.D.M., reference should be made to *The Proceedings of the Tellurometer Symposium*, (London: Tellurometer Ltd 1962). The Geodimeter and the Tellurometer systems are now considered separately since they differ in detail.

13.2. THE GEODIMETER (Model IV)

Detailed attention is confined to the Geodimeter Model IV, since to date (September 1964) this model is the most widely used, and its principles of operation are similar to the new Model VI. The following is a brief description of the instrument (figure 13.4). A beam of visible light emanates from a lamp,

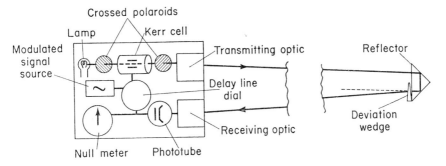

Figure 13.4

passes through a system of polaroids and a Kerr cell, and finally shines out from the transmitting optic towards a passive reflector at the other end of the line to be measured. The purpose of the polaroids and the Kerr cell is to superimpose on the visible carrier, the modulation pattern produced by the generator. The light is then reflected back to the receiving optic by a trihedral prism, and is retranslated into electrical impulses by a photoelectric cell. The phases of the outgoing and incoming signals are compared indirectly on a delay line dial, on which arbitrary units of up to 300 may be read. These units are converted to metres by tables specially constructed for each individual instrument. The final slant length of the survey line is obtained as detailed below, and corrections for meteorological conditions are applied if necessary. Two types of lamp are available; (1) the standard tungsten lamp, whose more modest power requirements can be supplied by battery, gives ranges of 1·5 km by day and 13 km by night provided visibility is good; and (2) the mercury vapour lamp, which requires a portable power generator, but which enables lines of 6 km by day and 25 km by night, to be measured (see table 13.1). The accuracy of the measured length is ±1 cm ±1/500 000 part of the distance.

13.2.1. REDUCTION OF READINGS TO SLANT LENGTH

The method of reducing the geodimeter readings to slant length is illustrated with reference to the example given in table 13.2, showing a typical field sheet.

The computation of the slant length falls into two parts: The derivation of quantities L_1, L_2, and L_3 from the delay line readings; and the derivation of the slant length from the quantities L_1, L_2, and L_3.

In this explanation, we consider the latter first.

13.2.2. SLANT LENGTH FROM L_1, L_2, AND L_3

The measurement of a length D can be compared to taping this length with a short tape of length x_1, such that

$$D = nx_1 + L_1 \tag{3}$$

where n is a whole number, and L_1 is a fractional part of x_1.

In taping the line the surveyor counts the number of tape lengths n, but the geodimeter cannot do so and merely records L_1. To find the value of n, the line is measured again with a second tape x_3 which is slightly shorter than x_1. Then

$$D = n_3 x_3 + L_3 \tag{4}$$

In the Geodimeter Model IV $x_1 = 5\cdot000$ metres and $x_3 = 20x_1/21 = 4\cdot761\ 90$ m. If the line D is less than $20x_1$, n_3 must equal n, since an extra length x_3 is only made up when the line is $20x_1 = 100$ metres or greater. Thus if $D < 100$ m, equation (4) becomes

$$D = nx_3 + L_3 \tag{5}$$

Equating D from (3) and (5)

$$L_3 - L_1 = n(x_1 - x_3) \tag{6}$$

$$= n\left(x_1 - \frac{20}{21}x_1\right) = \frac{nx_1}{21}$$

$$\therefore\ nx_1 = 21(L_3 - L_1)$$

substituting in (3)

$$D = 21(L_3 - L_1) + L_1 \tag{7}$$

Table 13.2. Geodimeter Booking Sheet [Model IV]

Geod. number NASM 4 _116_	Time on _12:05_ Time off _12:30_				Geod. Stn. _Stable Hill_ Reflector Stn. _Long Cross_							
Temp.° _56° F_ Press. _29·9 in Hg_ Ca^n. from nomogram $+0·149$	Geo. eccent. $+0·002$ Geo. const. $+0·220$ Refl. eccent. $-$ Refl. const. $-0·030$ At^m. corr. $+0·149$ Sum $\overline{+0·341}$				Height _293·60_ Date _2·1·64_ _feet_ Height _989·75_ _feet_				Observer _M. M._ Booker _J. H._ Type of reflector _Standard_			
Phase	F_1				F_2				F_3			
	C	S/O	R	S/O	C	S/O	R	S/O	C	S/O	R	S/O
1	54	S	57	O	60	S	65	S	54	S	34	S
2	58	O	50	S	64	O	74	O	60	O	44	O
3	54	S	57	O	60	S	65	S	54	S	34	S
4	58	O	50	S	64	O	74	O	60	O	44	O
(2)+(3)	112		107		124		139		114		78	
(1)+(4)	112		107		124		139		114		78	
Total	224		214		248		278		228		156	
Mean	56·0		53·5		62·0		69·5		57·0		39·0	
Mean, metres	0·750	S	0·721	O	0·798	S	0·890	SS	0·478	S	0·338	S

$L = R - C$ (If $-$ve add U and change S to O or O to S)	(2·500) 3·221 0·750 $\overline{2·471}$ SS	(2·494) 0·798 $\overline{0·092}$ SS	(2·381) 2·719 O 0·478 $\overline{2·241}$ OS
If SO or OS add U	(2·50) $\overline{2·471}$	(2·494) $\overline{0·092}$	(2·381) $\overline{48·622}$
$A = L_2 - L_1$ $B = L_3 - L_1$ (If $-$ve add 2U)	L_1 _2·471_	(4·988) L_2 _5·080_ $-L_1$ _2·471_ A _2·609_ $A \times 400$ _1043·6_	(4·762) L_3 _48·622_ $-L_1$ _2·471_ B _2·151_
$A \times 400 - B \times 21$	$A \times 400$ _1043·6_ $-B \times 21$ _45_ Nearest hundreds E _1000_		$+B \times 20$ _43·02_ $B \times 21$ _45·171_ Nearest multiple of 5 F _45_
$D' = E + F$	D' _1045_ m		

	D_1	D_2	D_3	
$D_1 = D' + L_1$ $D_2 = D' + L_2 - K_2$ $D_3 = D' + L_3 - K_3$ $K_2 = D' \times 0·0024936$ $K_3 = F \times 0·0476192$	D' 1045 $+L_1$ 2·471 D_1 1047·471	D' 1045 $+L_2$ 5·080 1050·080 $-K_2$ 2·606 D_2 1047·474	D' 1045 $+L_3$ 4·622 1049·622 $-K_3$ 2·143 D_3 1047·479	Mean D 1047·475 $+n \times 2000$ — Distance 1047·475 Total corr. $+0·341$ Corr. slope $-3·195$ Horizontal 1044·621

To feet multiply by 3·28084558

Remarks. _Not reduced to sea level_

Example: If $D = 47\cdot475 = 45 + 2\cdot475$
$$= (9 \times 5) + 2\cdot475$$

$n = 9$ and the Geodimeter would record $L_1 = 2\cdot475$ m. Then using the second 'tape' x_3, from (5)

$$D = (9 \times 4\cdot761\ 90) + L_3 = 47\cdot475$$

whence $L_3 = 4\cdot168$ m, which is the reading that would be obtained in this instance. Then from L_1 and L_3 the value of n is obtained from

$$L_3 = 4\cdot618$$
$$L_1 = 2\cdot475$$
$$\therefore\ L_3 - L_1 = 2\cdot143$$
$$\therefore\ nx_1 = 21(L_3 - L_1) = 45\cdot003$$

The small residual 0·003 is disregarded since it arises from the lower precision resulting from the multiplication of $(L_3 - L_1)$ by 21. In practice small errors in reading will produce larger discrepancies, and provided there is no ambiguity, we accept the nearest multiple of 5m for $21(L_3 - L_1)$. The amount by which $21(L_3 - L_1)$ differs from such a integral number of 5 metres is a good indication of the state of adjustment and alignment of the instrument. If D exceeds 100 m the above process enables us to determine the portion in excess of a whole number of units of 100 m only. The exact number of such units may often be estimated from a diagram. However, the instrument is equipped with a third 'tape' of length $x_2 = 400x_1/401 = 4\cdot987\ 53$ m which may be used in the same manner as x_3. Provided D is less than $400x_1$ (2 000 m), the subtraction of the L_2 and L_1 readings will give the number of units of 5 m contained in the line.

Example: If $D = 1\ 047.475$ m $= 1\ 045 + 2\cdot475 = 209x_1 + 2\cdot475$, n is 209 and L_1 is 2·475. Then L_2 is derived from

$$D = (209 \times 4\cdot987\ 53) + L_2 = 1\ 047\cdot475\ \text{m}$$

i.e. $L_2 = 5\cdot081$. Whence from the Geodimeter readings

$$L_2 = 5\cdot081 \quad \text{and} \quad L_1 = 2\cdot475$$

subtraction gives

$$L_2 - L_1 = 2\cdot606 \quad \text{and} \quad 401(L_2 - L_1) = 1\ 045$$

to the nearest 5 metres. Again in practice this multiplication will not produce an exact whole number of 5 m units. From all three readings L_1, L_2, and L_3 the line of length is obtained as follows:

Length from 0 to 5 metres—from L_1
Length from 5 to 100 metres—from $L_3 - L_1$
Length from 100 to 2 000 metres—from $L_2 - L_1$

e.g. in the example quoted
$$L_1 = 2\cdot 475 \text{ m}$$
$$21(L_3 - L_1) = 45$$
$$401(L_2 - L_1) = 1\ 045$$

In the geodimeter field, sheet table 13.2, $L_3 - L_1 = B$, and $(L_2 - L_1) = A$, $21B = F$ and $401A - F = E$ and $D' = E + F$. Many operators prefer to work only in terms of L_1, L_2, and L_3 and their multiplying factors 21, and 401.

For lines greater than 2 km an approximate length is required to an accuracy of about 1 km. This will be scaled from a map or from aerial photography or by estimation on the ground. If the above line had been 7 047·475, the L readings would be the same as given above. An estimate would give the length as say 7 km i.e. three units of 2 km plus a fraction to be measured by the geodimeter.

Further values of L_1 can be obtained from the L_2 and L_3 readings as follows: Since $nx_1 = 21(L_3 - L_1)$

$$L_1 = L_3 - \tfrac{1}{21}(21B)$$

e.g. in the example

$$L_1 = 4\cdot 618 - \tfrac{45}{21}$$
$$= 4\cdot 618 - 2\cdot 143 = 2\cdot 475$$

and $\quad L_1 = L_2 - \tfrac{1}{401}(401B) = 5\cdot 081 - \tfrac{1045}{401} = 5\cdot 081 - 2\cdot 606$
$$= 2\cdot 475$$

The mean of the three values of L_1 is accepted.

13.2.3. DERIVATION OF THE QUANTITIES L_1, L_2, AND L_3 FROM DELAY LINE READINGS

In fact the unit of measurement used in the Geodimeter is the period of oscillation T, and fractions of it θ. These readings are then converted to equivalent lengths x and L via the formula

$$D = \tfrac{1}{2}v_s w \tag{8}$$

where w is the transit time of the light signal over the double distance $2D$, and v_s the velocity of light under standard meteorological conditions. This conversion of readings to metres is carried out by *delay line tables* drawn up *empirically for each of the three frequencies of each individual geodimeter*. It is thus essential that the requisite instrument number is booked during measurement if an organization is operating several instruments.

Again, instead of the vector R figure 13.2(*a*) making a complete cycle, at the halfway stage a 180° phase reversal occurs which in effect returns the vector to zero, and the second half cycle is recorded on the same scale as the first. A phase angle θ may therefore represent either θ itself or $\theta + \tfrac{1}{2}T$. Obviously we must know which case is present. To do this note the relative movements (clockwise or anticlockwise) of the *delay line dial knob*, and the *pointer on the Null Meter* (see below §13.3.3). If these two movements are in the same (S) direction θ stands alone, and if they are opposed (O) a half cycle $\tfrac{1}{2}T$ has to be added to θ. These S and O indications must be booked for phase positions (1) and (3), and not for (2) and (4) which give the opposite indications. In practice, the extra half cycle is introduced after the readings have been converted to metres, i.e. $\frac{1}{2}x_1 = 2\cdot500$ m, $\frac{1}{2}x_2 = 2\cdot494$ m and $\frac{1}{2}x_3 = 2\cdot381$ m denoted by U_1, U_2, and U_3 respectively in the Geodimeter form, table 13.2.

Examples: (1) Consider a measurement of frequency F_1, $U_1 = 2\cdot500$. If $\theta = 56$ (S), from tables, $L_1 = 0\cdot750$, and since the indication is S accept $L_1 = 0\cdot750$. If the indication had been 56 (O) add U_1 and $L_1 = 3\cdot250$ m.

(2) Again, if measurement is on frequency F_2, $U_2 = 2\cdot494$ m; a reading of 61·75 converts via its tables to 0·798 m. If the indication is S, $L_2 = 0\cdot798$, but if it is O, $L_2 = 0\cdot798 + 2\cdot494 = 3\cdot292$ m.

13.2.4. REFLECTOR AND CALIBRATION READINGS

Since the position of the electrical centre of the instrument undergoes small changes during measurement, a differential measurement procedure is introduced to eliminate small errors from this source. Suppose the length to be measured is GR (figure 13.5) where G is the physical centre of the Geodimeter, and R the

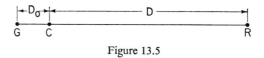

Figure 13.5

physical centre of the reflector. This distance GR will include a small instrumental error. A *calibration mirror or prism* C is then introduced in front of the Geodimeter at a known distance D_0 from G. The Geodimeter measurement of GC will also include the same instrumental error as GR if these two measurements are made in rapid succession. The distance GR—GC is then almost free of error, and finally GR is obtained by addition of the Geodimeter constant D_0 (usually about 0·2 m, but varies slightly from instrument to instrument).

Example: The following example gives the derivation of L_1 from the readings θ_R on the reflector, and θ_C on the calibration mirror. The mean readings on frequency F_1 are

$$\theta_R = 53\cdot75 \text{ (O)}, \qquad \theta_C = 56\cdot0 \text{ (S)}$$

which convert to metres to give 0·721 (O); 0·750 (S)

whence
$$\frac{2\cdot500}{3\cdot221} \qquad \frac{-}{0\cdot750}$$

and finally $L_1 = 3\cdot221 - 0\cdot750 = 2\cdot471$.

If θ_C is greater than θ_R the resultant L_1 is negative. This may be treated in the same manner as a positive quantity paying regard to the negative sign. Alternatively, as in the table 13.2, the result may be rendered positive by adding a whole length of 5·000 m.

13.2.5. PHASE READINGS

To eliminate small residual errors analogous to centering and collimation errors of a theodolite, readings on four phase positions are taken, the mean of which is almost free from errors. Readings on phases (1) and (3) should be similar, and those on (2) and (4) should be similar, though the (1) and (2) readings may differ by up to 10 divisions.

13.2.6. THE PASSIVE REFLECTORS

Although a plane mirror would suffice to reflect the light beam back to the Geodimeter, its alignment would be very critical over most distances. Over very short distances of up to 100 metres, reflecting Scotch tape can be used; this is especially useful in the survey of detail by radiation. For most distances, the reflector consists of a *trihedral prism* or combinations of such prisms. The

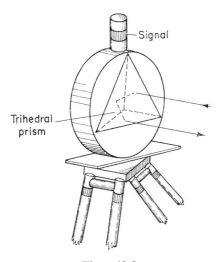

Figure 13.6

trihedral prism, used in cats-eyes, vehicle reflectors, etc, consists of the corner of a cube of glass, which has been cut away in a plane making an angle of 45° with the faces of the cube. The light is directed into the cut away face, i.e. that forming an equilateral triangle, as shown in figure 13.6, and is reflected by the inner surfaces of the prism which are highly silvered. Provided the angle of the incident light does not exceed about 20° to the normal to the front face of the prism, the reflected beam follows a path which is exactly parallel to the incident beam. Hence the alignment of the prism is not critical.

Due to a change of light velocity within the glass prism the effective optical centre of the prism does not coincide with the physical centre of the prism housing, which is centred over the survey mark: hence a small correction of -0.030 m has to be applied to the measured length for prisms in standard AGA housings. The correction for ordinary vehicle reflectors, sometimes used on medium distances, is negligible.

For optimum operational efficiency, the effective size of reflector is varied by increasing the number of prisms at the survey point, according to

(i) Length of line.
(ii) Weather conditions.
(iii) The type of lamp fitted to the instrument.

Prisms are mounted in holders in units of one, three, and seven, and a maximum of three holders can be fitted to a tripod. Figure 13.7 gives prism data for a variety of circumstances.

13.2.7. DEVIATION WEDGES

For short distances, a thin glass wedge is placed in front of the reflector prism covering one half of the front face of the prism. The object of this wedge is to deviate a beam of light from the reflector so that it enters the receiving optic

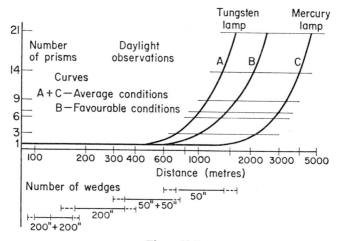

Figure 13.7

of the Geodimeter. Each wedge is housed in a circular ring which fits over the prism mount, and on the side of which is marked the wedge angle. Two angles 50" and 200" are available. The actual angle through which the light is deviated is about one half the wedge angle. Figure 13.7 shows the wedge sizes used over various distances. Two wedges can be fitted to each prism if necessary.

Table 13.1. Operating Distances (km) Favourable Conditions

Number of prisms	Standard tungsten lamp		Mercury vapour lamp	
	Day	Night	Day	Night
21	1·5	14	5	25
7	1·0	10	4	20
3	0·7	7	3	15
1	0·4	4	1·5	10

Care must be taken to ensure that the straight edge of the wedge is vertical. (The linear separation between the two optics of the Geodimeter is about 5 inches.)

13.2.8. ECCENTRICITY CORRECTION

If the top of the Geodimeter tripod is not horizontal, the centre of the instrument from which the measurement takes place is displaced with respect to the survey mark over which the tripod is plumbed, as shown in figure 13.8(a).

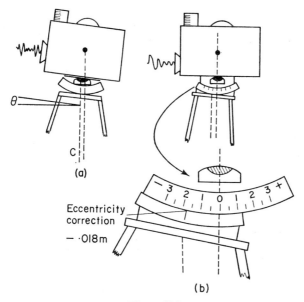

Figure 13.8

It will be noted that the correction depends solely on the position of the top of the tripod and not on the vertical angle at which the Geodimeter is pointed. The eccentricity correction to be applied to a measurement is read directly on a small scale at the base of the instrument, after the Geodimeter has been levelled with the aid of a small bubble mounted on its side (figure 13.8(b)).

13.2.9. METEOROLOGICAL CORRECTION

Allowance has to be made for variation in the velocity of light according to the atmospheric conditions of the measurement. The equation relating the distance D, the velocity of light *in vacuo* V_v, the index of refraction μ, and the transit time ω is

$$D = \frac{V_v}{\mu} \omega \qquad (9)$$

For a given measured value of ω and since V_v is constant, D is dependent on μ, which in turn depends on temperature, pressure, and humidity. From (9) by differentiation

$$\frac{dD}{D} = -\frac{d\mu}{\mu}$$

i.e. $$dD = -\frac{D}{\mu} d\mu \qquad (10)$$

For small changes in D, and since $\mu = 1\cdot000\ 3 \simeq 1$ we may put

$$dD = -D\, d\mu \qquad (11)$$

with ample accuracy. The general equation relating the refractive index of light μ_t at centigrade temperature t to μ_0 at 0°C is

$$\mu_t = 1 + \frac{\mu_0 - 1}{273 + t}\frac{P}{76} - \frac{55e10^{-7}}{1 + 0\cdot003\ 67t} \qquad (12)$$

where e is the vapour pressure. At 11°C, $e = 0\cdot013$ cm, whence the last term in (12) is negligible for most work, and it is ample to use

$$\mu_t = 1 + \frac{\mu_0 - 1}{273 + t}\frac{P}{76} \qquad (13)$$

The value of μ for the mean wavelength of light used by the Geodimeter is $1\cdot000\ 310\ 4$, which is the value at standard temperature $t_s\ (-6°C)$ and standard pressure P_s (76 cmHg), for dry air. Substituting these values in (13)

$$\mu_s = 1 + \frac{\mu_0 - 1}{267}$$

$$\therefore \mu_0 - 1 = 267(\mu_s - 1)$$

If $d\mu$ is the change in the refractive index from μ_s to μ_t

i.e. $$\mu_t = \mu_s + d\mu$$

$$\therefore -d\mu = \mu_s - \mu_t$$

$$= \mu_s - \left(1 + \frac{\mu_0 - 1}{273 + t}\frac{P}{76}\right)$$

$$= \mu_s - \left\{1 + \frac{(\mu_s - 1)267 \times P}{(273 + t)76}\right\}$$

$$= 1\cdot000\ 310\ 4 - 1 - \frac{(0\cdot000\ 310\ 4)(267)}{273 + t}\frac{P}{76}$$

$$\therefore -d\mu \times 10^6 = 310\cdot4 - \frac{82\ 880}{273 + t}\frac{P}{76}$$

whence from (11)

$$dD = \frac{D}{10^6}\left(310{\cdot}4 - \frac{82\,880}{273+t}\frac{P}{76}\right)$$

$$= \frac{D}{10^6} \times \text{correction}$$

For ease in calculation, this correction is read from a nomogram supplied by the manufacturers.

13.2.10. CORRECTION FOR LIGHT COLOUR

In very precise work, where accuracies of $1:10^6$ are sought, a correction is applied for the deviation of the particular colour of light on which an individual instrument operates. In most work this effect is neglected.

13.2.11. COMPLETE EXAMPLE

The reader should now study the abstraction of the readings given in table 13.2. Many surveyors will devise a method which involves less writing down and repetition. However, for the occasional Geodimeter operator, the form is preferable since it is self-explanatory.

The reduction of the slant lengths to the spheroid, and the subsequent computation of coordinates is considered in §13.9.

13.3. FIELD OPERATION OF THE GEODIMETER MODEL IV

The operation of the Geodimeter falls into three stages:

(i) Setting up the instrument and checking the operational levels.
(ii) The correct alignment of the instrument.
(iii) The actual measurement procedure.

The description of these stages is given with reference to figures 13.9 and 13.10. Meter X is a combined voltmeter-milliammeter which monitors the various operational levels of the instrument. It has a top scale on which volts are read, and a bottom scale on which current in milliamperes is read. The meter Y is the *null* or *zero* meter which acts as the zero mark of the micrometer reading system.

For efficient working of the instrument, the phototube must receive an optimum amount of light, which is indicated by a reading of between 0·8 and 1·0 milliamps on meter X. Before a delay line reading is made, the light sensitivity is adjusted to give this reading by moving the grey wedge plunger S_4 in or out very slightly.

To assist alignment, a rough sight is fitted to the left-hand side of the instrument, and for more accurate alignment subsequently, one may look through the monocular and through the transmitting optic and the Kerr cell plates. To view through the monocular it is pulled out, and the switch S_2 is in the position marked Adj. Actual measurement of distance takes place with the monocular pushed in, and S_2 at Meas. All switches have null points at the stages between the various setting positions to prevent carry over between circuits. The

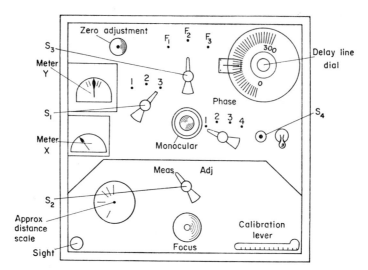

Figure 13.9

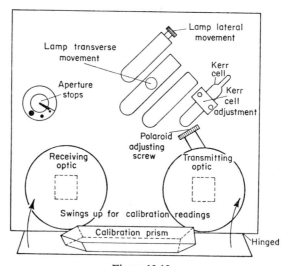

Figure 13.10

operator should therefore ensure that a switch is correctly positioned if no reading is obtained on meter X. The following is a description of the normal operational procedure which assumes that the instrument is functioning properly.

13.3.1. PART (A)

(1) Set up the instrument and reflector at the terminals of the line, and attach the power supply to the Geodimeter. The correct position of the converter switch is the 'blue' position.

(2) If the instrument is fitted with a mercury vapour lamp the arc has to be ignited by pressing the white starter button on the top cover for a few moments.

(3) Check the voltage on meter X top scale. With switch S_1 in position (1) the reading should be 6·3 V.

(4) Check current on the bottom scale of meter X. With switch S_1 in position (2) the reading should be between 30 and 40 mA.

(5) Check the operation of the phototube. With the calibration mirrors in the 'up' position and the light switched on, a reading of between 0·8 and 1·0 should be obtainable on the bottom scale of meter X by suitable manipulation of the grey wedge plunger. Switch S_1 in position (3).

(6) Check that the null meter Y is properly set to zero. The front of the instrument is closed up to exclude light.

Tungsten Lamp. Switch off the lamp, push in the grey wedge, and set the null meter pointer to read zero by the adjustment knob.

Mercury Vapour Lamp. Do not switch off the lamp as it will age rapidly if continually switched off and on. Turn switch S_2 to the 'Adj' position, and zero the null meter pointer.

13.3.2. PART (B)

(7) Set the aperture on the front of the instrument (figure 13.10) to the smallest aperture for daylight operation or to the largest for night work. An intermediate position may be required in dull day conditions to obtain the correct reading with the grey wedge.

(8) Align the instrument approximately with the small sight.

(9) Align through the monocular with S_2 in position Adj. The fine movement in azimuth and elevation is carried out by the slow motion screws on the base of the instrument. The reflector will probably require to be brought to focus. Through the monocular one sees the plates of the Kerr cell in the centre of the field of view, as in figure 13.11(*a*).

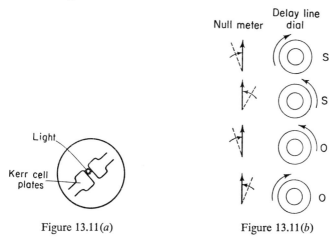

Figure 13.11(*a*)   Figure 13.11(*b*)

(10) To make the fine adjustment for optimum electronic sensitivity, proceed as follows: Adjust the grey wedge for a reading of between 0·8 and 1·0, and move the instrument slightly in elevation and/or azimuth until a maximum dip

of the needle to the left is obtained, a reading of about 0·6 will be obtained momentarily. When this maximum dip has been obtained, the alignment is complete. This fine adjustment is difficult to perform in daylight.

13.3.3. PART (C)

(11) A sequence of readings is made. It will be remembered that the grey wedge must be adjusted for a reading of from 0·8 to 1·0 on meter X prior to making any reading on the delay line dial. A reading is achieved by moving the delay line knob until the zero meter Y reads zero, i.e. when its pointer is vertical. If the Geodimeter instrument is correctly aligned the null meter pointer will respond quickly to changes in the delay knob. If its movement is sluggish, the alignment should be checked, or the Kerr cell adjustment checked (see §13.3.6). The directions of the final movements of the null pointer and the delay knob have to be noted since it is an important part of the reduction of observations. Therefore book S if they move finally in the same direction to achieve nullity, or O if they move in opposite directions (see figure 13.11(b)).

(12) The sequence of readings is:
With calibration mirror up, on frequency F_1 read four phases, and with calibration mirror down, on F_1 read four phases on the reflector. i.e. obtain

$$F_1 \quad P_1 \quad P_2 \quad P_3 \quad P_4 \quad \text{calibration}$$
$$F_1 \quad P_1 \quad P_2 \quad P_3 \quad P_4 \quad \text{reflector}$$

The process is repeated for frequencies F_2 and F_3, which completes the measurement.

(13) Pressure and temperature are booked if required in precise work, and the eccentricities of instrument and reflector if any.

13.3.4. MINOR INSTRUMENT ADJUSTMENTS

The following minor instrument adjustments may require to be carried out prior to a series of field measurements. Once they have been completed the instrument should stay in adjustment over a prolonged period provided it does not receive serious bumps in transportation.

13.3.4.1. *Lamp Alignment (Tungsten Type)*

For greatest efficiency, a maximum amount of light from the lamp should pass through the plates of the Kerr cell. Looking into the transmitting optic, the lamp is moved in its housing until this maximum is obtained.

13.3.4.2. *Lamp Alignment (Mercury Vapour Type)*

On no account should the light from the mercury vapour lamp be viewed with the naked eye, since severe burning may occur. The best way to adjust the lamp position is to project its image and that of the Kerr cell plates on to a piece of white paper through one telescope of a pair of binoculars placed with the objective lens over the transmitting optic. The lamp is moved in its housing to bring the image of the bright spot to the centre of the plates. This adjustment requires care, since the housing of the lamp becomes exceedingly hot within a few moments of being switched on.

13.3.5. POLAROID ADJUSTMENT

The polaroids (figure 13.4) are correctly set when their planes of polarization are mutually perpendicular. When the lamp is viewed from the front it should have a greeny-white (cold) colour if in correct adjustment, which contrasts with the orange-yellow (warm) colour of maladjustment. To achieve the correct adjustment, the front adjustable polaroid is rotated slightly in its housing. (see figure 13.10).

13.3.6. ADJUSTMENT OF THE KERR CELL

This check is to place the Kerr cell in the focal planes of both the transmitting and receiving optics, which are linked together with the focusing device. When stage (10) of above has been completed the observer should once again look through the monocular and note the relative positions of the reflector and the Kerr cell plates. If the reflector does not now lie at the centre of these plates, the instrument is operating inefficiently due to malalignment of the Kerr cell. To adjust the cell, a small light source at about 300 metres distance is used as if it were the illuminated reflector. This light source may be a highlight on some surface such as a car body. The instrument is aligned on this light as if on the reflector. The Kerr cell is then viewed through the monocular. If the light source appears to lie off the centre of the plates the cell is moved in its housing until the light appears to be central to the plates.

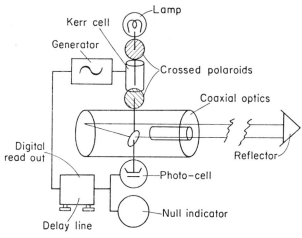

Figure 13.12

13.3.7. GEODIMETER MODEL VI

The latest production Model VI (see figure 13.12) has many improved features:

(i) A coaxial optic replaces the two optics of previous models, and thus eliminates the need for wedges.
(ii) With either type of lamp, the daylight range is improved on the Model IV by about 15%.

(iii) Due to transistorization, the weight of the basic instrument is reduced to about 30 lb.
(iv) A digital readout of the delay line dial is provided, but conversion tables are still required.
(v) A horizontal circle assists alignment.
(vi) An internal battery will give two hours operation of the standard lamp. Normal operation is by external battery or generator, the latter being necessary if the mercury lamp is used.
(vii) A special light weight generator has been designed weighing 30 lb.
(viii) All equipment, except the tripods, may be housed in two back packs for ease of working in difficult country.
(ix) The accuracy is the same as the Model IV.

A Model V was designed, but in view of the advantages of the Model VI, no production models were made.

13.4. THE TELLUROMETER

The Tellurometer system is a *microwave radio system* operating on carrier wavelengths of 10 cm (models (1) and (2),) and on 3 cm (model (3)). A reflected radio signal plays a similar role to the reflected light beam of the Geodimeter. We shall confine our detailed attention to the Model 1 MRA/1 (Master Remote Apparatus/One), and introduce the others later.

13.4.1. THE MRA/1 TELLUROMETER

Figure 13.13 shows the various components of the Tellurometer system. The apparatus requires two operators, one at each of its two components; the *master*, at which the time measurement is made, and the *remote* which acts as a 'radio reflector'. Both master and remote are combined radio transmitters and

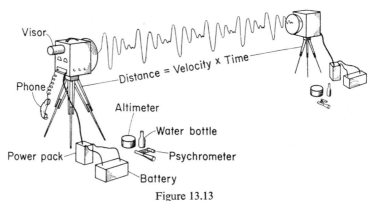

Figure 13.13

receivers of similar appearance, weighing about 30 lb each, and which stand on survey tripods or pillars. Electrical power is supplied to each from a 6 volt battery via an external *power pack* generating high tension (h.t.), low tension (l.t.), direct current (d.c.) and alternating current (a.c.). The current requirement is 8 A. A *duplex telephone* system is incorporated which enables the operators to communicate with each other as on the ordinary telephone. Time measurement is carried out with the instruments in the 'measure' position, and speech with them in the 'speak' position, though it is possible for the speech to be heard on one set while it is still in the measure setting. Transmission and reception of the radio waves is from antennae placed at the focus of a parabolic reflector, constructed in two halves to enable it to be transported in a folded position protecting the antennae (see figure 13.14). The normal range of the instrument is from 500 ft to 50 miles, with an optimum accuracy at 25 miles. (A line over 90 miles long has been measured.) Operation is possible by day and night, in haze, mist, or light rain. Normal accuracy is 5 cm ± 1/200 000 of the distance. A carrier frequency of 3 000 megs (megacycles per second)

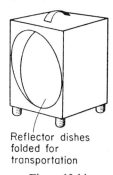

Reflector dishes
folded for
transportation

Figure 13.14

frequency modulated to produce an effective measurement frequency of about 10 megs is generated in the master, transmitted in a 10° wide beam to the remote where the phase of the modulation is converted to a one kilocycle (kc/s) pulse which is retransmitted instantaneously back to the master on a slightly different frequency. The phase of the master signal is also converted to a one kc/s signal which is displayed as a circular trace on a cathode ray tube (c.r.t.) as on a T.V. screen. The pulse from the remote is introduced into the c.r.t. where it blanks out the circular display at the correct phase angle with reference to a zero mark on a scale divided into 100 parts. In figure 13.21(c) the reading is 32, i.e. we read the beginning of the gap as it is approached in a clockwise manner round the circle. To assist reading, a visor fits over the c.r.t. to shield the display. Since the effective measurement frequency is 10 megs the period is $1/10 \times 10^6$ second or 0·1 microsecond. Since the scale is divided into 100 equal parts, these parts then represent $1/10^9$ of a second, one *milli-microsecond*, (1mμs) or one *nanosecond*, (1ns). From equation (1) the equivalent distance to 1 mμs of transit time is 0·3 m or about 1 foot. Since the transit time for the double distance is measured 1 mμs finally represents a measurement precision of 0·15 m or 6 inches in the length of the survey line.

13.4.2. MEASUREMENT OF THE TRANSIT TIME

In the geodimeter, the length of the line to be measured was found by using three different frequencies with which to time the interval taken by the impulses to traverse the double path. In the Tellurometer system four frequencies

Table 13.3. Tellurometer Field Sheet

TELLUROMETER FIELD SHEET

MASTER STN. _Somerton church tower_ INST. No. — OPERATOR _Allan_
REMOTE STN. _Lollover_ INST. No. — OPERATOR _Hough_
DATE _21/IV/63_ CONDITIONS _Continuous rain_
TIME _12:45 – 13:10_ _polythene covers in place_

COARSE READINGS

A+ 25   A+ 25   A+ 25   A+ 25   A+ 25   A+ 25   A+ 25   A+ 25
B 96   C 44   D 23   A– 74   B 97   C 46   D 23   A 74
Diff 29   81   02   51   28   79   02   51
Approx distance _28 025·5_   Course figure _28 025·5_ mµsec
3 miles

MET. READINGS	CRYSTAL TEMP	DRY OF BULB	WET °F BULB	DEPT °F	BAROMETER	CORRECTED BAR	BAROMETER NUMBERS
MASTER INITIAL	55	48·5	48·5	0	28·74	28·8	MASTER
FINAL	65	48	48	0	28·74	28·8	B 4402 +0·06
REMOTE INITIAL		47·5	47	0·5	28·32	28·3	REMOTE
FINAL		47	47	0	28·28	28·3	B 4386 +0·02
SUM		31·0					
MEAN	60	47·75		0·1		28·55	

FINE READINGS

Up from 3 (Master) by ½ turns

1	A+	25	A+R	76	100		10	A+	26	A+R	77	104
	A–	75	A–R	26				A–	73	A–R	26	
		50		50					53		51	
2	A+	25	A+R	77	101		11	A+		A+R		Mean of 40
	A–	75	A–R	26				A–		A–R		
		50		51								103·5
3	A+	26	A+R	77	105		12	A+		A+R		θ = 25·875
	A–	73	A–R	25				A–		A–R		
		53		52								
4	A+	25	A+R	77	105		13	A+		A+R		
	A–	73	A–R	24				A–		A–R		
		52		53								
5	A+	25	A+R	77	102		14	A+		A+R		
	A–	74	A–R	26				A–		A–R		
		51		51								
6	A+	26	A+R	77	105		15	A+		A+R		
	A–	73	A–R	25				A–		A–R		
		53		53								
7	A+	26	A+R	78	106		16	A+		A+R		
	A–	73	A–R	25				A–		A–R		
		53		53								
8	A+	26	A+R	77	104		17	A+		A+R		
	A–	74	A–R	25				A–		A–R		
		52		52								
9	A+	26	A+R	77	103		18	A+		A+R		
	A–	74	A–R	26				A–		A–R		
		52		51								

Uncorrected transit time   28 025·875
Crystal correction +1·8 p.p.m.   +0·050
                                  28 025·925 mµ sec

A, B, C, and D are used in a similar manner. In this system the time is actually recorded in millimicroseconds and converted by computation into distance. The technique of this time measurement is best explained with reference to an example. Consider the line of length L over which the transit time is 70 418 mµs.

Firstly a period of 100 mµs is used as the time measuring unit. This is the A reading unit. Since the dial is only capable of recording fractions of 100 mµs, the reading for this line will be 18 mµs i.e. $A = 18$ mµs. At this stage all we can say is that the transit time is an unknown integral number of units of 100 mμs plus 18.

To ascertain the number of units of 100 mμs up to 10, a second period D of length $\frac{10}{9}A$ i.e. with a frequency of 9 megs, is introduced. Since this unit of time is longer than the A unit by $\frac{10}{9}$, the reading on the same dial of 100 divisions will be $\frac{9}{10}$ of that recorded using the A units. Thus in the example if the total transit time were measurable we would find that there were 63 376·2 units of period D in the transit time (63 376·2 = $\frac{9}{10}$ or 70 418). Again since the dial cannot record more than 100 units, the D reading would be 76·2 in theory. In practice the reading would be 76, reading to a precision of 1 scale division. By subtracting the D reading from the A reading the number of units of 100 mμs up to ten contained in the line is found. For 70 418 − 63 376·2 = 7 041·8, which multiplied by 10 gives the whole transit time, and considering the actual readings that would be obtained $A - D = 18 - 76$, which is made positive by borrowing an additional A revolution of 100 mμs, i.e. $A - D = 118 - 76 = 42$. Whence 42 × 10 = 420, which an approximate value of the transit time under 1 000 mμs. Since it is known from the A reading alone that the last two figures are 18, we now know that the last three figures of the transit time are 418. This differential method is like using a vernier in reverse. If the zeros of the two scales are matched up (figure 13.15) their separation after 1 division on either

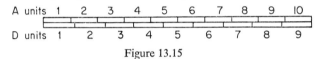

Figure 13.15

scale is 0·1 since the D scale division is $\frac{9}{10}$ of that of the A scale; after two divisions the separation is 0·2 and so on to 10 divisions of the A scale when both coincide. After eleven divisions on the A scale the separation is again 0·1 and so on.

In a similar manner two other periods C and B are introduced which are respectively $\frac{100}{99}A$ and 1 000A/999 respectively. The subtraction of the C reading from the A gives the number of revolutions of A (units of 100 mμs) in the line up to 100, and the A minus B reading gives the number up to 1000 revolutions, or 1 000 × 100 = 100 000 mμs. In the example the C reading is 99 × 70 418/100 = 69 713·82. Whence the actual C reading would be 14. $A - C = 18 - 14 = 04$, which shows that there are 04 units of 100 mμs in the transit time. The time is now known to be 0 418 mμs. The B reading would be 999 × 70 418/1 000 = 70 347·582 in theory and in practice would be 48. $A - B = 118 - 48 = 700$, hence there are 700 units of 100 mμs = 70 000 mμs in the time. Since there is only two digit precision at each subtraction, the complete transit time is built up as follows:

Using the A unit alone	1 8 mμs
From A–D	4 2
From A–C	0 4
From A–B	7 0
Thus the accepted time is	7 0 4 1 8 mμs

Which is best derived by considering the figures from the right to left.

In practice, since all readings are subject to small errors of reading the display, shaping, and centering of the display etc., the various pairs of numbers from which the coarse figure is constructed will not exhibit the exact agreement of this theoretical example. The precision of the second figure of each pair may be in doubt by two or three divisions. A likely set of readings would be

$$
\begin{array}{cccc}
A \ \ 18 & A \ \ 18 & A \ \ 18 & A \ \ 18 \\
B \ \ 45 & C \ \ 12 & D \ \ 75 & \\
A-B = 73 & & & \\
& A-C = 06 & & \\
& & A-D = 43 & \\
& & & A = 18
\end{array}
$$

Working from right to left the transit time is built up as follows:

$$
\begin{array}{lll}
A & 18 & \text{Precision of the } 8 \pm 3 \\
A-D & 43 & \text{Precision of the } 3 \pm 3 \\
A-C & 06 & \text{Precision of the } 6 \pm 3 \\
A-B & 73 & \text{Precision of the } 3 \pm 3 \\
\text{Coarse figure} & 704\ 18 &
\end{array}
$$

If the first figure of a pair, say $A-D$ differs from the second figure of the next pair $A-C$ by 5, an ambiguity is present. For example if the figures were

$$
\begin{array}{c}
18 \\
43 \\
09 \\
73
\end{array}
$$

It is not certain whether the coarse figure is 70 418 or 71 418, since the 09 could be either 04 or 14. Cases like this are very rare; provided the operator has adjusted his set properly an unresolved ambiguity of this kind should not arise. In some cases with a badly swinging line the operator may not be at fault (see §13.4.4).

If the transit time exceeds 100 000 mμs, which is equivalent to a distance of about 15 km or 10 miles, the approximate length of the line has to be known to within 10 miles and a unit of 100 000 mμs added for each 10 mile unit. For example, if the above line were about 27 miles the accepted transit time is 2 7 0 4 1 8 mμs. This process of obtaining the approximate transit time is known as 'taking the coarse figure'.

13.4.3. FINE READINGS

In taking the coarse figure the transit time is known to a precision of about 10 or 20 mμs only, since there are two sources of error present which have yet to be considered. These errors are

(i) *Instrumental*—due to small malcentring of the display, etc, and
(ii) *External*—due to the reflections of the radio beam from the intervening ground surface, etc.

To eliminate the effects of (i) a procedure similar to changing face and zero in

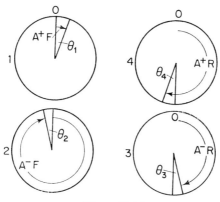

Figure 13.16

theodolite angular measurement is adopted. In figure 13.16 four values of the phase angle are shown in the forward (F) and reverse (R) positions and in the plus (+) and minus (−) positions. The zero mark for the forward readings is 0 and for the reverse readings is 50. The mean of angle θ read on each of the four positions shown is almost free from errors of centring and shaping the display. For ease of reading and abstraction, all four positions are read in a clockwise sence from the zero mark, i.e. the readings are taken as follows in the order A$^+$, A$^-$, A$^-$R, and A$^+$R.

$$A^+ = \theta_1 + 100 \qquad A^+R = 50 + \theta_4$$
$$A^- = 100 - \theta_2 \qquad A^-R = 50 - \theta_3$$

subtracting $(\theta_1 + \theta_2)$ + $(\theta_3 + \theta_4)$ Mean $= \frac{1}{4}(\theta_1 + \theta_2 + \theta_3 + \theta_4)$

They are booked in the shape of a 'U' and as shown, the subtraction gives the sums $(\theta_1 + \theta_2)$ and $(\theta_3 + \theta_4)$, and so the total sum and final mean are readily obtained.

A typical set of A readings would be as follows adding 100 to A$^+$ and/or A$^+$R as required:

$$\begin{array}{ll} A^+ \quad 17\cdot5 & A^+R \quad 68\cdot0 \\ A^- \quad 85\cdot5 & A^-R \quad 36\cdot5 \\ \hline 32\cdot0 & 31\cdot5 \quad \text{Mean} \quad \tfrac{1}{4}(63\cdot5) = 15\cdot9 \end{array}$$

It should be noted that the final mean value should be of the same order as the A$^+$ reading. In practice it will usually be apparent from the magnitude of each reading if a mistake has been made in the order of reading or booking. This is not so however when the A$^+$ reading is close to zero, in which case the operator should take greater care to ensure that the correct sequence is adopted.

13.4.4. GROUND SWING

To eliminate the external error the carrier frequency is varied by small amounts and a set of four *fine readings* is taken at each frequency. In this way the effect on the phase of the signal due to part of the beam being reflected from intervening ground surfaces is also varied. The frequency variation has been designed so that sufficient variation in phase will occur to reduce the ground reflection effects to small amounts as a rule. About twelve sets of readings are taken over frequencies varying from cavity tune setting of from 5 to 20. (The dial numbers are purely arbitrary.) Thus twelve values of the mean A reading are obtained. These values are plotted on a graph to ensure that a maximum and minimum value have been obtained, since the graph should theoretically approximate to a sine curve. The variation of the sets of fine readings is called 'ground swing' which in precise work should not exceed 10 mμs.

Lines with a ground swing of more than 10 mμs should be remeasured with the instruments in different positions or at different heights, or the line split in two parts. For a detailed analysis of the effects of ground swing the reader is referred to: R. C. Gardiner-Hill, *Proceedings of the Commonwealth Survey Officers' Conference*, 1963, Paper 30, (London: Dept. of Technical Cooperation and HMSO, 1964).

Readings of the B, C, and D patterns are not normally required at different carrier frequencies. Occasionally over lines with large ground swing, it is not possible to obtain a unique coarse figure in the first instance. In this event, coarse figures should be taken at those settings of the carrier which give the greatest and smallest A readings, one of which will usually give an acceptable coarse figure.

13.4.5. CORRECTION FOR FREQUENCY VARIATION

Since the absolute unit of measurement is the master A$^+$ frequency, maintained by the A$^+$ crystal whose nominal frequency is 10 megs, any deviation from this value will affect the result proportionally. The frequency of a crystal

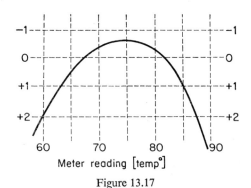

Figure 13.17

varies with its temperature, thus the crystal temperature has to be recorded and the necessary correction to the transit time obtained from a graph such as that shown in figure 13.17. Each crystal calibration graph should be checked at an electronics laboratory about every six months to ensure that it is still applicable.

13.4.6. METEOROLOGICAL EFFECTS

As with light waves, radio waves are affected by the density of the atmosphere through which they travel. The velocity of radio waves in air V_A is related to their velocity *in vacuo* V_V and the refractive index (μ) by the relationship

$$V_A = V_V/\mu$$

Since the distance $D = V_A \times$ Time (1)

$$D = V_V \times T/\mu$$

and the length of the survey line $L = \frac{1}{2}D$ is given by

$$L = \frac{1}{2}V_V \times T/\mu$$

The currently accepted value of $\frac{1}{2}V_V$ is 149 896·25 km/second which is used in conjunction with the following empirical formulae for the refractive index

$$(\mu - 1)10^6 = 4\ 730(P + E)/(459\cdot69 + t) \qquad (14)$$

$$E = 8\ 540\ e/(459\cdot69 + t) \qquad (15)$$

$$e = e' - 0\cdot000\ 367P(t - t')\{1 + (t' - 32)/1\ 571\} \qquad (16)$$

The refractive index μ is about 1·000 3 therefore $(\mu - 1)10^6 \simeq 300$. P is the barometric pressure in inches of mercury (see table 13.14), e' is the saturation vapour pressure in inches of mercury at the wet bulb temperature t' (°F). t is the air temperature (dry bulb) in °F. The values of both t and t' are obtained from readings of a psychrometer of the hand or mechanical type (see figures 13.18(a), (b)). In the case of the former fairly vigorous whirling with one's arm extended upwards is required since the empirical formulae above depend on a minimum velocity of air past the wet bulb being exceeded. For convenience in computation the above formulae (14), (15), and (16) have been rearranged as follows

$$(\mu - 1)10^6 = P \times 10^{-2}(2) + e(3)$$

$$e = e' - P \times 10^{-2}(t - t')(1)$$

Where the factors e', (1), (2), and (3) are tabulated as in table 13.4, and

$$(1) = 0\cdot036\ 7\{1 + (t' - 32)/1\ 571\}$$
$$(2) = 47\ 300/(459\cdot69 + t)$$
$$(3) = 4\ 730 \times 8\ 540/(459\cdot69 + t)$$

The argument for e' and (1) is the wet bulb temperature t', while that for factors (2) and (3) is the dry bulb temperature t.

In recording meteorological ('met') data it is desirable to obtain a record of the air conditions typical of the whole path traversed by the radio waves. Since met readings may be taken conveniently only at the line terminals it has to be assumed that such readings are representative of the line as a whole. This assumption will often be untrue, particularly in measurements across deep valleys or from high to low ground. A certain degree of air mixing in the area of the measurement is beneficial, though very strong gusty winds indicate that unstable met conditions are present. Temperature readings should be taken as high above the ground as possible since the first 50 ft of the atmosphere

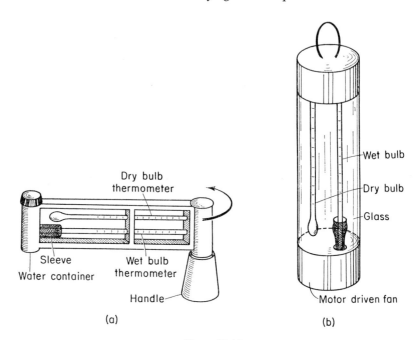

Figure 13.18

exhibits anomalies. Vegetation which is associated with its own climatic system will give rise to untypical met. Elevated thermocouples have been used to record met, though results appear to be inconclusive to date. It is essential that the thermometers are calibrated accurately as will be seen from table 13.5, which gives the magnitudes of the various errors affecting the Geodimeter and the Tellurometer. Generally speaking, measurement from towers with the lines grazing for most of their path should give the smallest errors due to untypical met readings. The problem however is by no means solved.

The tables for the determination of the refractive index of radio waves used here were devised at a time when the knowledge of the subject was not final, and although they are sufficiently accurate for most work, and have become well established by users of the Tellurometer, the International Union of Geodesy and Geophysics has recommended formulae by Essen and Froome of the National Physical Laboratory, England, for use in this connection, particularly for geodetic work. Tables suitable for the calculation of the refractive index have been constructed by N.P.L., and others more suitably cast for surveyors are available from the Directorate of Military Survey, England. The Geological Survey Department of the United States of America has designed and constructed a very useful protractor which enables the refractive index to be evaluated in a few moments, and is satisfactory either for the minor work, or for checking exact computation for gross error. The Tellurometer Company also supplied a suitable monogram on paper which may be used in a similar fashion to the protractor.

Table 13.4. Coefficients for Determining the Refractive Index of Radio Waves

t	e'	(1)	(2)	(3)	t'	e'	(1)	(2)	(3)
20	0·110	0·0364	986·1	175·6	70	0·739	0·0376	893·0	144·0
21	0·114		984·0	174·8	71	0·765		891·3	143·4
22	0·119		982·0	174·1	72	0·791		889·6	142·9
23	0·124	0·0365	979·9	173·4	73	0·818		887·9	142·4
24	0·130		977·9	172·7	74	0·846		886·3	141·8
25	0·135		975·9	171·9	75	0·875	0·0377	884·6	141·3
26	0·141		973·9	171·2	76	0·905		883·0	140·8
27	0·147		971·9	170·5	77	0·935		881·3	140·2
28	0·153	0·0366	969·9	169·8	78	0·967		879·7	139·7
29	0·160		967·9	169·1	79	0·999	0·0378	878·1	139·2
30	0·166		965·9	168·5	80	1·032		876·4	138·7
31	0·173		964·0	167·8	81	1·067		874·8	138·2
32	0·180	0·0367	962·0	167·1	82	1·102		873·2	137·7
33	0·188		960·0	166·4	83	1·138	0·0379	871·6	137·2
34	0·195		958·1	165·7	84	1·175		870·0	136·7
35	0·203		956·2	165·1	85	1·214		868·4	136·2
36	0·212	0·0368	954·2	164·4	86	1·253		866·8	135·7
37	0·220		952·3	163·7	87	1·294		865·2	135·2
38	0·229		950·4	163·1	88	1·335	0·0380	863·6	134·7
39	0·238		948·5	162·4	89	1·378		862·1	134·2
40	0·248		946·6	161·8	90	1·422		860·5	133·7
41	0·257	0·0369	944·7	161·1	91	1·467		858·9	133·2
42	0·268		942·8	160·5	92	1·514	0·0381	857·4	132·7
43	0·278		940·9	159·9	93	1·561		855·8	132·2
44	0·289		939·1	159·2	94	1·610		854·3	131·8
45	0·300	0·0370	937·2	158·6	95	1·661		852·7	131·3
46	0·312		935·4	158·0	96	1·712	0·0382	851·2	130·8
47	0·324		933·5	157·3	97	1·766		849·7	130·3
48	0·336		931·7	156·7	98	1·820		848·1	129·9
49	0·349	0·0371	929·8	156·1	99	1·876		846·6	129·4
50	0·362		928·0	155·5	100	1·933	0·0383	845·1	129·0
51	0·376		926·2	154·9	101	1·992		843·6	128·5
52	0·390		924·4	154·3	102	2·053		842·1	128·0
53	0·405	0·0372	922·6	153·7	103	2·115		840·6	127·6
54	0·420		920·8	153·1	104	2·179		839·1	127·1
55	0·436		919·0	152·5	105	2·244	0·0384	837·6	126·7
56	0·452		917·2	151·9	106	2·311		836·1	126·2
57	0·468		915·4	151·3	107	2·380		834·7	125·8
58	0·486	0·0373	913·7	150·7	108	2·450		833·2	125·3
59	0·503		911·9	150·1	109	2·523	0·0385	831·7	124·9
60	0·522		910·2	149·6	110	2·597		830·3	124·5
61	0·540		908·4	149·0	111	2·673		828·8	124·0
62	0·560	0·0374	906·7	148·4	112	2·751		827·4	123·6
63	0·580		904·9	147·9	113	2·831	0·0386	825·9	123·2
64	0·601		903·2	147·3	114	2·913		824·5	122·7
65	0·622		901·5	146·7	115	2·996		823·1	122·3
66	0·644	0·0375	899·8	146·2	116	3·082		821·6	121·9
67	0·667		898·1	145·6	117	3·170		820·2	121·5
68	0·690		896·4	145·1	118	3·261	0·0387	818·8	121·0
69	0·714		894·7	144·5	119	3·353		817·4	120·6
70	0·739	0·0376	893·0	144·0	120	3·448	0·0388	816·0	120·2

N.B.: Values of saturation vapour pressure e' and coefficient (1) correspond to wet bulb temperatures t'. Coefficients (2) and (3) correspond to dry bulb temperatures t.

Published by permission of Tellurometer (UK) Ltd.

Table 13.5. Comparative Accuracies. P.E of Mean from Various Sources of Error

GEODIMETER MODEL II 30 measures using all 3 frequencies on 3 nights		TELLUROMETER MRA 1 6 sets of 48 fine readings over 2 days	
Velocity	$\pm 1.3 \times 10^{-6}$ distance	$\pm 1.3 \times 10^{-6}$ distance	
Temperature	$\pm 1.0 \times 10^{-6}$ distance	$\pm 2.0 \times 10^{-6}$ distance	
Pressure	$\pm 0.1 \times 10^{-6}$ distance	$\pm 0.1 \times 10^{-6}$ distance	
Humidity	$\pm 0.05 \times 10^{-6}$ distance	$\pm 4.0 \times 10^{-6}$ distance	
Crystal	$\pm 0.15 \times 10^{-6}$ distance	$\pm 1.0 \times 10^{-6}$ distance	
Index	± 2.5 cm	± 5.0 cm	
Total ± 2.5 cm $\pm 1.65 \times D$		± 5.0 cm $\pm 5.0 \times 10^{-6} \times D$	

N.B. The index error of the Tellurometer MRA/3 is ± 2.0 cm and with both instruments, special observations can give even higher accuracies than those quoted above for average conditions.

13.4.7. INDEX ERROR

Recent investigations† have shown that the small index error of the Tellurometers in which readings are taken on a cathode ray tube, exhibit a periodic relationship to the mean A reading as follows:

$$\text{Error} = b_0 + b_2 \sin 2A + b_3 \sin A \sin 2A \qquad (17)$$

where $A = 2\pi$(Mean A reading)/100 and b_0, b_2 and b_3 are constants found empirically for each instrument. For one particular instrument tested by the Ordnance Survey $b_0 = 6.2$ cm $b_2 = 7.4$ cm and $b_3 = 7.8$ cm. It is therefore possible that at a particular A setting the index error could be almost nil for a particular instrument.

13.4.8. FIELD OPERATION OF THE MRA/1 TELLUROMETER

In all Tellurometer work it is essential that an exact and detailed routine procedure for operation is agreed between the two operators if confusion is to be avoided. The master is in charge as a rule, informing the remote to carry out various operations either by speaking on the telephone or by a flick of his speak/measure switch once a regular routine has been established. Since it is essential to conserve battery power at all times a time of switching on both sets is agreed, and it is also agreed that the sets should be switched on for only ten minutes in the first instance, and thereafter at intervals of say half an hour for periods of five minutes. These times will be varied according to circumstances. The following are the main stages of the measurement procedure:

(1) Master and remote take up position with the centre of the instruments plumbed over the survey stations and pointing towards each other. It will be noted that although measurement is from the focus of the parabolic reflector this is also the same as measuring from the directrix of the parabola which is arranged to coincide with the physical centre of the instrument (see figure 13.19). Each operator checks his Tellurometer to ensure that it is functioning correctly and that the battery is charged. (REG reading on the switched meter M

† R. C. Gardiner-Hill, *Proceedings of the Commonwealth Conference, 1963*, Paper 30, (London: Dept. of Technical Cooperation and H.M.S.O, 1964). Op. cit. 13.4.5.

(figures 13.20(a) and (b) should be between 30 and 80). Each operator peaks his crystal current, i.e. obtains a maximum reading on meter V by turning the reflector tune knob S_1. It is essential that all Tellurometer operation be carried out with a maximum crystal current. The remote operator sets, with knob S_2, his cavity tune dial D_1 to read a prearranged value, usually 5, and with the speak/measure switch at 'speak', awaits contact from the master.

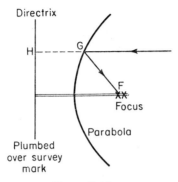

Figure 13.19

(2) The master sets his cavity tune reading a few divisions below that arranged for the remote, i.e. 3 in this case, and with his speak/measure switch also in the 'speak' position searches slowly for contact with the remote by gradually increasing his cavity tune reading. When contact is being made both Tellurometers should have switch S_3 in position A.V.C. (automatic voltage control). The needle of this meter will rise suddenly to a maximum in both instruments when contact is established, and speech is then possible. It is essential to make this contact with both sets on the 'speak' position to avoid making contact on a harmonic or side band of the measurement frequency. On instructions from the master, the remote moves his Tellurometer in azimuth until a maximum A.V.C. reading is obtained. His set is then correctly aligned on the master. The original alignment need only be to within about 45° for contact to be possible. This is a most useful characteristic of the Tellurometer in haze of cloudy conditions, or when working from towers. The master then aligns his set in a similar manner.

(3) Both operators switch to 'measure' and check that the modulation levels of their sets are correct. This is achieved with switch S_3 at 'mod'. With switch S_5 successively at positions A, B, C, and D, the readings on meter M should be respectively 40, 40, 40, and 36. If not, remove the cover marked 'Adjust modulation' and by turning the screws revealed with a non-metallic screwdriver these readings can be obtained. This adjustment should only rarely be necessary.

(4) The four crystals A, B, C, and D are then synchronized by the remote. He presses the small knob marked 'set crystals' located on the right side of the instrument and looks at his c.r.t. display. If the crystal in question is synchronized an ellipse is seen figure 13.21(d), if out of synchronization a rapidly moving series of figures-of-eight (Lissajous figures) is seen figure 13.21(e). In

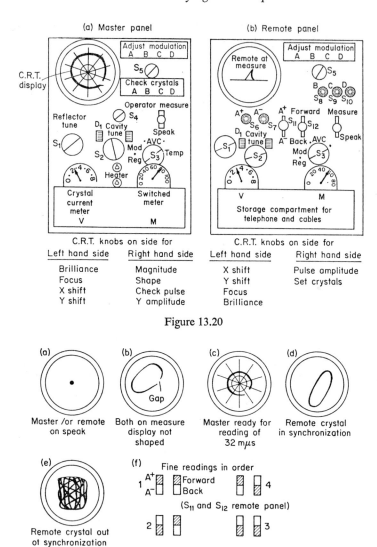

Figure 13.20

Figure 13.21

this latter case a small trimmer located below the frequency switch (S_6 for A^+ crystal $\cdots S_{10}$ for D of figure 13.20(*b*)) is turned until the relatively static ellipse is obtained. Whilst this adjustment is in progress the operators will hear a high pitched note on the telephone which varies as the remote alters his trimmer. At the correct synchronization position a characteristic note is obtained which is a useful rough indication of synchronization. The remote has five crystals to synchronize, viz. A^+, A^-, B, C, and D. Both Tellurometers are on the 'measure' position for this synchronization.

(5) The master has now to obtain a c.r.t. display which is a clear circle centred on the graticule of the c.r.t. as in figure 13.21(*c*). With both instruments

at 'measure' he suitably alters the knobs on the sides of his set to achieve this circle and a gap. A suitable gap will be obtained after fine tuning of his cavity tune, and possibly a little tuning with the reflector tune knob S_1 together with a correct balance between the illumination of the graticule and the intensity of the display. The remote operator has also the facility of varying the strength of the pulse from his Tellurometer to give a suitable break at the master. With experience an operator should never fail to produce a readable 'break' even if the quality of the display is very poor.

(6) The measurement is then carried out. Firstly, both operators record the met conditions which are booked by the master, as in table 13.3 which shows a typical Tellurometer field sheet. A booker is advisable if delays are not to occur while the master's eyes accustom themselves to the darkness of the reading visor after the bright page. As an alternative, a portable tape recorder is a useful means of booking if on one's own. A coarse figure is taken. Each Tellurometer is set on crystals A^+, A^-, B, C, and D in order, and in the measure position. The remote changes his crystal on orders from the master. Since the display at the remote alters from a horizontal line with a peak as in figure 13.20(b) to a dot figure 13.21(a) when the master changes from 'measure' to 'speak', a flick of this switch is sufficient to inform the remote to change crystal. The booker should check that the coarse figure 'breaks out' as indicated above in §13.4.2. If not, the readings should be checked and the Tellurometers re-synchronized if required.

(7) A series of fine readings is then taken. The remote operator moves his switches to A^+, A^-, A^-R and A^+R in order (see figure 13.21(f)) as instructed by the master by a flick of his speak/measure switch. After these four readings, the remote is instructed to move his cavity tune up a division say. This has the effect of destroying the contact between the Tellurometers. With both on the 'speak' position the master again establishes contact as described in §13.4.8 (2). This is a simple operation since he too will require to increase his cavity tune reading by one division. Another set of four fine readings is taken. This process is repeated until about twelve sets of fine readings is obtained. The booker should examine the means to ensure that a full cycle of swing is developed. If not, the full range of the instrument should be used. It is often found that a break is progressively more difficult to obtain at the higher cavity tune settings. A second coarse figure is taken as a check and a second series of met readings. This completes the measurement. Normally a period of about twenty minutes is required for the complete operation. Before finally closing down, the operators agree on the next stage of the survey.

13.5. PRACTICAL HINTS ON TELLUROMETER WORK

13.5.1. GROUND SWING

The ground swing will be reduced if the line clears the intervening ground by about 200 feet, and will be worst over a water surface. It will also cause an ambiguous coarse figure in some cases. If this arises, coarse figures should be taken at those cavity tune settings for which the extremes of the A^+ readings are obtained.

13.5.2. MET READINGS

To use the hand whirling psychrometer care should be taken to ensure that the minimum reading of the temperatures is obtained and that they are not affected by delay in reading or by the sun's rays. Care is also needed to ensure that the wet bulb is constantly moistened by the water from the holder. At low temperatures close to freezing point, some apparent anomalies can occur, in particular the wet bulb reading may exceed the dry. This occurs at the point when the water changes to ice releasing its latent heat. Correct readings will be obtained if the wet bulb is made to 'ice up' by touching it with some metal object.

13.5.3. BATTERIES

Batteries should be kept fully charged as far as possible, and if the Tellurometers are not in use for a prolonged period, the batteries should be used occasionally and recharged. The check on the specific gravity of the electrolyte is not a foolproof indication that the battery is serviceable. The surveyor should not grudge the relatively small expense of keeping good batteries solely for Tellurometer operation, and investing in a good charging motor to keep them fully operative. A log book should be kept for each battery in which is noted the operation times. It has been found that motor cycle 25 ampere-hour batteries are very effective, and since they weigh only $8\frac{3}{4}$ lb each they can be used in parallel in a specially designed box. Charging of four batteries is best carried out in series at 24 volts, 2 amps. At least one full set of spare batteries should be available for alternate use and charging.

13.5.4. EXAMPLE OF COMPUTATION OF TELLUROMETER SLANT LENGTH

Table 13.3 gives the necessary field data for the calculation of a Tellurometer line. The slant length is calculated in table 13.6 Part 1. The meteorological data and crystal corrections are obtained from table 13.4 and figure 13.17 respectively. The slant length is finally reduced to the spheroid as explained in §13.9.

13.6. OTHER MODELS OF THE TELLUROMETER

Production of the MRA/1 has been superseded by two other models, the MRA/2 and the MRA/3. Briefly the differences from the MRA/1 are:

13.6.1. MRA/2 TELLUROMETER

(i) Like the MRA/1 it operates on a carrier frequency of 10 cm. Each instrument is capable of acting either as master or remote while in the MRA/1 these functions are not interchangeable. This facility enables both operators to take a series of readings. It is also much more convenient to organize repairs etc. in a large organization operating several Tellurometers.

(ii) All crystals are thermostatically controlled at 110°F. No correction graph is required, nor has the remote operator to synchronize crystals.

Electromagnetic Distance Measurement

Table 13.6. Tellurometer Computation

PART 1. Computation of slant ranges   (Constants from tables of coefficients 13·4
MASTER: STATION _Somerton Church_ REMOTE: STATION _Lollover_
INST. ___M___   INST. ___R___
FIELD BOOK: _Table 13·3_   REMARKS. _Constant rain and wind_

MEASUREMENT No.			
	1.		
DRY TEMP. t (°F)	47·75		
WET TEMP. t' (°F)	47·65		
(1) $t-t'$	0·10		
(11) Pressure $P \times 10^{-2}$	0·2855	No second	
(111) Constant (1)	0·0370	measurement	
(1) × (11) × (111)	0·0011		
e'	0·333		
e	0·332		
Constant (2)	932·15		
Constant (3)	156·85		
(2) × (11)	266·13		
(3) × e	52·07		
μ	1·000	3182	1·000
(iv) Transit time mμs	28 025·925		
V/2 Half velocity × 10^{-6}		0·149 896 25	
Slant range $D=$			
$\frac{V/2 \times (IV)}{\mu}$	41 996·45 m		

Ground swing   1·5 mμs
Mean slant range   (Metres) ___41 996·45___

PART 2. REDUCTION TO SPHEROID   (Factors from Spheroid reduction tables)
Mean slant range _D 41 996·45_ Metres   Latitude __51°__   Azimuth __66°__
Slope correction C_1 ___-0·02__ Metres γ ___7·827___ From table _13·7_
Sea-level correction C_2 _-0·45_ $h_1 + h_2$ ___136___ Metres
Arc correction C_3 ___NIL___ h_1 ___90___ Metres
Spheroidal length __41 998·98__ h_2 ___46___ Metres
                                    $h_1 - h_2$ ___44___ Metres

C_1 Approx $- (h_1 - h_2)^2 / 2D$ $C_2 - \gamma (h_1 - h_2) D_1 \times 10^{-8}$
C_3 From table _13·8_

(iii) The power is supplied by a 12 or 24 V battery and the power pack is located inside the Tellurometer which is partly transistorized.

(iv) The operational panel is located on the side of the instrument.

(v) The parabolic reflector and the antennae are removed for transportation and fixed over the side operating panel.

(vi) The complete instrument in its storage position is fully waterproof.

(vii) A head telephone replaces the hand telephone as standard equipment.

(viii) The range and accuracy are the same as the MRA/1.

13.6.2. MRA/3 TELLUROMETER

(i) This instrument operates on a carrier of 3 cm and should therefore exhibit smaller errors due to ground swing.

(ii) The range is 30 metres to 100 km.

(iii) It is completely transistorized, has an internal power pack and an internal battery which will give two hours operation without recharging. Normal operation however is from external batteries.
(iv) Like the MRA/2 it can act as either master or remote.
(v) The antennae and reflector are permanently located under a fibreglass dome.
(vi) The effective measurement beam is 9°.
(vii) Two models are available, one giving transit time as with models MRA/1 and MRA/2, and the other which gives readings directly in metres. The standard refractive index used in the metric version is 1·000 325.
(viii) Readings on the metric system may be made on one of three possible alternative systems (*a*) on a full radial scale on which readings are made in conjunction with a null meter; (*b*) on a radial cathode ray tube display with the assistance of a cursor; (*c*) on a direct digital readout incorporating a null meter.

The digital readout system is not available with instruments reading in transit time.
(ix) The various systems are interchangeable except that an instrument designed for transit time readings will not operate with one designed for metric readout because of the different frequencies involved in the measurement.
(x) The overall accuracy of the instrument is ± 1 cm $\pm 3 \times 10^{-6}$ of the distance to be measured, provided the instruments have been accurately calibrated.

13.6.3. OTHER MODIFICATIONS TO THE TELLUROMETER

Experiments have been carried out in Canada with the reflectors separated from the Tellurometers and elevated on a long pole to enable measurements to be taken in forest areas. Results have been good, though the range of the instrument was reduced to about 10 miles in the first experiments due to power loss in the extension cables. This difficulty has been overcome by the use of wave guides. To increase the signal strength over long lines, a larger than standard reflector has been used with mixed results.

Special units for operation from air to ground called *Aerodist*, and from sea to ground, called *Hydrodist*, have been produced. Results have been satisfactory for the purposes intended.

13.7 OTHER DISTANCE MEASURING APPARATUS

13.7.1. MICROWAVE RADIO

The Cubic Corporation of U.S.A. has developed an instrument called *Electrotape*, operating on a carrier of 3 cm, with a digital read out, has a range of from 100 metres to 70 km and which gives results comparable to the Tellurometer.

Messrs Wild have developed the *Distomat* in conjunction with Albiswerk, Zurich. It has a 3 cm carrier, has a range of from 100 m to 50 km, each instrument consists of two units—a radio unit and a control unit—which enables the operator to be located some 20 m from the transmitter/receiver which can be elevated on a mast. There is also an analogue computer unit which produces the distance as a direct reading on a digital dial. Correction for met and ground swing is similar to the Tellurometer except that a standard refractive index is used, to which small corrections are applied, as with the Geodimeter. The equipment is much heavier than the Tellurometer. Results are comparable to the Tellurometer MRA/3.

Russian and Japanese versions of electromagnetic distance measurement equipment are also available.

13.7.2. TELLUROMETER MODEL 101

Recently the Tellurometer Company has manufactured the model 101 which has similar measurement characteristics to the MRA 3 but which is lighter and of a simpler design, since it is intended for the civilian user, instead of the military user which had imposed more stringent specifications hitherto.

13.7.3. FUTURE TELLUROMETERS

At the time of writing a prototype Model 4 has been exhibited in public. This instrument operates on a carrier wavelength of 8 mm and promises considerable reduction of index and reflection errors, as well as a narrower beam to enhance its performance in restricted areas. Other models operating on infrared carriers have also been envisaged.

13.7.4. ZEISS JENA ELECTRO-OPTICAL TELEMETER "EOS"

Operating on principles similar to the Geodimeter is the EOS of Carl Zeiss, Jena. This light wave instrument incorporates ultrasonic modulation of the light instead of a Kerr cell, and in consequence gives a more positive response in the null meter and a more efficient use of the light source. It operates from a 12v battery, will resolve a distance up to 3 km, has a daylight range of 7 km, and an accuracy quoted by the manufacturer as ± 0.5 cm ± 2 in 10^6 times distance. The major drawback of the instrument is its great weight.

13.7.5. LASERS

The firm of G & E Bradley Ltd, London is currently developing a ruby laser capable of measuring distances by reflecting a pulsed light beam from any prominent structure such as a building. Accuracies obtained to data are 1:1 000. Such an instrument may have great potential in detail surveys.

13.8 APPLICATION OF ELECTROMAGNETIC DISTANCE MEASUREMENT TO SURVEYING

Electromagnetic distance measurement (E.D.M.) has not merely supplanted other forms of distance measurement it has caused a change of emphasis and the introduction of new techniques. The following are some of the applications of E.D.M. to surveying:

  (i) Measurement of base lines in triangulation.
 (ii) Use in traverses of very large areas.
(iii) Fixation of control by trilateration, i.e. by linear measures only.
 (iv) Fixation of controls for aerial photography by radiation, i.e. by bearing and distance.
  (v) Setting out the important points such as bridge piers in engineering works.
 (vi) Measurement of the depths of mine shafts.
(vii) Fixation of a ship in inshore coastal surveys.
(viii) Three-dimensional trilateration from air to ground to fix controls in inaccessible terrain.

In addition, the duplex telephone of the microwave radio systems has improved survey organization and rendered practicable the observation of truly simultaneous reciprocal vertical angles which have increased the accuracy of trigonometrical heighting.

One of the most momentous effects of these instruments, capable of measuring great distances very accurately, is that the sides of triangulation systems have been checked with some quite surprising results. Some networks that were considered accurate to 1:100 000 have been found to be in error by as much as 1:20 000, while some low-order work has been proved to be very good.

The most serious objection to traversing, the tedium of obtaining accurate distances, has disappeared; with the result that a traverse will often now be preferred to a possible triangulation. Survey towers of heights ranging from 10 to 120 feet high are widely employed to give vision over obstacles, and street light maintenance lorries have been used to effect in township work.

Azimuth control of both traverse and trilateration is required at about every second station for the best results (see Chapter 7 for a detailed analysis of the propagation of error in traversing). Section 13.11.3 below gives a method of estimating the error propagated in trilateration.

When weather and visibility conditions prevent optical measurement of distances by Geodimeter, and also the observation of horizontal angles, the radio instruments can go ahead with the measurement of the sides of a network, leaving the angular work to be carried out later. Most purely trilateration networks have to be complex to reduce azimuth error, and to give a reasonable number of conditions in the network.

The scale of a trilateration is generally better than in a triangulation since the scale error of the former is not cumulative.

Generally, a combination of angular and linear measurement will ensure that both scale and azimuth are well maintained. Points on either side of a traverse are often fixed by trilateration and the network is called a *trilateration-traverse*; and if points on either side of the traverse are fixed by angular observations a *triangulation-traverse* results.

13.9. REDUCTION OF SLANT RANGES TO THE SPHEROID

In reducing slant measurements of E.D.M. instruments to the spheroid, the following corrections are applied:

(i) Curvature of signal path.
(ii) Mean terminal refraction.
(iii) Slope, C_1.
(iv) Height, C_2.
(v) Chord to arc.

The small corrections (i), (ii), and (v) are applied together as correction C_3 obtained from table 13.8.

13.9.1. CURVATURE OF SIGNAL PATH

The empirical formulae of §13.2.9 and §13.4.6 give values of the index of refraction μ for light and radio respectively, and enable the distance to be calculated from the transit time and a knowledge of the velocity *in vacuo*. Due to refraction, the path taken by the signal is curved. This curve is usually taken to be a circular arc of radius σ related to the radius of the Earth R by $k = R/2\sigma$ where k is the coefficient of refraction. If the angle subtended by the arc AB length l at the centre of curvature of the spheroid is θ, see figure 13.22(b), and the refraction angle at A (the angle between the curve and chord at A) is r, we have $r = k\theta = kl/R$

Also
$$l = 2r\sigma$$
then
$$k = R/2\sigma$$

Average values for k_L for light and k_R for radio are 0·065, ($\frac{1}{16}$) and 0·125, ($\frac{1}{8}$) respectively, giving the radii of curvature for light $\sigma_2 \simeq 8R$, and for radio

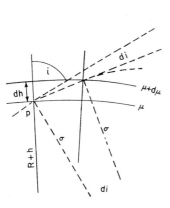

Figure 13.22(a)

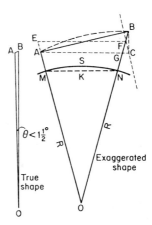

Figure 13.22(b)

$\sigma_R \simeq 4R$. The chord length of AB, D is given by $D = 2\sigma \sin r$

$$D = 2\sigma(r - r^3/3! + \cdots \text{etc.} \ldots)$$

Thus the correction for path curvature is $D - l = -l^3/24\sigma^2$

i.e. $$D - l = -l^3 k^2/6R^2$$

Second refraction correction

Under ideal conditions, the index of refraction μ may be considered constant over a spherical surface at height h whose radius is $R + h$. Conversely if the refractive index derived for the terminals of a line is taken to be typical of the line as a whole, it is assumed that the curvature of the signal path is constant and equals the radius of the earth R, neglecting the height above sea level. A correction has therefore to be applied to allow for the actual curvature σ and the consequent change in μ for the line as a whole. Figure 13.22(a) shows a small element dl of the signal path of radius σ passing from one spherical surface to another whose refractive indices differ by $d\mu$ and whose linear separation is dh, the height difference. The signal path makes an angle i with the normal at P. By the law of refraction $\mu \sin i = $ constant

$$\therefore \quad d\mu/\mu = -\cot i \, di \simeq -\cos i \, di \quad (\text{since } i \simeq 90)$$
$$= -dl \cos i/\sigma$$
$$= -dh/\sigma$$

The maximum separation between two circular arcs of radii R and σ and length l is

$$-\frac{l^2}{8}\left(\frac{1}{R} - \frac{1}{\sigma}\right) = -\frac{l^2}{8R}(1 - 2k)$$

and the mean separation for the two arcs is two thirds of this viz

$$-\frac{l^2}{12R}(1 - 2k). \quad (\text{Simpson's Rule})$$

The chord length D is calculated from the expression

$$D = \frac{1}{2} \cdot \frac{v}{\mu} \times (\text{transit time})$$

$$\therefore \quad dD/D = -d\mu/\mu$$

Thus the correction to be applied to D is

$$dD = D \, dh/\sigma = -\frac{D^3 k}{6R^2}(1 - 2k)$$

Slope correction C_1

In figure 13.22(b), the slant length AB of length D is to be reduced to the spheroidal arc S. The heights of A and B are respectively H_A and H_B and

$H_B - H_A = \Delta H$. Strictly $\Delta H = BG$ and not BC as assumed here. Since the angle subtended by D at the centre of the earth is never more than $1\frac{1}{2}°$, $BG = BC$ with ample accuracy for this reduction. The length of $AC = D_1$ is given by

$$D_1 = (D^2 - \Delta H^2)^{\frac{1}{2}}$$
$$= D - \Delta H^2/2D - \Delta H^4/8D^3 \text{ etc.}$$
$$= D - (\beta + \beta^2/2D) \quad \text{where} \quad \beta = \Delta H^2/2D$$
$$\therefore \quad D_1 = D - C_1$$

Tables of β are available in many survey departments, and C_1 can be obtained from them by successively entering the table with D and β as arguments.

Height correction C_2

It is considered here that the heights available are referred to the spheroid, though the separation between geoid and spheroid can introduce appreciable amounts. In figure 13.22(b) $AC = EF$, thus to obtain the chord length K, D_1 is reduced from the mean height $\frac{1}{2}(H_A + H_B) = H_M$

$$K = RD_1(R + H_M)^{-1}$$
$$= D_1 - D_1 H_M/R \quad \text{with ample accuracy.}$$

R is strictly the radius of curvature of the spheroid at latitude ϕ and azimuth α given by Euler's theorem. (See §4.1.6.) Whence the correction for height

$$C_2 = D_1 H_M(\cos^2 \alpha/\rho + \sin^2 \alpha/\nu)$$
$$= D_1(H_A + H_B)\gamma.$$

where γ is obtained from table 13.7 which is sufficiently precise for most work using any spheroid.

Chord to arc correction

The arc length S is obtained from the expression

$$S = K + K^3/24R^2 \cdots \text{etc.}$$

for $S = R\theta$,
$$K = 2R \sin \tfrac{1}{2}\theta = 2R(\tfrac{1}{2}\theta - \tfrac{1}{48}\theta^3 \cdots)$$
$$K \simeq S - S^3/24R^2$$
$$S = K + K^3/24R^2 \quad \text{with ample precision}$$

Combining this correction with the first two above we have a combined small correction C_3 where

$$C_3 = \frac{D^3}{24R^2}(1 - 2k)^2$$

which for radio $(k_R = \tfrac{1}{8})$

gives
$$C_3 = \frac{D^3}{43R^2}$$

Table 13.7. Reduction of Tellurometer Slant Ranges to the Spheroid

$$\gamma \text{ Factor} = \frac{1}{2R} \times 10^8, \ R \text{ in Metres}$$

(To determine the sea level correction c_2)

AZIMUTH (DEG.)	Latitude (deg)										AZIMUTH (DEG.)
	0	5	10	15	20	25	30	35	40	45	50	
0	7·893	7·892	7·890	7·887	7·883	7·878	7·873	7·866	7·860	7·852	7·846	180
5	7·892											175
10	7·891											170
15	7·889	7·888	7·887	7·884	7·880	7·875	7·870	7·864	7·857	7·851	7·844	165
20	7·887											160
25	7·883											155
30	7·879	7·879	7·877	7·875	7·871	7·867	7·863	7·857	7·852	7·846	7·840	150
35	7·875											145
40	7·871											140
45	7·866	7·866	7·864	7·862	7·860	7·856	7·852	7·848	7·844	7·839	7·835	135
50	7·861											130
55	7·857											125
60	7·852	7·852	7·851	7·850	7·848	7·845	7·843	7·839	7·836	7·833	7·829	120
65	7·849											115
70	7·845											110
75	7·843	7·843	7·842	7·841	7·839	7·837	7·835	7·833	7·830	7·828	7·825	105
80	7·841											100
85	7·840											95
90	7·839	7·839	7·838	7·837	7·836	7·835	7·832	7·830	7·828	7·826	7·824	90

This correction is tabulated in table 13.8, and it will be seen that it lies within the precision of present instruments.

The final spheroidal length S is given by

$$S = D - C_1 - C_2 + C_3 \qquad (18)$$

where the length used to evaluate each correction is that previously found.

13.9.2. ALTERNATIVE FORMULAE

S may be calculated via the small angle θ at the centre of curvature of the spheroid. This angle θ is computed from

$$\sin \tfrac{1}{2}\theta = \left(\frac{(s-a)(s-b)}{ab} \right)^{\frac{1}{2}} \qquad (19)$$

Table 13.8. Arc Correction c_3 Metres for Radio Instruments

L (km)	C (m)	L (km)	C (m)	L (km)	C (m)	L (km)	C (m)
0	0·000	21	0·005	35	0·025	49	0·068
8	0·000	22	0·006	36	0·027	50	0·072
9	0·000	23	0·007	37	0·029	51	0·076
10	0·001	24	0·008	38	0·032	52	0·081
11	0·001	25	0·009	39	0·034	53	0·086
12	0·001	26	0·010	40	0·037	54	0·091
13	0·001	27	0·011	41	0·040	55	0·096
14	0·002	28	0·013	42	0·043	56	0·101
15	0·002	29	0·014	43	0·046	57	0·107
16	0·002	30	0·016	44	0·049	58	0·112
17	0·003	31	0·017	45	0·052	59	0·118
18	0·003	32	0·019	46	0·056	60	0·124
19	0·004	33	0·021	47	0·060	61	0·131
20	0·005	34	0·023	48	0·064	62	0·137

where $\quad a = R + H_A, \quad b = R + H_B, \quad 2S = a + b + D$

whence $\quad\quad\quad\quad S = R\theta'' \sin 1''$ (20)

Although it is an attractive formula, the necessary computational precision can be obtained only with Andoyer's 15 figure tables in which interpolation is troublesome.

The chord length K may be written in terms of D, H_A, H_B, and R as follows

$$K = \left[\frac{(D - \Delta H)(D + \Delta H)}{(1 + H_A/R)(1 + H_B/R)} \right]^{\frac{1}{2}} \quad (21)$$

Computation with this formula needs care with significant figures on the calculating machine.

13.9.3. COMPUTATION OF COORDINATES IN A TRILATERATION NETWORK

Since the computation of triangulation and traverse coordinates is considered in Chapters 6 and 7, we shall confine our attention very largely to trilateration, although some mention is made of methods of computing with mixed angles and sides.

As with most surveys, computation may either be carried out on the surface of a reference spheroid using the mid-latitude or other formulae, or alternatively it may be carried out on a plane projection surface with the aid of the formulae of plane trigonometry. Generally speaking, the latter method is preferred by all but those having access to an electronic digital computer.

In precise work, the observed data—angles and sides—may be adjusted for consistency prior to the computation of the final coordinate values. Semi-rigorous or least squares adjustment may take place according to circumstances. Also, either the method of condition equations or of observation equations may be used. Generally, the method of observation equations and variation of coordinates is preferable because of its simplicity and versatility. The semigraphic method of combining the adjustment with the computation of coordinates has much to commend it, whilst Dr Jerie of Delft has produced an analogue computer to handle this problem in a manner similar to the semigraphic method.†

In spheroidal computation the spheroidal lengths are used in the computation directly, whereas in projection work the necessary scale factors have to be applied as described in Chapter 4.

13.9.4. COMPUTATION OF TRILATERATION BY THE METHOD OF VARIATION OF COORDINATES. (SEMIGRAPHIC AND LEAST SQUARES METHODS.)

In figure 13.23 the sides a and b have been measured, reduced to the spheroid and converted to grid lengths via the scale factors. We require to compute the coordinates of G from the two fixed points A and B whose coordinates are (E_A, N_A), and (E_B, N_B) respectively. The length of AB = g is computed from

† H. G. Jerie, Analogue computer for Calculating and Adjusting Trilateration Nets, *E.S.R.*, No. 126, October 1962.

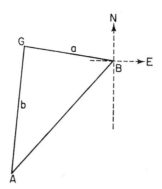

Figure 13.23

the coordinates. The computation is outlined with reference to the following data all in metres:

E_A 351 240·22 N_A 138 628·80
E_B 356 788·67 N_B 144 328·27
$a =$ BG $= 5$ 191·05 $b =$ AG $= 6$ 282·32

The length of $g =$ AB $= 7$ 954·20.

13.9.5. SEMI-GRAPHIC METHOD

The coordinates of A and B are plotted on a large scale diagram and the approximate position of G is found by swinging arcs equal to the lengths a and b. The approximate coordinates of G are

E_G' 351 590   and N_G' 144 880

From these approximate coordinates of G and the coordinates of A and B the computed lengths of a and b are obtained (see table 13.9). The computed lengths of b and a are 6 260·98 and 5 227·86, which differ from the observed lengths by the 'shifts' 21·34, and 36·81 respectively. In figure 13.24 the lines representing b and a are drawn at a convenient scale meeting at the approximate position of G, i.e. at G'. From G' the shifts are marked out at a convenient

Table 13.9. Fixation by Semigraphic Method
METRES

POINT	EASTING	NORTHING	LENGTH	POINT	EASTING	NORTHING	LENGTH
A	351 240·22	138 628·80	6 282·32	B	356 788·67	144 328·27	5 191·05
G'	351 590	144 880		G'	351 590	144 880	
Δ'	+349·78	+6 251·20	6 260·98	Δ'	−5 198·67	+551·73	5 227·86
1ˢᵗ calc.	+39·2	+19·0	+21·34	1ˢᵗ calc.	+39·2	+19·0	−36·81
Δ''	+388·98	+6 270·20	6 282·254	Δ''	−5 159·47	+570·73	5 190·940
2ⁿᵈ calc.	−0·10	+0·07	+0·066	2ⁿᵈ calc.	−0·10	+0·07	+0·110
Δ'''	+388·88	6 270·27	6 282·32	Δ'''	−5 191·57	+570·80	5 191·05
G	351 629·10	144 899·07		G	351 629·10	144 899·07	

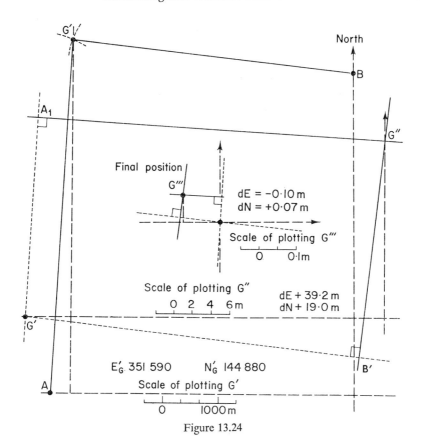

Figure 13.24

scale in their correct sense; e.g. since the measured length of b is longer than that computed from the provisional coordinates, the true position of G must lie further from A than does G'. In figure 13.24, the first position of G' plotted from A and B at the top left is moved down to enable G" to be plotted at a larger scale. Since the position of G lies at the intersection of the arcs centred on A and B and whose radii are equal to the measured lengths, its position will be located at the intersection of the perpendiculars drawn through the points A_1 and B_1 where $A_1G' = 21\cdot34$ m and $BG' = 36\cdot81$ m. This assumes that the departure of the tangent from the circular arc is negligible. This assumption will be more valid the closer that provisional position of G is to the correct one, other things equal. From the diagram, a second position of G is found at the intersection of the two perpendiculars from A_1 and B_1. At the same scale as the shifts, the corrections dE and dN are scaled from the diagram, e.g. in this example $dE = 39\cdot2$ m and $dN = 19\cdot0$ m.

Applying these corrections to the coordinates of G' a better estimate of the coordinates of G is obtained. In table 13.9 the new values of the ΔE and ΔN are obtained and the lengths of b and a recomputed to give second and smaller values of the shifts, i.e. of 0·066 m and 0·110 m respectively. This is an indication of the small computational error in the position of G'. To obtain a better position, these new shifts and perpendiculars are plotted to give the final position of G at their intersection point. As a final check the computed distances should agree exactly with the measured values.

If G is fixed from more than two points, the procedure is to compute the coordinates as above using two rays which give a good intersection. Thereafter the shifts are computed for all the other rays and the tangents (arcs) for each

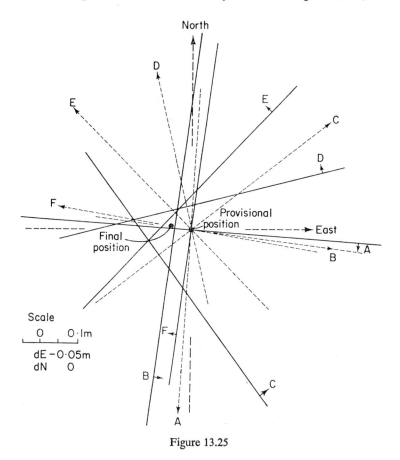

Figure 13.25

measured length are plotted on the final diagram. The final position of G is then obtained by inspection of the error figure produced, after any discrepant ray has been examined and corrected or rejected if necessary. Figure 13.25 shows the error figure for the fixation of G from points A, B, C, D, E, and F with the shifts obtained from table 13.10.

Since the observed angles may be similarly treated and represented on a diagram, the method may be used conveniently in the computation of co-ordinates of points fixed by both linear and angular measurements. In such problems it is advisable to use different colours to represent sides and angles. The provisional coordinates of the new point will also be derived most readily by analytical calculation, rather than by purely semi-graphic methods.

Table 13.10. Semi-graphic Trilateration Fixation of G by Multiple Rays Provisional Coordinates of G are 351 629.13E, 144 899.07N.

PROVISIONAL COORDINATES OF G ARE 351 629·13 E, 144 899·07N

NO	POINT	EASTING	NORTHING	DISTANCE	C = COMPUTED O = OBSERVED
1	A	351 240·22	138 628·80	6 282·32	O
2	Δ	+388·91	+6 270·27	6 282·32	C
3	$\frac{\Delta}{S}$ and d	+0·062	+0·998	0	O−C
4	B	356 788·67	144 328·27	5191·05	O
5	Δ	−5 159·54	+570·80	5191·01	C
6	$\frac{\Delta}{S}$ and d	−0·994	+0·110	+0·04	O−C
7	C	356 442·71	148 778·96	6 182·65	O
8	Δ	−4 813·58	−3 879·89	6 182·65	C
9	$\frac{\Delta}{S}$ and d	−0·779	−0·628	+0·09	O−C
10	D	350 044·25	150 752·70	6 064·34	O
11	Δ	+1 584·88	−5 853·63	6 064·39	C
12	$\frac{\Delta}{S}$ and d	+0·261	−0·965	−0·05	O−C
13	E	345 780·67	150 394·05	8 024·87	O
14	Δ	+5 848·46	−5 494·98	8 024·92	C
15	$\frac{\Delta}{S}$ and d	+0·729	−0·685	−0·05	O−C
16	F	347 490·50	145 480·79	4 179·31	O
17	Δ	+4 138·63	−581·72	4 179·31	C
18	$\frac{\Delta}{S}$ and d	+0·990	−0·139	0	O

13.9.6. ANALYTICAL METHOD

As an alternative to the graphical method, the corrections to the provisional coordinates of G may be found analytically. Consider the equation relating the length S to the corresponding differences in eastings and northings between its terminals (1) and (2), viz $S^2 = \Delta E^2 + \Delta N^2$.

Differentiating $\quad 2S\,dS = 2\Delta E\,\partial(\Delta E) + 2\Delta N\,\partial(\Delta N)$

i.e. $\quad dS = (\Delta E/S)\,\partial(\Delta E) + (\Delta N/S)\,\partial(\Delta N)$ \hfill (22)

where $\partial(\Delta E) = dE_2 - dE_1$ and $\partial(\Delta N) = \partial N_2 - \partial N_1$.

If one end of the line is fixed, say point (1), then $dE_1 = dN_1 = 0$. Equation (22) therefore reduces to

$$dS = (\Delta E/S)\,dE_2 + (\Delta N/S)\,dN_2 \qquad (23)$$

Since there are two variables dE_2 and dN_2 a second equation of the form of (23) is required for a solution to be obtained, i.e. at least two lines are required for a fixation. The value of dS (the shift) is obtained as for the semi-graphic method, and the coefficients $\Delta E/S$ and $\Delta N/S$ calculated from the best available values of ΔE, ΔN, and S. In table 13.11 the same data are used to calculate the first values of the shifts dS, and the coefficients $\Delta E/S$ and $\Delta N/S$ giving the equations

$$\left.\begin{array}{r} +0{\cdot}055\,9\,dE + 0{\cdot}998\,4\,dN = +21{\cdot}34 \\ -0{\cdot}994\,4\,dE + 0{\cdot}105\,5\,dN = -36{\cdot}81 \end{array}\right\} \qquad (24)$$

which on solution give $dE + 39{\cdot}05$ and $dN + 19{\cdot}18$. These values of dE and

Table 13.11. Analytical Fixation by Variation of Coordinates

POINT	EASTING	NORTHING	LENGTH	POINT	EASTING	NORTHING	LENGTH
A	351 240·22	138 628·80	6 282·32	B	356 788·67	144 328·27	5191·05
G'	351 590	144 880	6260·98	G'	351 590	144 880	5227·86
Δ'	+349·78	+6 251·20	+21·34	Δ'	−5 198·67	+551·73	−36·81
d'	+39·05	+19·18		d'	+39·05	+19·18	
Δ"	+388·83	+6270·38	6 282·42	Δ"	−5159·62	+570·91	5191·11
d"	+0·05	−0·10	−0·10	d"	+0·05	−0·10	−0·06
G	351 629·10	144 899·08		G	351 629·10	144 899·08	
Δ/S'	+0·0559	+0·9984	FIRST	Δ/S'	−0·9944	+0·1055	
Δ/S"	+0·062	+0·998	SECOND	Δ/S"	−0·994	+0·110	

EQUATIONS ARE OF THE FORM $(\Delta E/s)dE + (\Delta N/S)dN = O - C$

dN are added algebraically to the first values of ΔE and ΔN from which a better estimate of the computed length and therefore a smaller shift is obtained for each line. For greater accuracy in computation we form two further equations and solve. The second equations are

$$\left. \begin{array}{l} +0{\cdot}062\ dE + 0{\cdot}998\ dN = -0{\cdot}10 \\ -0{\cdot}994\ dE + 0{\cdot}110\ dN = -0{\cdot}06 \end{array} \right\} \quad (25)$$

and the solution is $dE + 0{\cdot}049$, $dN - 0{\cdot}103$. Whence the final values of ΔE and ΔN, and thence the final coordinates of G are obtained.

13.9.7. Least Squares Adjustment by Variation of Coordinates

In a complicated network or where more than the essential number of measurements is made to fix a point, an adjustment problem is present. For example in the example quoted above, in which G is fixed from points A, B, C, D, E, and F, the error figure 13.25 has to be solved to give the best position of G. If the problem is treated by the analytical method there is an equation for each line in terms of dE and dN, the corrections to the provisional coordinates. Let the most probable value of a length be V, and its observed value be O, then the residual v is given by $v = V - O$. Now let $V = C + dC$, where C is any provisional value of V and dC is the correction to be added to C to give V. In practice dC should be as small as possible for accurate results, i.e. the closer the provisional coordinates are to the correct values the better. Then

$$v = dC + (C - O) \quad (26)$$

For the length S, it has been shown that the relationship between a change dS in S and in the coordinates of its terminals (E_1, N_1) and (E_2, N_2) is

$$dS = (\Delta E/S)(dE_2 - dE_1) + (\Delta N/S)(dN_2 - dN_1) \quad (22)$$

whence substituting for $dC = dS$ in (26) gives the observation equation for the measured length S

$$v = (\Delta E/S)(dE_2 - dE_1) + (\Delta N/S)(dN_2 - dN_1) + (C - O) \quad (27)$$

The term (*computed minus observed*) or $(C - O)$ is the *shift* of the semi-graphic method. An equation of the form (27) is written for each measured length, and from them are formed the normal equations as described in Chapter 5. Table 13.10 gives the data for the formation of the observation equations of the example in lines (3), (6), (9), (12), (15), and (18). The observation equations are

$$\left.\begin{array}{ccc} dE & dN & \text{Abs } (C - O) \\ +0\cdot062 & +0\cdot998 & 0\cdot00 \\ -0\cdot994 & +0\cdot110 & +0\cdot04 \\ -0\cdot779 & -0\cdot628 & +0\cdot09 \\ +0\cdot261 & -0\cdot965 & -0\cdot05 \\ +0\cdot729 & -0\cdot685 & -0\cdot05 \\ +0\cdot990 & -0\cdot139 & 0\cdot00 \end{array}\right\} \quad (28)$$

The normal equations formed from these observation equations (28) are

$$\begin{array}{ccc} dE & dN & \text{Abs} \\ +3\cdot178 & -0\cdot448 & +0\cdot159 \\ & +2\cdot822 & -0\cdot030 \end{array}$$

which on solution give $dE - 0\cdot050$, and $dN + 0\cdot003$. Whence the final coordinates of G are 351 629·08 E, 144 899·07 N.

13.9.8. THE LEAST SQUARES ADJUSTMENT OF BOTH ANGULAR AND LINEAR MEASUREMENTS BY VARIATION OF COORDINATES

In adjusting a mixed network, the linear and directional (or angle) observation equations are in dissimilar terms—lengths and angles. To effect the combined analytical adjustment it is necessary to convert *either* the angles to linear equivalents *or* the sides to angular equivalents, and thereafter proceed in the normal manner. The latter is currently used for this type of adjustment, though there may be some advantage in adopting the former.

For example, the equation (23) may be written

$$\frac{dS}{S} = \left(\frac{\Delta E}{S^2}\right) dE + \left(\frac{\Delta N}{S^2}\right) dN \quad (29)$$

in which dS/S can be considered as an angle in radians, which on conversion to seconds of arc $\Delta S''$

gives
$$\Delta S'' = (\Delta E/S^2 \sin 1'') \, dE + (\Delta N/S^2 \sin 1'') \, dN \quad (30)$$

This is then comparable to the *direction equation* for the same line

$$d\alpha'' = (\Delta N/S^2 \sin 1'') \, dE - (\Delta E/S^2 \sin 1'') \, dN \quad (31)$$

(see §6.11.14). Observation equations are written for all directions (or angles), and for all measured sides, fixed quantities being dealt with as described in

Chapter 6. Thereafter, the normal equations are formed and the adjustment completed in the usual way. Finally the length corrections in angular terms are converted to linear units,

For $\quad \Delta S'' \sin 1'' = dS/S$, i.e. $dS = S \Delta S''/206\ 265$

13.10. THE LEAST SQUARES ADJUSTMENT OF MEASURED LENGTHS BY CONDITION EQUATIONS

As an alternative to adjusting measured lengths by variation of coordinates and observation equations, they may also be adjusted by the condition equations method, though in the opinion of the author, the former is to be preferred in most instances on account of its simplicity. Again, if computation is finally on a projection system, provisional coordinates are required for $t - T$ and scale factors, even if the data is adjusted on the spheroid, and therefore the variation of coordinates method is more suitable. The method of variation of geographical coordinates on the spheroid is practicable only for an organization with access to an electronic digital computer. For further discussion of these and other points, the reader should refer to Rainsford.[†] However, since the authors' opinion is not shared by many surveyors, and in the adjustment of relatively simple networks the condition equations method is preferable, and in important work one method may be used to check the other, an outline of the method is given in the following sections.

13.10.1. CONDITION EQUATIONS IN TRILATERATION

Apart from that arising from a fixed length, the condition equations which the measured sides of a trilateration net must satisfy are ultimately reduced to conditions that certain angles computed from the sides must satisfy, or to certain conditions that areas must satisfy. Since the consideration of areas has certain serious limitations and is less convenient computationally, we shall consider angles only.

13.10.2. FIXED SIDES

Suppose a length a has to equal a fixed value A, then the condition equation to be satisfied is

$$a + da - A = 0 \qquad (32)$$

This does not involve angles.

13.10.3. FIGURAL CONDITIONS

In the polygon ABCD $\cdots$ H with centre point 0 (figure 13.26) the geometrical condition to be satisfied is that the angles θ should add up to 360° or 2π. Let the misclosure on 360° be $d\theta$. The adjustment is achieved when the points D and H coincide. The condition equation is

$$d\theta_1 + d\theta_2 + \cdots + d\theta_7 + d\theta = 0 \qquad (33)$$

[†] H. F. Rainsford, Survey Adjustments and Least Squares, London, Constable.

Electromagnetic Distance Measurement

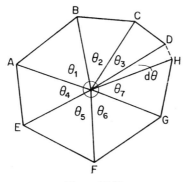

Figure 13.26

13.10.4. BEARING CONDITIONS

The bearings of AB and FH are fixed (figure 13.27). Let their difference be α, then the condition is that $\theta_1 - \theta_2 + \theta_3 - \theta_4 + \theta_5$ should equal α, introducing 180° if required. If the misclosure is $d\theta$ the condition equation is

$$d\theta_1 + d\theta_3 + d\theta_5 - (d\theta_2 + d\theta_4) + d\theta = 0 \tag{34}$$

13.10.5 POSITION CONDITIONS

In figure 13.27 points $A(y_A, x_A)$ and $F(y_F, x_F)$ are fixed. i.e.

$$y_F - y_A = \Delta Y \quad \text{a constant}$$

and

$$x_F - x_A = \Delta X \quad \text{a constant}$$

Denote the respective bearings and lengths of AC, CD, DE, and EF by $\beta_1, \beta_2, \beta_3, \beta_4$ and S_1, S_2, S_3, S_4. Then the condition equation for easting y is

$$S_1 \sin \beta_1 + S_2 \sin \beta_2 + S_3 \sin \beta_3 + S_4 \sin \beta_4 = \Delta Y$$

Differentiating with respect to the bearings and lengths

$$S_1 \cos \beta_1 \, d\beta_1 + S_2 \cos \beta_2 \, d\beta_2 + S_3 \cos \beta_3 \, d\beta_3 + S_4 \cos \beta_4 \, d\beta_4$$
$$+ \sin \beta_1 \, dS_1 + \sin \beta_2 \, dS_2 + \sin \beta_3 \, dS_3 + \sin \beta_4 \, dS_4 + dy = 0 \tag{35}$$

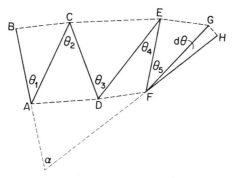

Figure 13.27

where dy is the misclosure. Remembering that

$$\beta_1 = \theta_1 + \text{constant}; \quad \beta_2 = \theta_1 - \theta_2 + \text{constant, etc.}$$

and therefore

$$d\beta_1 = d\theta_1; \quad d\beta_2 = d\theta_1 - d\theta_2; \quad d\beta_3 = d\theta_1 - d\theta_2 + d\theta_3;$$
$$d\beta_4 = d\theta_1 - d\theta_2 + d\theta_3 - d\theta_4$$

this equation (35) reduces to

$$d\theta_1(S_1 \cos \beta_1 + S_2 \cos \beta_2 + S_3 \cos \beta_3 + S_4 \cos \beta_4)$$
$$- d\theta_2(S_2 \cos \beta_2 + S_3 \cos \beta_3 + S_4 \cos \beta_4) + d\theta_3(S_3 \cos \beta_3 + S_4 \cos \beta_4)$$
$$- d\theta_4(S_4 \cos \beta_4) + [\sin \beta \, dS]_1^4 + dy = 0 \quad (36)$$

Putting $\Delta x = S \cos \beta$ the final condition equation for easting is

$$+d\theta_1 \, {}_A\Delta x_F - d\theta_2 \, {}_C\Delta x_F + d\theta_3 \, {}_D\Delta x_F - d\theta_4 \, {}_E\Delta x_F$$
$$+ [\sin \beta \, dS]_1^4 + dy = 0 \quad (37)$$

The similar equation for northings x is

$$-d\theta_1 \, {}_A\Delta y_F + d\theta_2 \, {}_C\Delta y_F - d\theta_3 \, {}_D\Delta y_F + d\theta_4 \, {}_E\Delta y_F$$
$$+ [\cos \beta \, dS]_1^4 + dx = 0 \quad (38)$$

13.10.6. BASIC EQUATION FOR A TRIANGLE

Thus it is seen that, with the exception of a fixed side, the condition equations involve angles θ computed from the measured lengths, and that accurate values of these angles are required to assess the misclosures. Since the observed values are not angles but sides, the condition equations must be written in terms of variations to the sides, i.e. the $d\theta$'s have yet to be related to the dS's.

Consider the triangle ABC with measured sides a, b, c. Suppose A is the required computed angle given by

$$\sin \tfrac{1}{2} A = \left[\frac{(s-b)(s-c)}{bc} \right]^{\frac{1}{2}}$$

where
$$2s = a + b + c$$

then
$$2 \log \sin \tfrac{1}{2} A = \log(s-b) + \log(s-c) - \log b - \log c \quad (39)$$

Differentiation of equation (39) gives the required relationship between the computed angle A and the measured sides a, b, and c. This differentiation will now be effected in two ways giving two different methods of computing the differential equation, both of which are computationally simple. The object of using two methods is to enable the work to be checked.

13.10.7. METHOD ONE: FOR MACHINE COMPUTATION AND TABLES OF NATURAL TRIGONOMETRICAL FUNCTIONS

Differentiating (39) in the normal way we have

$$2 \cot \tfrac{1}{2} A \cdot \tfrac{1}{2} dA = \frac{d(s-b)}{s-b} + \frac{d(s-c)}{s-c} - \frac{db}{b} - \frac{dc}{c} \quad (40)$$

Electromagnetic Distance Measurement

now
$$s - b = \tfrac{1}{2}(a + b + c - 2b) = \tfrac{1}{2}(a - b + c)$$
$$\therefore \; d(s - b) = \tfrac{1}{2}(da - db + dc)$$
also
$$d(s - c) = \tfrac{1}{2}(da + db - dc)$$

Substituting in (40) gives

$$\cot \tfrac{1}{2}A \cdot dA = \frac{1}{2(s - b)}(da - db + dc)$$

$$+ \frac{1}{2(s - c)}(da + db - dc) - \frac{db}{b} - \frac{dc}{c}$$

$$\therefore \; 2\cot \tfrac{1}{2}A \cdot dA = da\left(\frac{1}{s - b} + \frac{1}{s - c}\right)$$

$$- db\left(\frac{1}{s - b} - \frac{1}{s - c} + \frac{2}{b}\right) - dc\left(\frac{1}{s - c} - \frac{1}{s - b} + \frac{2}{c}\right)$$

And putting $R = \tfrac{1}{2} \tan \tfrac{1}{2}A$

$$dA = R\, da\left(\frac{1}{s - b} + \frac{1}{s - c}\right) - R\, db\left(\frac{1}{s - b} - \frac{1}{s - c} + \frac{2}{b}\right)$$

$$- R\, dc\left(\frac{1}{s - c} - \frac{1}{s - b} + \frac{2}{c}\right) \quad (41)$$

Example: Consider triangle ABC whose measured sides are given in the table 13.12, together with the computation of the necessary factors for the formation of the differential equation.

Table 13.12

a 69 847·62			s − a 29 209·695	Sum	8·3033
b 83 587·77	$10^5 \times 2/b$ 2·3926		s − b 15 469·545	$10^5 \times 1/s - b$	6·4643
c 44 679·24	$10^5 \times 2/c$ 4·4764		s − c 54 378·075	$10^5 \times 1/s - c$	1·8390
2s 198 114·63			s 99 057·315	Diff.	+4·6253
s 99 057·315			Checks.		
$\sin \tfrac{1}{2} A$ 0·474 59858					
$\tfrac{1}{2} A$ 28° 19′ 59″·58		$\tan \tfrac{1}{2} A$ 0·5392	$R = 0·2696$		
A 56° 39′ 59″·16					

The differential equation is then

$$dA = +2·238\,6\, da - 1·892\,0\, db + 0·040\,1\, dc \quad (42)$$

where dA is in units of (radians $\times 10^{-5}$).

13.10.8. METHOD TWO: FOR LOGARITHMIC COMPUTATION

We now differentiate equation (39) by the numerical method of tabular logarithmic differences, which is commonly used in the formation of the side equation in a triangulation figure. d(log b) may be expressed as $\Delta_b \cdot db$ where Δ_b is the tabular difference of log b for a variation of one unit of db. Any convenient unit may be chosen, e.g. the difference may be for 1 foot as in the example below. Denoting the tabular log differences of $\sin \frac{1}{2} A$; $s-b$, $s-c$, b, and c by respectively Δ_A, Δ_{s-b}, Δ_{s-c}, Δ_b, and Δ_c the differential equation from (39) is

$$2\Delta_A \, dA = \Delta_{s-b} \, d(s-b) + \Delta_{s-c} \, d(s-c) - \Delta_b \, db - \Delta_c \, dc$$

Substituting for $d(s-b)$ and $d(s-c)$ as before, we have after some rearrangement

$$2\Delta_A \, dA = da(\Delta_{s-b} + \Delta_{s-c}) - db(\Delta_{s-b} - \Delta_{s-c} + 2\Delta_b)$$
$$- dc(\Delta_{s-c} - \Delta_{s-b} + 2\Delta_c)$$

and denoting $1/2\Delta_A$ by R the final form of the equation is

$$dA = R \, da(\Delta_{s-b} + \Delta_{s-c}) - R \, db(\Delta_{s-b} - \Delta_{s-c} + 2\Delta_b)$$
$$- R \, dc(\Delta_{s-c} - \Delta_{s-b} + 2\Delta_c) \quad (43)$$

where dA is in units of seconds of arc or some other chosen multiple, e.g. in the example we choose units of 10 seconds of arc. Converting dA to units of radians 10^{-5} gives an equation which checks (42) to three significant figures, which is sufficient for most work.

Example: The same example as above example in §13.10.7 is worked in the table 13.13 in the logarithmic form, and gives the degree of agreement expected. The differential equation is then

$$dA = +0{\cdot}461 \, 1 \, da - 0{\cdot}389 \, 8 \, db + 0{\cdot}008 \, 2 \, dc \quad (44)$$

and converting to units of radians 10^{-5} gives

$$dA = +2{\cdot}235 \, da - 1{\cdot}890 \, db + 0{\cdot}040 \, dc \quad (45)$$

which checks (42) with sufficient accuracy.

It will be apparent that an example such as the fixation of G from the six points A to F in §13.9.6 will be very much more involved if adjusted by condition equations than by the variation of coordinates method. For this reason the conditions method is less convenient than the latter in all but very simple examples seldom encountered in practice.

Table 13.13

	Δ tool	2Δ	Sum Δ	360·6
log b 4·922 1427	52·0	104·0	log (s−b) 4·189 4775	+280·7
log c 4·650 1057	97·2	194·4	log (s−c) 4·735 4238	+ 79·9
9·572 2484			Diff. Δ	+200·8
			8·924 9013	
½ A 28° 19′ 59·″58			9·572 2484	
A 56° 39′ 59″·16			9·352 6529	Δ for 10″
			log sin ½ A 9·676 32645	+391
			R = 0·001279	

13.10.9. THE LEAST SQUARES ADJUSTMENT OF COMBINED ANGULAR AND LINEAR MEASUREMENTS BY THE METHOD OF CONDITION EQUATIONS

As with a trilateration, it is usually simpler to adjust a combined network by variation of coordinates on the plane as described in §13.9.7. The problem may however be solved by condition equations as follows.

Consider a triangle ABC whose spherical excess is ϵ in which all three sides a, b, and c, and all three angles A, B, and C have been measured. Then the conditions to be satisfied are that

$$A + B + C = 180 + \epsilon \tag{46}$$

and
$$a/\sin A' = b/\sin B' \tag{47}$$

and
$$a/\sin A' = c/\sin C' \tag{48}$$

where A', B', and C' are the Legendre angles, i.e.

$$A' = A - \tfrac{1}{3}\epsilon, \quad B' = B - \tfrac{1}{3}\epsilon, \quad C' = C - \tfrac{1}{3}\epsilon$$

Differentiating (47)

$$da/a - db/b + \cot B' \, dB' - \cot A' \, dA' = 0 \tag{49}$$

Normally the equations such as (47) will not be satisfied exactly. Let the misclosure on unity be k_1, i.e. $1 + k_1 = (a/\sin A')(\sin B'/b)$. The condition equation to be satisfied is then

$$da/a - db/b + \cot B' \, dB' - \cot A' \, dA' + k_1 = 0 \tag{50}$$

where dA' and dB' are in radians. Converting to seconds of arc we have

$$da/a \sin 1'' - db/b \sin 1'' + \cot B'(dB')'' - \cot A'(dA')'' + k_1/\sin 1'' = 0 \tag{51}$$

Now da/a can be considered to be an angle in radians, whose value in seconds of arc is $\Delta a''$, i.e. $da/a = \Delta a'' \sin 1''$. Similarly we put $db/b = \Delta b'' \sin 1''$. Equation (51) may then be written

$$\Delta a'' - \Delta b'' + \cot B'(dB')'' - \cot A'(dA')'' + k_1 206\,265 = 0 \tag{52}$$

Similarly the equation for condition (48) may be written

$$\Delta a'' - \Delta c'' + \cot C'(dC')'' - \cot A'(dA')'' + k_2\, 206\,265 = 0 \tag{53}$$

From (46) the plane angle condition is

$$(dA')'' + (dB')'' + (dC')'' + k_0 = 0 \tag{54}$$

where k_0 is the triangle misclosure on 180°.

Example: A simple example of a tertiary triangle will serve to illustrate the method. In triangle ABC, $A = 69° 21' 00''$, $B = 71° 04' 32''$, $C = 39° 34' 13''$; $a = 8\,259 \cdot 01$ ft, $b = 8\,349 \cdot 53$ ft, $c = 5\,622 \cdot 30$ ft. Spherical excess is negligible.

The angle equation is
$$dA + dB + dC - 15 = 0 \tag{55}$$
$$(a \sin B / b \sin A) - 1 = k_1 = -0{\cdot}000\ 064\ 5$$
$$\therefore\ k_1 \times 206\ 265 = -13{\cdot}2$$

and
$$(a \sin C / c \sin A) - 1 = k_2 = +0{\cdot}000\ 020\ 7$$
$$\therefore\ k_2 \times 206\ 265 = +4{\cdot}5$$
$$\cot A = +0{\cdot}377,\quad \cot B = +0{\cdot}343,\quad \cot C = +1{\cdot}210$$

Whence the condition equations are

dA	dB	dC	Δa	Δb	Δc	Abs.
+1	+1	+1				−15·0
−0·377	+0·343		+1	−1		−13·2
−0·377		+1·210	+1		−1	+4·5

which on solution by the method of correlatives give
$$dA = +5''{\cdot}4;\quad dB = +9''{\cdot}6,\quad dC = -0''{\cdot}1,\quad \Delta a = +3''{\cdot}19,$$
$$\Delta b = -8''{\cdot}74,\quad \Delta c = +5''{\cdot}55$$

whence $da = +0{\cdot}13$ ft, $db = -0{\cdot}35$ ft, $dc = +0{\cdot}15$ ft and the final adjusted sides and angles are

$$\begin{array}{llll}
A & 69°\ 21'\ 05''{\cdot}4 & a & 8\ 259{\cdot}14 \\
B & 71°\ 04'\ 41''{\cdot}6 & b & 8\ 349{\cdot}18 \\
C & 39°\ 34'\ 13''{\cdot}0 & c & 5\ 622{\cdot}45
\end{array}$$

The final check is that $a/\sin A = b/\sin B = c/\sin C$, viz.

$$\frac{a}{\sin A} = 8\ 826{\cdot}12;\quad \frac{b}{\sin B} = 8\ 826{\cdot}12,\quad \frac{c}{\sin C} = 8\ 826{\cdot}12$$

13.11. THE COMPUTATION OF TRILATERATION COORDINATES

In addition to the method of variation of coordinates described above, two other methods of computing the coordinates of a trilateration network are worthy of mention.

13.11.1. BY SOLUTION OF TRIANGLES

In this method the values of the angles of the triangle are derived from the sides from the formula

$$\sin \tfrac{1}{2} A = \left[\frac{(s-b)(s-c)}{bc}\right]^{\frac{1}{2}} \quad \text{where}\ \ 2s = a + b + c \tag{56}$$

If the spherical excess ϵ'' is significant it is derived from

$$\epsilon'' = bc \sin A / 2 \rho v \sin 1'' \tag{57}$$

whence the spheroidal angles are obtained by adding $\tfrac{1}{3}\epsilon''$ to the plane 'Legendre'

13.11.2. BY DIRECT SOLUTION OF COORDINATES

In triangle ABC if the coordinates of points A and B are known and sides a and b are measured the coordinates of C are given by the formulae

$$E_c = \tfrac{1}{2}(E_A + E_B) + \frac{1}{2c^2}(a^2 - b^2)(E_A - E_B) - \frac{2\Delta}{c^2}(N_A - N_B) \quad (58)$$

$$N_c = \tfrac{1}{2}(N_A + N_B) + \frac{1}{2c^2}(a^2 - b^2)(N_A - N_B) + \frac{2\Delta}{c^2}(E_A - E_B) \quad (59)$$

where $\quad \Delta = [s(s-a)(s-b)(s-c)]^{\frac{1}{2}}$

The convention of sign is that A, B, and C are in a clockwise order. The formulae are proved as follows: By the cotangent formula for intersection §6.9.3

$$E_c = (E_A \cot B - E_B \cot A - N_A + N_B)/(\cot A + \cot B) \quad (60)$$

and, remembering that

$$\cos A = \frac{1}{2bc}(b^2 + c^2 - a^2) \quad \text{and} \quad \cos B = \frac{1}{2ac}(a^2 + c^2 - b^2)$$

and $\quad \Delta = \tfrac{1}{2}bc \sin A = \tfrac{1}{2}ac \sin B$

$$\sin A = 2\Delta/bc, \quad \sin B = 2\Delta/ac$$

therefore

$$\cot A = \frac{1}{4\Delta}(b^2 + c^2 - a^2) \quad \text{and} \quad \cot B = \frac{1}{4\Delta}(a^2 + c^2 - b^2)$$

Substituting for $\cot A$, and $\cot B$ in equation (60) gives the formula (58). The formula for northings is proved in a similar manner by substitution in

$$N_c = (N_A \cot B + N_B \cot A + E_A - E_B)/(\cot A + \cot B)$$

It will be remembered that projection scale factors may be required to convert spheroidal into grid lengths before computing coordinates.

13.11.3. THE ERROR IN POSITION OF A LINEAR FIXATION

In a similar way to that for an angular fixation, the error of a linear fixation may be estimated as follows: Consider the point C, fixed by linear intersection from A, and B by measured sides a and b, subject to errors of $\pm da$ and $\pm db$ respectively (see figure 13.28). C lies on position lines centred on A and B

Table 13.14. Tables of Equivalent Pressure Units

inches mercury	millibars	millimetres mercury	millibars	feet altitude	millibars	metres altitude	millibars
19·0	643	490	653	−1000	1050	−250	1044
19·2	650	495	660	−800	1043	−200	1037
19·4	657			−600	1035	−150	1031
19·6	664	500	667	−400	1028	−100	1025
19·8	671	505	673	−200	1020	−50	1019
		510	680				
20·0	677	515	687	0	1013	0	1013
20·2	684	520	693	200	1005	50	1006
20·4	691			400	998	100	1000
20·6	698	525	700	600	990	150	994
20·8	704	530	707	800	983	200	988
		535	713				
21·0	711	540	720	1000	976	250	982
21·2	718	545	727	1200	969	300	977
21·4	725			1400	962	350	971
21·6	731	550	733	1600	955	400	965
21·8	738	555	740	1800	948	450	959
		560	747				
22·0	745	565	753	2000	941	500	953
22·2	752	570	760	2200	934	550	948
22·4	759			2400	927	600	942
22·6	765	575	767	2600	920	650	936
22·8	772	580	773	2800	913	700	930
		585	780				
23·0	779	590	787	3000	907	750	925
23·2	786	595	793	3200	900	800	919
23·4	792			3400	893	850	914
23·6	799	600	800	3600	887	900	908
23·8	806	605	807	3800	880	950	903
		610	813				
24·0	813	615	820	4000	874	1000	897
24·2	820	620	827	4200	868	1050	892
24·4	826			4400	861	1100	887
24·6	833	625	833	4600	855	1150	881
24·8	840	630	840	4800	849	1200	876
		635	847				
25·0	847	640	853	5000	842	1250	871
25·2	853	645	860	5200	836	1300	865
25·4	860			5400	830	1350	860
25·6	867	650	867	5600	824	1400	855
25·8	874	655	873	5800	818	1450	850
		660	880				
26·0	880	665	887	6000	812	1500	845
26·2	887	670	893	6200	806	1550	840
26·4	894			6400	800	1600	835
26·6	901	675	900	6600	794	1650	830
26·8	908	680	907	6800	788	1700	825
		685	913				
27·0	914	690	920	7000	783	1750	820
27·2	921	695	927	7200	777	1800	815
27·4	928			7400	771	1850	810
27·6	935	700	933	7600	766	1900	805
27·8	941	705	940	7800	760	1950	800
		710	947				
28·0	948	715	953	8000	754	2000	795
28·2	955	720	960	8200	749	2100	786
28·4	962			8400	743	2200	776
28·6	969	725	967	8600	738	2300	767
28·8	975	730	973	8800	732	2400	758
		735	980				
29·0	982	740	987	9000	727	2500	749
29·2	989	745	993	9200	722	2600	740
29·4	996			9400	716	2700	731
29·6	1002	750	1000	9600	711	2800	722
29·8	1009	755	1007	9800	706	2900	713
		760	1013				
30·0	1016	765	1020	10 000	701	3000	705
30·2	1023	770	1027	10 400	691	3100	696
30·4	1029			10 800	681	3200	688
30·6	1036	775	1033	11 200	671	3300	680
30·8	1043	780	1040	11 600	661	3400	672
		785	1047				
31·0	1050	790	1053	12 000	651	3500	664

Derived from Smithsonian Physical Tables
Smithsonian Institution, Washington D.C., 1954.

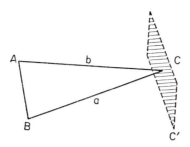

Figure 13.28

consisting of circular arcs which can be approximated to straight lines for small errors. The shaded portion of figure 13.28 represents the error figure of the position of C. At worst the error in position of C is CC' and at best is nil. If the angle C is right, CC' is reduced to a minimum for given values of da and db, an estimate of the angular error at A, dA, is given by

$$dA = da/b \quad \text{in radians}$$

or in general, the angular error at A, σ_A, is given by

$$\sigma_A^2 = \frac{1}{b^2}[\cot^2 C(\sigma_a^2 + \sigma_b^2) + \sigma_a^2] + \frac{1}{c^2}\sigma_c^2 \cot^2 B$$

where σ_a, σ_b and σ_c are the standard errors in a, b and c respectively.

The positional error propagated through a chain can then be estimated as for a traverse with angular errors $(\sigma_A)2^{\frac{1}{2}}$, and linear errors σ_c, etc (see §7.12.2). It must be stressed that this procedure will only give a rough estimate of the error propagated. For an accurate estimate the corrections of the least squares solution of an actual case have to be analysed as described in Rainsford.†

† H. F. Rainsford, Survey Adjustments and Least Squares (London: Constable, 1957).

14 *The Determination of Astronomical Azimuth*

14.1. INTRODUCTION

In all but very small isolated surveys the direction of at least one line has to be known accurately so that the survey may be oriented correctly with respect to other surveys of adjacent areas, and that the directional errors of a survey may be prevented from becoming too large. Traverses and trilateration particularly require azimuth control. The direction of a survey line with respect to true North—the *azimuth* of the line—is determined from theodolite observations to the stars or to the Sun, which although is itself a star, presents special problems when observed from the Earth. The quantities observed are:

  (i) The *horizontal angle* between the star (Sun) and a reference object (R.O.) placed at some suitable point of the survey: and *either*:
  (ii) The *vertical angle* to the star (Sun) at the same instant as the horizontal angle; *or*
  (iii) The *accurate time* of the horizontal observation.

If the vertical angle is observed, an approximate time of the observation is required to enable data to be extracted from an almanac. The accuracy with which these quantities are observed depends on the accuracy required of the final azimuth, the particular method used, and the position of the observer on the Earth. Space does not permit of a full treatment of these questions in this book, and for further information the reader should refer to Biddle†. The

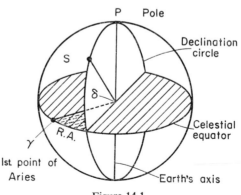

Figure 14.1

† C. A. Biddle, *The Text Book of Field Astronomy*, (London: H.M.S.O., 1958).

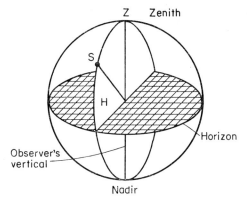

Figure 14.2

methods described here are normally sufficient for all but geodetic work, though observations to Polaris can achieve very high accuracy.

14.1.1. THEORY

Due to the axial rotation of the Earth, the period of which defines the *day*, the stars appear to rotate round the Earth. The Earth also revolves round the Sun in an elliptical orbit in a period of time—*the year*. Since the stars are so distant from the Earth, both the Earth and its orbit can be considered as a point in space, and the stars can be considered as fixed to the inside surface of a huge sphere—the *celestial sphere*—at whose centre is the Earth and its orbit. Only directions on the Earth have significance. The two fundamental directions in field astronomy are

(i) The *axis* of the Earth.
(ii) The direction of the plumb line at the observer's station—*the vertical*.

Figures 14.1 and 14.2 show the celestial sphere as seen from a point outside it and above the North Pole.

14.1.2. THE AXIS OF THE EARTH DEFINES THE FOLLOWING

(i) The North and South *celestial poles* about which the stars appear to revolve.
(ii) The plane of the *celestial equator* which is perpendicular to the axis and passes through the centre of the Earth.
(iii) The *declination* δ of a star (Sun)—the angle between the equator and the star, measured in the great circle passing through the star and the pole—*the declination circle*. A South declination is negative.
(iv) The *right ascension* (R.A.) of the star; the angle measured *anticlockwise* along the celestial equator from a reference point γ—the *first point of Aries*—to the declination circle through the star. The directions *clockwise* and *anticlockwise* are as viewed from outside the celestial sphere and above the North Pole.

The quantities declination δ and right ascension (R.A.) are tabulated in almanacs from data obtained by direct observations at observatories. The almanac specially designed for land surveyors for all but geodetic work is *The Star Almanac for Land Surveyors*, (London: H.M.S.O., Published annually). The data used in this book are obtained from the *Star Almanac for Land Surveyors*, contracted here to the S.A. Both declination and right ascension change very slowly throughout the year, though for the sun the change is more rapid than for the other stars. The argument used in the S.A. for various quantities is generally *Greenwich Mean Time* (G.M.T.), which is now called *Universal Time* (U.T.).

14.1.3. THE OBSERVER'S VERTICAL DEFINES THE FOLLOWING (figure 14.2)

(i) The *zenith* and *nadir*—the points vertically above and below the observer respectively.
(ii) The *horizon*—the plane through the observer perpendicular to the vertical.
(iii) The *altitude* (H) of a star—the angle between the horizon and the star measured in the great circle through the star and the zenith—the *vertical circle* through the star.

14.1.4. BOTH SYSTEMS DEFINED BY THE EARTH'S AXIS AND THE VERTICAL ARE COMBINED TOGETHER VIA

(i) The great circle containing the zenith and the poles—*the meridian*.
(ii) The angle between the Earth's axis and the vertical—the *co-latitude c*. The latitude $\phi = 90 - c$, is the angle between the vertical and the equator, measured in the meridian (see figure 14.3). It will be seen that the altitude of the elevated pole equals the observer's latitude.

Figure 14.4 shows the two systems combined together, with the plane of the meridian in the plane of the paper, and the pole to the right of the zenith. If P

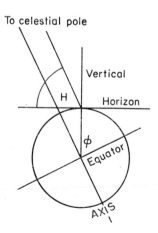

Figure 14.3

The Determination of Astronomical Azimuth

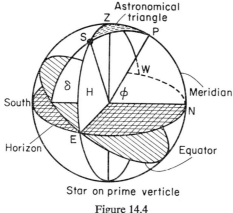

Figure 14.4

is the North Pole, the North, and South *cardinal points* are defined by the intersection of the meridian with the horizon, and points East and West are at right angles to this North–South line.

14.1.5

When a star crosses the meridian it is said to *transit*, and since it does so twice in one revolution of the Earth, we distinguish *upper transit* when the star transits on the same side of the pole as the zenith, and *lower transit* when the transit is on the opposite side of the pole to the zenith.

14.1.6. THE ASTRONOMICAL TRIANGLE

The spherical triangle formed by the zenith Z, the pole P, and the star S is the *astronomical triangle* ZPS (Figure 14.5). The angle Z is called the *azimuth angle*, the angle h the *hour angle*, and the angle S the *parallactic angle*. When the azimuth angle Z is 90° the star is said to be on the *prime vertical* (P.V.)

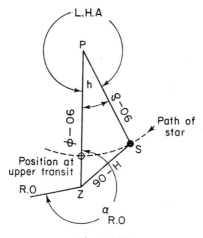

Figure 14.5

and when the parallactic angle $S = 90°$, the star is at *elongation*. A star cannot elongate unless PS < PZ i.e. unless $90 - \delta < 90 - \phi$ or $\delta > \phi$.

14.1.7

The *azimuth* α of a star is the angle between the meridian plane and the plane of the vertical circle through the star, *reckoned clockwise from North*, even in the Southern hemisphere. Hence for an eastern star, $Z = \alpha$ and for a western star, $Z = 360 - \alpha$.

14.1.8

The *Local Hour Angle* (L.H.A.) of a star is the angle between the plane of the meridian and the plane of the declination circle through the star, measured *clockwise from* the star's position *at upper transit*. Hence for an eastern star $h = 24$ hours $-$ L.H.A. and for a western star $h =$ L.H.A. Hour angles are considered in units of time, and have to be converted to units of arc to obtain the values of their trigonometrical functions, unless tables with the argument in time are available (see §2.4.3 below).

The sides of the astronomical triangle are

$$PZ = 90 - \phi$$
$$PS = 90 - \delta$$
$$ZS = 90 - H$$

14.2. AZIMUTH BY SIMULTANEOUS OBSERVATION OF HORIZONTAL AND VERTICAL ANGLES

Applying the cosine formula (see §2.8.3.1) to the astronomical triangle gives

$$\cos(90 - \delta) = \cos(90 - \phi)\cos(90 - H) + \sin(90 - \phi)\sin(90 - H)\cos Z$$
$$\therefore \quad \cos Z = (\sin \delta - \sin \phi \sin H)/\cos \phi \cos H \qquad (1)$$

From which Z and therefore α is derived. The angle θ to the R.O. is also known, and thence the azimuth of the R.O. is obtained.

14.2.1. OBSERVATION OF ALTITUDE H

In star or Sun observations the telescope is pointed in such a manner that the star will appear to cross the diaphragm at the required point without the observer requiring to move the slow motion screws at the time of the observation. When both horizontal and vertical angles are required, this procedure is difficult, and at least one slow motion screw has usually to be moved to bring about simultaneous bisection of both cross hairs. Care must be exercised in observing the Sun since the direct light passing through the telescope can cause serious damage to the eye if viewed even momentarily without the dark glass in position. If such a glass is not available, the image of both cross hairs and the Sun can be focused on a piece of white card and the observation carried out viewing the card. When picking up the sun in the field of view of the telescope, the best way is to view the shadows of the open sights of the telescope: when they are coincident the Sun is in the field of view. Picking up a star is facilitated if the open sights are illuminated by holding a torch high up behind one's head and just illuminating the sights. In all astro work involving altitudes, the vertical index error of the theodolite should be determined by observing a fixed object on both faces, so that if necessary the results can be computed for the separate observations on each face. Great care must be taken to level the theodolite properly and to ensure that the vertical circle bubble is level when taking vertical angles.

14.2.2. CORRECTIONS TO AN OBSERVED ALTITUDE H_0

14.2.2.1. If the star is observed on one face, the index error of the theodolite has to be applied. If both faces are used—as is normal in azimuth work—the mean of both faces is correct provided the two observations are carried out within a period of not more than five minutes, otherwise a correction for the curvature of the stellar path may have to be applied in precise work.

14.2.2.2. Due to the refracting or bending of the light path from the star to the observer, the theodolite vertical angle is too high and a correction amounting to about $58''$ cot H_0 has to be subtracted from the observed altitude H_0. A better value of this refraction correction is obtained from tables in the S.A which allow for the effects of atmospheric pressure P and the temperature t. For example, if the observed altitude $H_0 = 26° 27' 32''$, $P = 29.9$ inches of mercury, and $t = 75°F$; the S.A. gives the refraction correction $r = r_0 f$ where f is obtained from a table entering with P and t, and in this case $f = 0.95$ (*Star Almanac for Surveyors*, page 61). The angle r_0 is obtained from a critical table on page 60, the argument being H_0. The necessary critical entries in this case are

$$H_0 = 26° 26'$$
$$26° 38' \quad r_0 = 116''$$

whence $\quad r = 110'' = 01' 50''$

The final altitude corrected for refraction is then

$$H = 26° 27' 32'' - 01' 50'' = 26° 25' 42''$$

Since the refraction correction increases rapidly as H_0 decreases, the altitude should not be less than 30° if possible, though in Sun observations at high latitudes, altitudes lower than this may have to be observed in winter.

14.2.2.3. A parallax correction of $+8''\!.9 \cos H_0$ is applied to all Sun altitudes if considered significant. In this example $\cos H_0 = 0.89$ and the correction is $+8''$.

14.2.2.4. Since the Sun appears so large in the field of the telescope, accurate bisection is not possible without special attachments. Observations are made with the Sun tangential to the cross hairs as shown if figure 14.6. On face left the limbs of the Sun are placed tangentially to the cross hairs in the top left-hand quadrant, and on face right in the quadrant diametrically opposite (or *vice versa*). The means of the horizontal and of the vertical angles give the position of the centre of the Sun. If an observation is made in this manner on one face only, a correction for the Sun's semi-diameter (s.d.) is applied. The s.d. is tabulated for each month in the S.A. and is about 16 minutes of arc.

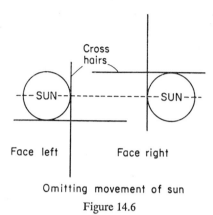

Omitting movement of sun
Figure 14.6

An attachment which enables the centre of the Sun to be bisected is the Roelofs' solar prism. This prism splits the Sun into four images when viewed in the telescope, the overlapping positions of which form a bright star-shape which is bisected to make the observation (figure 14.7). This prism attachment permits of more accurate work in Sun observations than the method of limbs. Corrections have to be applied if observations are made on one face only using the prism.

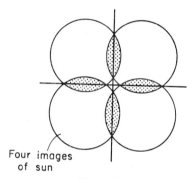

Figure 14.7

14.2.3. CORRECTIONS TO AN OBSERVED HORIZONTAL READING

(1) If the observation is made on one face, the horizontal collimation error of the theodolite has to be applied.

(2) If the observation is on one face to the limb of the Sun the correction for the semi-diameter is $16' \sec H_0$ which become excessively large as $H_0 \rightarrow 90°$. Observations on both faces eliminate errors from this source if both values of H_0 are sensibly the same.

(3) If the vertical axis of the theodolite is tilted, the correction to be applied to the horizontal reading to the elevated star is $+i \tan H_0$. This effect is not eliminated by changing face. To determine i, the plate bubble is read when making a pointing to the star. Both ends of the bubble are read (i.e. left L and right R as viewed from the observer's position).

The Determination of Astronomical Azimuth

For example the bubble readings are as follows

	Face left		Face right	
	Bubble L	Bubble R	Bubble L	Bubble R
	+2·5	+2·5	+3·0	+2·0

Then i is given by $i = (\sum L - \sum R)\frac{1}{4}D$, both in magnitude and sign, where D is the value of one division of the particular bubble in question. In the example if $D = 20''$, $H_0 = 26° 27'$, the complete correction is $+(\sum L - \sum R)\frac{1}{4}D \tan H_0$,

e.g. $\quad\quad\quad \frac{1}{4}(+2·5 + 3·0 - 2·5 - 2·0)20'' \times 0·497 = +2''$

which is negligible in this Sun observation. The effect of axis tilt is small if H_0 is kept low, but this conflicts with the requirement that H_0 has to be kept above 30° to avoid large refraction corrections to the vertical angle. Thus if vertical angles are observed, the value of H_0 is kept between 30° and 60° as a compromise. In any case, the theodolite should be levelled as accurately as possible to avoid the axis tilt error.

14.2.4. COMPLETE EXAMPLE

In table 14.1(a) a complete example of a Sun observation is shown, together with the calculation of the final azimuth of the r.o. in table 14.1(b). Normally at least three such observations are made and the resultant azimuths should

Table 14.1(a). Sun Azimuth by Altitudes: Field Book:

Theodolite _Tavistock Mk II_ Temp° _75°F_ At station __1__
Observer __A.L.A.__ Pressure _29·9 in Hg_ R.O. at __2__
Booker __D.J.M.__ Value of one division of plate bubble _20"_
Date ___24 Sept 1964___

NOTES	HORIZONTAL READING	BUBBLE L R	VERTICAL READING	APPROX U.T.
① Face R R.O.	271° 40' 04"			
F.R. Sun ⚹	58 11 40	3·0 2·0	62° 56' 47"	14h 39m
F.L. Sun ⚹	238 02 47	2·5 2·5	297 22 20	
F.L. R.O.	91 39 16	5·5 4·5		
② F.L. R.O.	00 10 28			
F.L. Sun ⚹	147 43 47	2·5 2·5	296 49 16	14h 46m
F.R. Sun ⚹	328 43 40	3·0 2·0	63 54 12	
F.R. R.O.	180 11 18	5·5 4·5		

agree to within a minute of arc at the outside. The value of the latitude was scaled from a map, and the declination is obtained from the *Star Almanac*. The tabular entries in the table for the Sun are

24th September U.T. 12 h δ = South 00° 34'·8
$\quad\quad\quad\quad\quad\quad\quad$ 18 h δ = South 00° 40'·6

Table 14.1(b). Computation of Sun Azimuth Observations of Table 14.1(a).

Formula: $\cos Z = \dfrac{\sin \delta - \sin H \sin \phi}{\cos H \cos \phi}$

$\phi = 51° 35' 30'' N$
$\sin \phi = 0.783\,603$
$\cos \phi = 0.621\,262$

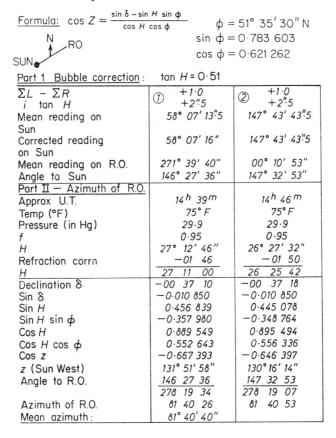

Part 1 Bubble correction: $\tan H = 0.51$

	①	②
$\Sigma L - \Sigma R$	+1·0	+1·0
$i \tan H$	+2″5	+2″5
Mean reading on Sun	58° 07' 13″5	147° 43' 43″5
Corrected reading on Sun	58° 07' 16″	147° 43' 43″5
Mean reading on R.O.	271° 39' 40″	00° 10' 53″
Angle to Sun	146° 27' 36″	147° 32' 53″
Part II – Azimuth of R.O.		
Approx U.T.	14h 39m	14h 46m
Temp (°F)	75°F	75°F
Pressure (in Hg)	29·9	29·9
f	0·95	0·95
H	27° 12' 46″	26° 27' 32″
Refraction corrn	−01 46	−01 50
H	27 11 00	26 25 42
Declination δ	−00 37 10	−00 37 18
Sin δ	−0·010 850	−0·010 850
Sin H	0·456 839	0·445 078
Sin H sin φ	−0·357 980	−0·348 764
Cos H	0·889 549	0·895 494
Cos H cos φ	0·552 643	0·556 336
Cos z	−0·667 393	−0·646 397
z (Sun West)	131° 51' 58″	130° 16' 14″
Angle to R.O.	146 27 36	147 32 53
	278 19 34	278 19 07
Azimuth of R.O.	81 40 26	81 40 53
Mean azimuth:	81° 40' 40″	

Since the approximate time of the observation was 14 h 46 m U.T. the interpolated declination is South 00° 37′·3 which is treated as a negative quantity in the computation, i.e.

$$\delta = -00° 37' 18''$$

14.2.5. POSITION OF STAR OR SUN FOR OBSERVATION

If the azimuth is computed from observed altitudes the best results are obtained from stars at elongation; or with stars which cannot elongate, from stars observed close to the prime vertical, i.e. at azimuths of 90° or 270°. The altitude of the star should be between 30° and 60° for the reasons mentioned in §14.2.2 and 3.

To eliminate the effects of errors in the refraction corrections and of the value of latitude scaled from the map, observations should be balanced East and West. In star work an observation is made to a western prime vertical star and to an eastern P.V. star balanced for altitude to within five degrees. The mean azimuth computed from these two balanced stars is accepted. If more observations are made, an equal number of East and West stars should be used to give the final results. To balance Sun observations, morning and afternoon observations are made at equal times before and after transit. Observations for azimuth should not be made at a transit if the vertical angle is observed, since the altitude barely changes for a considerable change in azimuth, the star appearing to move horizontally.

14.3. THE DETERMINATION OF AZIMUTH FROM OBSERVED HORIZONTAL ANGLES AND RECORDED TIME

This method of determining azimuth is potentially more accurate than that described above in §14.2, since the observation involves the bisection of only the vertical crosshair (horizontal angle), vertical refraction does not affect the method, and the stars may be observed low down close to the horizon thus reducing the errors due to axis tilt.

14.3.1. THEORY

From the observed time the hour angle h is calculated as shown in §14.3.2. The declination δ is obtained from the S.A. and the latitude and longitude of the observer are scaled from a map or obtained from the survey coordinates. Applying the cot formula to the astronomical triangle (see §2.8.3) gives

$$\cos h \cos (90 - \phi) = \sin (90 - \phi) \cot (90 - \delta) - \sin h \cot Z$$

$$\therefore \quad \cot Z = (\cos \phi \tan \delta - \cos h \sin \phi)/\sin h \qquad (2)$$

whence Z and therefore the azimuth α is calculated.

14.3.2. DERIVATION OF THE HOUR ANGLE h

An accurate chronometer is required for this method. The chronometer error on U.T. is found by comparison with a radio time signal (such as the B.B.C. six pips, or M.S.F. Rugby, which station transmits signals at intervals of one second almost continuously throughout the day and night), or from the G.P.O. speaking clock. A stop watch enables an accurate determination of the chronometer error to be made. Consider as an example, an error determination using the B.B.C. pips. The sixth pip of a series is the exact time of the signal. For example if the signal is transmitted at 09 h 00 m 00·0 s U.T. the first pip is emitted at 08 h 59 m 54·0 s. As the pips are heard, the observer beats time with them keeping the stop watch ready. On the sixth pip he starts the watch, which now theoretically continues the exact U.T. of the signal. The watch is then stopped at some convenient exact second on the clock; thus the U.T. of this chronometer second is the U.T. of the signal plus the watch reading, and the chronometer error on U.T. is thus known. For example, the U.T. of the signal is 09 h 00 m 00·0 s usually written $09^h\ 00^m\ 00^s\cdot0$, the stop watch reading is say $23^s\cdot7$ when the chronometer reading was $09^h\ 10^m\ 10^s\cdot0$, i.e. the U.T. corresponding to the chronometer time of $09^h\ 10^m\ 10^s\cdot0$ is $09^h\ 00^m\ 23^s\cdot7$, and the chronometer is fast on U.T. by $10^m\ 00^s\cdot0 - 23^s\cdot7 = 09^m\ 46^s\cdot3$.

The chronometer time of an observation is obtained in a similar manner using the stop watch. When the star crosses the vertical hair the observer starts the watch, and later stops it on an exact second of the chronometer. The chronometer time of the observation is therefore this chronometer time minus the reading on the watch. For example, if the watch reading is $11^s \cdot 8$ when stopped on a chronometer reading of $11^h\ 23^m\ 20^s \cdot 0$, the chronometer time of the observation is $11^h\ 23^m\ 08^s \cdot 2$. The addition of the chronometer correction on U.T. to this observation time gives the U.T. of the observation. For example, supposing the error on U.T. is fast $09^m\ 46^s \cdot 3$ the correction is $-09^m\ 46^s \cdot 3$ and the U.T. of the observation is $11^h\ 23^m\ 08^s \cdot 2 - 09^m\ 46^s \cdot 3 = 11^h\ 13^m\ 21^s \cdot 9$.

Once the U.T. of the observation is calculated, the hour angle is obtained differently for the Sun as for the stars.

14.3.2.1. *Hour Angle of the Sun*

The Greenwich hour angle (G.H.A.) is first obtained from

$$\text{G.H.A.} = \text{U.T.} + E \tag{3}$$

where E is obtained from the S.A. For example if the U.T. is $14^h\ 22^m\ 43^s \cdot 8$ the requisite portion of the table for the Sun is

24th September 12 h U.T. $E = 12^h\ 08^m\ 02^s \cdot 3$

18 h U.T. $E = 12^h\ 08^m\ 07^s \cdot 5$

and the interpolated value of E for the observed time is

$$E = 12^h\ 08^m\ 04^s \cdot 3$$

Thus the G.H.A. $= 26^h\ 30^m\ 48^s \cdot 1 = 02^h\ 30^m\ 48^s \cdot 1$. The local hour angle (L.H.A.) is given by

$$\text{L.H.A.} = \text{G.H.A.} + \lambda$$

where λ is the longitude of the observer with respect to Greenwich, reckoned positive eastwards and negative westwards. In the example λ is $2^s \cdot 5$ West, i.e. $\lambda = -2^s \cdot 5$ whence the L.H.A. (Sun) is $02^h\ 30^m\ 48^s \cdot 1 - 2^s \cdot 5$. Since the Sun is in the West, $h = \text{L.H.A.} = 02^h\ 30^m\ 45^s \cdot 6$.

14.3.2.2. *Hour Angle of a Star*

To find the L.H.A. of a star the Greenwich Sidereal Time (G.S.T) is first found from

$$\text{G.S.T.} = \text{U.T.} + R \tag{4}$$

where R is tabulated in the *Star Almanac*. For the U.T. $21^h\ 45^m\ 28^s \cdot 7$ on the 24th September the relevant tabular values of the S.A. are from the table for the sun:

U.T. R

18 h $00^h\ 14^m\ 22^s \cdot 3$

24 h $00^h\ 15^m\ 21^s \cdot 5$

from which the interpolated value of R is $00^h\ 14^m\ 59^s \cdot 3$ and the G.S.T. $= 21^h\ 45^m\ 28^s \cdot 7 + 00^h\ 14^m\ 22^s \cdot 3 = 22^h\ 00^m\ 28^s \cdot 0$. The G.H.A of the star is then obtained from the G.S.T. and the right ascension R.A. of the star obtained from the S.A. from

$$\text{G.H.A.} = \text{G.S.T.} - \text{R.A.} \tag{5}$$

Suppose the R.A. is $11^\text{h}\ 20^\text{m}\ 11^\text{s}\cdot 0$, the G.H.A. is then

$$22^\text{h}\ 00^\text{m}\ 28^\text{s}\cdot 0 - 11^\text{h}\ 20^\text{m}\ 11^\text{s}\cdot 0 = 10^\text{h}\ 40^\text{m}\ 17^\text{s}\cdot 0$$

The L.H.A. is then found from the G.H.A. as for the Sun, i.e. from

$$\text{L.H.A.} = \text{G.H.A.} + \lambda$$

The *Star Almanac* contains tables to assist in the interpolation of the various quantities E, R, δ, and R.A.

14.3.3. TRIGONOMETRICAL FUNCTIONS OF h

The trigonometrical functions of h are best obtained from tables with the argument in time such as the H.M.S.O. seven figure tables. If these are not available, h has to be converted to units of arc via:

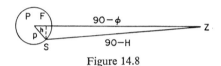

Figure 14.8

$360° = 24$ h, $15° = 1$ h, $15' = 1$ m, $15'' = 1$ s, or via tables designed to facilitate this conversion such as that given in the *Star Almanac*.

Practical Example: Table 14.2(*a*) gives the observed angles, time and chronometer data for the computation of the azimuth of the R.O. in table 14.2(*b*). The error of the chronometer was determined from the G.P.O. speaking clock which is accurate to $\frac{1}{20}$ second, and may be conveniently used throughout the United Kingdom.

14.3.4. POSITION FOR OBSERVATION

As with the altitude method, the stars should be balanced East and West, but the altitude at which the star is observed should be kept low. Good results may be obtained at almost any azimuth, though the position of elongation is also best in this method since the star then appears to move in a vertical direction only. Good results are obtainable from close circumpolar stars such as Polaris in the northern hemisphere or σ Octantis in the southern. These stars appear to move very slowly in azimuth and because of their proximity to the poles special simplified computation is possible using specially designed tables, such as the *Pole Star Tables of the* S.A.

14.4. AZIMUTH FROM CIRCUMPOLAR STARS

In figure 14.8 the path of a close circumpolar star is shown, and PZS is the astronomical triangle. The polar distance of the star is $p = 90 - \delta$. Then

$$PZ = 90 - \phi \simeq 90 - H + p \cos h \tag{6}$$

Table 14.2(a). Sun Azimuth by Hour Angle: Field Book:

Theodolite _Tavistock Mk II_ Temp° _____ At station __(1)__
Observer ___A.L.A._____ Pressure ___ in Hg. R.O. at __(2)____
Booker ____D.J.M._____ Value of one division of plate bubble _20"_
Date __24/Sept/1964__

NOTES		HORIZONTAL READING	BUBBLE L. R.	CHRONOMETER TIME	STOP WATCH	CHRON. TIME OF OBS
③ F.L.	R.O.	00° 02' 31"				
R.L. Sun ☉		142° 07 09	2·5 2·5	14ʰ 08ᵐ 10ˢ	09ˢ·4	14ʰ 08ᵐ 00ˢ·6
F.R. Sun ☉		323 12 20	3·0 2·0	14 09 50	06·9	14 09 43·1
F.R.	R.O	180 03 24	5·5 4·5			
Approx. altitude Sun 27°						
④ F.R.	R.O.	271 40 07				
F.R. Sun ☉		55 45 08	3·0 2·0	14 13 30	08·0	14 13 22·0
F.L. Sun ☉		235 30 50	3·0 2·0	14 15 10	11·5	14 14 58·5
F.L.	R.O.	91 39 19	6·0 4·0			
⑤ F.L.	R.O	91 39 18				
F.L. Sun ☉		236 23 59	3·0 2·0	14 18 40	11·6	14 18 28·4
F.R. Sun ☉		57 26 15	3·5 1·5	14 20 10	06·9	14 20 03·1
F.R.	R.O.	271 40 02	6·5 3·5			

Chronometer data: U.T. Signal   14ʰ 11ᵐ 10ˢ·0
                   Stop watch   42·0
                                       14 11 52·0
                   Chron: 13 58 00·0
                   Chron. slow on U.T. 13 52·0

Table 14.2(b). Computation of Sun Azimuth Observations of Table 14.2a

Formula: $\cot Z = \dfrac{\cos\phi \tan\delta - \sin\phi \cos h}{\sin h} = A/\sin h$

Latitude $\phi = 51° 35' 30"$ Longitude $\lambda = 02^S·5$ West
$\sin\phi = 0.783\ 603$ $\cos\phi = 0.621\ 262$

Part I Bubble correction:	(3)	(4)	(5)
L − R	+1·0	+2·0	+3·0
i tan H	+2"·5	+5"·0	+8"·0
Mean reading on Sun	142° 39' 44"·5	55° 37' 59"	236° 55' 07"
Corrected reading on Sun	142 39 47	55 38 04	236 55 15
Mean reading on R.O.	00 02 58	271 39 43	91 39 40
Angle to Sun	142 36 49	143 58 21	145 15 35
Part II Azimuth:			
Chron. Time obs.	14ʰ 08ᵐ 51ˢ·8	14ʰ 14ᵐ 10ˢ·2	14ʰ 19ᵐ 15ˢ·8
Chron. corrn.	+13 52·0	+13 52·0	+13 52·0
U.T. obs.	14 22 43·8	14 28 02·2	14 33 07·8
E at 12 h	12 08 02·3	12 08 02·3	12 08 02·3
Δ	+02·0	+02·1	+02·2
G.H.A (Sun)	26 30 48·1	26 36 06·6	26 41 12·3
Longitude λ	−2·5	−2·5	−2·5
L.H.A. (Sun) = h	02 30 45·6	02 36 04·1	02 41 09·8
h arc	37° 41' 24"	39° 01' 01"	40° 17' 27"
Sin h	0·611 389	0·629 550	0·646 668
Cos h	0·791 330	0·776 960	0·762 772
Declination δ	−00° 37' 00"	−00° 37' 12"	−00° 37' 12"
Tan δ	−0·010 763	−0·010 821	−0·010 821
Cos φ tan δ	−0·006 687	−0·006 723	−0·006 723
Sin φ cos h	0·620 089	0·608 828	0·597 710
A	−0·626 776	−0·615 551	−0·604 433
cot Z	−1·025 167	−0·977 764	−0·934 688
Z	135° 42' 43"	134° 21' 21"	133° 04' 00"
Angle to R.O.	142 36 49	143 58 21	145 15 35
	278 19 32	278 19 42	278 19 35
Azimuth of R.O.	81 40 28	81 40 18	81 40 25
Mean of R.O.	81° 40' 35"		

The Determination of Astronomical Azimuth

Applying Napier's rules for circular parts to the right angled triangle ZSF gives

$$\sin(p \sin h) = \cos(90 - Z) \cos H$$

$$\therefore \sin Z = \sin(p \sin h) \sec H$$

$$Z'' \simeq p'' \sin h \sec H$$

putting

$$p \sin h = x, \; p \cos h = y$$

$$\therefore Z'' = x \sec(\phi + y) \quad \text{(from (6))}$$

$$= x \sec \phi + xy \sec \phi \tan \phi \quad \text{(Taylor)}$$

$$= \sec \phi (p \sin h + p^2 \sin h \cos h \sin 1'' \tan \phi)$$

or say

$$Z'' = \sec \phi (b_0 + b_1 + b_2)$$

The quantities b_0, b_1, and b_2 are given in the Pole Star tables of the *Star Almanac*,

Table 14.3. Observations on Polaris for Azimuth: Field Book:

NIGHT: 25/26 June 1963 Observer: J.C.
Station: Parridge House Booker: J.C.
Pressure 29.93 Temp. 38°F Tavistock VO 41397
One division of plate bubble 25" R.O. No. 4

	NOTES	FACE	HORIZONTAL READING	MEAN HOR.	CHRONOMETER READING	WATCH	BUBBLE L. R.
①	R. O.	L	00° 00' 00"	179 59 33		13ˢ.4	3·5 0·5
	Star	L	61 38 26	61 38 30	02ʰ 09ᵐ 25ˢ	15·5	1·5 2·5
	Star (Cloud)	R	241 38 34	61 38 57	02 21 25	14·45	5·0 3·0
	R.O	R	179 59 06		02 15 25		+2·0
				Chron. time obs.	02 15 10·55		
②	R. O.	L	00 00 10	00 00 18		17·3	3·5 0·5
	Star	L	61 40 24	61 40 39	02 42 25	14·4	1·5 2·75
	Star	R	241 40 54	61 40 21	02 48 50	15·85	5·0 3·25
	R.O.	R	180 00 26		02 46 37·5		+1·75
				Chron. time obs.	02 46 21·65		
③	R. O.	L	90 00 20	90 00 34·5		11·2	3·5 0·5
	Star	L	151 40 18	151 40 30·5	03 14 25	10·8	1·25 2·75
	Star	R	331 40 43	61 39 56·0	03 18 25	11·0	4·75 3·25
	R.O.	R	270 00 49		03 16 25		+1·50
				Chron. time obs.	03 16 14·0		

Table 14.4. Chronometer Data:

DATE 1963	U.T. SIGNAL	CHRON: READING	WATCH	CHRON. TIME SIGNAL	ERROR ON U.T.
25 June	08ʰ 00ᵐ 00ˢ	08ʰ 57ᵐ 30ˢ	13ˢ.0	08ʰ 57ᵐ 17ˢ.0	57ᵐ 17ˢ.0 FAST
26 "	No signal				
27 "	10ʰ 00ᵐ 00ˢ	10ʰ 57ᵐ 00ˢ	11·3	10ʰ 56ᵐ 48ˢ.7	56ᵐ 48ˢ.7 FAST
	Δ 50 hrs			Δ	28·3
		Chron. rate =	$\frac{28·3}{50} = 0·566$ sec/hr losing		

the argument for which is L.S.T. (see table 14.6). Based on (6) the latitude ϕ can be obtained from the Pole Star tables from

$$\phi = H - (a_0 + a_1 + a_2)$$

where the argument for the tabular quantities a_0, a_1, and a_2 is again the L.S.T.

In this method of determining azimuth the time is required to only 10 seconds giving a precision equivalent to that of the tables, which is 6 seconds of arc in the final azimuth.

Table 14.5. Computation of Azimuth by Pole Star Tables from Observations of Table 14.3.

Latitude 51° 16′ 04″ N   Longitude 02° 35′ 49″ W arc
Night 25/26 June 1963 $-0^h\ 10^m\ 23^s\!.2$ time
Bubble divn. 25″

Part I Bubble correction: $= i \tan H$. $\tan H = 1{\cdot}25 = \tan \phi$

$\Sigma L - \Sigma R$	+2·0	+1·75	+1·50
$i \tan H$	+0′.3	+0′.2	+0′.2
Angle to R.O.	61° 38′.9	61° 40′.3	61° 39′.9
Corrected R.O.	61° 39′.2	61° 40′.5	61° 40′.1

Part II Azimuth computation:

Chron time Obs:	$02^h\ 15^m\ 10^s\!.6$	$02^h\ 46^m\ 21^s\!.7$	$03^h\ 16^m\ 14^s\!.0$
Chron correction	−57 08	−57 08	−57 08
U.T. Obs:	01 18 03	01 49 14	02 19 06
R at 0 h	18 13 35·8	18 13 35·8	18 13 35·8
Δ	12·8	17·9	22·9
G.S.T.	19 31 51·6	20 03 17·7	20 33 04·7
Longitude λ	−10 23·2	−10 23·2	−10 23·2
L.S.T.	19 21 28·4	19 52 54·5	20 22 41·5

From Pole Star table: $a = (b_0 + b_1 + b_2) \sec \phi$ (See table 14.6)
Argument L.S.T. $\sec \phi = 1{\cdot}599$

b_0	+53′.7	+54′.5	+54′.3
b_1	—	—	—
b_2	+0′.2	+0′.2	+0′.2
Sum	+53·9	+54·7	+54·5
Z STAR *	+86′.2	+87′.5	+87′.2
a STAR	361° 26′.2	361° 27′.5	361° 27′.2
Angle to R.O.	61 39·2	61 40·2	61 40·1
Azimuth R.O.	299 47·0	299 47·0	299 47·1

* If Z is positive Polaris is East of the pole,
if Z is negative Polaris is West of the pole.

Practical Example: Tables 14.3, 14.4 give the necessary field book and chronometer data, from which the azimuth of the R.O. is calculated in table 14.5, the Pole Star data being given in table 14.6. It will be noted that the chronometer error is interpolated between two values. The change in the chronometer error per unit time is the *rate* of the chronometer.

Table 14.6. Extracts from Pole Star Tables: To be Used with Table 14.5.

L.S.T.	19h b_0	20h b_0	
21m	+53·7	+54·3	
24m	+53·8	+54·2	
			etc
51m	+54·5		
54m	+54·5		
LATITUDE	b_1	b_1	
45°	0	0	
50°	0	0	
55°	0	0	
MONTH	b_2	b_2	
May	+0·1	+0·1	
June	+0·2	+0·2	
July	+0·2	+0·3	

Reproduced from the *Star Almanac* by permission of the controller of H.M. Stationery Office.

14.5. IDENTIFICATION OF STARS

The identification of stars is best carried out in the field with the aid of a star chart, such as the companion volume to the *Star Almanac*, and which indicates those stars tabulated in the S.A. Other suitable charts are Norton's *Star Atlas*†, or *A New Popular Star Atlas*‡. With these latter, care has to be exercised not to observe a star for which the right ascension and declination are not tabulated in the *Star Almanac*. The procedure of identifying a star is to work systematically over the heavens from a known familiar constellation such as the Plough, Orion, etc, until the particular star is reached, and identified.

As a safeguard against mis-identification the approximate time of an altitude observation should be noted, or the approximate altitude of the star when time is being recorded. This enables both the R.A. and δ to be evaluated from

$$\sin \delta = \sin \phi \sin H + \cos \phi \cos H \cos Z$$

whence δ;

and
$$\sin h = \cos H \sin Z / \cos \delta$$

whence h, and therefore the R.A. from equation (5) and a knowledge of the longitude λ.

† Norton, *Star Atlas* (Edinburgh: Gall and Inglis, 1964).
‡ *A New Popular Star Atlas* (Edinburgh: Gall and Inglis, 1961).

Index

ABSTRACTS, 33
Acceleration in vertical curves, 553
Access to land, 34
Accuracy,
 definition, 201
 plottable, 4
Adjustment,
 angles and sides, 661
 heights, 416
 least squares, 221
 traverses, 345
 traverse networks, 357
 weighted observations, 239, 360
Adjustment of instruments,
 level, 145
 theodolite, 118
Aerial photography, 1
Alidade,
 simple, 1
 telescopic, 462
 theodolite, 82, 136
Alignment correction, 327
Almanac, Star, 668
Altimeter, 426
 Baromec, Paulin, Wallace & Tiernan, 422
Altimetry, practical, 428
Altitude of star, 668
Anallactic telescope, 447, 468
Aneriod barometer, 422
Angle measurement, 134, 261
Angle conditions, 284
Apex distance,
 compound combined curves, 503
 transition and combined curves, 500
Arc to Chord correction on projection, 184, 191, 197
Areas, 592
 by planimeter, 598
 from rectangular coords, 601
 subdivision of, 604
Aries, first point of, 667
Astronomical triangle, 669
Automatic level, 142
 adjustment, 148
Auxiliary bearing, 78
 in traversing, 344
Axes of theodolite, 82, 98, 119, 131, 132
 strain, 82, 127, 130

Axis of the earth, 667
 tilt, 673
Azimuth, astronomical, 666
 by altitude, 670
 by hour angle, 675
 geodetic and astronomical, 151
 grid, 176
Azimuths, frequency,
 Cadastral traverses, 342
 Precise traverses, 374

BACKLASH, 262
Backsight, 386
Balance line, Mass Haul, 582
Balancing of stars, 674
Barograph, 424
Barometric formulae, 424
Barometric heighting, 387, 421
 metrological factors, 426
Barycentric resection, 293
Base line, 4
Barometer, mercury, 421
Batteries, 640
Beacons, 252
Beaman stadia arc, 462
Bearing, change of, 74, 355
Bearings, 78
 adjustment, 344
 grid, 172
Bench marks, 11, 27
Bessel's interpolation formula, 68
Bessel's, method of resection, 8, 299
Bilby Towers, 260
Booking, 32
 Cadastral Surveys, 591
 Chain Surveys, 444
 levelling, 403
 tacheometry, 454
 taping, 340
 triangulation, 267
Borrowing in earthworks, 583
Bosshardt tacheometer, 465
Boundary posts, 589
Boundaries,
 geographical, 164
 re-establishment, 593
Bowditch adjustment, 345

683

Bromides, 19
Building line, 437
Buildings,
    representation of, 438
    cadastral surveys, 594
Bulking in earthworks, 586
Butt jointed plates, 478

CADASTRAL SURVEYING, 588
Cadastral traverse, 341
Cairns, 590
Calculating Machines, 37
    areas, 603
    square roots, 47
    types, 51
Calibration readings Geodimeter, 616
Capsule of barometer, 423
Catenary taping, 323
    sag correction, 328
    booking, 341
Cardinal points, 669
Carrier frequency, 611
Cassini projection, 168, 171, 172
Cathode ray tube, 638
Cavity Tune, 637
Celestial Sphere, 667
Centering of towers, 260
Central scale factor, 179
Chain Survey, 441
Chainage in setting out, 481
Cholesky solution, 233
Chord to arc linear correction, 647
Chronometer, 675, 680
Circle, theodolite, 84, 104
    eccentricity, 128
    graduation errors, 130
Circles, in spherical trig, 58
Circular curves, 482
    calculation of components, 490
    degree of, 484
    setting out, 508
Class interval, 205
Clinometer heighting, 14
Clip screw, 83, 123
Clothoid-transition curve, 484
    polar coordinates, 488
    rectangular coordinates, 486
    setting out, 516
Coarse figure, 630
Coincidence bubble reader, 100, 140
Co-latitude, 668
Collimation, error of level, 146, 397
    height of, in level booking, 392
    horizontal, in theodolite adjustment, 121
Collin's Point resection, 8, 299

Combined Curves,
    calculation of components, 501
    definition, 482
Compass, box, 2
Completion of detail surveys, 477
Compilation, 4, 19
Compound curves, 482
    circular, 492
Compound combined curves, 501
Computation, 37
    number of figures to use, 54
    logarithmic, 57
Cone constant, 195
Conical projection, simple, 194
Condition equations, 222, 242
Conditions,
    bearing, 657
    figural, 656
    in network, number of, 284
    of position, 657
    for triangle, 658
Conformal projections, 166, 168
Consolidation, in earthworks, 586
Contour datum, 10
Contour geometry, 559
Contouring, 10
Contours, earthwork volumes from, 559
Control for Cadastral Surveys, 590
Conventional signs, 4, 438
Convergence, Cassini projection, 170
    conical projection, 196
    of meridians, 159
    Transverse Mercator projection, 174
Coordinates,
    transformation of, 73
Correlatives, 242
Corrections to measured distances, 324
Critical tables, 56
Cross hairs, 90
Cross sections, areas of, 567
    calculation of eccentricity, 577
    curvature correction, 576
    earthwork volumes from, 567, 579
Crystals, Tellurometer, 638
Cubic parabola, as transition curve, 487
Cubic parabola, as vertical curve, 548
Curvature correction for earthworks, 576
Curvature of earth, 400, 411
Curvature, mean radius of, 155
Curves circular, see 'Circular curves'
Curve design, horizontal, 531
    vertical, 552, 554
Curve Setting out, see 'Setting out'
Curves, transition, see 'Transition curves'
Curves for wide roads and dual carriageways, 505

Curves, obstructed, 521
Curve ranging, circular curves,
    chain and tape, 510
    theodolite, 512
Curve ranging, obstructed curves, 521
Curve ranging, transition curves, 517
Curve widening, 507
Cutting points on projection, 171
Cuttings,
    See also 'cross sections', 561, 579

DALBY'S THEOREM, 159
Danger circle, 7
Declination, 667
Deed, plan for, 592
Deflection angles for transition curves, 488, 518
Degree of curve, 484
Delay line, 615, 624
Demarcation of boundaries, 589
Details, 10
Detail Survey, 437
    completion, 477
Deviation curve, 207
Deviation, in traverse, 356
Diagonal scale, 6
Diagram circle, 458
Diurnal wave, 428
Dislevelment correction, 136
Distance curve, 459
Distomat, 643
Double base altimetry, 430
Double entry tables, 71
Drawing mediums, 4
Dual carriageway roads, 505
Dumpy level, 140, 147
Dyeline, 20
Dynamic heights, 406

EARTHWORK VOLUMES, SEE 'VOLUMES OF EARTHWORKS'
Easting false, 181
Eccentricity,
    in theodolite, 128
    in earthwork volumes, 577
    of ellipse, 150
Electronic computer, 52, 164, 166, 249
Electromagnetic waves, 608
Electrotape, 643
Elongation, 670
Embankments
    See also 'cross sections', 561, 579
End area rule for earthwork volumes, 558
Equal Shifts, method of adjustment, 277
Equations,
    condition, 222

normal, 228
observation, 222
Equator, celestial, 667
Equipotential surface, 406
Error, average, 209
    gross, 202
    logarithmic differences method, 318
    of mean, 217
    probable, 210
    random, 202, 204, 216
    standard, 208, 210
    mean square, 208
    systematic, 202
    true, 201
Errors in Tacheometry,
    with vertical staff, 456
    with horizontal staff, 474
Errors in Traversing, 330
    location, 353
    propagation, 379
Errors in Triangulation,
    propagation, 319
Errors in Subtense Measurement, 366
Euler's theorem, 152, 154
Excess and Deficiences, 594
External focusing telescope, 91, 448
Eyepieces, 88
    diagonal, 95

FACE—CHANGE OF, 262
Ferrero's formula, 321
Field books, 32
Field examination, 477
Fine readings, 630
Fixations on projection, 189
Flattening, 150
Foresight, 386
Free haul, 585
Frequency, 609
Frequency variation, 632

GAUSS-KRUGER PROJECTION, 199
Gauss-Doolittle method of solution, 233
Geodetic factors, 163
Geodetic levelling, 402
Geodetic level, 140
Geodimeter, 608, 611, 624
Geodimeter model 6, 625
Geopotential numbers, 409
Geoid, 149
Grade staff reading, 573
Grades, 104
Gradienter, 404
Gradient—rule of change, 540, 553
Graduation errors in circle, 130

Graphical methods in detail survey, 478
Graticule, 167
Graticule tables, UTM, 181
Great circle, 58
Greenwich hour angle, 676
Greenwich sidereal time, 676
Grey wedge, Geodimeter, 623
Grid, 167
Grid of levels, 12
Grid, plotting of, 15
Ground marks, 337, 339
Ground swing, 631, 640
Ground taping, 323, 340

HAMMER SELF REDUCING TACHEOMETER, 458
Haul, 579
Haul distance, 583
Height control points for photogrammetry, 12
Height transfers by theodolite, 419
Height traversing, 420
Heighting, accuracies in, 388
Heighting, precision of, 415
Heights, Air Photo, distances for, 420
Helios, 256
Highway spirals, see 'transition curves',
Histogram, 205
Horizon, 668
Horizontal angle measurement, 134, 261
Horizontal collimation, 121, 133
Hour angle, 670, 676
Hydrostatic levelling, 385
Hysteresis, 423

INACCESSIBLE BASE, 10, 315
Illuminated beacons, 256
Index error,
    altimeter, 430
    tellurometer, 636
    theodolite, 123, 134
Indian Clinometer, 2, 12
Indicatrix, 154
Indicatory posts, 589
Inking in, 16
Intersection, 5
Intersection, computation of, 288
    in detail survey, 478
    semi graphic, 301
    variation of coordinates, 308
Internal focusing telescope, 92, 447
Interpolation, 67, 68, 70, 71
Intervisibility, 418
Invar, 325

JERIE ANALOGUE COMPUTER, 649

KERN PLANE TABLE EQUIPMENT RK, 463, 464
Kerr Cell, 625

LABOUR, 35
Lagrangian multiplier, 242
Lambert's Conical Orthomorphic projection, 194
Lambert's projection, 196
Lamps, beacon, 256
Laplace azimuth, 151
Laser, 644
Latitude, 668
    geodetic, 150
    astronomical, 151
Layouts, 594
Leap frog altimetry, 430
Least squares adjustment, 221
    of linear and angular measures, 655
    of traverse, 357
Least squares, principle of, 213
Legendre's theorem, 156
Lehmann's rules, 7
Lemniscate as transition curve, 535
Lettering, 17
Level, see 'Surveyor's level',
Level booking, 389
Level, 96
    sensitivity, 101
Levelling, errors in, 393
    geodetic, 402
    hydrostatic, 385
Licenced surveyor, 588
Linear measurement, 323, 332, 340, 341
Logarithmic differences, 659
    error propagation, 318
Longitude, 151, 676
Lower plate, 84, 134

MATHEMATICAL TABLES, 53, 54, 55
Mass-haul diagrams, 579
Master Tellurometer, 637
Marks, permanent, reference, witness, 25
Mean sea level, 11
Meridian, 668
    ellipse, 149
Meridional distances, 155
Meteorological correction, Geodimeter, 619
    Tellurometer, 632
Meteorological readings, 640
Metes and bounds, 592
Micrometer microscope, 109, 110, 111
Micrometer, optical, 111
Mid latitude formulae, 158, 160
Millibar, 422, 664
Millieme, 104

Millimicrosecond, 627
Mistake, 202
Modulation, 610
Mosaics, air photo, 1, 20
Mounting, paper, 5
Multiple base altimetry, 430
Multiplying constant, tacheometry, 447

NADIR, 668
Nanosecond, 627
Normal distribution, 207
Normal equations, 228
    correlative, 244
    solution of, 233
Normal and vertical staff (tacheometry), 451

OBLIQUE PROJECTIONS, 198
Observation equations, 222, 225
Observation of horizontal angles, 261
Obstructed curves, 521
Octantis, sigma, 677
Offsets, 441
Omitted measurements (traverses), 353
On line points, 591
Opaque beacons, 253
Optical micrometer, 111, 115, 129
Optical scale, 107
Optical square, 443
Ordnance Survey, 27, 476, 592
Orientation, 5
Origin, change of, 75
Orthometric heights, 406
Orthomorphic, 168
Overhaul, in earthworks, 585

PANORAMIC SKETCH, 22
Parabola, cubic—as transition curve, 487
    —as vertical curve, 548
Parabola, simple—as vertical curve, 541
Parallactic angle, 669
Parallax, Sun's, 671
Parallax, in telescope, 94, 134
    effect on measured angles, 331
Parallel, length of, 155
Parallel plate, in optical micrometer, 112
    micrometer, 141
Parallel, setting out a, 165
Paper traverse, 594
Phase angle, 609
Phase error on beacons, 254
Phase readings (Geodimeter), 617
Photo points, 29
Plane rectangular coordinates, 77
Plane table 1, 462
Planimeter, 598
Plate level, 98, 119, 120

Plumbing fork, 3
Plus measurements, 443
Polaris, 677
Polaroid (Geodimeter), 625
Poles, North and South, 667
Pole Star Tables, 677
Porro's anallactic telescope, 448
Pothonot-Snellius resection, 297
Precision, 201, 207
Pre-marking of photo points, 30
Pressure, measurement of, 421
Prime vertical, 669
Prismoidal, correction, 575
    rule for earthwork volumes, 557
Probability, 204
Projection, map, 164, 167
    Cassini, 168
    conformal, 166
    Lambert, 194
    survey-computation on, 182
    Tables, 174
    Transverse Mercator, 166
Psychrometer, 633

QUADRILATERAL, ADJUSTMENT OF, 275

RADIANS TO SEXAGESIMAL VALUES, 57
Radiation, 441, 445, 6
Radii of curvature of spheroid, 152
Railways—see roads.
Random error, law of, 207
Ranging of curves—see curve ranging.
Ramsden eyepiece, 88
Rate, chronometer, 681
Recce description of stations, 20
Reconnaissance, 19, 20, 22
Record keeping, 33
Reduction of tacheometric observations, 455
Reduction to centre, 316
Re-establishment of boundaries, 593
Reference object (R.O.), 264
Reflector, Geodimeter, 617
Refraction, coefficient of, 415
    (levelling), 400
    (trigonometrical heighting), 412
    index of (Geodimeter), 619
    index of (Table of), 635
    index of (Tellurometer), 633
    second correction, 646
    tables, 671
Rejection of observations, 213
Relative shift—for compound combined curves, 501
Remote Tellurometer, 637
Repere marks, 5
Report writing, 35

Resection, 7
    Bessel's or Collins' point, 299
    Barycentric or Tienstra, 293
    computation of, 293
    in detail surveying, 478
    Pothonot-Snellius, 297
    semi-graphic, 303
    two point, 10, 315
    variation of coordinates, 310
Residual, 201
Reticule, 90
Reverse curves, 496, 482
Right Ascension, 667
Rise and Fall booking of levelling, 391
Roads, design of horizontal curves, 531
    earthwork volumes, 561, 567, 579
    mass haul diagrams for, 579
    vertical curve design, 552
    widening, 505
Roeloffs' prism, 672

SAG CORRECTION IN LINEAR MEASUREMENTS, 328, 333
Satellite stations, 316
Scale, map, 4
Scale change, in coordinates, 73, 75, 76
Scale error, Transverse Mercator, 179
Scale factor, 167, 180
    Lambert projection, 197
    line, 183, 184
Self aligning level, see 'automatic level'
Self reducing tacheometer with horizontal staff, 465, 471, 476
Self reducing tacheometer with vertical staff, 458, 462
Semi-diameter of Sun, 671
Semi-graphic computation of linear measures, 650
    of angular measures, 301
Setting out circular curves, 508, 530
Setting out in cadastral surveys, 591, 595
Setting out obstructed curves, 521
Setting out transition curves, clothoid and cubic parabola, 516
    lemniscate, 537
Setting out works, general, 480
Shift, for transition curves, 499
    relative- for compound curves, 501
Side conditions, 285
Side equation in quadrilateral, 279
Sidereal Time, Greenwich, 676
Sight distance—vertical curves, 554
Signals, 252, 334, 335
Single base altimetry, 430
Simpson's rule, 597

Slope correction, 326, 332, 647
Slope staking, 572
Speed, in design of horizontal curves, 531
    in design of vertical curves, 553, 554
Spherical excess, 65, 156
Spherical triangles, solution of, 60 to 65
Spherical triangle, 58, 60
    polar, 63
    right angled, 64
Spheroid, 149
Spirals—highway, see 'transition curves'
Spirit level, 96, 137
Spirit levelling, 385, 388
Spot heights—earthwork volumes from, 564
Stabilizer in automatic level, 144
Stadia marks, 90, 446
Staff, levelling, 389
Standard parallel, 194
Star Almanac for Land Surveyors, 668, 673, 676, 679
Star identification, 681
Station adjustment, 270
Station descriptions, 23, 337, 338
Station yards, 583
Strain in axis of rotation, 82, 127, 130
Striding level, 100, 126, 136
Subdivision, 604
Subtense bar, 363
    traversing, 360, 362, 366, 368
Sun azimuth, 670, 673
Surveyor's level, 137 to 147
Summit—in vertical curves, 541
Super elevation of roads and railways, 531

$t - T$ CORRECTION, LAMBERT PROJECTION, 197
    transverse Mercator projection, 184, 189, 191
Tacheometers, self reducing-, see 'self reducing tacheometers'
Tacheometry, 446, 449, 453, 456, 476
Tangent lengths, 490, 492, 496, 503
Tangent points, for cubic vertical curves, 550
    for parabolic vertical curves, 544
Taping, 340, 341
Telescope, 83, 85, 88, 91, 92, 94, 137
Telescopic alidade, 462
Telephone, Tellurometer duplex, 626
Tellurometer, 608, 626, 636, 639, 641, 643
Temporary marks, 591
Theodolite, adjustment of, 118
    component parts of, 81
    errors, 330, 331
    maladjustment, 134
    malconstruction, 127, 134
    types of, 138

Three tripod equipment, 333
Tienstra resection, 293
Time signal, 675
Time, Universal, 675
Topographical maps, 1
Towers, survey, 260
Town survey control, 341
Transformation of coordinates, 73, 75, 76, 77
Transit axis, 83, 125
Transit method of traverse adjustment, 346, 351
Transition curves, definition of, 481
    derivation of equations, 484
    lemniscate, 535
    rectangular coordinates, 486
    setting out, 516
    polar coordinates, 488
    wide roads and dual carriageways for, 505
Transit, of star, 669
Transit time (E.D.M.), 608, 627
Transverse Mercator projection, 166, 172, 174
Transverse projections, 198
Trapezoidal rule, 597
Traverses, 323
    plane table, 6
    tacheometric, 455, 476
Triangle, adjustment of, 271
    solution of, 80
Triangulation, 5, 252
Tribrach, 82, 333, 334
Trigonometrical heighting, 387, 410
Trihedral prism, 617
Trilateration, 6, 644, 649, 656, 662
Trivet stage, 82
Trunnion axis, see 'transit axis'
Tunnels, earthwork volumes of, 561
Two point resection, 315

UNIVERSAL TIME (U.T.), 675
Universal Transverse Mercator Projection (U.T.M.), 181
Upper plate, 84
Urban layouts, 594

VALLEY, IN VERTICAL CURVES, 541

Value, most probable, 201
    true, 200
Variance, 209
Variation of coordinates, angles, 300
Variation of coordinates—lengths, 300, 649, 653
Vernier, 106
Versine, 326
Vertical acceleration, 553
Vertical angle, 124
Vertical circle of theodolite, 105, 106
    in astronomy, 668
Vertical collimation, 123
Vertical curves, design constants, 552
Vertical curves, 540
Vertical, the—definition of, 668
    the—deviation of, 151
Vertical interval of contour, 11
Vertical, the 'Prime', 669
Vision distance-vertical curves, 554
Volumes of earthworks—curvature correction for, 576
    'end area' rule, 558
    from contours, 559
    from cross-sections, 567, 579
    from spot heights, 564
    prismoidal correction, 575
    prismoidal rule, 557

WASTING—IN EARTHWORKS, 583
Wedge-Geodimeter, 618
Wedge, travelling- in micrometer, 112
Weighted observations, adjustment of, 239, 244
Weights, 201, 217, 218
Wide roads, 505
Wild R.D.H. tacheometer, 466, 470, 471
Wild R.D.S. tacheometer, 458, 461
Working plans, 594

YOUNG'S MODULUS OF ELASTICITY, 326

ZEISS JENA E.O.S., 643
Zenith, the, 668
Zero, change of, 130, 263, 342
Zero curve (tacheometer), 459
Zones, projection, 181